Springer Finance

Textbooks

Springer Finance Textbooks

Springer Finance is a programme of books addressing students, academics and practitioners working on increasingly technical approaches to the analysis of financial markets. It aims to cover a variety of topics, not only mathematical finance but foreign exchanges, term structure, risk management, portfolio theory, equity derivatives, and financial economics.

This subseries of Springer Finance consists of graduate textbooks.

More information about this series at http://www.springer.com/series/11355

Robert A. Jarrow

Continuous-Time Asset Pricing Theory

A Martingale-Based Approach

Springer

Robert A. Jarrow
Samuel Curtis Johnson Graduate School
Cornell University
Ithaca
New York, USA

ISSN 1616-0533 ISSN 2195-0687 (electronic)
Springer Finance
Springer Finance Textbooks
ISBN 978-3-030-08549-0 ISBN 978-3-319-77821-1 (eBook)
https://doi.org/10.1007/978-3-319-77821-1

Mathematics Subject Classification (2010): 90C99, 60G99, 49K99, 91B25

Printed on acid-free paper

This Springer imprint is published by the registered company Springer International Publishing AG part of Springer Nature.
The registered company address is: Gewerbestrasse 11, 6330 Cham, Switzerland

This book is dedicated to my wife, Gail.

Preface

> The fundamental paradox of mathematics is that abstraction leads to both simplicity and generality. It is a paradox because generality is often thought of as requiring complexity, but this is not true. This insight explains both the beauty and power of mathematics.

Philosophy

My philosophy in creating models for practice and for understanding is based on two simple principles:

1. always impose the least restrictive set of assumptions possible to achieve maximum generality, and
2. when choosing among assumptions, it is better to impose an assumption that is observable and directly testable versus an assumption that is unobservable and only indirectly testable.

This philosophy affects the content of this book.

The Key Topics

Finance's asset pricing theory has three topics that uniquely identify it.

1. Arbitrage pricing theory, including derivative valuation/hedging and multiple-factor beta models.
2. Portfolio theory, including equilibrium pricing.
3. Market informational efficiency.

These three topics are listed in order of increasing structure (set of assumptions), from the general to the specific. In some sense, topic 3 requires less structure than topic 2 because market efficiency only requires the existence of an equilibrium, not a characterization of the equilibrium.

The more assumptions imposed, the less likely the structure depicts reality. Of course, this depends crucially on whether the assumptions are true or false. If the assumptions are true, then no additional structure is being imposed when an assumption is added. But in reality, all assumptions are approximations, therefore all assumptions are in some sense "false." This means, of course, that the less assumptions imposed, the more likely the model is to be "true."

The Key Insights

There are at least nine important insights from asset pricing theory that need to be understood. These insights are obtained from the three fundamental theorems of asset pricing. The insights are enriched by the use of preferences, characterizing an investor's optimal portfolio decision, and the notion of an equilibrium. These nine insights are listed below.

1. The existence of a state price density or an equivalent local martingale measure (First Fundamental Theorem).
2. Hedging and exact replication (Second Fundamental Theorem).
3. The risk-neutral valuation of derivatives (Third Fundamental Theorem).
4. Asset price bubbles (Third Fundamental Theorem).
5. Spanning portfolios (mutual fund theorems) (Third Fundamental Theorem).
6. The meaning of Arrow–Debreu security prices (Third Fundamental Theorem).
7. The meaning of systematic versus idiosyncratic risk (Third Fundamental Theorem).
8. The meaning of diversification (Third Fundamental Theorem and the Law of Large Numbers).
9. The importance of the market portfolio (Portfolio Optimization and Equilibrium).

Insight 1 requires the first fundamental theorem. Insight 2 requires the second fundamental theorem. Insights 3–8 require the first and third fundamental theorems of asset pricing. Insight 8 also requires the law of large numbers. Insight 9 requires the notion of an equilibrium with heterogeneous traders. There are three important aspects of insights 1–9 that need to be emphasized. The first is that all of these insights are derived in incomplete markets, including markets with trading constraints. The second is that all of these insights are derived for *discontinuous sample path processes*, i.e. asset price processes that contain jumps. The third is that all of these insights are derived in models where traders have heterogeneous beliefs, and in certain subcases, differential information as well. As such, these insights are very robust and relevant to financial practice. All of these insights are explained in detail in this book.

The Martingale Approach

The key topics of asset pricing theory have been studied, refined, and extended for over 40 years, starting in the 1970s with the capital asset pricing model (CAPM), the notion of market efficiency, and option pricing theory. Much knowledge has been accumulated and there are many different approaches that can be used to present this material. Consistent with my philosophy, I choose the most abstract, yet the simplest and most general approach for explaining this topic. This is the martingale approach to asset pricing theory—the unifying theme is the notion of an equivalent local martingale probability measure (and all of its extensions). This theme can be used to understand and to present the known results from arbitrage pricing theory up to, and including, portfolio optimization and equilibrium pricing. The more restrictive historical and traditional approach based on dynamic programming and Markov processes is left to the classical literature.

Discrete Versus Continuous Time

There are three model structures that can be used to teach asset pricing.

1. A static (single period) model,
2. discrete-time and multiple periods, or
3. continuous-time.

Static models are really only useful for pedagogical purposes. The math is simple and the intuition easy to understand. They do not apply in practice/reality. Consistent with my philosophy, this reduces the model structure choice to two for this book, between discrete-time multiple periods and continuous-time models. We focus on continuous-time models in this book because they are the better model structure for matching reality (see Jarrow and Protter [103]).

Trading in continuous time better matches reality for three reasons. One, a discrete-time model implies that one can only trade on the grid represented by the discrete time points. This is not true in practice because one can trade at any time during the day. Second, trading times are best modeled as a finite (albeit very large) sequence of random times on a continuous time interval. It is a very large finite sequence because with computer trading, the time between two successive trades is very small (milli- and even microseconds). This implies that the limit of a sequence of random times on a continuous time interval should provide a reasonable approximation. This is, of course, continuous trading. Three, continuous-time has a number of phenomena that are not present in discrete-time models—the most important of which are strict local martingales. Strict local martingales will be shown to be important in understanding asset price bubbles.

Mean-Variance Efficiency and the Static CAPM

As an epilogue to Part III of this book, its last chapter studies the static CAPM. The static CAPM is studied after the dynamic continuous-time model to emphasize the omissions of a static model and the important insights obtained in dynamic models. This is done because the static model is not a good approximation to actual security markets. This book only briefly discusses the mean-variance efficient frontier. Consequently, an in depth study of this material is left to independent reading (see Back [5], Duffie [52], Skiadas [171]). Generalizations of this model in continuous time—the intertemporal CAPM due to Merton [137] and the consumption CAPM due to Breeden [22]—are included as special cases of the models presented in this book.

Stochastic Calculus

Finance is an application of stochastic process and optimization theory. Stochastic processes because asset prices evolve randomly across time. Optimization because investors trade to maximize their preferences. Hence, this mathematics is essential to developing the theory. This book is not a mathematics book, but an economics book. The math is not emphasized, but used to obtain results. The emphasis of the book is on the economic meaning and implications of assumptions and results. The proofs of most results are included within the text, except those that require a knowledge of functional analysis. Most of the excluded proofs are related to "existence results," examples include the first fundamental theorem of asset pricing and the existence of a saddle point in convex optimization. For those proofs not included, references are provided. The mathematics assumed is that obtained from a first level graduate course in real analysis and probability theory. Sources of this knowledge include Ash [3], Billingsley [13], Jacod and Protter [75], and Klenke [123]. Excellent references for stochastic calculus include Karatzas and Shreve [117], Medvegyev [136], Protter [151], Roger and Williams [157], Shreve [169], while those for optimization include Borwein and Lewis [19], Guler [66], Leunberger [134], Ruszczynski [162], and Pham [149].

ASSET PRICING THEORY: TRADITIONAL VERSUS MARKET MICROSTRUCTURE

Asset Pricing Theory

1. continuous/discrete-time
2. frictionless/frictions
3. equal/differential beliefs
4. equal/differential information

(Traditional)

competitive markets $\Longleftrightarrow$ Walrasian equilibrium

(Market Microstructure)

competitive markets $\Longleftrightarrow$ Nash equilibrium or zero expected profit

Traditional Asset Pricing Theory versus Market Microstructure

Although the distinction between traditional asset pricing theory and market microstructure is not "black and white," one useful classification of the difference between these two fields is provided in the previous Table. In this classification, traditional asset pricing theory and market microstructure have common the structures (1)–(4). They differ in the meaning of a competitive market, in particular, the notion of an equilibrium. Traditional asset pricing uses the concept of a Walrasian equilibrium (supply equals demand, price-takers) whereas market microstructure uses Nash equilibrium or a zero expected profit condition (strategic traders, non-price-takers). This difference is motivated by the questions that each literature addresses.

Asset pricing abstracts from the mechanism under which trades are executed. Consequently, it assumes that investors are price-takers whose trades have no quantity impact on the price. This literature focuses on characterizing the price process, optimal trading strategies, and risk premium. In contrast, the market microstructure literature seeks to understand the trade execution mechanism itself, and its impact on market welfare. This alternative perspective requires a different equilibrium notion, one that explicitly incorporates strategic trading. This book presents asset pricing theory using the traditional representation of market clearing. For a book that reviews the market microstructure literature, see O'Hara [147].

Themes

The themes in this book differ from those contained in most other asset pricing books in four notable ways. First, the emphasis is on price processes that include jumps, not just continuous diffusions. Second, stochastic optimization is based on martingale methods using convex analysis and duality, and not diffusion processes with stochastic dynamic programming. Third, asset price bubbles are an important consideration in every result presented herein. Fourth, the existence and characterization of economic equilibrium is based on the use of a representative trader. Other excellent books on asset pricing theory, using the more traditional approach to the topic, include Back [5], Bjork [14], Dana and Jeanblanc [42], Duffie [52], Follmer and Schied [63], Huang and Litzenberger [72], Ingersoll [74], Karatzas and Shreve [118], Merton [140], Pliska [150], and Skiadas [171].

Acknowledgements I am grateful for a lifetime of help and inspiration from family, colleagues, and students.

Ithaca, NY, USA Robert A. Jarrow

Contents

Part III Equilibrium

List of Notation

For easy reference, this section contains the notation used consistently throughout the book. Notation that is used only in isolated chapters is omitted from this list, but complete definitions are included within the text.

$\underset{n\times 1}{x} = (x_1, \ldots, x_n)' \in \mathbb{R}^n$, where the prime denotes transpose, is a column vector.

$t \in [0, T]$ represents time in a finite horizon and continuous-time model.

$(\Omega, \mathscr{F}, (\mathscr{F}_t), \mathbb{P})$ is a filtered probability space on $[0, T]$ with $\mathscr{F} = \mathscr{F}_T$, where Ω is the state space, $\mathscr{F}$ is a σ-algebra, $(\mathscr{F}_t)$ is a filtration, and $\mathbb{P}$ is a probability measure on Ω.

$E[\cdot]$ is expectation under the probability measure $\mathbb{P}$.

$E^{\mathbb{Q}}[\cdot]$ is expectation under the probability measure $\mathbb{Q}$ given $(\Omega, \mathscr{F}, \mathbb{Q})$, where $\mathbb{Q} \neq \mathbb{P}$.

$\mathbb{Q} \sim \mathbb{P}$ means that the probability measure $\mathbb{Q}$ is equivalent to $\mathbb{P}$.

r_t is the default-free spot rate of interest.

$\mathbb{B}_t = e^{\int_0^t r_s ds}$, $\mathbb{B}_0 = 1$ is the value of a money market account.

$\mathbb{S}_t = (\mathbb{S}_1(t), \ldots, \mathbb{S}_n(t))' \geq 0$ represents the prices of n of risky assets (stocks), semimartingales, adapted to $\mathscr{F}_t$.

$B_t \equiv \frac{\mathbb{B}_t}{\mathbb{B}_t} = 1$ for all t represents the normalized value of the money market account.

$S_t = (S_1(t), \ldots, S_n(t))' \geq 0$ represents prices when normalized by the value of the money market account, i.e. $S_i(t) = \frac{\mathbb{S}_i(t)}{\mathbb{B}(t)}$.

$(S, (\mathscr{F}_t), \mathbb{P})$ is a market.

$\mathscr{B}(0, \infty)$ is the Borel σ-algebra on $(0, \infty)$.

$L^0 \equiv L^0(\Omega, \mathscr{F}, \mathbb{P})$ is the space of all $\mathscr{F}_T$-measurable random variables.

$L^0_+ \equiv L^0_+(\Omega, \mathscr{F}, \mathbb{P})$ is the space of all nonnegative $\mathscr{F}_T$-measurable random variables.

$L^1_+(\mathbb{P}) \equiv L^1_+(\Omega, \mathscr{F}, \mathbb{P})$ is the space of all nonnegative $\mathscr{F}_T$-measurable random variables X such that $E[X] < \infty$.

$\mathscr{O}$ is the set of optional stochastic processes.

$\mathscr{L}(S)$ is the set of predictable processes integrable with respect to S.

$\mathfrak{L}(\mathbb{B})$ is the set of optional processes that are integrable with respect to $\mathbb{B}$.

$\mathscr{L}^0$ is the set of adapted, right continuous with left limit existing (cadlag) stochastic processes.
$\mathscr{L}^0_+$ is the set of adapted, right continuous with left limit existing (cadlag) stochastic processes that are nonnegative.
$\mathscr{A}(x) = \{(\alpha_0, \alpha) \in (\mathscr{O}, \mathscr{L}(S)) :\ X_t = \alpha_0(t) + \alpha_t \cdot S_t,\ \exists c \leq 0,$
$X_t = x + \int_0^t \alpha_u \cdot dS_u \geq c,\ \forall t \in [0, T]\}$
is the set of admissible, self-financing trading strategies.
$\mathfrak{M} = \{\mathbb{Q} \sim \mathbb{P} :\ S \text{ is a } \mathbb{Q}\text{-martingale}\}$
is the set of martingale measures.
$\mathfrak{M}_l = \{\mathbb{Q} \sim \mathbb{P} :\ S \text{ is a } \mathbb{Q}\text{-local martingale}\}$
$= \{\mathbb{Q} \sim \mathbb{P} :\ X \text{ is a } \mathbb{Q}\text{-local martingale},\ X = 1 + \int \alpha \cdot dS,\ (\alpha_0, \alpha) \in \mathscr{A}(1)\}$
is the set of local martingale measures.
$\mathfrak{M}_s = \{\mathbb{Q} \sim \mathbb{P} :\ S \text{ is a } \mathbb{Q}\text{-supermartingale}\}$
is the set of supermartingale measures.
$\mathscr{D}_l = \big\{Y \in \mathscr{L}^0_+ : Y_0 = 1,\ XY \text{ is a } \mathbb{P}\text{-local martingale},$
$X = 1 + \int \alpha \cdot dS,\ (\alpha_0, \alpha) \in \mathscr{A}(1)\big\}$
is the set of local martingale deflator processes.
$D_l = \{Y_T \in L^0_+ : \exists Z \in \mathscr{D}_l,\ Y_T = Z_T\}$
is the set of local martingale deflators.
$\mathscr{M}_l = \left\{Y \in \mathscr{D}_l : \exists \mathbb{Q} \sim \mathbb{P},\ Y_T = \frac{d\mathbb{Q}}{d\mathbb{P}}\right\}$
is the set of local martingale deflator processes generated by a probability density with respect to $\mathbb{P}$.
$M_l = \big\{Y_T \in L^0_+ : \exists Z \in \mathscr{M}_l,\ Y_T = Z_T\big\}$
is the set of local martingale deflators that are probability densities with respect to $\mathbb{P}$.
$\mathscr{M} = \{Y \in \mathscr{L}^0_+ :\ Y_T = \frac{d\mathbb{Q}}{d\mathbb{P}},\ Y_t = E[Y_T | \mathscr{F}_t],\ \mathbb{Q} \in \mathfrak{M}\}$
$= \{Y \in \mathscr{L}^0_+ :\ Y \in \mathscr{M}_l,\ Y_T = \frac{d\mathbb{Q}}{d\mathbb{P}},\ \mathbb{Q} \in \mathfrak{M}\}$
is the set of martingale deflator processes generated by martingale measures.
$M = \{Y_T \in L^0_+ :\ Y_T = \frac{d\mathbb{Q}}{d\mathbb{P}},\ \mathbb{Q} \in \mathfrak{M}\}$
$= \{Y \in L^0_+ :\ \exists Z \in \mathscr{M},\ Y_T = Z_T\}$
is the set of martingale deflators generated by martingale measures.
$\mathscr{N}(x) = \{(\alpha_0, \alpha) \in (\mathscr{O}, \mathscr{L}(S)) :\ X_t = \alpha_0(t) + \alpha_t \cdot S_t,$
$X_t = x + \int_0^t \alpha_u \cdot dS_u \geq 0,\ \forall t \in [0, T]\big\}$
is the set of nonnegative wealth, self-financing trading strategies.
$\mathscr{D}_s = \big\{Y \in \mathscr{L}^0_+ : Y_0 = 1,\ XY \text{ is a } \mathbb{P}\text{-supermartingale},$
$X = 1 + \int \alpha \cdot dS,\ (\alpha_0, \alpha) \in \mathscr{N}(1)\big\}$
is the set of supermartingale deflator processes.
$D_s = \{Y_T \in L^0_+ : \exists Z \in \mathscr{D}_s,\ Y_T = Z_T\}$
$= \{Y_T \in L^0_+ : Y_0 = 1,\ \exists (Z_n(T))_{n \geq 1} \in M_l,\ Y_T \leq \lim_{n \to \infty} Z_n(T)\ a.s.\}$
is the set of supermartingale deflators.
$\mathscr{X}(x) = \left\{X \in \mathscr{L}^0_+ :\ \exists (\alpha_0, \alpha) \in \mathscr{N}(x),\ X_t = x + \int_0^t \alpha_u \cdot dS_u,\ \forall t \in [0, T]\right\}$
is the set of nonnegative wealth processes generated by self-financing trading strategies.

$\mathscr{X}(x) = \left\{X \in \mathscr{L}_+^0 : \exists(\alpha_0, \alpha) \in \mathscr{N}(x),\ x + \int_0^t \alpha_u \cdot dS_u \geq X_t,\ \forall t \in [0, T]\right\}$
is the set of nonnegative wealth processes dominated by the value process of a self-financing trading strategy.
$\mathscr{C}(x) = \{X_T \in L_+^0 : \exists(\alpha_0, \alpha) \in \mathscr{N}(x),\ x + \int_0^T \alpha_t \cdot dS_t = X_T\}$
$= \{X_T \in L_+^0 : \exists Z \in \mathscr{X}(x),\ X_T = Z_T\}$
is the set of nonnegative random variables generated by self-financing trading strategies.
$\mathscr{C}(x) = \left\{X_T \in L_+^0 : \exists(\alpha_0, \alpha) \in \mathscr{N}(x),\ x + \int_0^T \alpha_t \cdot dS_t \geq X_T\right\}$
is the set of nonnegative random variables dominated by the value process of a self-financing trading strategy.
$\beta_t = S_t - E^{\mathbb{Q}}[S_T \,|\mathscr{F}_t]$
is an asset's price bubble with respect to the equivalent local martingale measure $\mathbb{Q}$.
$p(t, T)$ is the time t price of a default-free zero-coupon bond paying \$1 at time T with $t \leq T$.
$f(t, T) = -\frac{\partial \log(p(t,T))}{\partial T}$
is the time t default-free (continuously compounded) forward rate for date T with $t \leq T$.
$L(t, T) = \frac{1}{\delta}\left[\frac{p(t,T)}{p(t,T+\delta)} - 1\right]$
is the time t default-free discrete forward rate for the time interval $[T, T + \delta]$ with $t \leq T$.
$D(t, T)$ is the time t price of a risky zero-coupon bond paying \$1 at time T with $t \leq T$.
$U_i(x, \omega) : (0, \infty) \times \Omega \to \mathbb{R}$
is the state dependent utility function of wealth for investor $i = 1, \ldots, \mathscr{I}$.
$\left\{((\mathscr{F}_t), \mathbb{P}), (N_0, N), \left(\mathbb{P}_i, U_i, \left(e_0^i, e^i\right)\right)_{i=1}^{\mathscr{I}}\right\}$ is an economy.
$U(x, \omega) : (0, \infty) \times \Omega \to \mathbb{R}$ is the aggregate utility function of wealth for a representative trader.

Part I
Arbitrage Pricing Theory

Overview

The key results of finance that are successfully used in practice are based on the three fundamental theorems of asset pricing. Part I presents the three theorems. The applications of these three theorems are also discussed, including state price densities (Arrow–Debreu prices), systematic risk, multiple-factor beta models, derivatives pricing, derivatives hedging, and asset price bubbles. All of these implications are based on the existence of an equivalent local martingale measure.

The three fundamental theorems of asset pricing relate to the existence of an equivalent local martingale measure, its uniqueness, and its extensions. Roughly speaking, the first fundamental theorem of asset pricing equates no arbitrage with the existence of an equivalent local martingale measure. The second fundamental theorem relates market completeness to the uniqueness of the equivalent local martingale measure. The third fundamental theorem states that there exists an equivalent martingale measure, without the prefix "local," if and only if there is no arbitrage and no dominated assets in the economy.

There are three major models used in derivatives pricing: the Black–Scholes–Merton (BSM) model, the Heath–Jarrow–Morton (HJM) model, and the reduced form credit risk model. These models are discussed in this part. Other extensions and refinements of these models exist in the literature. However, if you understand these three classes of models, then their extensions and refinements are easy to understand. These models are divided into three cases: complete markets, "extended" complete markets, and incomplete markets.

In complete markets, there is unique pricing of derivatives and exact hedging is possible. The two model classes falling into this category are the BSM and the HJM model. There are two models for studying credit risk: structural and reduced form models. Structural models assume that the markets are complete. Reduced form models, depending upon the structure imposed, usually (implicitly) assume that the markets are incomplete.

In reduced form models, market incompleteness is due to the use of inaccessible stopping times to model default (jump processes). "Extended" complete markets contain the reduced form class of models. This class of models is called "extended" complete because to obtain unique pricing in such a model, one assumes that the market studied is embedded in a larger market that is complete and therefore the equivalent local martingale measure is unique. This extended complete market usually includes the trading of derivatives (e.g. call and put options with different strikes and maturities) on the primary traded assets (e.g. stocks, zero-coupon bonds). In this case the local martingale measure never needs to be explicitly identified for pricing. It is important to note that in this circumstance, however, exact hedging of credit risk is impossible without the use of traded derivatives. The primary use of these models is for pricing and static hedging using derivatives, and not dynamic hedging using the primary traded assets (risky and default-free zero-coupon bonds). In incomplete markets, which are not "extended" complete, exact pricing and hedging of assets is (usually) impossible. In this case upper and lower bounds for derivative prices are obtained by super- and sub-replication.

Chapter 1
Stochastic Processes

We need a basic understanding of stochastic processes to study asset pricing theory. Excellent references are Karatzas and Shreve [117], Medvegyev [136], Rogers and Williams [157], and Protter [151]. This chapter introduces some terminology, notation, and key theorems. Few proofs of the theorems are provided, only references for such. The basics concepts from probability theory are used below without any detailed explanation (see Ash [3] or Jacod and Protter [75] for this background material).

1.1 Stochastic Processes

We consider a continuous-time setting with time denoted $t \in [0, \infty)$. We are given a filtered probability space $(\Omega, \mathscr{F}, (\mathscr{F}_t)_{0 \le t \le \infty}, \mathbb{P})$ where Ω is the state space with generic element $\omega \in \Omega$, $\mathscr{F}$ is a σ-algebra representing the set of events, $(\mathscr{F}_t)_{0 \le t \le \infty}$ is a filtration, and $\mathbb{P}$ is a probability measure defined on $\mathscr{F}$. A *filtration* is a collection of σ-algebras which are increasing, i.e. $\mathscr{F}_s \subseteq \mathscr{F}_t$ for $0 \le s \le t \le \infty$.

A *random variable* is a mapping $Y : \Omega \to \mathbb{R}$ such that Y is $\mathscr{F}$-measurable, i.e. $Y^{-1}(A) \in \mathscr{F}$ for all $A \in \mathscr{B}(\mathbb{R})$ where $\mathscr{B}(\mathbb{R})$ is the Borel σ-algebra on $\mathbb{R}$, i.e. the smallest σ-algebra containing all open intervals (s, t) with $s \le t$ for $s, t \in \mathbb{R}$ (see Ash [3, p. 8]).

A *stochastic process* is a collection of random variables indexed by time, i.e. a mapping $X : [0, \infty) \times \Omega \to \mathbb{R}$, denoted variously depending on the context, $X(t, \omega) = X(t) = X_t$. It is *adapted* if X_t is $\mathscr{F}_t$-measurable for all $t \in [0, \infty)$.

A *sample path* of a stochastic process is the graph of $X(t, \omega)$ across time t keeping ω fixed.

We assume that the filtered probability space satisfies the usual hypotheses. The *usual hypotheses* are that $\mathscr{F}_0$ contains the $\mathbb{P}$-null sets of $\mathscr{F}$ and that the filtration $(\mathscr{F}_t)_{t \ge 0}$ is right continuous. Right continuous means that $\mathscr{F}_t = \cap_{u > t} \mathscr{F}_u$ for all

R. A. Jarrow, *Continuous-Time Asset Pricing Theory*, Springer Finance,
https://doi.org/10.1007/978-3-319-77821-1_1

$0 \le t < \infty$. Letting $\mathscr{F}_0$ contain the $\mathbb{P}$-null sets of $\mathscr{F}$ facilitates the measurability of various events, random variables, and stochastic processes. Right continuity implies the important result that given a random variable $\tau : \Omega \to [0, \infty]$, $\{\tau(\omega) \le t\} \in \mathscr{F}_t$ for all t if and only if $\{\tau(\omega) < t\} \in \mathscr{F}_t$ for all t, see Protter [151, p. 3]. This fact will be important with respect to the mathematics of stopping times, which are introduced below. One can think of right continuity as implying that the information at time t^+ is known at time t, see Medvegyev [136, p. 9].

A stochastic process is said to be *cadlag* if it has sample paths that are right continuous with left limits existing a.s. $\mathbb{P}$.

It is said to be *caglad* if its sample paths are left continuous with right limits existing a.s. $\mathbb{P}$.

Both of these stochastic processes allow sample paths that contain jumps, i.e. a sample path which exhibits at most a countable number of discontinuities (jumps) over any compact interval (see Medvegyev [136, p. 5]). An interval in the real line is compact if and only if it is closed and bounded.

A stochastic process is said to be *predictable* if it is measurable with respect to the predictable σ-algebra. The predictable σ-algebra is the smallest σ-algebra generated by the processes that are caglad and adapted, see Protter [151, p. 102].

A stochastic process is said to be *optional* if it is measurable with respect to the optional σ-algebra. The optional σ-algebra is the smallest σ-algebra generated by the processes that are cadlag and adapted, see Protter [151, p. 102].

It can be shown that the predictable σ-algebra is always contained in the optional σ-algebra (see Medvegyev [136, p. 27]). Hence, we get the following relationship among the three types of stochastic processes,

$$\text{predictable} \subseteq \text{optional} \subseteq \text{adapted}.$$

A stochastic process is said to be *continuous* if its sample paths are continuous a.s. $\mathbb{P}$, i.e. it is both cadlag and caglad.

Definition 1.1 (Nondecreasing Process) Let X be a cadlag process. X is *a nondecreasing* process if the paths $X_t(\omega)$ are nondecreasing in t for all $\omega \in \Omega$ a.s. $\mathbb{P}$.

Definition 1.2 (Finite Variation Process) Let X be a cadlag process. X is a *finite variation process* if the paths $X_t(\omega)$ are of finite variation on compact intervals for all $\omega \in \Omega$ a.s. $\mathbb{P}$.

A real-valued function $f : \mathbb{R} \to \mathbb{R}$ being of finite variation on a compact interval means that the function can be written as the difference of two nondecreasing (monotone) real-valued functions, see Royden [160, p. 100]. From Royden [160, Lemma 6, page 101] we get the next lemma.

Lemma 1.1 (Lebesgue Integrals) *Let Y be a cadlag and adapted process such that*

$$X_t(\omega) = \int_0^t Y_s(\omega)ds$$

exists for all $t \in [0, T]$ and for all $\omega \in \Omega$ a.s. $\mathbb{P}$.

Then, X_t is a continuous and adapted process of finite variation on $[0, T]$.

Lemma 1.2 (Increasing Functions of Continuous Finite Variation Processes) *Let X_t be a continuous and adapted process of finite variation on $[0, T]$.*

Let $f : \mathbb{R} \to \mathbb{R}$ be strictly increasing (or strictly decreasing) and differentiable with f' continuous.

Then, $f(X_t)$ is a continuous and adapted process of finite variation on $[0, T]$.

Proof The continuity of $f(X_t)$ follows trivially, and the continuity of f implies $f(X_t)$ is adapted. Consider a partition of the time interval $[0, T]$, denoted $t_0, \cdots, t_n$ where $\max[t_i - t_{i-1}] \to 0$ as $n \to \infty$.

Fix $\omega \in \Omega$. Note that $\sum_{i=1}^n \left|f(X_{t_i}) - f(X_{t_{i-1}})\right| = \sum_{i=1}^n f'(\xi_i)\left|X_{t_i} - X_{t_{i-1}}\right|$ for some $\xi_i \in (X_{t_i}, X_{t_{i-1}})$ by the mean value theorem. Since X_t is continuous, $I \equiv [\min\{X_t; t \in [0, T]\}, \max\{X_t; t \in [0, T]\}]$ is a compact interval on the real line. Since f' is continuous, there exists a $\xi \in I$ such that $f'(\xi_i) \leq f'(\xi)$ for all $\xi_i \in I$.

Hence, $\sum_{i=1}^n f'(\xi_i)\left|X_{t_i} - X_{t_{i-1}}\right| \leq f'(\xi)\sum_{i=1}^n \left|X_{t_i} - X_{t_{i-1}}\right|$.

Since X_t is of finite variation on $[0, T]$, taking the supremum across all such partitions of time gives $\sup \sum_{i=1}^n \left|X_{t_i} - X_{t_{i-1}}\right| < \infty$, which implies

$\sup \sum_{i=1}^n \left|f(X_{t_i}) - f(X_{t_{i-1}})\right| < \infty$. This completes the proof.

Definition 1.3 (Martingales) A stochastic process X is a *martingale* with respect to $(\mathscr{F}_t)_{0 \leq t \leq \infty}$ if

(i) X is cadlag and adapted,
(ii) $E[|X_t|] < \infty$ all t, and
(iii) $E[X_t \,|\mathscr{F}_s] = X_s$ a.s. for all $0 \leq s \leq t < \infty$.

It is a *submartingale* if (iii) is replaced by $E[X_t \,|\mathscr{F}_s] \geq X_s$ a.s.

It is a *supermartingale* if (iii) is replaced by $E[X_t \,|\mathscr{F}_s] \leq X_s$ a.s.

For the definition of an expectation and a conditional expectation, see Ash [3, Chapter 6]. Within the class of martingales, uniformly integrable martingales play an important role (see Protter [151, Theorem 13, p. 9]).

Definition 1.4 (Uniformly Integrable Martingales) A stochastic process X is a *uniformly integrable martingale* with respect to $(\mathscr{F}_t)_{0 \leq t \leq \infty}$ if

(i) X is a martingale,
(ii) $Y = \lim_{t \to \infty} X_t$ a.s. $\mathbb{P}$ exists, $E[|Y|] < \infty$, and
(iii) $E[Y \,|\mathscr{F}_t] = X_t$ a.s. for all $0 \leq t < \infty$.

Remark 1.1 (Uniformly Integrable Martingales) Suppose we are given a filtered probability space $(\Omega, \mathscr{F}, (\mathscr{F}_t)_{0\leq t\leq T}, \mathbb{P})$ for a finite time horizon $T < \infty$ with $X : [0, T] \times \Omega \rightarrow \mathbb{R}$, where $\mathscr{F} = \mathscr{F}_T$. Then, if X is a martingale, we have $E[X_T | \mathscr{F}_s] = X_s$ a.s. $\mathbb{P}$ for all $s \in [0, T]$. This implies that all martingales on a finite horizon are uniformly integrable. This completes the remark.

Definition 1.5 (Stopping Time) A random variable $\tau : \Omega \rightarrow [0, \infty]$ is a *stopping time* if $\{\omega \in \Omega : \tau(\omega) \leq t\} \in \mathscr{F}_t$ for all $t \in [0, \infty]$.

Note that $+\infty$ is included in the range of the stopping time.

Definition 1.6 (Stopping Time σ-Algebra) Let τ be a stopping time. The *stopping time σ-algebra* is

$$\mathscr{F}_\tau \equiv \{A \in \mathscr{F} : A \cap \{\tau \leq t\} \in \mathscr{F}_t \text{ for all } t\}.$$

Let τ be a stopping time. Then, the *stopped process* is defined as

$$X_{t\wedge\tau} \equiv \begin{cases} X_t & \text{if } t < \tau \\ X_\tau & \text{if } t \geq \tau. \end{cases}$$

Definition 1.7 (Local Martingales) A stochastic process X is a *local martingale* with respect to $(\mathscr{F}_t)_{0\leq t\leq\infty}$ if

(i) X is cadlag and adapted,
(ii) there exists a sequence of stopping times (τ_n) such that $\lim_{n\to\infty} \tau_n = \infty$ a.s. $\mathbb{P}$, where $X_{t\wedge\tau_n}$ is a martingale for each n, i.e.

$$X_{s\wedge\tau_n} = E[X_{t\wedge\tau_n} | \mathscr{F}_s] \text{ a.s. } \mathbb{P}$$

for all $0 \leq s \leq t < \infty$.

Remark 1.2 (Finite Horizon Local Martingales) Suppose we are given a filtered probability space $(\Omega, \mathscr{F}, (\mathscr{F}_t)_{0\leq t\leq T}, \mathbb{P})$ for a finite time horizon $T < \infty$ with $X : [0, T] \times \Omega \rightarrow \mathbb{R}$, where $\mathscr{F} = \mathscr{F}_T$. In the definition of a local martingale, condition (ii) is modified to the existence of a sequence of stopping times (τ_n) such that $\lim_{n\to\infty} \tau_n = T$ a.s. $\mathbb{P}$, where $X_{t\wedge\tau_n}$ is a martingale for each n. This completes the remark.

Remark 1.3 (Local Submartingales and Supermartingales) The notion of a local process extends to both submartingales and supermartingales. Indeed, in the definition of a local martingale replace the word "martingale" with either "submartingale" or "supermartingale." This completes the remark.

Lemma 1.3 (Sufficient Condition for a Local Martingale to be a Supermartingale) *Let X be a local martingale that is bounded below, i.e. there exists a constant $a > -\infty$ such that $X_t \geq a$ for all t a.s. $\mathbb{P}$.*

Then, X is a supermartingale.

Proof $X_t \geq a$ for all t implies $Z_t = X_t - a \geq 0$ a.s. $\mathbb{P}$. Note that Z is a local martingale. Hence, without loss of generality we can consider only nonnegative processes.

By definition of a local martingale, let the sequence of stopping times $(\tau_n) \uparrow \infty$ be such that $E[X_{t\wedge\tau_n} | \mathscr{F}_s] = X_{s\wedge\tau_n}$. Keeping s, t fixed, taking limits of both the left and right sides gives $\lim_{n\to\infty} E[X_{t\wedge\tau_n} \big| \mathscr{F}_s] = \lim_{n\to\infty} X_{s\wedge\tau_n} = X_s$. Now, by Fatou's lemma

$$\lim_{n\to\infty} E[X_{t\wedge\tau_n} \big| \mathscr{F}_s] \geq E[\lim_{n\to\infty} X_{t\wedge\tau_n} | \mathscr{F}_s] = E[X_t | \mathscr{F}_s] \ .$$

Combined these give $X_s \geq E[X_t | \mathscr{F}_s]$. This completes the proof.

Lemma 1.4 (Sufficient Condition for a Local Martingale to be a Martingale) *Let X be a local martingale.*

Let Y be a martingale such that $|X_t| \leq |Y_t|$ *for all t a.s.* $\mathbb{P}$.

Then, X is a martingale.

Proof For a fixed T, by Remark 1.1, Y_t is a uniformly integrable martingale on $[0, T]$.

By Medvegyev [136, Proposition 1.144, p. 107], the set

$$\{Y_\tau : \tau \text{ is a finite-valued stopping time}\}$$

is of class D. Hence

$$\{X_\tau : \tau \text{ is a finite-valued stopping time}\}$$

is of class D.

This follows because $\limsup_{n\to\infty} \sup_\tau \int_{\{|Y_\tau| \geq n\}} |Y_\tau| \, d\mathbb{P} = 0$ implies $\limsup_{n\to\infty} \sup_\tau \int_{\{|X_\tau| \geq n\}} |X_\tau| \, d\mathbb{P} = 0$ (by the definition of uniform integrability Protter [151, p. 8]).

By Medvegyev [136, Proposition 1.144, p. 107], again, X is a uniformly integrable martingale on $[0, T]$.

Since this is true for all T, X is a martingale. This completes the proof.

Remark 1.4 (Bounded Local Martingales are Martingales) Let X be a local martingale that is bounded, i.e. there exists a constant $k > 0$ such that $|X_t| \leq k$ for all t a.s. $\mathbb{P}$. Then, $Y_t \equiv k$ for all t is a (uniformly integrable) martingale. Applying Lemma 1.4 shows that X is a martingale. This completes the remark.

Definition 1.8 (Semimartingales) A stochastic process X is a *semimartingale* with respect to $(\mathscr{F}_t)_{0\leq t\leq\infty}$ if it has a decomposition

$$X_t = X_0 + M_t + A_t,$$

where

(i) $M_0 = A_0 = 0$,
(ii) A is adapted, cadlag, and of finite variation on compact intervals of the real line, and
(iii) M is a local martingale (hence cadlag).

Semimartingales are important because they are the class of processes for which one can construct stochastic integrals (see Protter [151, Chapter 2]).

Definition 1.9 (Independent Increments) A stochastic process X has *independent increments* with respect to $(\mathscr{F}_t)_{0\le t\le\infty}$ if

(i) $X_0 = 0$,
(ii) X is cadlag and adapted, and
(iii) whenever $0 \le s < t < \infty$, $X_t - X_s$ is independent of $\mathscr{F}_s$.

By *independence* of $\mathscr{F}_s$ we mean that $\mathbb{P}(X_t - X_s \in A \,|\mathscr{F}_s) = \mathbb{P}(X_t - X_s \in A)$ for all $A \in \mathscr{B}(\mathbb{R})$.

Definition 1.10 (Poisson Process) Let X_t be an adapted process with respect to $(\mathscr{F}_t)_{0\le t\le\infty}$ taking values in the set $\{0, 1, 2, \ldots\}$ with $X_0 = 0$.

It is a *Poisson process* if

(i) for any s, t with $0 \le s < t < \infty$, $X_t - X_s$ is independent of $\mathscr{F}_s$ and
(ii) for any s, t, u, v with $0 \le s < t < \infty$, $0 \le u < v < \infty$, $t - s = v - u$, then the distribution of $X_t - X_s$ is the same as that of $X_v - X_u$.

Remark 1.5 (Poisson Process Distribution) It can be shown (see Protter [151, p. 13]) that for all $t > 0$,

$$\mathbb{P}(X_t = n) = \frac{e^{-\lambda t}(\lambda t)^n}{n!}, \qquad n = 0, 1, 2, \cdots$$

for some constant $\lambda \ge 0$, where $E(X_t) = \lambda t$ and $\text{Var}(X_t) \equiv E([X_t - E(X_t)]^2) = \lambda t$.

A Poisson process is a discontinuous sample path process, i.e. its sample paths have at most a countable number of jumps. This completes the remark.

Definition 1.11 (Brownian Motion) Let X_t be an adapted process with respect to $(\mathscr{F}_t)_{0\le t\le\infty}$ taking values in $\mathbb{R}$ with $X_0 = 0$.

It is a 1-dimensional *Brownian motion* if

(i) for any s, t with $0 \le s < t < \infty$, $X_t - X_s$ is independent of $\mathscr{F}_s$ and
(ii) for $0 < s < t$, $X_t - X_s$ is normally distributed with $E(X_t - X_s) = 0$ and $\text{Var}(X_t - X_s) = (t - s)$.

$X_t - X_s$ being *normally distributed* with mean zero and variance $(t - s)$ means that

$$\mathbb{P}(X_t - X_s \le x) = \int_{-\infty}^{x} \frac{1}{\sqrt{2\pi(t-s)}} e^{-\frac{1}{2(t-s)}z^2} dz$$

for all $x \in \mathbb{R}$.

Remark 1.6 (Continuous Sample Path Brownian Motions) It can be shown that conditions (i) and (ii) imply that a Brownian motion process X_t always has a modification that has continuous sample paths a.s. $\mathbb{P}$ (see Protter [151, p. 17]). A modification of a stochastic process X is another stochastic process Y that is equal to X a.s. $\mathbb{P}$ for each t (see Protter [151, p. 3]). When discussing Brownian motions, without loss of generality, we will always assume that the Brownian motion process has continuous sample paths. This completes the remark.

Definition 1.12 (Levy Process) A stochastic process X is a *Levy process* if

(i) it has independent increments with respect to $(\mathscr{F}_t)_{0 \le t \le \infty}$ and
(ii) the distributions of $X_{t+s} - X_t$ and $X_s - X_0$ are the same for all $0 \le s < t < \infty$.

Remark 1.7 (Examples of Levy Processes) Both Brownian motions and Poisson processes are examples of Levy Processes (see Medvegyev [136, Chapter 7]). This completes the remark.

Definition 1.13 (Cox Process) Let X_t be an adapted process with respect to $(\mathscr{F}_t)_{0 \le t \le \infty}$ taking values in the set $\{0, 1, 2, \ldots\}$ with $X_0 = 0$.

Let Y_t be an adapted process with respect to $(\mathscr{F}_t)_{0 \le t \le \infty}$ taking values in $\mathbb{R}^d$.

Denote by $\mathscr{F}_t^Y \equiv \sigma(Y_s : 0 \le s \le t)$ the σ-algebra generated by Y up to and including time t and by $\mathscr{F}_\infty^Y \equiv \vee_{t=0}^{\infty} \mathscr{F}_t^Y$ the smallest σ-algebra containing $\mathscr{F}_t^Y$ for all $t \ge 0$.

Let $\lambda : [0, \infty) \times \mathbb{R}^d \longrightarrow [0, \infty)$, denoted $\lambda_t(y) \ge 0$, be jointly Borel measurable with $\int_0^t \lambda_u(Y_u)du < \infty$ for all $t \ge 0$ a.s. $\mathbb{P}$.

X is a *Cox process* if for all $0 \le s < t$ and $n = 0, 1, 2, \cdots$

$$\mathbb{P}\left(X_t - X_s = n | \mathscr{F}_\infty^Y \vee \mathscr{F}_s\right) = \frac{e^{-\int_s^t \lambda_u(Y_u)du} \left(\int_s^t \lambda_u(Y_u)du\right)^n}{n!},$$

where $\mathscr{F}_\infty^Y \vee \mathscr{F}_s$ is the smallest σ-algebra containing both $\mathscr{F}_\infty^Y$ and $\mathscr{F}_s$.

This is sometimes called a doubly stochastic process or a conditional Poisson process (see Bremaud [23, p. 21], Bielecki and Rutkowski [12, p. 193]). Intuitively, conditioned on the entire history of Y over $[0, \infty)$, X is a Poisson process.

1.2 Stochastic Integration

This section introduces the notion of a stochastic integral. It is based on Protter [151]. We define two integrals in this section. The first, the Ito–Stieltjes integral, is with respect to a finite variation process. The second, the (Ito) stochastic integral, is with respect to a semimartingale.

Definition 1.14 (Ito–Stieltjes Integrals) Let X be an adapted process of finite variation.

Let Y be an adapted cadlag process. Then,

$$\int_0^t Y_s(\omega) dX_s(\omega),$$

the pathwise Lebesgue–Stieltjes integral exists for all $\omega \in \Omega$ a.s. $\mathbb{P}$ (Medvegyev [136, Proposition 2.9, p. 115]). This pathwise integral is called the *Ito–Stieltjes integral.*

For a definition of the Lebesgue–Stieltjes integral, see Royden [160, p. 263].

For future use, let $\mathfrak{L}(X)$ denote the set of *optional* processes that are Ito–Stieltjes integrable with respect to X. We next start the process of defining a stochastic integral for semimartingales.

Definition 1.15 (Simple Predictable Processes) A stochastic process $\alpha_t(\omega)$ is a *simple predictable process* if it has a representation

$$\alpha_t = \alpha_0 + \sum_{i=1}^{n} \alpha_i 1_{(T_i, T_{i+1}]}(t),$$

where $0 = T_1 \leq \cdots \leq T_{n+1} < \infty$ is a finite sequence of stopping times and α_i is $\mathscr{F}_{T_i}$-measurable for all $i = 1, \ldots, n$.

Let the set of simple predictable processes be denoted by $\mathfrak{S}$. Note that this process is adapted and left continuous with right limits existing.

Definition 1.16 (Stochastic Integrals) Let X be a semimartingale.

For $\alpha \in \mathfrak{S}$, the *stochastic integral* is defined by

$$\int_0^t \alpha_s dX_s \equiv \alpha_0 X_0 + \sum_{i=1}^{n} \alpha_i (X_{T_{i+1}} - X_{T_i}) 1_{[T_{i+1}, \infty)}(t)$$

for all t.

We need to extend these integrands to a larger class of stochastic processes. First, we consider the set of all adapted, left continuous with right limits existing processes, denoted $\alpha \in \mathbb{L}$. We endow this set of processes with the uniformly on compacts in probability (ucp) topology (see Protter [151, p. 57]). We note that the space $\mathfrak{S}$ is dense in $\mathbb{L}$ under the ucp topology. Hence, given any $\alpha \in \mathbb{L}$ there exists a sequence $\alpha^n \in \mathfrak{S}$ such that $\alpha^n \to \alpha$. Also, endow the set of cadlag and adapted processes with the ucp topology. The stochastic integral defined above is in this set.

We can now define the stochastic integral for a semimartingale X and $\alpha \in \mathbb{L}$.

Definition 1.17 (Stochastic Integrals) Let X be a semimartingale.

For $\alpha \in \mathbb{L}$, choose a sequence $\alpha^n \in \mathfrak{S}$ such that it converges to α, then the *stochastic integral* is defined by

$$\int_0^t \alpha_s dX_s \equiv \text{ucp-}\lim_{n\to\infty} \int_0^t \alpha_s^n dX_s.$$

In this notation, we interpret $\int_0^t \alpha_s dX_s$ as a stochastic process defined on $t \in [0, \infty)$. We need to extend this stochastic integral to an even larger class of integrands, the set of predictable processes. Let the set of predictable processes be denoted by $\alpha \in \mathfrak{P}$.

This stochastic integral is extended from $\mathbb{L}$ to the class $\mathfrak{P}$, again, by taking limits. We sketch this construction. The construction is rather complicated. It proceeds by first restricting the set of semimartingales for which the stochastic integral is defined. To obtain this restriction, we introduce the $\mathscr{H}^2$ norm on the set of semimartingales and consider the set of semimartingales with finite norm (see Protter [151, p. 154]). This space of semimartingales is a Banach space. This norm induces a topology on the set of semimartingales. This gives the appropriate notion of limits in the space of semimartingales.

Next, we consider the predictable processes that are bounded, i.e. $\alpha \in \mathfrak{P}$ such that $|\alpha(w, t)| \leq K$ for all t a.s. $\mathbb{P}$, where K is a constant. We denote this set by $\alpha \in b\mathfrak{P}$. Continuing, we fix a semimartingale $X \in \mathscr{H}^2$ for which we will construct the stochastic integral. Using this semimartingale, we define a distance function $d_X(\alpha^1, \alpha^2)$ on the set of bounded predictable processes $\alpha^1, \alpha^2 \in b\mathfrak{P}$ (see Protter [151, p. 155]). This distance function induces a topology on the space of bounded predictable processes. This gives the appropriate notion of limits in the space of bounded predictable processes. Denote the set of bounded, adapted, and left continuous processes by $\alpha \in b\mathbb{L}$. We note that the space of bounded, adapted, and left continuous processes is dense in $b\mathfrak{P}$ (see Protter [151, p. 156]). Hence, given any $\alpha \in b\mathfrak{P}$ there exists a sequence $\alpha^n \in b\mathbb{L}$ such that $\alpha^n \to \alpha$.

We can now define the stochastic integral for $X \in \mathscr{H}^2$ and $\alpha \in b\mathfrak{P}$.

Definition 1.18 (Stochastic Integrals) Let $X \in \mathscr{H}^2$ be a semimartingale.

For $\alpha \in b\mathfrak{P}$, choose a sequence $\alpha^n \in b\mathbb{L}$ such that it converges to α, then

$$\int_0^t \alpha_s dX_s \equiv \mathscr{H}^2\text{-}\lim_{n\to\infty} \int_0^t \alpha_s^n dX_s.$$

As before, here we interpret $\int_0^t \alpha_s dX_s$ as a stochastic process defined on $t \in [0, \infty)$. Finally, this integral can be extended to a semimartingale X and a predictable process $\alpha \in \mathfrak{P}$ using a localization argument via a sequence of stopping times τ_n approaching infinity where the stopped processes $X_{\min(t,\tau_n)}$, $\alpha_{\min(t,\tau_n)}$ are in $\mathscr{H}^2$ and $b\mathfrak{P}$, respectively. We leave a description of this localization argument to Protter [151, p. 163]. This completes the construction. We define $\mathscr{L}(X) \subset \mathfrak{P}$ to be the set of predictable processes where the stochastic integral with respect to X exists (after the localization argument).

Not all stochastic integrals are local martingales.The following lemma gives sufficient conditions for a stochastic integral to be a local martingale.

Lemma 1.5 (Sufficient Condition for a Stochastic Integral to be a Local Martingale) *Let $H \in \mathscr{L}(X)$, where X is a local martingale.*

Consider

$$Y_t = Y_0 + \int_0^t H_s dX_s$$

for all t.

Let Y be bounded below, i.e. there exists a constant $c > -\infty$ such that $Y_t \geq c$ for all t a.s. $\mathbb{P}$.

Then, Y is a local martingale.

Proof By Protter [151, Theorem 89, p. 234], Y is a σ-martingale. Since Y is bounded below, by Ansel and Stricker [2], Y is a local martingale. This completes the proof.

1.3 Quadratic Variation

Definition 1.19 (Quadratic Variation and Quadratic Covariation) Let X, Y be semimartingales.

The *quadratic variation* of X, denoted $[X, X]_t$, is defined by

$$[X, X]_t = X_t^2 - 2\int_0^t X_{s-} dX_s.$$

The *quadratic covariation* of X, Y, denoted $[X, Y]_t$, is defined by

$$[X, Y]_t = X_t Y_t - \int_0^t X_{s-} dY_s - \int_0^t Y_{s-} dX_s,$$

where $X_{s-} = \lim_{t \to s,\, t<s} X_t$ and $Y_{s-} = \lim_{t \to s,\, t<s} Y_t$, and we use the convention that $X_{0-} = Y_{0-} = 0$.

Remark 1.8 (Decomposition) It can be shown that for a general semimartingale,

$$[X, X]_t = \left[X^c, X^c\right]_t + \sum_{0 \le s \le t} (\Delta X(s))^2 = [X, X]_t^c + \sum_{0 \le s \le t} (\Delta X(s))^2,$$

where Y^c denotes the continuous part of the stochastic process Y (see Medvegyev [136, p. 245]) and $\Delta X(t) = X(t) - X(t-)$. Note that $\Delta X(t)$ has only a countable number of nonzero values over $[0, t]$ (see Medvegyev [136, p. 5]), so the sum is well defined. This completes the remark.

Remark 1.9 (Alternative Characterization) The quadratic variation of X is a non-decreasing, adapted, and cadlag process where

$$[X, X]_0 = X_0^2, \ \Delta [X, X]_t = (\Delta X_t)^2,$$

$$X_0^2 + \sum_{i=1}^{n} \left(X_{T_{i+1}^n} - X_{T_i^n}\right)^2 \to [X, X]_t$$

under the uniformly on compacts in probability (ucp) topology and where $T_0^n \le T_1^n \le \cdots \le T_{k_n}^n$ is a sequence of stopping times such that $\lim_{n\to\infty} \sup_k T_k^n = t$ a.s. and $\sup_k \left|T_{K+1}^n - T_k^n\right| \to 0$ as $n \to \infty$ a.s. $\mathbb{P}$ (see Protter [151, p. 66]). This completes the remark.

Remark 1.10 (Conditional Quadratic Variation) The conditional quadratic variation of X, denoted $\langle X, X\rangle$, is the compensator of $[X, X]$, i.e. the unique finite variation predictable process such that $[X, X] - \langle X, X\rangle$ is a local martingale (see Protter [151, p. 122]). This implies that when X is continuous (i.e. $X = X^c$), the quadratic variation is equal to the conditional quadratic variation, i.e.

$$[X, X] = \left[X^c, X^c\right] = [X, X]^c = \langle X, X\rangle .$$

See Protter [151, p. 123], for a proof. This completes the remark.

Lemma 1.6 (Quadratic Variation of a Finite Variation Process) *Let X be a continuous, adapted process of finite variation.*

Then, $[X, X]$ is a constant process equal to X_0^2.

Let Y be a semimartingale.

Then, $[X, Y]$ is a constant process equal to $X_0 Y_0$.

Proof X a continuous process of finite variation implies that X is a quadratic pure jump semimartingale (Protter [151, Theorem 26, p. 71]). Then, Protter [151, Theorem 28, p. 75], implies both results, given that $\Delta X_s = 0$ for all s because X is continuous. This completes the proof.

1.4 Integration by Parts

The next theorem enables integration-by-parts (see Protter [151, p. 68]). The new result in this theorem is that the product of two semimartingales is again a semimartingale.

Theorem 1.1 (Integration by Parts) *Let X and Y be semimartingales. Then, X_tY_t is a semimartingale and*

$$X_tY_t = \int_0^t X_{s-}dY_s + \int_0^t Y_{s-}dX_s + [X, Y]_t .$$

1.5 Ito's Formula

An important theorem is Ito's formula (see Protter [151, p. 81]). For Ito's formula, we restrict our attention to continuous semimartingales.

Theorem 1.2 (Ito's Formula) *Let $X \equiv (X^1, \ldots, X^n)$ be an n-tuple of continuous semimartingales.*

Let $f : \mathbb{R}^n \to R$ be twice continuously differentiable.

Then,

$$f(X_t) - f(X_0) = \sum_{i=1}^{n} \int_0^t \frac{\partial f(X_s)}{\partial x_i} dX_s^i + \frac{1}{2} \sum_{i=1}^{n} \sum_{j=1}^{n} \int_0^t \frac{\partial^2 f(X_s)}{\partial x_i \partial x_j} d\left\langle X^i, X^j \right\rangle_s .$$

1.6 Girsanov's Theorem

The next theorem (see Protter [151, p. 141]) will prove useful in pricing derivatives.

Theorem 1.3 (Girsanov's Theorem) *Let W be a standard Brownian motion on $(\Omega, \mathscr{F}, (\mathscr{F}_t)_{0 \le t \le \infty}, \mathbb{P})$.*

Let $H \in \mathscr{L}(W)$ be a stochastic process.

Let $\mathbb{Q}$ defined by $\frac{d\mathbb{Q}}{d\mathbb{P}} = \exp\left\{-\frac{1}{2}\int_0^T H_s^2 ds - \int_0^T H_s dW_s\right\} > 0$ be a probability measure for some $T > 0$.

Then,

$$X_t = \int_0^t H_s ds + W_t$$

is a standard Brownian motion under $\mathbb{Q}$ for $0 \le t \le T$.

Remark 1.11 (Equivalent Probability Measures) The probability measure $\mathbb{Q}$ defined in Girsanov's theorem is equivalent to $\mathbb{P}$, written $\mathbb{Q} \sim \mathbb{P}$. This means that $\mathbb{Q}$ agrees with $\mathbb{P}$ on zero probability events, i.e. $\mathbb{P}(A) = 0 \Leftrightarrow \mathbb{Q}(A) = 0$ for all $A \in \mathscr{F}$. This completes the remark.

1.7 Essential Supremum

The following theorem (see Pham [149, p. 174]; note the proof here does not depend on S being a continuous process) will be important in super- and sub-replication.

Theorem 1.4 (Essential Supremum) *Let $X_T \geq 0$ be $\mathscr{F}_T$-measurable.*
Let S be an $\mathbb{R}^n$-valued semimartingale.
Then, the cadlag modification of the process

$$X_t = \operatorname*{ess\ sup}_{\mathbb{Q}\in\mathfrak{M}_l} E^{\mathbb{Q}}[X_T \,|\mathscr{F}_t\,]$$

is a supermartingale for any $\mathbb{Q} \in \mathfrak{M}_l$, where $\mathfrak{M}_l \equiv \{\mathbb{Q} \sim \mathbb{P} : S \text{ is a } \mathbb{Q}\text{-local martingale}\}$.

1.8 Optional Decomposition

The next theorem (see Follmer and Kabanov [62]) will also prove to be useful in super- and sub-replication.

Theorem 1.5 (Optional Decomposition) *Let S be an $\mathbb{R}^n$-valued semimartingale.*
Let $\mathfrak{M}_l \neq \emptyset$.
Let X be a local supermartingale with respect to any $\mathbb{Q} \in \mathfrak{M}_l$.
Then, there exists a nondecreasing cadlag adapted process C with $C_0 = 0$ and a predictable integrand $\alpha \in \mathscr{L}(S)$ such that

$$X_t = X_0 + \int_0^t \alpha_s dS_s - C_t.$$

1.9 Martingale Representation

The following theorem (see Protter [151, p. 187]) will prove useful in hedging derivatives.

Theorem 1.6 (Martingale Representation) *Let $W \equiv (W^1, \ldots, W^n)$ be an n-dimensional Brownian motion on $(\Omega, \mathscr{F}, (\mathscr{F}_t)_{t\in[0,T]}, \mathbb{P})$ with the filtration $(\mathscr{F}_t)_{t\in[0,T]}$ its completed natural filtration, where $\mathscr{F} = \mathscr{F}_T$.*
Let Z_T be $\mathscr{F}_T$-measurable with $E\,[|Z_T|] < \infty$.

Then, there exist predictable processes α^i *in* $\mathcal{L}(W^i)$ *with* $\int_0^T (\alpha_s^i)^2 ds < \infty$ *for* $i = 1, \ldots, n$ *such that*

$$Z_T = E[Z_T] + \sum_{i=1}^{n} \int_0^T \alpha_s^i dW_s^i$$

and where

$$Z_t = E[Z_T] + \sum_{i=1}^{n} \int_0^t \alpha_s^i dW_s^i$$

is a martingale.

1.10 Equivalent Probability Measures

The next theorem characterizes equivalent probability measures on a Brownian filtration. It will subsequently prove useful to understand arbitrage-free markets in a Brownian motion market.

Theorem 1.7 (Equivalent Probability Measures on a Brownian Filtration) *Let* $W \equiv (W^1, \ldots, W^n)$ *be an n-dimensional Brownian motion on* $(\Omega, \mathscr{F}, (\mathscr{F}_t)_{t\in[0,T]}, \mathbb{P})$ *with the filtration* $(\mathscr{F}_t)_{t\in[0,T]}$ *its completed natural filtration, where* $\mathscr{F} = \mathscr{F}_T$.

$\mathbb{Q} \sim \mathbb{P}$ *if and only if*

(1) there exist predictable processes α^i *in* $\mathcal{L}(W^i)$ *with* $\int_0^T (\alpha_s^i)^2 ds < \infty$ *for* $i = 1, \ldots, n$ *such that*
(2) $Z_t = e^{-\sum_{i=1}^n \int_0^t \alpha_s^i dW_s^i - \frac{1}{2}\sum_{i=1}^n \int_0^t (\alpha_s^i)^2 ds}$,
(3) $Z_T = \frac{d\mathbb{Q}}{d\mathbb{P}} > 0$, *and*
(4) $E[Z_T] = 1$ *where* $E[\cdot]$ *is expectation under* $\mathbb{P}$.

Proof ($\Leftarrow$) This direction is trivial. Assuming the hypotheses (1)–(4), the measure $\mathbb{Q}$ defined by condition (2) is a probability measure equivalent to $\mathbb{P}$.

($\Rightarrow$) Assume $\mathbb{Q} \sim \mathbb{P}$. Then, define $Z_T = \frac{d\mathbb{Q}}{d\mathbb{P}} > 0$. This gives $E[Z_T] = 1$. Protter [151, Corollary 4, p. 188], gives conditions (1) and (2). This completes the proof.

1.11 Notes

Excellent references for stochastic calculus include Karatzas and Shreve [117], Medvegyev [136], Protter [151], and Roger and Williams [157]. For books that present both the basics of stochastic calculus and its application to finance, see

Baxter and Rennie [10], Jeanblanc, Yor, and Chesney [113], Korn and Korn [124], Lamberton and Lapeyre [129], Mikosch [141], Musiela and Rutkowski [145], Shreve [169], and Sondermann [172].

Chapter 2
The Fundamental Theorems

This chapter presents the three fundamental theorems of asset pricing. These theorems are the basis for pricing and hedging derivatives, understanding the risk return relations among assets including the notion of systematic risk, portfolio optimization, and equilibrium asset pricing.

2.1 The Set-Up

We consider a continuous-time setting with a finite horizon $[0, T]$ where trading takes place at times $t \in [0, T)$ and the outcome of all trades are realized at time T. We assume that no trades take place at time $t = T$. This is because at time T all traded assets are liquidated and their proceeds distributed.

We are given a complete filtered probability space $(\Omega, \mathscr{F}, (\mathscr{F}_t)_{t\in[0,T]}, \mathbb{P})$ where the filtration $(\mathscr{F}_t)_{t\in[0,T]}$ satisfies the usual hypotheses and $\mathscr{F} = \mathscr{F}_T$. Here $\mathbb{P}$ is the *statistical probability measure*. By the statistical probability measure we mean that the probability $\mathbb{P}$ is that measure from which historical time series data are generated (drawn by nature). Hence, standard statistical methods can be used to estimate the probability measure $\mathbb{P}$ from historical time series data. For simplicity of notation, we adopt the convention that all of the subsequent equalities and inequalities given are assumed to hold almost surely (a.s.) with respect to $\mathbb{P}$, unless otherwise noted.

Traded in this market are a money market account and n risky assets. The markets are assumed to be *frictionless* and *competitive*. By frictionless we mean that there are no transaction costs, no differential taxes, shares are infinitely divisible, and there are no trading constraints, e.g. short sales restrictions, borrowing limits, or margin requirements. By competitive we mean that traders act as price-takers, i.e. they can trade any quantity of shares desired without affecting the market price. Alternatively stated, there is no liquidity risk. Liquidity risk is when there is a quantity impact from trading on the price. In Part IV of this book we will relax

R. A. Jarrow, *Continuous-Time Asset Pricing Theory*, Springer Finance,
https://doi.org/10.1007/978-3-319-77821-1_2

the no trading constraints assumption. However, the competitive market assumption is maintained throughout this book.

Let $\mathbb{B}_t = e^{\int_0^t r_s ds}$ denote the time t value of a money market account (mma), initialized with a dollar investment, $\mathbb{B}_0 = 1$. By construction it is continuous in time t and of finite variation on compact intervals (use Lemma 1.1 and Lemma 1.2 in Chap. 1). Here r_t is the default-free spot rate of interest, adapted to $\mathscr{F}_t$, and integrable with $\int_0^t |r_s| ds < \infty$. Note that $\mathbb{B}_t > 0$ for all t.

Let $\mathbb{S}_t = (\mathbb{S}_1(t), \ldots, \mathbb{S}_n(t))' > 0$ denote the time t prices of the n risky assets, which are strictly positive semimartingales with respect to $\mathscr{F}_t$. The prime denotes transpose. With a slight loss of generality, we assume that all risky asset prices are strictly positive, so that returns are well defined. We also assume that the risky assets have no cash flows (dividends) over $[0, T)$. The time T value of the risky assets can be viewed as a *liquidating cash flow (dividend)*. Recall that there is no trading at time T. Surprisingly perhaps, it is shown in Sect. 2.3 below that the assumption of no cash flows over $[0, T)$ is without loss of generality. It is important to note that the risky asset price processes can be discontinuous (with jumps). All of the subsequent results apply to these general price processes unless otherwise noted.

Remark 2.1 (Risky Assets May Include Derivatives) In the collection of risky assets, some could be derivatives. For example, $(\mathbb{S}_1(t), \ldots, \mathbb{S}_k(t))$ for $k < n$ could be stocks, and $(\mathbb{S}_{k+1}(t), \ldots, \mathbb{S}_n(t))$ could be European call options on stocks (e.g. $\mathbb{S}_{k+1}(t)$ could be a European call option on $\mathbb{S}_1(t)$). This completes the remark.

A *trading strategy* is defined to be a specification of the *number of units* of the mma and risky assets held for all state-time pairs $(\omega, t) \in \Omega \times [0, T]$, i.e. the specification of the stochastic processes $(\alpha_0(t), \alpha_t)' \in \mathbb{R}^{n+1}$ where $\alpha_t = \alpha(t) = (\alpha_1(t), \ldots, \alpha_n(t))' \in \mathbb{R}^n$ and the stochastic processes $(\alpha_0(t), \alpha_t)$ are adapted to $\mathscr{F}_t$. The trading strategy $(\alpha_0(t), \alpha_t)$ being adapted means that when constructing these trading strategies at time t, only information known at time t can be used in the construction. Alternatively stated, these trading strategies cannot be constructed using information from the future.

Let x be the initial value or wealth invested in the trading strategy. Then,

$$x = \mathbb{X}_0 = \alpha_0(0)\mathbb{B}_0 + \alpha_0 \cdot \mathbb{S}_0, \tag{2.1}$$

where $z \cdot y = \sum_{j=1}^n z_j y_j$ for $z, y \in \mathbb{R}^n$. This equals the number of shares in the trading strategy multiplied by the price per share and then summed across all traded assets. The trading strategy's value at time $t \in [0, T]$ is

$$\mathbb{X}_t = \alpha_0(t)\mathbb{B}_t + \alpha_t \cdot \mathbb{S}_t. \tag{2.2}$$

Define $\mathfrak{L}(\mathbb{B})$ to be the set of optional processes that are integrable with respect to $\mathbb{B}$, and $\mathscr{L}(\mathbb{S})$ to be the set of predictable processes that are integrable with respect to $\mathbb{S}$, see Sect. 1.2 for the formal definition of a stochastic integral. We assume that the holdings in the mma $\alpha_0 \in \mathfrak{L}(\mathbb{B})$ and the holdings in the risky asset $\alpha \in \mathscr{L}(\mathbb{S})$. Note the difference in measurability requirements on the position in the mma versus the risky assets.

From a mathematical perspective, this difference is due to the restrictions needed to guarantee the existence of the relevant integrals. This distinction also has an important economic interpretation. Recall that a predictable process is a stochastic process generated by the set of adapted and caglad (left continuous with right limits existing) processes. This means that when forming the position in the risky assets, only past information and current information generated *by the flow of time from the left* can be utilized. In contrast (this will be more clearly seen after normalizing prices by the value of the mma below), the position in the mma is a "residual" decision. The mma position is completely determined by the dynamic holdings in the risky assets (expression (2.4)) and the value of the portfolio at any time (expression (2.2)). This implies that its measurability requirements are determined by the position in the risky assets $\alpha \in \mathscr{L}(\mathbb{S})$ and the risky asset price process $\mathbb{S}_t$ via expression (2.2). Since $\mathbb{S}_t$ is an adapted and cadlag (right continuous with left limits existing) process (not necessarily predictable), and $\alpha \in \mathscr{L}(\mathbb{S})$ can be a constant (a buy and hold), this implies the more general measurability of the position in the mma (an optional process) is needed.

As defined, a trading strategy may require cash infusions or generate cash outflows after time 0 and before time T. To understand why, consider the trading strategy consisting of buying 1 unit of the mma at time 0 and purchasing another unit of the mma at time $\frac{T}{2}$, i.e.

$$(\alpha_0(t), \alpha_t) = \begin{cases} (1, (0, 0, \ldots, 0)) \text{ if } 0 \leq t < \frac{T}{2} \\ (2, (0, 0, \ldots, 0)) \text{ if } \frac{T}{2} \leq t \leq T. \end{cases}$$

At time $\frac{T}{2}$, the change in the value of this trading strategy is $\Delta\mathbb{X}_{\frac{T}{2}} = 2\mathbb{B}_{\frac{T}{2}} - \mathbb{B}_{\frac{T}{2}} = \mathbb{B}_{\frac{T}{2}} > 0$. To finance this change in value, a cash infusion of $\mathbb{B}_{\frac{T}{2}}$ dollars is needed.

Trading strategies that require no cash outflows or inflows between $(0, T)$ are called self-financing.

Definition 2.1 (Self-financing Trading Strategy (s.f.t.s.)) A trading strategy $(\alpha_0, \alpha) \in (\mathfrak{L}(\mathbb{B}), \mathscr{L}(\mathbb{S}))$ is *self-financing* if

$$\begin{aligned} \mathbb{X}_t &= x + \int_0^t \alpha_0(u) d\mathbb{B}_u + \int_0^t \alpha_u \cdot d\mathbb{S}_u \\ &= x + \int_0^t \alpha_0(u) r_u \mathbb{B}_u du + \int_0^t \alpha_u \cdot d\mathbb{S}_u \quad \textit{or} \end{aligned} \tag{2.3}$$

$$d\mathbb{X}_t = \alpha_0(t) r_t \mathbb{B}_t dt + \alpha_t \cdot d\mathbb{S}_t \tag{2.4}$$

for all $t \in [0, T]$.

As seen by expression (2.4), the change in the trading strategy's value is completely determined by the change in the value of the underlying assets' prices. There are no cash flows in or out of the trading strategy, hence the terminology.

Not all s.f.t.s.'s are reasonable. With continuous trading, without further restrictions on the trading strategy, it is possible to construct an s.f.t.s. that generates positive values for sure from zero initial wealth. These s.f.t.s.'s are called doubling strategies (see the example following the next definition). To exclude doubling strategies, we add the following restriction on an s.f.t.s.

Definition 2.2 (Admissible s.f.t.s.'s) An s.f.t.s. $(\alpha_0, \alpha) \in (\mathfrak{L}(\mathbb{B}), \mathscr{L}(\mathbb{S}))$ with initial value x and value process X_t is called *admissible* if there exists a constant $c \leq 0$ such that

$$X_t = x + \int_0^t \alpha_0(u) d\mathbb{B}_u + \int_0^t \alpha_u \cdot d\mathbb{S}_u \geq c$$

for all $t \in [0, T]$.

The additional restriction imposed by admissibility is that the value of the trading strategy cannot fall below some fixed negative level $c \leq 0$. This fixed lower bound imposes an implicit *borrowing constraint* on the trading strategy, which excludes doubling strategies. A doubling trading strategy and the purpose of the admissibility condition are illustrated in the following example.

Example 2.1 (Doubling Strategies) This example has two purposes. The first is to illustrate an s.f.t.s. that is a doubling strategy. The second is to show that the doubling strategy is not admissible. Hence the set of admissible s.f.t.s.'s will exclude such doubling strategies.

Fix a sequence of dates $\{t_i\}_{i=0,1,2,\ldots} \to T$ with $t_0 = 0$. Consider a risky asset whose market price at time $t_0 = 0$ is $\mathbb{S}_0 = 1$, and at each subsequent time t_i its price changes discretely

$$\mathbb{S}_{t_i} = \begin{cases} 2\mathbb{S}_{t_{i-1}} & \text{with probability } \frac{1}{2} \\ \frac{1}{2}\mathbb{S}_{t_{i-1}} & \text{with probability } \frac{1}{2}. \end{cases}$$

Let the stock price process be right continuous with left limits existing. This asset price process is strictly positive and discontinuous.

Let the mma be identically one, i.e. $\mathbb{B}_t \equiv 1$.

We define the trading strategy recursively. The trading strategy starts with zero investment, holding one share of the risky asset and shorting the money market account to finance the position. It liquidates the first time the risky asset price increases. Otherwise, it retrades and "doubles-up" as follows.

At time $t_0 = 0$, $\alpha(0) = 1$ and $\alpha_0(0) = -1 < 0$ is chosen so that

$$\mathbb{X}_0 = \alpha_0(0) + \alpha(0)\mathbb{S}_0 = -1 + 1 \cdot 1 = 0.$$

At time t_i for $i \geq 1$, the stock price $\mathbb{S}_{t_i}$ is known.

If $\mathbb{S}_{t_i} = 2\mathbb{S}_{t_{i-1}}$, then liquidate the position. The liquidated value is

$$\mathbb{X}_{t_i} = \alpha_0(t_{i-1}) + \alpha(t_{i-1})2\mathbb{S}_{t_{i-1}}.$$

If $\mathbb{S}_{t_i} = (\frac{1}{2})\mathbb{S}_{t_{i-1}}$, then retrade. The value at the start of time t_i before retrading is

$$\mathbb{X}_{t_i} = \alpha_0(t_{i-1}) + \alpha(t_{i-1})\left(\frac{1}{2}\right)^i.$$

This is because the asset price has never increased (otherwise the position was liquidated strictly before t_i).

We want to buy more shares in the risky asset $\alpha(t_i)$ and short additional units of the mma $\alpha_0(t_i)$ keeping the strategy self-financing. Hence, $(\alpha_0(t_i), \alpha(t_i))$ are chosen so that

$$\mathbb{X}_{t_i} = \alpha_0(t_{i-1}) + \alpha(t_{i-1})\left(\frac{1}{2}\right)^i = \alpha_0(t_i) + \alpha(t_i)\left(\frac{1}{2}\right)^i.$$

In addition, choose $\alpha(t_i)$ so that if the asset price jumps up at the next time t_{i+1}, the value of risky asset position is strictly greater than the short position in the mma, i.e.

$$\alpha(t_i)2\left(\frac{1}{2}\right)^i > -\alpha_0(t_i) = -\alpha_0(t_{i-1}) + [\alpha(t_i) - \alpha(t_{i-1})]\left(\frac{1}{2}\right)^i.$$

This is always possible. Indeed, simplifying this inequality, $\alpha(t_i)$ just needs to satisfy

$$\begin{aligned}\alpha(t_i) &> -2^i\alpha_0(t_{i-1}) - \alpha(t_{i-1}) \\ &= -2^i\left[\alpha_0(t_{i-1}) + (\tfrac{1}{2})^i\alpha(t_{i-1})\right] = -2^i\mathbb{X}_{t_i}.\end{aligned} \tag{2.5}$$

Note that if the risky asset jumps up at the next time t_{i+1}, then

$$\mathbb{X}_{t_{i+1}} = \alpha_0(t_i) + \alpha(t_i)2\left(\frac{1}{2}\right)^i > 0.$$

This is an "arbitrage opportunity" (to be defined later) because it starts with zero investment and it is terminated with a strictly positive value with probability one. A jump up occurs with probability one because the realizations of the stock price across time are independent and identically distributed, and there are an infinite number of realizations possible before time T.

We now prove by induction that this strategy's value has a strictly positive probability of falling below any negative constant $c < 0$. Hence it is not admissible. Thus, the admissibility condition will remove such doubling strategies.

Step ($i = 1$). Suppose we are at time t_1 with an asset price equal to $\mathbb{S}_{t_1} = \frac{1}{2}$. The value is

$$\mathbb{X}_{t_1} = \alpha_0(0) + \alpha(0)\frac{1}{2} = -\frac{1}{2} < 0.$$

Step ($i + 1$). Suppose that we enter this time step with $\mathbb{X}_{t_i} < 0$ where the asset price is $\mathbb{S}_{t_{i+1}} = (\frac{1}{2})^i(\frac{1}{2})$. Such a consecutive sequence of down jumps occurs with positive probability for all $i \geq 1$. The value of the trading strategy is

$$\mathbb{X}_{t_{i+1}} = \alpha_0(t_i) + \alpha(t_i)\left(\frac{1}{2}\right)^i\left(\frac{1}{2}\right).$$

First, we show that $\mathbb{X}_{t_{i+1}} - \mathbb{X}_{t_i} < (\frac{1}{2})\mathbb{X}_{t_i} < 0$, i.e. the loss in value is greater than half the absolute value of the position's previous value at each time step.

Proof

$$\begin{aligned}\mathbb{X}_{t_{i+1}} - \mathbb{X}_{t_i} &= \alpha(t_i)\left(\frac{1}{2}\right)^i\left(\frac{1}{2}\right) + \alpha_0(t_i) - \alpha(t_i)\left(\frac{1}{2}\right)^i - \alpha_0(t_i)\\ &= -\alpha(t_i)\left(\frac{1}{2}\right)^i\left(\frac{1}{2}\right) < 0.\end{aligned}$$

By expression (2.5),

$$\alpha(t_i) > -2^i\mathbb{X}_{t_i} > 0$$

if and only if

$$\alpha(t_i)\left(\frac{1}{2}\right)^i\left(\frac{1}{2}\right) > -\left(\frac{1}{2}\right)^i\left(\frac{1}{2}\right)2^i\mathbb{X}_{t_i} = -(\frac{1}{2})\mathbb{X}_{t_i} > 0.$$

Combining the two previous inequalities completes the first proof.

Second, we show that this implies that there is a sequence of $i + 1$ bad draws such that $\mathbb{X}_{t_{i+1}} < c$ with positive probability, proving the claim.

Proof Note that $\mathbb{X}_{t_{i+1}} - \mathbb{X}_{t_i} < (\frac{1}{2})\mathbb{X}_{t_i}$ implies that

$$\mathbb{X}_{t_{i+1}} < \left(\frac{3}{2}\right)\mathbb{X}_{t_i} < \left(\frac{3}{2}\right)^2\mathbb{X}_{t_{i-1}}\cdots < \left(\frac{3}{2}\right)^{i-2}\mathbb{X}_{t_1} = \left(\frac{3}{2}\right)^{i-2}\left(-\frac{1}{2}\right).$$

If we choose i such that $(\frac{3}{2})^{i-2}(-\frac{1}{2}) < c$, then the proof is completed.

Simple algebra shows that an $i > \frac{\log(-2c)}{\log 3 - \log 2} + 2$ satisfies this condition.
Note that this sequence of bad draws occurs with probability $\left(\frac{1}{2}\right)^{i+1} > 0$.
This completes the second proof.

Hence, this doubling strategy is not admissible. This completes the example.

The set of admissible s.f.t.s.'s for a given initial wealth x is denoted by

$$\mathscr{A}(x) = \{(\alpha_0, \alpha) \in (\mathfrak{L}(\mathbb{B}), \mathscr{L}(\mathbb{S})) : \mathbb{X}_t = \alpha_0(t)\mathbb{B}_t + \alpha_t \cdot \mathbb{S}_t, \ \exists c \leq 0,$$

$$\mathbb{X}_t = x + \int_0^t \alpha_0(u) d\mathbb{B}_u + \int_0^t \alpha_u \cdot d\mathbb{S}_u \geq c, \ \forall t \in [0, T]\Big\}.$$

Remark 2.2 (Admissibility and (Naked) Short Selling) If a risky asset's price process $\mathbb{S}_i(t)$ is unbounded above, i.e. for any $c > 0$, $\mathbb{P}(\sup_{t \in [0,T]} \mathbb{S}_i(t) > c) > 0$, the admissibility condition excludes (naked) short selling of the risky asset from the set of admissible s.f.t.s.'s $\mathscr{A}(x)$ where the initial wealth is $x = -\mathbb{S}_i(0) < 0$. Indeed, (naked) short selling of the ith risky asset is the s.f.t.s.

$$(\alpha_0(t), \alpha_1(t), \ldots, \alpha_{i-1}(t), \alpha_i(t), \alpha_{i+1}(t), \ldots \alpha_n(t)) = (0, 0, \ldots, 0, -1, 0, \ldots, 0)$$

for all $t \in [0, T]$. Because the value process of this trading strategy is

$$\mathbb{X}(t) = \alpha_0(t)\mathbb{B}_t + \alpha_t \cdot \mathbb{S}_t = -\mathbb{S}_i(t)$$

for all t, we see that for any $c > 0$,

$$\mathbb{P}(\inf_{t \in [0,T]} \mathbb{X}(t) < -c) = \mathbb{P}(\sup_{t \in [0,T]} \mathbb{S}_i(t) > c) > 0,$$

violating the admissibility condition. This completes the remark.

For subsequent usage, a *market* is defined to be a collection $((\mathbb{B}, \mathbb{S}), (\mathscr{F}_t), \mathbb{P})$ representing the stochastic processes for the traded assets, the market's information set, and the statistical probability measure. *A market is always assumed to be frictionless (unless otherwise indicated) and competitive.* The underlying state space and σ-algebra, $(\Omega, \mathscr{F})$, are always implicit in this collection and not included.

Remark 2.3 (Frictionless and Competitive Market Assumptions) The assumption of frictionless markets implicitly appears in the definition of the set of admissible s.f.t.s.'s $\mathscr{A}(x)$ in that the accumulated value of the trading strategy has no additional adjustments for frictions, e.g. transaction costs, taxes, indivisible shares, short sale constraints, margin requirements.

The assumption of competitive markets implicitly appears in the definition of $\mathscr{A}(x)$ because the price processes $(\mathbb{B}_u, \mathbb{S}_u)$ do not depend on the trading strategy (α_0, α). That is, the trader is a price-taker in that there is no quantity impact on the price processes from trading.

Although the convention in the literature is to label the above market structure as frictionless, this is somewhat of a misnomer. Indeed, the admissibility condition is, in fact, a trading constraint imposed on the aggregate value of the s.f.t.s. Because admissibility is needed to exclude doubling strategies, its imposition is thought to be very mild. It is also standard in portfolio optimization problems, in the context of a frictionless market, to impose an analogous constraint that a trader's wealth is always nonnegative (see Part II of this book).

It is important to keep this misnomer in mind when using the phrase "frictionless markets" in the remainder of this book. Doing so one can more easily understand why asset price bubbles often appear. For example, in the asset price bubbles Chap. 3 this clarifies why asset price bubbles exist in a frictionless and competitive market where there are no arbitrage opportunities (to be defined). Second, in the portfolio optimization Chaps. 10–12 this clarifies how an optimal wealth and consumption path can exist in the presence of asset price bubbles in a frictionless and competitive market. And finally, in Chaps. 13–16 studying economic equilibrium this also clarifies how asset price bubbles can exist in a frictionless and competitive market rational equilibrium. This completes the remark.

2.2 Change of Numeraire

Normalization by the money market account, which is a change of numeraire, simplifies the notation and is almost without loss of generality. The lost of generality is that the set of trading strategies $\mathscr{A}(x)$ after the change of numeraire may differ from the set of trading strategies before due to the modified integrability conditions needed to guarantee that the relevant integrals exist. This section presents the new notation and the evolutions for the mma and the risky assets under this change of numeraire.

Let $B_t \equiv \frac{\mathbb{B}_t}{\mathbb{B}_t} = 1$ for all t. This represents the normalized value of the money market account (mma).

Let $S_t = (S_1(t), \ldots, S_n(t))' \geq 0$ represent the risky asset prices when normalized by the value of the mma, i.e. $S_i(t) = \frac{\mathbb{S}_i(t)}{\mathbb{B}_t}$. Then,

$$\frac{dB_t}{B_t} = 0 \qquad \text{and} \tag{2.6}$$

$$\frac{dS_t}{S_t} = \frac{d\mathbb{S}_t}{\mathbb{S}_t} - r_t dt. \tag{2.7}$$

Proof Using the integration by parts formula Theorem 1.1 in Chap. 1, one obtains (dropping the t's)

$$d\left(\frac{\mathbb{S}}{\mathbb{B}}\right)=\frac{1}{\mathbb{B}}d\mathbb{S}+\mathbb{S}d\left(\frac{1}{\mathbb{B}}\right)=\frac{d\mathbb{S}}{\mathbb{S}}\frac{\mathbb{S}}{\mathbb{B}}-\frac{\mathbb{S}}{\mathbb{B}}\frac{d\mathbb{B}}{\mathbb{B}}.$$

The first equality uses $d\left[\mathbb{S},\frac{1}{\mathbb{B}}\right]=0$, since $\mathbb{B}$ is continuous and of finite variation (use Lemmas 1.2 and 1.6 in Chap. 1).

Substitution yields

$$dS=\frac{d\mathbb{S}}{\mathbb{S}}S-S\frac{d\mathbb{B}}{\mathbb{B}}.$$

Algebra completes the proof.

Recall that $\mathscr{L}(S)$ is the set of predictable processes integrable with respect to S and $\mathscr{O}$ is the set of optional processes. We have that the set of *admissible self-financing trading strategies* is

$$\begin{aligned}\mathscr{A}(x)=\{(\alpha_0,\alpha)\in(\mathscr{O},\mathscr{L}(S)):\ &X_t=\alpha_0(t)+\alpha_t\cdot S_t,\ \exists c\leq 0,\\ &X_t=x+\textstyle\int_0^t\alpha_u\cdot dS_u\geq c,\ \forall t\in[0,T]\Big\}.\end{aligned}$$

The normalized value of a trading strategy is

$$\frac{\mathbb{X}_t}{\mathbb{B}_t}=\alpha_0(t)+\alpha_t\cdot\frac{\mathbb{S}_t}{\mathbb{B}_t}$$

or

$$X_t=\alpha_0(t)+\alpha_t\cdot S_t. \tag{2.8}$$

The value process evolution

$$d\mathbb{X}_t=\alpha_0(t)r_t\mathbb{B}_tdt+\alpha_t\cdot d\mathbb{S}_t$$

is equivalent to

$$\begin{aligned}X_T&=X_0+\textstyle\int_0^T\alpha_t\cdot dS_t \text{ or}\\ dX_t&=\alpha_t\cdot dS_t.\end{aligned} \tag{2.9}$$

As easily seen, these expressions are obtainable from the non-normalized evolutions by setting $\mathbb{B}_t\equiv 1$ and $r_t\equiv 0$.

Proof All of the next steps are reversible.

$$dX = \alpha \cdot dS$$

$$d\left(\frac{\mathbb{X}}{\mathbb{B}}\right) = \alpha \cdot d\left(\frac{\mathbb{S}}{\mathbb{B}}\right).$$

The integration by parts formula Theorem 1.1 in Chap. 1, using the fact that $d\left[\mathbb{X}, \frac{1}{\mathbb{B}}\right] = 0$ since $\mathbb{B}$ is continuous and of finite variation (use Lemmas 1.2 and 1.6 in Chap. 1), yields

$$d\left(\frac{\mathbb{X}}{\mathbb{B}}\right) = \frac{1}{\mathbb{B}}d\mathbb{X} - \mathbb{X}d\left(\frac{1}{\mathbb{B}}\right) = \frac{d\mathbb{X}}{\mathbb{B}} - \frac{\mathbb{X}}{\mathbb{B}}\frac{d\mathbb{B}}{\mathbb{B}} = \alpha \cdot \left(\frac{d\mathbb{S}}{\mathbb{B}} - \frac{\mathbb{S}}{\mathbb{B}}\frac{d\mathbb{B}}{\mathbb{B}}\right).$$

This equality uses

$$\mathbb{X} = \alpha_0\mathbb{B} + \alpha \cdot \mathbb{S}.$$

$$d\mathbb{X} - Xrdt = \alpha \cdot (d\mathbb{S} - \mathbb{S}rdt).$$

Again using

$$\mathbb{X} = \alpha_0\mathbb{B} + \alpha \cdot \mathbb{S}$$

we get

$$d\mathbb{X} = (\alpha_0\mathbb{B} + \alpha \cdot \mathbb{S})rdt + \alpha \cdot (d\mathbb{S} - \mathbb{S}rdt).$$

Simplifying yields

$$d\mathbb{X} = \alpha_0 r\mathbb{B}dt + \alpha \cdot d\mathbb{S}.$$

This completes the proof.

A *normalized market*, frictionless (unless otherwise noted) and competitive, is defined to be the collection $(S, (\mathscr{F}_t), \mathbb{P})$.

2.3 Cash Flows

This section shows that the assumption of no cash flows to the risky assets over $[0, T)$ is without loss of generality. Indeed, it is shown that if there are cash flows, a simple transformation of the risky asset price processes generates an "equivalent" market with no cash flows. All the theorems can be generated for this equivalent market with no cash flows and then the transformation reversed to obtain the results in the original market. Of course, the time T value of the risky assets can be viewed as a liquidating cash flow since there is no trading at that date.

We are given a *normalized market* $(S, (\mathscr{F}_t), \mathbb{P})$. For simplicity, assume that the market consists of just one risky asset with price process denoted by S_t. Let this risky asset have a *cumulative cash flow* process which is a nonnegative, nondecreasing, cadlag, adapted stochastic process that equals 0 at time 0, i.e.

$$\mathscr{G}_t \geq 0 \quad \text{with} \quad \mathscr{G}_0 = 0.$$

Since we are using the normalized market representation, the cumulative cash flow process is in units of the mma. These conditions imply that the cash flow process is of finite variation.

To transform the market with cash flows into an equivalent market without cash flows, we need to reinvest all cash flows into either the mma or the risky asset. We consider both possibilities.

2.3.1 *Reinvest in the MMA*

Denote the new risky asset's price with reinvestment in the mma by S_t^*. The transformation is

$$S_t^* = S_t + \mathscr{G}_t.$$

Note that this transformation is additive. A useful special case of this transformation is where for each $\omega \in \Omega$, $\mathscr{G}_t$ is absolutely continuous, i.e. there exists a nonnegative, cadlag, adapted stochastic process $G_t \geq 0$ such that

$$\mathscr{G}_t = \int_0^t G_t dt.$$

In this case,

$$dS_t^* = dS_t + G_t dt.$$

Here, the price change in the risky asset with reinvestment in the mma over $[t, t+dt]$ equals the price change in the risky asset plus the cash flows received.

2.3.2 Reinvest in the Risky Asset

Denote the new risky asset's price with reinvestment in the risky asset by S_t^*. The transformation is

$$S_t^* = S_t + S_t \int_0^t \frac{1}{S_u} d\mathcal{G}_u.$$

To understand this expression, note that at time u a cash flow of $d\mathcal{G}_u$ occurs. This enables the purchase of $\frac{1}{S_u} d\mathcal{G}_u$ shares of the risky asset. The accumulated value of this position at time t is the right side of this expression.

A useful special case of this transformation is where there exists a nonnegative, nondecreasing, cadlag, adapted stochastic process $G_t \geq 0$ with $G_0 = 1$ such that

$$d\mathcal{G}_t = S_t dG_t.$$

Substitution yields

$$S_t^* = S_t + S_t \int_0^t dG_u = S_t(G_t - 1).$$

Note that this transformation is multiplicative. Define g_t by the following expression

$$1 + G_t = e^{g_t}.$$

Then,

$$S_t^* = S_t e^{g_t}.$$

Assume, in addition, that g_t is a continuous process. Then,

$$\frac{dS_t^*}{S_t^*} = \frac{dS_t}{S_t} + dg_t.$$

In this case, the return on the risky asset with cash flows reinvested over $[t, t + dt]$ equals the return on the risky asset plus the cash flows received.

Proof Using the integration by parts Theorem 1.1 in Chap. 1, we obtain

$$dS_t^* = dS_t \cdot e^{g_t} + S_t \cdot e^{g_t} dg_t = \frac{dS_t}{S_t} \cdot S_t e^{g_t} + S_t \cdot e^{g_t} \cdot dg_t.$$

This uses the fact that $[S, g] = 0$ since g is continuous and of finite variation (Lemma 1.6 in Chap. 1). Dividing by S_t^* completes the proof.

As just proved, a market $(S_t, \mathcal{G}_t)$ with cash flows over $[0, T)$ can be transformed into an equivalent market $(S_t^*, 0)$ with no cash flows over $[0, T)$ using either of the transformations given above. All the analysis can be done in the equivalent market without cash flows, then the transformation reversed to get the results in the original market with cash flows. For the remainder of this book, without loss of generality, we will only consider the equivalent market with no cash flows.

2.4 The First Fundamental Theorem

This section presents the First Fundamental Theorem of asset pricing. Before presenting the First Fundamental Theorem, however, there are a number of concepts that need to be introduced, so that the theorem is well understood. We start with a *normalized market* $(S, (\mathcal{F}_t), \mathbb{P})$.

2.4.1 No Arbitrage (NA)

We start with a definition of an arbitrage opportunity.

Definition 2.3 (No Arbitrage (NA)) An admissible s.f.t.s. $(\alpha_0, \alpha) \in \mathcal{A}(x)$ with value process X is a (simple) *arbitrage opportunity* if

(i) $X_0 = x = 0$, (zero investment)
(ii) $X_T = \int_0^T \alpha_t \cdot dS_t \geq 0$ with $\mathbb{P}$ probability one, and
(iii) $\mathbb{P}(X_T > 0) > 0$.

A market satisfies no arbitrage (NA) if there are no trading strategies that are arbitrage opportunities.

A simple arbitrage opportunity requires zero wealth at time 0. It is called a *zero investment* trading strategy. In addition, an arbitrage opportunity never loses any wealth and with strictly positive probability it generates positive wealth at time T. This is equivalent to getting a free lottery ticket, where the lottery is yet to occur.

Remark 2.4 (Other Simple Arbitrage Opportunities) Two other types of simple arbitrage opportunities, that are also excluded by NA, need to be understood. The first is where the admissible s.f.t.s. $(\alpha_0, \alpha) \in \mathcal{A}(x)$ for $X_0 = x$ is not zero investment, but it generates an immediate positive cash flow. In particular, it satisfies

(i) $X_0 = x < 0$, and
(ii) $X_T = x + \int_0^T \alpha_t \cdot dS_t \geq 0$ with $\mathbb{P}$ probability one.

Using the mma, one can easily transform this type of arbitrage opportunity into that used in the definition. This is done by immediately purchasing x units of the mma with the positive cash flow generated by condition (i), so that the time 0 value of the transformed trading strategy is zero. Note that a negative value at time 0, as in condition (i), implies a positive cash flow. With this reinvestment, condition (ii) then implies that the trading strategy's terminal value will be greater than or equal to $\int_0^T \alpha_t \cdot dS_t$, which is strictly greater than zero with $\mathbb{P}$ probability one. Hence, this transformed trading strategy is now an arbitrage opportunity as in Definition 2.3.

The second type of alternative arbitrage trading strategy is where the admissible s.f.t.s. $(\alpha_0, \alpha) \in \mathscr{A}(x)$ for $X_0 = x$ is not zero investment, and it generates payoffs in excess of its initial investment. In particular, it satisfies

(i) $X_0 = x > 0$,
(ii) $X_T = x + \int_0^T \alpha_t \cdot dS_t \geq x$ with $\mathbb{P}$ probability one, and
(iii) $\mathbb{P}(X_T > x) > 0$.

Using the mma, one can easily transform this type of arbitrage opportunity into that used in the definition. This is done by immediately shorting x units of the mma, and adding this to the initial s.f.t.s.

Hence, because of these possibilities, restricting our attention to zero investment trading strategies as in Definition 2.3 is without-loss-of-generality. This completes the remark.

Remark 2.5 (Numeraire Invariance) It is important to note that the definition of an arbitrage opportunity is invariant to normalization by the mma. Indeed, since $\mathbb{B}_t > 0$ all t,

$$\begin{aligned} \mathbb{X}_0 > 0 \quad &\Leftrightarrow \quad X_0 = \tfrac{\mathbb{X}_0}{\mathbb{B}_0} > 0 \\ \mathbb{X}_T \geq 0 \quad &\Leftrightarrow \quad X_T = \tfrac{\mathbb{X}_T}{\mathbb{B}_T} \geq 0 \\ \mathbb{P}(\mathbb{X}_T > 0) > 0 &\Leftrightarrow \mathbb{P}(X_T > 0) = \mathbb{P}\left(\tfrac{\mathbb{X}_T}{\mathbb{B}_T} > 0\right) > 0 \end{aligned}$$

This completes the verification and the remark.

Remark 2.6 (Invariance With Respect to a Change in Equivalent Probability Measures) The admissible s.f.t.s.'s in the definitions of NA are invariant with respect to a change in equivalent probability measures. This follows because the definitions only depend on events of probability zero or strictly positive probability. This completes the remark.

Example 2.2 (Asset Price Processes Violating NA) This example illustrates asset price processes that violate NA.

1. (*Reflected Brownian Motion*)

Let the market consist of a single risky asset and an mma with the risky asset price process given by

$$S_1(t) = S_0 \left| W_t \right|,$$

where $S_0 > 0$ and W_t is a standard Brownian motion with $W_0 = 1$. This process is known as a reflected Brownian motion.

This price process admits a simple arbitrage opportunity. Indeed, consider the following trading strategy. Buy the risky asset the first time after time 0 that $S_t = 0$. Let $\tau > 0$ correspond to this first hitting time of 0. Note that $P(\tau < T) > 0$. Then, sell the risky asset the first time after τ that $S_t > \varepsilon$, where $\varepsilon > 0$ is a constant. Note that $\mathbb{P}(S_t > \varepsilon | S_\tau = 0, \tau < T) > 0$ for all $t \in (\tau, T]$. After selling, invest the proceeds in the mma until time T.

Let X_t denote the value process of this trading strategy. Note that at time 0, $X_0 = 0$ because no position is taken in the risky asset or the mma. At time τ, $X_\tau = 1 \cdot S_\tau = 0$ and

$$X_T = \begin{cases} S_\tau = \varepsilon \text{ if } \tau < T \\ 0 \quad \text{ if } \tau = T. \end{cases}$$

This trading strategy is self-financing and it is admissible since $X_t \geq 0$ for all $t \in [0, T]$. Finally, since $\mathbb{P}(X_T = \varepsilon > 0) > 0$ and $\mathbb{P}(X_T \geq 0) = 1$, it is a simple arbitrage opportunity.

This example can be generalized to any risky asset price process that hits zero at some $\tau \in (0, T]$ with strictly positive probability and after hitting zero the value process has a strictly positive probability of becoming strictly positive again before time T. The same admissible s.f.t.s. works to create an arbitrage opportunity. Hence, we get the following fact.

Assume NA. Given a risky asset price process S_t, if $S_\tau = 0$ for any $\tau > 0$, then $S_t = 0$ for all $t \in [\tau, T]$.

2. (*Redundant Assets with Different Drifts*)

Let the market consist of two risky assets and an mma with the risky asset price processes given by

$$S_1(t) = S_0 e^{R_1 t + \sigma W_t}, \quad \text{and}$$
$$S_2(t) = S_0 e^{R_2 t + \sigma W_t},$$

where $S_0 > 0$, $R_1 > R_2 \geq 0$ are constants, and W_t is a standard Brownian motion with $W_0 = 0$. Both of these risky asset prices follow geometric Brownian motions.

These price processes admit a simple arbitrage opportunity. Indeed, consider the following trading strategy. At time 0, buy risky asset 1 and short risky asset 2. Hold these positions until time T. Next, let X_t denote the value process of this trading strategy. Note that at time 0, $X_0 = S_0 - S_0 = 0$, and

$$X_T = S_0(e^{R_1 T} - e^{R_2 T}) > 1 \ a.s. \ \mathbb{P}.$$

This strategy is admissible since $X_t \geq 0$ for all $t \in [0, T]$, it is self-financing and, therefore, it is a simple arbitrage opportunity. This completes the example.

Unfortunately, this intuitive definition of no arbitrage is not strong enough to obtain the First Fundamental Theorem of asset pricing. The reason is because an admissible s.f.t.s.'s value is characterized as a stochastic integral, and stochastic integrals are themselves defined as limits of sums. Hence, approximate arbitrage opportunities that become arbitrage opportunities in the limit must also be excluded. We study an additional type of admissible s.f.t.s. that characterizes these "approximate arbitrage opportunities."

2.4.2 No Unbounded Profits with Bounded Risk (NUPBR)

An additional type of admissible s.f.t.s. relates to a limiting type of arbitrage opportunity called no unbounded profits with bounded risk (NUPBR). This definition is from Karatzas and Kardaras [116, p. 465].

Definition 2.4 (No Unbounded Profits with Bounded Risk (NUPBR)) A sequence of admissible s.f.t.s.'s $(\alpha_0, \alpha)_n \in \mathscr{A}(x)$ with value processes X^n generate *unbounded profits with bounded risk (UPBR)* if

(i) $X_0^n = x > 0$, (positive investment)
(ii) $X_t^n \geq 0$ a.s. $\mathbb{P}$ for all $t \in [0, T]$ (always nonnegative), and
(iii) $(X_T^n)_n \geq 0$ are unbounded in probability, i.e. $\lim_{m \to \infty} \left(\sup_n \mathbb{P}\left(X_T^n > m\right)\right) > 0$.
If no such sequence of trading strategies exists, then the market satisfies NUPBR.

A sequence of admissible s.f.t.s.'s whose value processes are always nonnegative with an initial value of $x > 0$ (condition (i)) generates UPBR if it never loses more than the strictly positive initial investment of x units (condition (ii)), and it generates unboundedly large profits at time T with strictly positive probability (condition (iii)). Note, that this admissible s.f.t.s. can incur losses, hence it has risk. But, the risk is bounded because the loss is limited to the initial investment of $x > 0$.

Without-loss-of-generality, one can replace $x > 0$ with $x = 1$ in Definition 2.4. Indeed, if the admissible s.f.t.s. $(\alpha_0, \alpha)_n \in \mathscr{A}(1)$ satisfies all of the conditions of Definition 2.4, then the admissible s.f.t.s. $(\tilde{\alpha}_0, \tilde{\alpha})_n = (x\alpha_0, x\alpha)_n \in \mathscr{A}(x)$ satisfies Definition 2.4 for an arbitrary $x > 0$.

Remark 2.7 (No Arbitrage of the First Kind) Because of condition (iii), the definition of UPBR is difficult to understand. The definition of an UPBR can be clarified by relating it to the notion of an arbitrage opportunity of the first kind. This definition is from Kardaras [121, p. 653].

Definition (No Arbitrage of the First Kind) An $\mathscr{F}_T$-measurable random variable χ generates an *arbitrage opportunity of the first kind* if

(i) $\mathbb{P}(\chi \geq 0) = 1$,
(ii) $\mathbb{P}(\chi > 0) > 0$, and

(iii) for all $x > 0$ there exists an admissible s.f.t.s. $(\alpha_0, \alpha) \in \mathscr{A}(x)$ with value process $X^{x,\alpha} \geq 0$ such that $X_T^{x,\alpha} \geq \chi$.
The market satisfies *no arbitrage of the first kind* if there exists no arbitrage opportunities of the first kind.

As defined, no arbitrage of the first kind excludes a related, but different type of admissible s.f.t.s. Indeed, to be an arbitrage opportunity, there needs to exist a random variable χ satisfying conditions (i) and (ii), and a zero investment ($x = 0$) admissible s.f.t.s. $(\alpha_0, \alpha) \in \mathscr{A}(0)$ such that $X_T^{0,\alpha} \geq \chi$. Here, however, the admissible s.f.t.s. must have a positive initial investment $x > 0$ with a value process that is nonnegative at time T. Key here is that the initial investment can be made arbitrarily small since condition (ii) must be true for all $x > 0$.

The fact that the initial investment must be strictly positive is crucial. It implies that a "small" loss on the trading strategy is possible, although the loss is bounded by the initial value of x. Note also that the magnitude of this random variable χ can be made unboundedly large by multiplying both it and the initial investment x by an arbitrary constant $c > 0$ and letting $c \to \infty$.

An amazing result, shown by Kardaras [121], is that NUPBR is equivalent to no arbitrage of the first kind. This completes the remark.

Remark 2.8 (Invariance With Respect to a Change in Equivalent Probability Measures) It is important to note that the admissible s.f.t.s.'s in the definition of NUPBR are invariant with respect to a change in equivalent probability measures. This follows because the definitions only depend on events of probability zero or strictly positive probability. This completes the remark.

Let us denote by $\mathscr{L}_+^0$ the set of adapted and cadlag (right continuous with left limits existing) stochastic processes $Y(t, \omega)$ on $[0, T] \times \Omega$ such that $Y_t \geq 0$. Next, we define a subset of $\mathscr{L}_+^0$ that are needed to give an alternative characterization NUPBR.

$$\begin{aligned} \mathscr{D}_l \equiv \big\{ Y \in \mathscr{L}_+^0 : Y_0 = 1,\ XY \text{ is a } \mathbb{P}\text{-local martingale}, \\ X = 1 + \textstyle\int \alpha \cdot dS,\ (\alpha_0, \alpha) \in \mathscr{A}(1) \big\}. \end{aligned} \tag{2.10}$$

A stochastic process in this set is called a *local martingale deflator* process. It is called this because when the value process of an admissible s.f.t.s. with unit initial wealth is multiplied by a $Y \in \mathscr{D}_l$ (a deflator or change of numeraire) the resulting product XY is a $\mathbb{P}$-local martingale. This set of local martingale deflator processes depends on the probability measure $\mathbb{P}$. We can now present the characterization of NUPBR, which is stated without proof (see Takaoka and Schweizer [173] or Kardaras [121]).

Theorem 2.1 (Probability Characterization of NUPBR) *The market satisfies NUPBR if and only if $\mathscr{D}_l \neq \emptyset$, i.e. there exists a local martingale deflator process.*

This result will prove useful in the portfolio optimization Chap. 11 below when we study portfolio optimization in an incomplete market. The notion of NUPBR in conjunction with NA is sufficient to obtain the First Fundamental Theorem of asset pricing. Before stating and proving this theorem, however, it is necessary to note some useful properties of the set of local martingale deflator processes.

Associated with the set of local martingale deflator (stochastic) processes is the set of local martingale deflators (random variables)

$$D_l = \{Y_T \in L^0_+ : \exists Z \in \mathscr{D}_l,\ Y_T = Z_T\}.$$

2.4.3 Properties of $\mathscr{D}_l$

A number of observations are important with respect to the set of local martingale deflator processes $\mathscr{D}_l$.

1. Note that buying and holding only the mma generates an admissible s.f.t.s. with $X_0 = 1$ and $X_t = 1$ for all t. This implies that any $Y \in \mathscr{D}_l$ is a $\mathbb{P}$-local martingale. And, since it is bounded below by zero, $Y \in \mathscr{D}_l$ is a $\mathbb{P}$-supermartingale (see Lemma 1.3 in Chap. 1). Finally, since $Y_0 = 1$, this implies $E[Y_T] \le 1$.
2. Because the admissible s.f.t.s. X is bounded below and $Y \ge 0$, $Y_t X_t$ is bounded below. $Y_t X_t$ being a $\mathbb{P}$-local martingale implies that for any $Y \in \mathscr{D}_l$, $Y_t X_t$ is a $\mathbb{P}$-supermartingale, i.e. $Y_t X_t \ge E[Y_T X_T | \mathscr{F}_t]$ for all t (see Lemma 1.3 in Chap. 1).
3. A special subset of the set of local martingale deflator processes are those $Y \in \mathscr{D}_l$ such that $Y_T = \frac{d\mathbb{Q}}{d\mathbb{P}} > 0$ for some probability measure $\mathbb{Q}$ equivalent to $\mathbb{P}$, i.e. Y_T is a probability density with respect to $\mathbb{P}$. In this case, $E[Y_T] = E\left[\frac{d\mathbb{Q}}{d\mathbb{P}}\right] = 1 = Y_0$ and $Y_t = E[Y_T | \mathscr{F}_t]$, implying that Y is a $\mathbb{P}$-martingale, i.e. $Y_t = E[Y_T | \mathscr{F}_t]$. Indeed, suppose not. Then, $Y_t > E[Y_T | \mathscr{F}_t]$ with positive probability for some t since Y is a $\mathbb{P}$-supermartingale. Taking expectations, $E[Y_t] > E[E[Y_T | \mathscr{F}_t]] = E[Y_T] = 1$. But, $1 = Y_0 \ge E[Y_t]$ yielding the contradiction. We denote this subset of local martingale deflator processes by

$$\mathscr{M}_l = \left\{Y \in \mathscr{D}_l : \exists \mathbb{Q} \sim \mathbb{P},\ Y_T = \frac{d\mathbb{Q}}{d\mathbb{P}}\right\}.$$

   This set of stochastic processes generates the associated set of local martingale deflators (random variables)

$$M_l = \{Y_T \in L^0_+ :\ \exists Z \in \mathscr{M}_l,\ Y_T = Z_T\}.$$

Given these properties, we can prove the next lemma.

Lemma 2.1 (**X is a $\mathbb{Q}$-local martingale**) *Let $\mathbb{Q}$ be such that for some $Y_T \in M_l$, $\frac{d\mathbb{Q}}{d\mathbb{P}} = Y_T$. Then,*

X is a $\mathbb{Q}$-local martingale for $X_t = 1 + \int_0^t \alpha_u \cdot dS_u$, $(\alpha_0, \alpha) \in \mathscr{A}(1)$.

Proof Since $Y_T \in M_l$, $Y_t X_t$ is a local martingale under $\mathbb{P}$ where $Y_t = E[Y_T | \mathscr{F}_t]$. Hence, there exists a sequence of stopping times (τ_n) such that $\lim_{n\to\infty} \tau_n = \infty$ a.s. $\mathbb{P}$ and $Y_{t\wedge\tau_n} X_{t\wedge\tau_n}$ is a martingale for each n. To prove this lemma it is sufficient to show that $X_{t\wedge\tau_n}$ is a martingale under $\mathbb{Q}$, i.e. $E\left[\frac{Y_T}{Y_t} X_{T\wedge\tau_n} | \mathscr{F}_t\right] = X_t$ for $\tau_n \geq t$ since $\frac{d\mathbb{Q}}{d\mathbb{P}}|_t = \frac{Y_T}{Y_t}$ (see Karatzas and Shreve [117, Lemma, p. 193]) implies that $E^{\mathbb{Q}}\left[X_{T\wedge\tau_n} | \mathscr{F}_t\right] = X_t$. But,

$$E\left[\frac{Y_T}{Y_t} X_{T\wedge\tau_n} | \mathscr{F}_t\right] = E\left[E\left[\frac{Y_T}{Y_t} X_{T\wedge\tau_n} | \mathscr{F}_{T\wedge\tau_n}\right] | \mathscr{F}_t\right] = E\left[E\left[Y_T | \mathscr{F}_{T\wedge\tau_n}\right] \frac{1}{Y_t} X_{T\wedge\tau_n} | \mathscr{F}_t\right].$$

Since Y is a uniformly integrable martingale under $\mathbb{P}$, by Doob's Optional Sampling Theorem (Protter [151, p. 9]),

$$E\left[Y_T | \mathscr{F}_{T\wedge\tau_n}\right] = Y_{T\wedge\tau_n}.$$

Substitution yields

$$E\left[\frac{Y_{T\wedge\tau_n}}{Y_t} X_{T\wedge\tau_n} | \mathscr{F}_t\right] = \frac{1}{Y_t} E\left[Y_{T\wedge\tau_n} X_{T\wedge\tau_n} | \mathscr{F}_t\right] = \frac{Y_t X_t}{Y_t} = X_t,$$

which completes the proof.

The set of equivalent probability measures $\mathbb{Q}$ generated by the local martingale deflators $Y_T \in M_l$ is equivalent to the set of *equivalent local martingale measures*, denoted

$$\mathfrak{M}_l = \{\mathbb{Q} \sim \mathbb{P} : \ S \text{ is a } \mathbb{Q}\text{-local martingale}\},$$

as the next lemma proves.

Lemma 2.2 (Characterization of $\mathfrak{M}_l$)

$$\mathfrak{M}_l = \left\{\mathbb{Q} : \exists Y_T \in M_l, \ Y_T = \frac{d\mathbb{Q}}{d\mathbb{P}}\right\}$$

$$= \left\{\mathbb{Q} \sim \mathbb{P} : \ X \text{ is a } \mathbb{Q}\text{-local martingale}, \ X = 1 + \int \alpha \cdot dS, \ (\alpha_0, \alpha) \in \mathscr{A}(1)\right\}.$$

Proof Define
$\mathfrak{N} = \left\{\mathbb{Q} \sim \mathbb{P} : \ X \text{ is a } \mathbb{Q}\text{-local martingale}, \ X = 1 + \int \alpha \cdot dS, \ (\alpha_0, \alpha) \in \mathscr{A}(1)\right\}$.

(Step 1) Show $\mathfrak{N} \subset \mathfrak{M}_l$.

Choose $\mathbb{Q} \in \mathfrak{N}$. Consider a buy and hold trading strategy in a single asset with a unit holding in that asset. This is an admissible s.f.t.s. For this buy and hold trading strategy $X_t = \frac{S_t}{S_0}$ is a $\mathbb{Q}$-local martingale by the definition of $\mathfrak{N}$. Hence, S is a $\mathbb{Q}$-local martingale, implying $\mathbb{Q} \in \mathfrak{M}_l$.

(Step 2) Show $\mathfrak{M}_l \subset \mathfrak{N}$.

Choose $\mathbb{Q} \in \mathfrak{M}_l$. Since S is a $\mathbb{Q}$-local martingale, and given that $X = 1 + \int \alpha \cdot dS$ is bounded below by admissibility of the s.f.t.s., by Lemma 1.5 in Chap. 1, X is a $\mathbb{Q}$-local martingale. This implies $\mathbb{Q} \in \mathfrak{N}$.

(Step 3) The first equality follows from the definitions of the various sets and Lemma 2.1.

This completes the proof.

The subset of equivalent probability measures making S a $\mathbb{Q}$-martingale, and not just a local martingale, will also prove to be an important set. This is called the set of *equivalent martingale measures* and denoted by

$$\mathfrak{M} = \{\mathbb{Q} \sim \mathbb{P} : \ S \text{ is a } \mathbb{Q}\text{-martingale}\} .$$

Associated with the set of equivalent martingale measures is the set of martingale deflator processes

$$\mathscr{M} = \{Y \in \mathscr{L}_+^0 : \ Y_T = \frac{d\mathbb{Q}}{d\mathbb{P}}, \ Y_t = E\left[Y_T \,|\mathscr{F}_t\right], \ \mathbb{Q} \in \mathfrak{M}\}$$

$$= \{Y \in \mathscr{L}_+^0 : \ Y \in \mathscr{M}_l, \ Y_T = \frac{d\mathbb{Q}}{d\mathbb{P}}, \ \mathbb{Q} \in \mathfrak{M}\}$$

and the set of martingale deflators

$$M = \{Y_T \in L_+^0 : \ Y_T = \frac{d\mathbb{Q}}{d\mathbb{P}}, \ \mathbb{Q} \in \mathfrak{M}\}$$

$$= \{Y_T \in L_+^0 : \ \exists Z \in \mathscr{M}, \ Y_T = Z_T\}.$$

For easy reference, we summarize the relationships among the various sets.

probability measures :	$\mathfrak{M} \subset \mathfrak{M}_l \subset \{\mathbb{Q} : \ \mathbb{Q} \sim \mathbb{P}\}$
stochastic processes :	$\mathscr{M} \subset \mathscr{M}_l \subset \mathscr{D}_l \subset \mathscr{L}_+^0$
random variables :	$M \subset M_l \subset D_l \subset L_+^0$

Remark 2.9 (Properties of $\mathfrak{M}$) We note that unlike the result that holds for local martingale measures as given in Lemma 2.2,

$$\mathfrak{M} \neq \left\{\mathbb{Q} \sim \mathbb{P} : X \text{ is a } \mathbb{Q}\text{-martingale}, \ X = 1 + \int \alpha \cdot dS, \ (\alpha_0, \alpha) \in \mathscr{A}(1)\right\}.$$

This is true because given a $\mathbb{Q} \sim \mathbb{P}$, $X = 1 + \int \alpha \cdot dS$ for an arbitrary $(\alpha_0, \alpha) \in \mathscr{A}(1)$ is in general only a local martingale and not a martingale. Hence for a given martingale measure $\mathbb{Q}$, to ensure that X is also a martingale under $\mathbb{Q}$, additional restrictions on the trading strategies $(\alpha_0, \alpha) \in \mathscr{A}(1)$ are needed. For a set of sufficient conditions on both the martingale measure $\mathbb{Q}$ (therefore S) and the stochastic process X (therefore the trading strategies $(\alpha_0, \alpha) \in \mathscr{A}(1)$) that guarantee that X is a martingale under $\mathbb{Q}$, see Theorem 2.7 below. This completes the remark.

2.4.4 No Free Lunch with Vanishing Risk (NFLVR)

Finally, we give the strengthening of NA needed to prove the First Fundamental Theorem of asset pricing. The appendix to this chapter shows that this definition is equivalent to that contained in Delbaen and Schachermayer [44].

Definition 2.5 (No Free Lunch with Vanishing Risk (NFLVR)) A *free lunch with vanishing risk* (FLVR) is a sequence of admissible s.f.t.s.'s $(\alpha_0, \alpha)_n \in \mathscr{A}(x)$ with initial value $x \geq 0$, value processes X_t^n, lower admissibility bounds $0 \geq c_n$ where $\exists c \leq 0$ with $c_n \geq c$ for all n, and an $\mathscr{F}_T$-measurable random variable $\chi \geq x$ with $\mathbb{P}(\chi > x) > 0$ such that $X_T^n \to \chi$ in probability.

The market satisfies NFLVR if there exist no FLVR trading strategies.

As seen, a market satisfies FLVR if there is a sequence of admissible s.f.t.s.'s $(\alpha_0, \alpha)_n \in \mathscr{A}(x)$ that approaches a simple arbitrage opportunity in the limit. Note that the sequence of admissible s.f.t.s.'s have a uniform lower admissibility bound $c \leq 0$.

It is easy to see that a simple arbitrage opportunity is an FLVR. Indeed, consider the simple arbitrage opportunity $(\alpha_0, \alpha) \in \mathscr{A}(0)$ with value process X_t and admissibility bound c that satisfies $X_0 = 0$, $X_T \geq 0$, and $\mathbb{P}(X_T > 0) > 0$. Define the (constant) sequence of admissible s.f.t.s.'s using the simple arbitrage opportunity, i.e. $(\alpha_0, \alpha)_n \equiv (\alpha_0, \alpha)$ whose time T value process $X_T^n = X_T$ approaches (equals) $\chi \equiv X_T$ where $\chi \geq 0$ with $\mathbb{P}(\chi > 0) > 0$. This sequence $(\alpha_0, \alpha)_n$ is an FLVR. Hence, NFLVR excludes both simple and approximate arbitrage opportunities.

Remark 2.10 (Equivalent Definitions of NFLVR) Equivalent, but different formulations of NFLVR appear in the literature (see Karatzas and Kardaras [116, p. 466]). Two of these are worth mentioning. In the first, one can replace $x > 0$ with $x = 1$. To see this, suppose $(\alpha_0, \alpha)_n \in \mathscr{A}(1)$ has an initial value 1, value processes X_t^n, lower admissibility bounds $0 \geq c_n$ where $\exists c \leq 0$ with $c_n \geq c$ for all n, and an $\mathscr{F}_T$-measurable random variable $\chi \geq 1$ with $\mathbb{P}(\chi > 1) > 0$ such that $X_T^n \to \chi$ in probability. This sequence of admissible s.f.t.s.'s can be transformed into an FLVR. Indeed, for an arbitrary $x > 0$, define the admissible s.f.t.s. $(\tilde{\alpha}_0, \tilde{\alpha})_n = (x\alpha_0, x\alpha)_n \in \mathscr{A}(x)$. A straightforward verification shows that this is an FLVR with $x > 0$.

In the second, one can replace $c \leq 0$ with $c = 0$. However, in this case one must restrict the initial wealth $x > 0$. To see this, suppose $(\alpha_0, \alpha)_n \in \mathscr{A}(x)$ has an initial value $x > 0$, value processes X_t^n, lower admissibility bounds $c_n \geq 0$ for all n, and an $\mathscr{F}_T$-measurable random variable $\chi \geq x$ with $\mathbb{P}(\chi > x) > 0$ such that $X_T^n \to \chi$ in probability. This sequence of admissible s.f.t.s.'s can be transformed into an FLVR. Indeed, let $c < 0$ be an arbitrary uniform lower bound. If $x + c \geq 0$, then a straightforward verification shows that $(\tilde{\alpha}_0, \tilde{\alpha})_n = (\alpha_0 + c, \alpha)_n \in \mathscr{A}(x+c)$ is an FLVR with the uniform lower admissibility bound equal to $c < 0$. If $x + c < 0$, then first transform $(\alpha_0, \alpha)_n \in \mathscr{A}(x)$ to $(\tilde{\alpha}_0, \tilde{\alpha})_n = (\frac{\tilde{x}}{x}\alpha_0, \frac{\tilde{x}}{x}\alpha)_n \in \mathscr{A}(\tilde{x})$ where $\tilde{x} + c \geq 0$, which is an FLVR, and then apply the previous modification with $\tilde{x} + c \geq 0$ to complete the argument. This completes the remark.

An alternative characterization of NFLVR is given in the next theorem, which is stated without proof (see Delbaen and Schachermayer [44, Corollary 3.8]).

Theorem 2.2 (Characterization of NFLVR) *The market satisfies NA and NUPBR if and only if NFLVR.*

This theorem characterizes NFLVR as equivalent to both NA and NUPBR. As shown, NFLVR is a stronger assumption than either of NUPBR or NA alone.

Remark 2.11 (Invariance With Respect to a Change in Equivalent Probability Measures) It is important to note that the admissible s.f.t.s.'s in the definitions of NFLVR are invariant with respect to a change in equivalent probability measures. This follows because the definitions only depend on events of probability zero or strictly positive probability. This completes the remark.

Example 2.3 (Asset Price Processes Violating NFLVR) This example illustrates a collection of asset price processes that violate NFLVR. Consider the Brownian motion market of Sect. 2.7 below. Theorem 2.9 gives necessary and sufficient conditions on the asset price processes such that NFLVR is violated. This completes the example.

NFLVR is the basis of the First Fundamental Theorem of asset pricing to which we now turn.

2.4.5 The First Fundamental Theorem

Given the economic notion of NFLVR, to prove theorems, we need a probabilistic characterization. This is given by the First Fundamental Theorem of asset pricing.

Theorem 2.3 (The First Fundamental Theorem) *NFLVR if and only if $\mathfrak{M}_l \neq \emptyset$, i.e. there exists a $\mathbb{Q} \sim \mathbb{P}$ such that S is a $\mathbb{Q}$-local martingale.*

$\mathbb{Q}$ is called an *equivalent local martingale measure*.

Proof We only prove sufficiency.

($\mathfrak{M}_l \neq \emptyset \Rightarrow$ NFLVR) This is a proof by contradiction. Assume $\mathfrak{M}_l \neq \emptyset$. Let $\mathbb{Q}$ be such an equivalent local martingale measure. This proof uses the equivalent definition of NFLVR given in the appendix. There are two cases to consider.

(Case 1) Let there exist an arbitrage opportunity $(\alpha_0, \alpha) \in \mathscr{A}(x)$ with value process X_t with initial value $X_0 = x = 0$ such that $X_T = \int_0^T \alpha_t \cdot dS_t \geq 0$ and $\mathbb{P}(X_T > 0) > 0$. Since $\mathbb{Q}$ is a local martingale measure, X_t is a local martingale under $\mathbb{Q}$ (see Lemma 1.5 in Chap. 1). X_t is also a supermartingale under $\mathbb{Q}$ by Lemma 1.3 in Chap. 1 because X_t is bounded below by the admissibility condition. Thus, $0 = X_0 \geq E^{\mathbb{Q}}[X_T] > 0$, which is the contradiction.

(Case 2) Let there exist an FLVR with initial value $x = 0$. Then there exists a sequence of admissible s.f.t.s.'s $(\alpha_0, \alpha)_n \in \mathscr{A}(0)$ with lower admissibility bounds (c_n), values (X_t^n), and an $\mathscr{F}_T$-measurable random variable $f \geq 0$ with $\mathbb{P}(f > 0) > 0$ such that $c_n \to 0$ and $X_T^n \to f$ in probability.

For the limiting trading strategy, note that there exists a subsequence such that $X_T^n \to f$ a.s. $\mathbb{P}$ (see Jacod and Protter [75, p. 141]). By Fatou's lemma,

$$\lim_{n\to\infty} E^{\mathbb{Q}}[X_T^n] \geq E^{\mathbb{Q}}[\lim_{n\to\infty} X_T^n] = E^{\mathbb{Q}}[f] > 0.$$

Since $\mathbb{Q}$ is a local martingale measure, X_t^n is a local martingale under $\mathbb{Q}$ (see Lemma 1.5 in Chap. 1). X_t^n is also a supermartingale under $\mathbb{Q}$ by Lemma 1.3 in Chap. 1 because X_t^n is bounded below by the admissibility condition. Thus, $0 = X_0 \geq E^{\mathbb{Q}}[X_T^n]$ for all n, which implies $0 \geq \lim_{n\to\infty} E^{\mathbb{Q}}[X_T^n] > 0$. This is the contradiction.

(NFLVR $\Rightarrow \mathfrak{M}_l \neq \emptyset$) is given in Delbaen and Schachermayer [46]. The equivalence result states that S is a σ-martingale under $\mathbb{Q}$, but since $S \geq 0$, every nonnegative σ-martingale is a local martingale, see the proof of Lemma 1.5 in Chap. 1. This completes the proof.

Remark 2.12 (An Infinite Number of Traded Risky Assets) The statement NFLVR $\Rightarrow \mathfrak{M}_l \neq \emptyset$ depends on the fact that there is only a finite number of risky assets trading.

The statement $\mathfrak{M}_l \neq \emptyset \Rightarrow$ NFLVR holds more generally for an arbitrary number of risky assets (see the proof of the First Fundamental Theorem). This collection could be a continuum (an uncountably infinite set). However, in this extension, any trading strategy must still consist of only a finite number of traded assets, even though an infinite number of risky assets are available to choose the finite number included in the trading strategy. This completes the remark.

Using Lemma 1.3 in Chap. 1, this theorem has an important corollary.

Corollary 2.1 *If NFLVR, then S_t is a supermartingale under $\mathbb{Q}$, i.e.*

$$S_t \geq E^{\mathbb{Q}}[S_T | \mathscr{F}_t] \text{ for all } t. \tag{2.11}$$

Note that the asset price process is not a martingale. The difference between the price of the asset and the conditional expectation on the right side of expression (2.11), if nonzero, will be seen (in Chap. 3) to represent an asset price bubble.

2.4.6 Equivalent Local Martingale Measures

This section generates some observations about equivalent local martingale measures useful in subsequent chapters. Fix a $\mathbb{Q} \in \mathfrak{M}_l$. Because $\mathbb{Q}$ is mutually absolutely continuous (equivalent) with respect to $\mathbb{P}$, there exists a unique strictly positive Radon–Nikodym derivative (see Ash [3, p. 63])

$$Y_T = \frac{d\mathbb{Q}}{d\mathbb{P}} > 0 \in M_l \tag{2.12}$$

such that

$$E[1_A Y_T] = E^{\mathbb{Q}}[1_A] = \mathbb{Q}(A) \tag{2.13}$$

for all $A \in \mathscr{F}_T$.

This strictly positive Radon–Nikodym derivative uniquely determines a stochastic process $Y \in \mathscr{L}^0_+$ defined by

$$Y_t = E[Y_T | \mathscr{F}_t]. \tag{2.14}$$

As seen, Y is a uniformly integrable martingale. Using this stochastic process, we can define conditional expectations of indicator functions with respect to $\mathbb{Q}$, i.e.

$$E^{\mathbb{Q}}[1_A | \mathscr{F}_t] = E\left[1_A \frac{Y_T}{Y_t} \middle| \mathscr{F}_t \right] \tag{2.15}$$

for all $A \in \mathscr{F}_T$. Dividing by Y_t on the right side of this equality is necessary to make $E^{\mathbb{Q}}[1_\Omega | \mathscr{F}_t] = 1$.

Last, given an $\mathscr{F}_T$-measurable random variable X_T that is integrable with respect to $\mathbb{Q}$, we have

$$E^{\mathbb{Q}}[X_T] = E[Y_T X_T] \tag{2.16}$$

and

$$E^{\mathbb{Q}}[X_T | \mathscr{F}_t] = E\left[X_T \frac{Y_T}{Y_t} \middle| \mathscr{F}_t \right]. \tag{2.17}$$

2.4.7 The State Price Density

We define the state price density using non-normalized prices.

Definition 2.6 (State Price Density) Let $Y \in \mathscr{M}_l$. The *state price density* $H \in \mathscr{L}_+^0$ is

$$H_t = \frac{Y_t}{\mathbb{B}_t} = \frac{E[Y_T | \mathscr{F}_t]}{\mathbb{B}_t}. \tag{2.18}$$

Lemma 2.3 *Assume NFLVR.* $(\mathbb{S}_t H_t)$ *is a local martingale under* $\mathbb{P}$.

Proof Note that $\mathbb{S}_t H_t = S_t Y_t$. We have that S_t is a local martingale under $\mathbb{Q}$. This is equivalent to $S_t Y_t$ being a local martingale under $\mathbb{P}$. Indeed, consider a sequence of stopping times $(\tau_n) \uparrow \infty$. Using expression (2.16), because the stopped process $S_{t \wedge \tau_n}$ under $\mathbb{Q}$ is a martingale, the stopped process $S_{t \wedge \tau_n} Y_{t \wedge \tau_n}$ is a martingale under $\mathbb{P}$. This completes the proof.

By Lemma 1.3 in Chap. 1, this implies that the nonnegative process $(\mathbb{S}_t H_t)$ is a supermartingale under $\mathbb{P}$, i.e.

$$\mathbb{S}_t \geq E\left[\mathbb{S}_T \frac{H_T}{H_t} \middle| \mathscr{F}_t \right]. \tag{2.19}$$

For subsequence use, the following lemma will be important.

Lemma 2.4 *Assume NFLVR. Consider an admissible s.f.t.s.* $(\alpha_0, \alpha) \in \mathscr{A}(x)$ *with initial value* $x > 0$ *and value process* $d\mathbb{X}_t = \alpha_0(t) r_t \mathbb{B}_t dt + \alpha(t) \cdot d\mathbb{S}_t$.

Then, $(\mathbb{X}_t H_t)$ *is a local martingale under* $\mathbb{P}$.

Proof Note that $\mathbb{X}_t H_t = X_t Y_t$, where $dX_t = \alpha_t \cdot dS_t$ is a local martingale under $\mathbb{Q}$. Consider a sequence of stopping times $(\tau_n) \uparrow \infty$. We have that expression (2.16) holds for the stopped process $X_{t \wedge \tau_n} Y_{t \wedge \tau_n}$ under $\mathbb{P}$. This completes the proof.

Since $\mathbb{X}_t$ is bounded below by the admissibility condition, again by Lemma 1.3 in Chap. 1, the nonnegative process $(\mathbb{X}_t H_t)$ is a supermartingale under $\mathbb{P}$, i.e.

$$\mathbb{X}_t \geq E\left[\mathbb{X}_T \frac{H_T}{H_t} \middle| \mathscr{F}_t\right]. \tag{2.20}$$

2.5 The Second Fundamental Theorem

This section presents the Second Fundamental Theorem of asset pricing, which relates the notion of a complete market to local martingale measures. We are given a *normalized market* $(S, (\mathscr{F}_t), \mathbb{P})$, and as before, we start with some definitions. Let

$L_+^0 \equiv L_+^0(\Omega, \mathscr{F}, \mathbb{P})$ be the space of all nonnegative $\mathscr{F}_T$-measurable random variables, and

$L_+^1(\mathbb{Q}) \equiv L_+^1(\Omega, \mathscr{F}, \mathbb{Q})$ be the space of all nonnegative $\mathscr{F}_T$-measurable random variables X such that $E^{\mathbb{Q}}[X] < \infty$.

The set of random variables L_+^0 should be interpreted as the set of "derivative securities" and $L_+^1(\mathbb{Q})$ should be interpreted as the set of "derivative securities whose payoffs are integrable with respect to the probability measure $\mathbb{Q}$."

Define

$$\mathscr{C}(x) = \left\{X_T \in L_+^0 : \exists(\alpha_0, \alpha) \in \mathscr{A}(x),\ x + \int_0^T \alpha_t \cdot dS_t = X_T\right\}.$$

This is the set of nonnegative random variables generated by all admissible s.f.t.s.'s starting with an initial value of $x \geq 0$.

The collection $\underset{x\geq 0}{\cup} \mathscr{C}(x)$ is the set of securities that can be obtained—*synthetically constructed*—via the use of admissible s.f.t.s.'s from nonnegative initial values. These securities are called attainable.

Definition 2.7 (Attainable Securities) A random variable $Z_T \in L_+^0$ is *attainable* if $Z_T \in \underset{x\geq 0}{\cup} \mathscr{C}(x)$, i.e. there exists an $x \geq 0$ and an $(\alpha_0, \alpha) \in \mathscr{A}(x)$ such that

$$x + \int_0^T \alpha_t \cdot dS_t = Z_T. \tag{2.21}$$

An important market is one satisfying NFLVR in which the set of attainable securities, integrable with respect to a local martingale measure $\mathbb{Q} \in \mathfrak{M}_l$, generate value processes that are $\mathbb{Q}$-martingales. Such a market is called complete.

Definition 2.8 (Complete Market with Respect to $\mathbb{Q} \in \mathfrak{M}_l$) Given $\mathfrak{M}_l \neq \emptyset$, i.e. NFLVR.

Choose a $\mathbb{Q} \in \mathfrak{M}_l$.

The market is *complete with respect to* $\mathbb{Q}$ if given any $Z_T \in L_+^1(\mathbb{Q})$, there exists an $x \geq 0$ and $(\alpha_0, \alpha) \in \mathscr{A}(x)$ such that

$$x + \int_0^T \alpha_u \cdot dS_u = Z_T,$$

and the value process defined by

$$X_t \equiv x + \int_0^t \alpha_u \cdot dS_u$$

for all $t \in [0, T]$ is a $\mathbb{Q}$-martingale, i.e. $X_t = E^{\mathbb{Q}}[Z_T | \mathscr{F}_t]$.

The definition of a complete market depends on the market satisfying NFLVR and the choice of a probability measure $\mathbb{Q} \in \mathfrak{M}_l$. Unlike the definitions of NA, NUPBR, and NFLVR, the definition of a complete market is not invariant with respect to a change in equivalent probability measures. This follows because the set of integrable random variables $L^1_+(\mathbb{Q})$ depends on $\mathbb{Q}$. The definition of a complete market is formulated in this way because of its subsequent use for pricing derivatives in an NFLVR market using conditional expectations. For a definition of a complete market that is independent of both NFLVR and a particular $\mathbb{Q} \in \mathfrak{M}_l$, see Battig and Jarrow [9].

Remark 2.13 (Attainable Securities and $\mathbb{Q}$-Martingales) The definition of a complete market with respect to $\mathbb{Q} \in \mathfrak{M}_l$ requires that the admissible s.f.t.s. satisfies $X_T = x + \int_0^T \alpha_u \cdot dS_u = Z_T$ at time T. Hence, the security Z_T is attainable. But, the definition of a complete market is stronger. In addition, the definition requires that the admissible s.f.t.s makes the value process $X_t = x + \int_0^t \alpha_u \cdot dS_u$ a $\mathbb{Q}$-martingale.

This is a stronger condition than attainability because the stochastic integral $\int_0^t \alpha_u \cdot dS_u$, being bounded below (here by zero), is a $\mathbb{Q}$-local martingale (see Lemma 1.5 in Chap. 1). It need not, however, be a $\mathbb{Q}$-martingale. The next corollary gives a sufficient condition on Z_T such that the value process X is a $\mathbb{Q}$-martingale. An alternative set of sufficient conditions for X to be a $\mathbb{Q}$-martingale are provided in Theorem 2.7 later in this chapter. This completes the remark.

Corollary 2.2 (Attainability and Boundedness) *Given $\mathfrak{M}_l \neq \emptyset$.*

Let $Z_T \geq 0$ be attainable, i.e. there exists an $x \geq 0$ and $(\alpha_0, \alpha) \in \mathscr{A}(x)$ with value process

$$X_t \equiv x + \int_0^t \alpha_u \cdot dS_u$$

such that

$$x + \int_0^T \alpha_u \cdot dS_u = Z_T.$$

Let X_t be bounded above, i.e. $X_t \leq k$ for some $k > 0$. Then, X_t is a $\mathbb{Q}$-martingale, i.e. $X_t = E^{\mathbb{Q}}[Z_T | \mathscr{F}_t]$.

Proof By admissibility X_t is bounded below, hence by Lemma 1.5 in Chap. 1, X_t is a $\mathbb{Q}$-local martingale. By Remark 1.4 in Chap. 1, X_t is a $\mathbb{Q}$-martingale. This completes the proof.

Remark 2.14 (Non-Unique Triplet $\{x, (\alpha_0, \alpha)\}$ Generating Z_T) The initial investment and admissible s.f.t.s, $\{x, (\alpha_0, \alpha)\}$, generating a derivative $Z_T \in L^1_+(\mathbb{Q})$ may not be unique. This occurs, for example, if S is not a martingale under $\mathbb{Q}$, see Remark 2.15 below. However, given an initial investment x, the s.f.t.s. (α_0, α) generating Z_T is unique, if the traded risky assets are non-redundant. By non-redundant we mean that the stochastic process $S_j(t)$ with initial value $S_j(0)$ cannot be generated by an admissible s.f.t.s. using all the other risky assets ($i = 1, \cdots, n$ and $i \neq j$). Assuming that the risky assets are non-redundant, the admissible s.f.t.s. triplet $\{x, (\alpha_0, \alpha)\}$ generating $Z_T \in L^1_+(\mathbb{Q})$ is uniquely determined because the initial wealth $x = X_0 = E^{\mathbb{Q}}[Z_T]$ is given and fixed as part of the definition of a complete market. This completes the remark.

Remark 2.15 (Strict Local Martingales and Complete Markets) To simplify the discussion, consider a market consisting of the mma and only one risky asset with price process $S_1(t)$. In this case, we omit the subscript "1" on the risky asset's price. It may be the case that for a local martingale measure $\mathbb{Q} \in \mathfrak{M}_l$, S is not a $\mathbb{Q}$-martingale. Here, we say that S is a *strict* $\mathbb{Q}$-local martingale. Since $S \geq 0$, it is also a (strict) supermartingale by Lemma 1.3 in Chap. 1, i.e. $E^{\mathbb{Q}}[S_T | \mathscr{F}_t] < S_t$ for $0 \leq t < T$.

When S is a *strict* $\mathbb{Q}$-local martingale, a buy and hold trading strategy in S (say with a unit holding) generates the value process

$$X_t \equiv 1 \cdot S_t = S_0 + \int_0^t 1 \cdot dS_u.$$

Note that $X_T = S_T$ and $X_0 = S_0$. The value process X_t is *not* a $\mathbb{Q}$-martingale because although

$$X_T = S_T, E^{\mathbb{Q}}[X_T] = E^{\mathbb{Q}}[S_T] < S_0 = X_0.$$

Nonetheless, the market could still be complete with respect to $\mathbb{Q} \in \mathfrak{M}_l$. An explicit example of such a market is provided in the asset price bubbles Chap. 3, Example 3.1. If the market is complete with respect to $\mathbb{Q} \in \mathfrak{M}_l$, then there exists an *alternative* admissible s.f.t.s. $(\tilde{\alpha}_0, \tilde{\alpha}) \in \mathscr{A}(\tilde{x})$, where $\tilde{x} + \int_0^T \tilde{\alpha}_u dS_u = S_T$ and

$$\tilde{X}_t \equiv \tilde{x} + \int_0^t \tilde{\alpha}_u dS_u = E^{\mathbb{Q}}[X_T | \mathscr{F}_t] = E^{\mathbb{Q}}[S_T | \mathscr{F}_t]$$

for all $t \in [0, T]$. This trading strategy must be dynamic with $(\tilde{\alpha}_0, \tilde{\alpha})$ changing across time, since the value process for a buy and hold trading strategy in the risky asset (shorting) was shown not to be a $\mathbb{Q}$-martingale.

To determine the exact admissible s.f.t.s. $(\tilde{\alpha}_0, \tilde{\alpha}) \in \mathscr{A}(\tilde{x})$, the method used in Sect. 2.6.3 below in the synthetic construction of a derivative can be employed with the "derivative's" time t value being $\tilde{X}_t = E^{\mathbb{Q}}[S_T | \mathscr{F}_t]$. At this time, however, we can still say something qualitative about the trading strategy $(\tilde{\alpha}_0, \tilde{\alpha})$ and how it differs from buying and holding the risky asset. The self-financing condition is that $\tilde{\alpha}_0(t) + \tilde{\alpha}_t S_t = \tilde{x} + \int_0^t \tilde{\alpha}_u dS_u$, and therefore $\tilde{\alpha}_0(t) + \tilde{\alpha}_t S_t = E^{\mathbb{Q}}[S_T | \mathscr{F}_t] < S_t$ for all $t \in [0, T)$. This implies that if the position in the mma is strictly positive $\tilde{\alpha}_0(t) > 0$ at time t, then the number of shares held in the risky asset is less than one, i.e. $\tilde{\alpha}_t < 1$. Conversely, if $\tilde{\alpha}_0(t) < 0$ at time t, then $\tilde{\alpha}_t > 1$.

This insight yields another implication. It implies that by using $(\tilde{\alpha}_0, \tilde{\alpha}) \in \mathscr{A}(\tilde{x})$, an investor can obtain the same time T payoff as generated by the risky asset S_T, but at a cheaper cost of $\tilde{x}$ dollars at time 0 versus S_0 dollars. It is cheaper because

$$\tilde{x} = \tilde{X}_0 = E^{\mathbb{Q}}[S_T | \mathscr{F}_0] = E^{\mathbb{Q}}[S_T] < S_0.$$

In this case the risky asset S is said to be a dominated security (see Definition 2.9 below). This remark documents the link between the existence of S being a dominated security and S being a strict local martingale. In the asset price bubbles Chap. 3, we will connect both of these conditions to the existence of asset price bubbles.

It is natural to wonder whether shorting the risky asset and going long the admissible s.f.t.s. $(\tilde{\alpha}_0, \tilde{\alpha}) \in \mathscr{A}(\tilde{x})$ generates an arbitrage opportunity, i.e. violates NA. In general, if the risky asset's price process is unbounded above, then the answer is no due to the admissibility condition on this combined s.f.t.s. Indeed, this combined s.f.t.s. would necessarily be unbounded below due to the lower bound on $(\tilde{\alpha}_0, \tilde{\alpha}) \in \mathscr{A}(\tilde{x})$ and the unbounded short position in the risky asset, thereby violating the admissibility condition (see Remark 2.2 above). This completes the remark.

Remark 2.16 (Indicator Functions) Note that if the market is complete with respect to $\mathbb{Q}$, then $1_A \in L^1_+(\mathbb{Q})$ is attainable for all $A \in \mathscr{F}_T$ and the value process of the admissible s.f.t.s. that generates it is a martingale. These indicator random variables will be identified as Arrow–Debreu securities in Chap. 4 below. This completes the remark.

We now state and prove the Second Fundamental Theorem of asset pricing.

Theorem 2.4 (Second Fundamental Theorem) *Assume $\mathfrak{M}_l \neq \emptyset$, i.e. NFLVR.*

If the market is complete with respect to $\mathbb{Q} \in \mathfrak{M}_l$, then $\mathfrak{M}_l$ is a singleton, i.e., the equivalent local martingale measure is unique.

Conversely, assume $\mathfrak{M} \neq \emptyset$, i.e. there exists an equivalent martingale measure. If $\mathfrak{M}$ is a singleton, then the market is complete with respect to $\mathbb{Q} \in \mathfrak{M}$.

Proof ($\mathfrak{M}_l \neq \emptyset$. Completeness $\Rightarrow$ $\mathfrak{M}_l$ a singleton) Consider two measures $\mathbb{Q}^1, \mathbb{Q}^2 \in \mathfrak{M}_l$. Choose an arbitrary $A \in \mathscr{F}_T$. Consider $Z_T = 1_A \geq 0$.

Since the market is complete with respect to both $\mathbb{Q}^1, \mathbb{Q}^2 \in \mathfrak{M}_l$, there exists an $x^i \geq 0$ and $(\alpha_0^i, \alpha^i) \in \mathscr{A}(x^i)$ such that

$$x^i + \int_0^T \alpha_u^i \cdot dS_u = 1_A$$

and

$$X_t^i \equiv x^i + \int_0^t \alpha_u^i \cdot dS_u = E^{\mathbb{Q}^i}[1_A | \mathscr{F}_t]$$

for all t and for $i = 1, 2$. We note that for each i, the value process X_t^i is bounded below by zero and above by 1 for all t.

We claim that $x^1 = x^2$. We prove this by contradiction. Suppose, without loss of generality, that $x^1 > x^2$ at time 0. Then, NFLVR is violated (in fact NA is violated), which contradicts the hypothesis of the theorem. To see that NFLVR is violated, consider the following trading strategy: short $\{x^1 \geq 0$ and $(\alpha_0^1, \alpha^1) \in \mathscr{A}(x^1)\}$, go long $\{x^2 \geq 0$ and $(\alpha_0^2, \alpha^2) \in \mathscr{A}(x^2)\}$, and invest $\{x^1 - x^2 > 0$ in a buy and hold position in the mma$\}$. Since $-X_t^1$ is bounded below by -1 for all t, the trading strategy is admissible. It is trivially self-financing. The initial value of this admissible s.f.t.s. is 0 and its time T value is $x^1 - x^2 > 0$ with probability one. This is a (simple) arbitrage opportunity, which proves the claim.

Hence, we have

$$\begin{aligned}
x^1 &= E^{\mathbb{Q}^1}[X_T] = E^{\mathbb{Q}^1}[1_A] = \mathbb{Q}^1(A), \\
x^2 &= E^{\mathbb{Q}^2}[X_T] = E^{\mathbb{Q}^2}[1_A] = \mathbb{Q}^2(A), \text{ and} \\
x^1 &= x^2, \text{ which yields} \mathbb{Q}^1(A) = \mathbb{Q}^2(A).
\end{aligned}$$

Since this is true for all $A \in \mathscr{F}_T$, we have $\mathbb{Q}^1 = \mathbb{Q}^2$.
($\mathfrak{M} \neq \emptyset$. $\mathfrak{M}$ a singleton $\Rightarrow$ Completeness)

This proof is contained in Harrison and Pliska [67]. We prove the theorem under the additional hypothesis that $S_1, \ldots, S_n$ are $\mathscr{H}_0^1(\mathbb{Q})$-martingales.

(Step 1) Prove a martingale representation theorem.

First, we need various definitions and a version of the Jacod and Yor Theorem.

Let $\{X_1, \ldots, X_n\}$ be $\mathscr{H}_0^1(\mathbb{Q})$-martingales. We note that these are uniformly integrable martingales because they are martingales on $[0, T]$.

Define $\chi = \text{span}\{X_1, \ldots, X_n\}$ and stable(χ) to be the smallest closed linear subspace containing χ such that if $X_t \in \chi$, then $X_{t \wedge \tau} \in \chi$ for all stopping times τ (see Protter [151, p. 178]).

By Medvegyev [136, Proposition 5.44, p. 341], this is equivalent to closure under stochastic integration, i.e. if $X \in$ stable (χ) and there exists an $\alpha \in L(X)$ such that $\alpha \bullet X \in \mathscr{H}_0^1(\mathbb{Q})$, then $\alpha \bullet X \in$ stable (χ) where $(\alpha \bullet X)_t \equiv \int_0^t \alpha \cdot dX$.

We say that χ has martingale representation if stable $(\chi) = \mathcal{H}_0^1(\mathbb{Q})$, see Medvegyev [136, p. 329].

Define

$$\mathfrak{M}_H(\chi) = \{\mathbb{P} \sim \mathbb{Q} : X \in \chi \text{ implies } X \in \mathcal{H}_0^1(\mathbb{P})\}.$$

These are the equivalent probability measures such that X is an $\mathcal{H}_0^1(\mathbb{P})$-martingale. Note that $\mathfrak{M}_H(\chi) \neq \emptyset$ since $\mathbb{Q} \in \mathfrak{M}_H(\chi)$. Also note that $\mathfrak{M}_H(\chi) \subset \mathfrak{M}$, the equivalent probability measures $\mathbb{P}$ such that $X \in \chi$ are $\mathbb{P}$-martingales. Finally, $\mathfrak{M}_H(\chi)$ is a convex set.

A version of the Jacod and Yor Theorem (Medvegyev [136, p. 341]) states

$\chi \subset \mathcal{H}_0^1(\mathbb{Q})$ has martingale representation if and only if $\mathbb{Q}$ is an extremal point of $\mathfrak{M}_H(\chi)$.

We can now prove the following lemma.

Lemma *If* $\mathbb{Q} \in \mathfrak{M}$ *is unique, then* stable$(\chi) = \mathcal{H}_0^1(\mathbb{Q})$.

Proof $\mathbb{Q} \in \mathfrak{M}$ is unique implies that $\mathbb{Q} \in \mathfrak{M}_H(\chi)$ is unique. This trivially implies $\mathbb{Q}$ is an extremal point of $\mathfrak{M}_H(\chi)$.

By the Jacod and Yor Theorem given above, $\chi \subset \mathcal{H}_0^1(\mathbb{Q})$ has martingale representation. Thus, by definition, stable $(\chi) = \mathcal{H}_0^1(\mathbb{Q})$. This completes the proof of the Lemma.

(Step 2) Definition of a complete market (revisited).

The market is complete with respect to $\mathbb{Q}$ if

$$L_+^1(\mathbb{Q}) = \{Z_T \in L_+^1(\mathbb{Q}) : \exists x \geq 0, \exists(\alpha_0, \alpha) \in \mathcal{A}(x),$$
$$X \equiv x + \alpha \bullet S \text{ is a } \mathbb{Q}\text{-martingale with } X_T = Z_T\}.$$

Rewritten in terms of stochastic processes, define

$$\mathcal{G}^1 \equiv \left\{ Z \in \mathcal{L}_+^0 : \exists x \geq 0, \exists(\alpha_0, \alpha) \in \mathcal{A}(x), Z \equiv x + \alpha \bullet S \text{ is a } \mathbb{Q}\text{-martingale} \right\}$$

and

$$\mathcal{G}^2 \equiv \left\{ Z \in \mathcal{L}_+^0 : Z \text{ is a } \mathbb{Q}\text{-martingale} \right\}.$$

The market is complete with respect to $\mathbb{Q}$ if $\mathcal{G}^1 = \mathcal{G}^2$.

Without loss of generality we can restrict consideration to the sets

$$\mathcal{G}_0^1 = \mathcal{G}^1 \cap \left\{ Z \in \mathcal{L}_+^0 : Z_0 = 0 \right\}$$

and

$$\mathscr{G}_0^2 = \mathscr{G}^2 \cap \left\{ Z \in \mathscr{L}_+^0 : Z_0 = 0 \right\}.$$

Indeed, given $Z \in \mathscr{G}^1$, this $\mathbb{Q}$-martingale can be generated by the $X \in \mathscr{G}_0^1$ given by $X = Z - Z_0$. The admissible s.f.t.s. determined by $\mathscr{G}_0^1$ that attains X can be used to attain Z by adding Z_0 additional units in the mma, which retains admissibility and nonnegativity.

Hence, the market is complete with respect to $\mathbb{Q}$ if $\mathscr{G}_0^1 = \mathscr{G}_0^2$.

(Step 3) Show the market is complete if $\mathbb{Q} \in \mathfrak{M}$ is unique.

First, by the definitions, we have $\mathscr{G}_0^1 \subset \mathscr{G}_0^2$. We need to show that $\mathscr{G}_0^2 \subset \mathscr{G}_0^1$.

By hypothesis, $X_i(t) \equiv S_i(t) - S_i(0)$ are uniformly integrable $\mathscr{H}_0^1(\mathbb{Q})$-martingales.

Hence $\{X_1, \ldots, X_n\} \in \mathscr{G}_0^1$, where each X_i represents a buy and hold trading strategy in the mma and the risky asset. This is an admissible s.f.t.s.

Next, $\left(\text{span}\{X_1, \ldots, X_n\} \cap \mathscr{L}_+^0\right) \subset \mathscr{G}_0^1$. These are buy and hold trading strategies in the mma and all the risky assets whose value processes are nonnegative, hence admissible s.f.t.s.

Finally, $\left(\text{stable}(\chi) \cap \mathscr{L}_+^0\right) \subset \mathscr{G}_0^1$, because stable$(\chi)$ is closed under stochastic integration. Hence, this set corresponds to all s.f.t.s. that generate nonnegative-valued processes, which are admissible.

Then, by Lemma in (Step 1), $\left(\mathscr{H}_0^1(\mathbb{Q}) \cap \mathscr{L}_+^0\right) \subset \mathscr{G}_0^1$, i.e. all $\mathscr{H}_0^1(\mathbb{Q})$ martingales are in $\mathscr{G}_0^1$.

We note that given any admissible (nonnegative) s.f.t.s. in $\mathscr{H}_0^1(\mathbb{Q}) \cap \mathscr{L}_+^0 = \text{stable}(\chi) \cap \mathscr{L}_+^0$, stopping the trading strategy at a stopping time τ and investing the proceeds in the mma until time T is included within the set. Stopping an admissible s.f.t.s. is an admissible s.f.t.s. Thus, by considering a sequence of stopping times τ_n with $\tau_n \to T$, the set of admissible s.f.t.s.'s using $\{X_1, \ldots, X_n\}$ generates the set of all $\mathbb{Q}$-local martingales with nonnegative values. Hence, since $\mathbb{Q}$-martingales are $\mathbb{Q}$-local martingales, the set of admissible s.f.t.s.'s using $\{X_1, \ldots, X_n\}$ generates $\mathscr{G}_0^2$, thus $\mathscr{G}_0^2 \subset \mathscr{G}_0^1$, which completes the proof of (Step 3) and the theorem.

In a market satisfying NFLVR, the Second Fundamental Theorem relates the notion of a complete market to the non-empty set of local martingale measures $\mathfrak{M}_l$. The first statement in the theorem is that if the market is complete with respect to any $\mathbb{Q} \in \mathfrak{M}_l$, then the local martingale measure is unique. For the converse, the hypothesis needs to be strengthened to the existence of a martingale measure $\mathbb{Q} \in \mathfrak{M}$. If the martingale measure is unique, then the market is complete with respect to $\mathbb{Q} \in \mathfrak{M}$. This additional hypothesis for the converse makes it important to understand, in economic terms, the relation between the two sets of equivalent probability measures $\mathfrak{M}_l$ and $\mathfrak{M}$. This economics behind these two sets of measures is studied in the next section.

Remark 2.17 (Infinite Number of Local Martingale Measures) In an incomplete market, if the set of equivalent local martingale measures is not a singleton, then the cardinality $|\mathfrak{M}_l| = \infty$, i.e. the set contains an infinite number of elements. This follows because if not a singleton, the set contains at least two distinct elements and all convex combinations of these two elements. This completes the remark.

Remark 2.18 (Discontinuous Price Processes and Market Completeness) For continuous sample path price processes, the market can be complete or incomplete, depending on properties of the filtration and the evolutions of the risky asset price processes. When the filtrations and the risky asset price processes are generated by a finite-dimension Brownian motion, then the market is complete if and only if the volatility matrix of the risky asset return processes satisfies a non-singularity condition for all times with probability one (see Theorem 2.12 in Sect. 2.7 below). This condition fails, for example, when the risky asset price processes have stochastic volatility (see Eisenberg and Jarrow [56]).

In contrast, however, for discontinuous sample path price processes the market is almost always incomplete. This occurs because when a jumps occurs, the distribution for the change in the price process is usually not just discrete with a finite number of jump amplitudes where the number of jump amplitudes is less then the number of traded assets (see Cont and Tankov [34, Chapter 9.2]). In this case the number of traded assets will be insufficient to hedge the different jump magnitudes possible (in mathematical terms, the martingale representation property fails for the value processes of admissible s.f.t.s.'s). This completes the remark.

2.6 The Third Fundamental Theorem

This section presents the Third Fundamental Theorem of asset pricing. The Third Fundamental Theorem is the "work horse" of the three, yielding as a corollary the key tool used to value and hedge derivatives, called risk neutral valuation. To obtain this theorem, as in the previous sections, we are given a *normalized market* $(S, (\mathscr{F}_t), \mathbb{P})$. We start with some definitions. First, we introduce the notion of no dominance (ND) initially employed by Merton [138] to study the properties of option prices.

Definition 2.9 (No Dominance) The ith asset $S_i(t)$ is *undominated* if there exists no admissible s.f.t.s. $(\alpha_0, \alpha) \in \mathscr{A}(x)$ with an initial value $x = S_i(0)$ such that

$$\mathbb{P}\{x + \int_0^T \alpha_t \cdot dS_t \geq S_i(T)\} = 1 \quad \text{and} \quad \mathbb{P}\{x + \int_0^T \alpha_t \cdot dS_t > S_i(T)\} > 0.$$

A market $(S, (\mathscr{F}_t), \mathbb{P})$ satisfies no dominance (ND) if each S_i, $i = 0, \ldots, n$, is undominated.

ND states that it is not possible to find an admissible s.f.t.s. with initial investment equal to an asset's initial value such that the trading strategy's payoffs at time T dominate the payoffs to the traded asset. The definition of ND is invariant with respect to a change in equivalent probability measures.

A dominated asset need not imply the existence of an arbitrage opportunity. This is due to the admissibility condition in the definition of a trading strategy, which precludes shorting an asset whose price is unbounded above (see Remark 2.2 above). NFLVR and ND are distinct conditions, although both imply the simpler no arbitrage NA condition. Indeed, that NFLVR implies NA follows directly from Theorem 2.2 above. Second, since ND implies that the mma ($B_t \equiv 1$) is undominated, making the following identifications in the definition of ND,

$$(x = S_i(0) = S_i(T) = B_t = 1$$

and

$$X_T = 1 + \int_0^T \alpha_t \cdot dS_t),$$

yields the definition of NA.

ND is a condition related to supply equalling demand. To understand why, note that an NFLVR is a mispricing opportunity that any single trader can exploit in unlimited quantities. As an NFLVR is exploited, prices adjust to eventually eliminate the mispricing. Hence, NFLVR is independent of the aggregate market supply or demand. In contrast, ND compares two different investment alternatives for obtaining the same time T payoff. Of these two investment alternatives, the first, an s.f.t.s., dominates the second, holding the asset directly. A trader, if they desire the time T payoff from holding this position in their optimal portfolio, will never hold the asset. Since all traders see the same dominance, no one in the market will hold the asset. Hence, the supply for the asset (if positive) will exceed market demand. Positive supply can only equal demand if ND holds.

Example 2.4 (NFLVR But Not ND) The finite-dimension Brownian motion market in Sect. 2.7 below can be used to generate examples of risky asset price evolutions that satisfy NFLVR but not ND. To obtain these examples, use Theorem 2.9 to guarantee that the market satisfies NFLVR. Then, use the insights from Theorem 2.13 to obtain a price process that does not satisfy Novikov's condition and is a strict local martingale. An illustration of such a risky asset price process is contained in the asset price bubbles Chap. 3, Example 3.1. Another example is given in the asset price bubbles Chap. 3, Example 3.2, for a CEV price process with $\alpha > 1$. This completes the example.

We now state and prove the Third Fundamental Theorem of asset pricing. Recall the notation for the set of equivalent martingale measures

$$\mathfrak{M} = \{\mathbb{Q} \sim \mathbb{P} : \ S \text{ is a } \mathbb{Q}\text{-martingale}\}$$

and the set of equivalent local martingale measures

$$\mathfrak{M}_l = \{\mathbb{Q} \sim \mathbb{P} : \ S \text{ is a } \mathbb{Q}\text{-local martingale}\},$$

where $\mathfrak{M} \subset \mathfrak{M}_l$, often a strict subset.

Theorem 2.5 (Third Fundamental Theorem) *NFLVR and ND if and only if $\mathfrak{M} \neq \emptyset$, i.e. there exists an equivalent probability $\mathbb{Q}$ such that S is a $\mathbb{Q}$ martingale.*

Proof (NFLVR and ND $\Rightarrow \mathfrak{M} \neq \emptyset$) By the First Fundamental Theorem of asset pricing, NFLVR $\Rightarrow \mathfrak{M}_l \neq \emptyset$. Hence, we only need to show that ND $\Rightarrow \mathfrak{M} \neq \emptyset$. The proof of this is based on Jarrow and Larsson [96]. This proof requires the definition of a maximal trading strategy and two lemmas. First the definition.

An admissible s.f.t.s. $(\alpha_0(t), \alpha_t) \in \mathscr{A}(0)$ with zero initial wealth is *maximal* if for every other admissible s.f.t.s. $(\beta_0(t), \beta_t) \in \mathscr{A}(0)$ such that $\int_0^T \beta_t \cdot dS_t \geq \int_0^T \alpha_t \cdot dS_t$, $\int_0^T \beta_t \cdot dS_t = \int_0^T \alpha_t \cdot dS_t$ holds.

Next, the two lemmas.

Lemma 2.5 *(Delbaen and Schachermayer [46, Theorem 5.12]). $(\alpha_0(t), \alpha_t) \in \mathscr{A}(0)$ is maximal if and only if there exists a $\mathbb{Q} \in \mathfrak{M}_l$ such that $\int_0^t \alpha_s \cdot dS_s$ is a $\mathbb{Q}$-martingale.*

This theorem in Delbaen and Schachermayer [46] applies to our market because local martingales and σ-martingales coincide when the risky asset prices are nonnegative.

Lemma 2.6 *(Delbaen and Schachermayer [45, Theorem 2.14]) Finite sums of maximal strategies are again maximal.*

This theorem in Delbaen and Schachermayer [45] applies to our market because the proof in Delbaen and Schachermayer [45] never uses the local boundedness assumption.

For all t, define the trading strategy

$$\left(\gamma_0^i(t), \gamma_t^i\right) \equiv (-S_i(0), (0, \ldots, 1, \ldots, 0))$$

with a 1 in the ith place of the risky assets. This is buying and holding the ith risky asset and shorting and holding $-S_i(0)$ units of the mma. This s.f.t.s. (because it is a buy and hold trading strategy) has zero initial value because

$$\gamma_0^i(0) + \gamma_t^i \cdot S_0 = -S_i(0) + S_i(0) = 0.$$

It is admissible because the time t value of this trading strategy is

$$\int_0^t \gamma_s^i \cdot dS_s = -S_i(0) + S_i(t) \geq -S_i(0)$$

for all t.

First, note that ND implies that for all i, $\left(\gamma_0^i(t), \gamma_t^i\right) \in \mathscr{A}(0)$ is maximal.
Define

$$(\gamma_0(t), \gamma_t) = \sum_{i=1}^{n} \left(\gamma_0^i(t), \gamma_t^i\right) = \left(-\sum_{i=1}^{n} S_i(0),\ (1, \ldots, 1)\right).$$

This is an admissible s.f.t.s. with zero initial value. The time t value of this trading strategy is $\int_0^t \gamma_s \cdot dS_s$.

By Lemma 2.6, we get $(\gamma_0(t), \gamma_t) \in \mathscr{A}(0)$ is maximal.

By Lemma 2.5, there is a $\mathbb{Q}$ making $\int_0^t \gamma_s \cdot dS_s$ a martingale.

But, for all i, $\int_0^t \gamma_s \cdot dS_s = \sum_{i=1}^n (S_i(t) - S_0(t)) \geq S_i(t) - S_i(0)$ for all t.

Using Lemma 1.4 in Chap. 1, we have that a local martingale bounded by a martingale is itself a martingale. This completes the proof.

($\mathfrak{M} \neq \emptyset \Rightarrow$ NFLVR and ND).

Let $\mathbb{Q} \in \mathfrak{M}$.

First, $\mathfrak{M} \neq \emptyset$ implies $\mathfrak{M}_l \neq \emptyset$, hence NFLVR by the First Fundamental Theorem of asset pricing.

Second, we prove ND holds by contradiction. Suppose ND is violated. Then there exists an $(\alpha_0, \alpha) \in \mathscr{A}(x)$ with an initial value $x = S_0$ such that for some risky asset i,

$$S_i(0) + \int_0^T \alpha_t \cdot dS_t \geq S_i(T) \qquad \text{and} \qquad \mathbb{P}\{S_i(0) + \int_0^T \alpha_t \cdot dS_t > S_i(T)\} > 0.$$

Taking expectations, we obtain

$$S_i(0) + E^{\mathbb{Q}}\left[\int_0^T \alpha_t \cdot dS_t\right] > E^{\mathbb{Q}}[S_i(T)] = S_i(0).$$

The last equality follows by the definition of $\mathbb{Q}$ being a martingale measure.

Thus, $E^{\mathbb{Q}}\left[\int_0^T \alpha_t \cdot dS_t\right] > 0$. But, using Lemma 2.4 we have that $X_T = \int_0^T \alpha_t \cdot dS_t$ is a supermartingale, thus, $E\left[\int_0^T \alpha_t \cdot dS_t\right] \leq 0$. This is the contradiction, which completes the proof.

The Third Fundamental Theorem gives economic meaning to the difference between markets for which $\mathfrak{M}_l \neq \emptyset$ versus $\mathfrak{M} \neq \emptyset$. The first is equivalent to the market satisfying NFLVR. The second is equivalent to the market satisfying both NFLVR and ND. As emphasized previously ND is a much stronger condition, related to supply equalling demand.

Under NFLVR and ND, let $\mathbb{Q} \in \mathfrak{M}$ and $Y_T = \frac{d\mathbb{Q}}{d\mathbb{P}} > 0$ be its Radon–Nikodym derivative. Then, the risky asset price processes are $\mathbb{Q}$-martingales, i.e.

$$S_t = E^{\mathbb{Q}}[S_T \,|\mathscr{F}_t] = E\left[S_T \frac{Y_T}{Y_t}\middle|\mathscr{F}_t\right]. \tag{2.22}$$

Alternatively stated, the asset's price equals its expected (discounted, due to the normalization by the mma) value using the probability $\mathbb{Q}$. Notice that multiplying the risky asset's liquidating cash flow by the Radon–Nikodym derivative generates the "certainty equivalent" of the cash flow. Indeed, the asset's present value equals the expected (discounted) value of $S_T \frac{Y_T}{Y_t}$ under the probability $\mathbb{P}$. This shows that we can interpret $\frac{Y_T}{Y_t}$ *as an adjustment for risk.* This is a key economic insight that will be used repeatedly below.

Remark 2.19 (An Infinite Number of Risky Assets) The statement NFLVR and ND $\Rightarrow \mathfrak{M} \neq \emptyset$ depends on the fact that there is only a finite number of risky assets trading.

The statement $\mathfrak{M} \neq \emptyset \Rightarrow$ NFLVR and ND holds more generally for an arbitrary number of risky assets (see the proof of the Third Fundamental Theorem). This collection could be a continuum (an uncountably infinite set). However, in this extension, the trading strategy must still consist of only a finite number of the traded assets selected from the infinite collection that trade. This extension will subsequently be used in both Ross' Arbitrage Pricing Theory (APT) in the multiple-factor model Chap. 4 and the Heath–Jarrow–Morton model in Chap. 6. This completes the remark.

2.6.1 Complete Markets

Using ND, we get a modification of the Second Fundamental Theorem of asset pricing, which is the original theorem given by Harrison and Pliska [67, 69].

Theorem 2.6 (Characterization of Complete Markets Under NFLVR and ND) *Assume NFLVR and ND, i.e.* $\mathfrak{M} \neq \emptyset$.

The market is complete with respect to $\mathbb{Q} \in \mathfrak{M}$ *if and only if* $\mathfrak{M}$ *is a singleton, i.e. the equivalent martingale measure is unique.*

Proof (Completeness $\Rightarrow \mathfrak{M}$ *a singleton)* By the Second Fundamental Theorem 2.4 of asset pricing, completeness implies $\mathfrak{M}_l$ is a singleton. Since $\mathfrak{M} \subset \mathfrak{M}_l$, we get this implication.

($\mathfrak{M}$ a singleton $\Rightarrow$ Completeness)

This is the second statement within the Second Fundamental Theorem 2.4 of asset pricing. This completes the proof.

The difference between the Theorems 2.6 and 2.4 is the hypothesis that an equivalent martingale measure exists, i.e. $\mathfrak{M} \neq \emptyset$. As noted earlier, there exist complete markets that satisfy NFLVR but where an equivalent martingale measure does not exist (see Remark 2.15). Complete markets satisfying NFLVR and violating ND are important because, as shown in the next chapter, they are the markets that contain asset price bubbles.

2.6.2 Risk Neutral Valuation

Given the Third Fundamental Theorem, we can now discuss risk neutral valuation.

Theorem 2.7 (Risk Neutral Valuation) *Assume there exists a* $\mathbb{Q} \in \mathfrak{M}$ *(i.e. NFLVR and ND holds).*

Let $E^{\mathbb{Q}}\left[[S_i, S_i]_T^{\frac{1}{2}}\right] < \infty$ *for* $i = 1, \ldots, n$.

Given any attainable claim $Z_T \in \bigcup_{x \geq 0} \mathscr{C}(x) \subset L_+^0$ *with* $E^{\mathbb{Q}}\left[[X, X]_T^{\frac{1}{2}}\right] < \infty$, *where* X_t *is the value process of the admissible s.f.t.s.* $(\alpha_0, \alpha) \in \mathscr{A}(x)$ *that generates* Z_T, *then*

$$X_t = E^{\mathbb{Q}}[Z_T | \mathscr{F}_t] \tag{2.23}$$

for all $t \in [0, T]$ *with* $x = X_0$.

Proof The integrability condition with respect to $\mathbb{Q} \in \mathfrak{M}$ makes S_i an $\mathscr{H}^1$-martingale (see Protter [151, p. 193 and p. 238, Ex 17]).

Since Z_T is attainable, there exists an $x \geq 0$ and $(\alpha_0, \alpha) \in \mathscr{A}(x)$ such that $x + \int_0^T \alpha_t \cdot dS_t = Z_T$. Define the value process of this s.f.t.s. $X_t \equiv x + \int_0^t \alpha_s \cdot dS_s$. Note $X_T = Z_T$. Then, $E^{\mathbb{Q}}\left[[X, X]_T^{\frac{1}{2}}\right] < \infty$ implies that $E^{\mathbb{Q}}\left[\sup_{t\in[0,T]} X(t)\right] < \infty$ (see Protter [151, Theorem 48, p. 193]). This imposes an implicit restriction on $(\alpha_0, \alpha) \in \mathscr{A}(x)$, which implies that $\int_0^t \alpha_u \cdot dS_u$ is an $\mathscr{H}^1$-martingale (see Medvegyev [136], p. 341). Thus, X_t is an $\mathscr{H}^1$-martingale with $X_0 = x$, which completes the proof.

Note that in the statement of this theorem, expression (2.23) gives the time t value for a derivative with payoff Z_T at time T. The time t value is equal to the payoff's conditional expectation using the equivalent martingale probability measure $\mathbb{Q} \in \mathfrak{M}$. This theorem is called risk neutral valuation because it provides the valuation formula that would exist in equilibrium (see Part III of this book) in an economy populated by risk neutral investors whose beliefs are all equal to $\mathbb{Q}$.

It is important to note that the above theorem does not require the market to be complete with respect to $\mathbb{Q} \in \mathfrak{M}$. Consequently, there could exist many equivalent martingale probability measures. Nonetheless, all attainable claims (if properly integrable) have the same price under any of these equivalent martingale probability measures because they generate martingales with the same terminal and initial values.

Remark 2.20 (An Alternative Sufficient Condition for Risk Neutral Valuation) An alternative sufficient condition that yields risk neutral valuation is first to assume there exists a $\mathbb{Q} \in \mathfrak{M}$. Let $Z_T \in \bigcup_{x \geq 0} \mathscr{C}(x) \subset L_+^0$ be attainable with $(\alpha_0, \alpha) \in \mathscr{A}(x)$ and value process X_t. If X_t is bounded above, i.e. $X_t \leq k$ for some $k > 0$ for all t, then expression (2.23) applies.

This statement is true because the value process X is a $\mathbb{Q}$-local martingale by Lemma 2.1. And, any bounded local martingale is a martingale by Remark 1.4 in Chap. 1. This completes the remark.

In a complete market, we get the following useful theorem. This theorem is the basis for the valuation methodologies used in the Black–Scholes–Merton (Chap. 5) and Heath–Jarrow–Morton (Chap. 6) models for pricing derivatives.

Theorem 2.8 (Risk Neutral Valuation in a Complete Market) *Assume $\mathfrak{M} \neq \emptyset$ (i.e. NFLVR and ND holds).*

Let the market be complete with respect to $\mathbb{Q} \in \mathfrak{M}$.

Then, risk neutral valuation works for any $Z_T \in L^1_+(\mathbb{Q})$, i.e. *let X_t be the value process of the admissible s.f.t.s. $(\alpha_0, \alpha) \in \mathscr{A}(x)$ that generates Z_T, then*

$$X_t = E^{\mathbb{Q}}[Z_T \mid \mathscr{F}_t]$$

for all $t \in [0, T]$ with $x = X_0$.

This theorem follows because in a complete market, by Theorem 2.6, $\mathbb{Q} \in \mathfrak{M}$ is unique. And, by definition of a complete market, all derivatives integrable with respect to $\mathbb{Q}$ are attainable, i.e. $Z_T \in L^1_+(\mathbb{Q})$ implies that $Z_T \in \bigcup_{x \geq 0} \mathscr{C}(x)$, and the value process is a $\mathbb{Q}$-martingale. Thus, there exists an admissible s.f.t.s. $(\alpha_0, \alpha) \in \mathscr{A}(x)$ such that $X_t = E^{\mathbb{Q}}[Z_T \mid \mathscr{F}_t]$ with $X_0 = x$. Note that we do not need to assume the additional integrability conditions as in Theorem 2.7 on either S or Z_T with respect to $\mathbb{Q}$.

2.6.3 *Synthetic Derivative Construction*

Given the risk neutral valuation formula for pricing any derivative, we now discuss synthetic construction of the derivative's payoffs. Assuming NFLVR and ND, i.e. $\mathfrak{M} \neq \emptyset$, plus assuming that the market is complete with respect to $\mathbb{Q} \in \mathfrak{M}$, we know for a given "traded derivative" $Z_T \in L^1_+(\mathbb{Q})$, there exists an initial investment $x > 0$ and an admissible s.f.t.s. $(\alpha_0, \alpha) \in \mathscr{A}(x)$ such that

$$x + \int_0^t \alpha_u \cdot dS_u \equiv X_t = E^{\mathbb{Q}}[Z_T \mid \mathscr{F}_t] \quad \text{for all } t \qquad \text{where} \tag{2.24}$$

$$x = X_0 \qquad \text{and} \qquad X_t = \alpha_0(t) + \alpha_t \cdot S_t. \tag{2.25}$$

This trading strategy with value process X_t is called the "synthetic derivative" because its time T payoffs Z_T are identical to those of the "traded derivative."

For practical applications, we need to identify this admissible s.f.t.s., i.e. we need to be able to determine both the initial investment x and the trading strategy itself, $(\alpha_0(t), \alpha_t)$. We now explain how this is done. First, taking expectations under $\mathbb{Q}$, we get the initial investment

$$x = X_0 = E^{\mathbb{Q}}[Z_T]. \tag{2.26}$$

In words, the initial cost of constructing the synthetic derivative is equal to the expected value of the derivative's time T payoffs under the equivalent martingale measure.

Second, once we know the holdings in the risky assets α_t, then we can compute the position in the mma $\alpha_0(t)$ by solving expression (2.25). This leaves only the determination of α_t. For a large class of processes one can determine α_t using Malliavin calculus and the generalized Clark–Ocone formula (see Nunno, Oksendal, and Proske [51]). Here, however, we add some additional structure and use more elementary methods. This additional structure is often available in practical applications.

In this regard, we assume that S is a *continuous* price process and that we can write the value process X_t as a $C^{1,2}$ function (continuously differentiable in the first argument and twice continuously differentiable in the second argument) f of time t and the risky asset prices S_t, i.e.

$$X_t = E^{\mathbb{Q}}[Z_T \,|\mathscr{F}_t] = f(t, S_t). \tag{2.27}$$

Applying Ito's formula (see Theorem 1.2 in Chap. 1) to the function in expression (2.27) determines α_t. Indeed, applying Ito's formula yields the expression

$$dX_t = \tfrac{\partial f}{\partial t}dt + \tfrac{1}{2}\textstyle\sum_{i=1}^{n}\sum_{j=1}^{n}\tfrac{\partial^2 f}{\partial S_{ij}^2}\left\langle dS_i(t), dS_j(t)\right\rangle + \tfrac{\partial f}{\partial S_t}\cdot dS_t \quad \text{or}$$

$$X_0 + \textstyle\int_0^T \tfrac{\partial f}{\partial t}dt + \tfrac{1}{2}\int_0^T\sum_{i=1}^{n}\sum_{j=1}^{n}\tfrac{\partial^2 f}{\partial S_{ij}^2}\left\langle dS_i(t), dS_j(t)\right\rangle + \int_0^T \tfrac{\partial f}{\partial S_t}\cdot dS_t = X_T, \tag{2.28}$$

where $\frac{\partial f}{\partial S_t} = \left(\frac{\partial f}{\partial S_1}, \ldots, \frac{\partial f}{\partial S_n}\right)' \in \mathbb{R}^n$. Identifying the integrands of dS_t across Eqs. (2.24) and (2.28) gives the self-financing trading strategy

$$\alpha_t = \frac{\partial f}{\partial S_t}. \tag{2.29}$$

These holdings are called the derivative's deltas.

2.7 Finite-Dimension Brownian Motion Market

This section illustrates an application of the fundamental theorems of asset pricing in a market where the randomness is generated by a finite-dimension Brownian motion process. Given is a normalized market $(S, (\mathscr{F}_t), \mathbb{P})$ where the money market account $B_t \equiv 1$.

2.7.1 The Set-Up

We assume that

$$\begin{aligned} dS_i(t) &= S_i(t)\left((b_i(t)-r_t)dt+\textstyle\sum_{d=1}^{D}\sigma_{id}(t)dW_d(t)\right) \qquad \text{or} \\ S_i(t) &= S_i(0)e^{\int_0^t (b_i(u)-r_u)du-\frac{1}{2}\int_0^t\left(\sum_{d=1}^{D}\sigma_{id}^2(u)\right)du+\int_0^t\sum_{d=1}^{D}\sigma_{id}(u)dW_d(u)} \end{aligned} \tag{2.30}$$

for $i = 1, \ldots, n$, where $W_t = W(t) = (W_1(t), \ldots, W_D(t))' \in \mathbb{R}^D$ are independent Brownian motions with $W_d(0) = 0$ for all $d = 1, \ldots, D$ that generate the filtration $(\mathscr{F}_t)$, $S_0 = S(0) = (S_1(0), \ldots, S_n(0))' \in \mathbb{R}^n$ is a vector of strictly positive constants, r_t is $\mathscr{F}_t$-measurable with $\int_0^T |r_t|\,dt < \infty$, $b_t = b(t) = (b_1(t), \ldots, b_n(t))' \in \mathbb{R}^n$ is $\mathscr{F}_t$-measurable (adapted) with $\int_0^T \|b_t\|\,dt < \infty$, where $\|x\|^2 = x \cdot x = \sum_{i=1}^n x_i^2$ for $x \in \mathbb{R}^n$, and

$$\sigma_t = \sigma(t) = \underset{n\times D}{\begin{bmatrix} \sigma_{11}(t) & \cdots & \sigma_{1D}(t) \\ \vdots & & \vdots \\ \sigma_{n1}(t) & \cdots & \sigma_{nD}(t) \end{bmatrix}} \tag{2.31}$$

is $\mathscr{F}_t$-measurable (adapted) with $\sum_{j=1}^n \sum_{d=1}^D \int_0^T \sigma_{jd}^2 dt < \infty$.

In vector notation, we can write the evolution of the stock price process as

$$\frac{dS_t}{S_t} = (b_t - r_t\mathbf{1})dt + \sigma_t dW_t, \tag{2.32}$$

where $\frac{dS_t}{S_t} = \left(\frac{dS_1(t)}{S_1(t)}, \ldots, \frac{dS_n(t)}{S_n(t)}\right)' \in \mathbb{R}^n$ and $\mathbf{1} = (1, \ldots, 1)' \in \mathbb{R}^n$.

This assumption implies that S has continuous sample paths. The quadratic variation is

$$\begin{aligned} \langle dS_i(t), dS_j(t)\rangle &= \left\langle \sum_{d=1}^D S_i(t)\sigma_{id}(t)dW_d(t), \sum_{k=1}^D S_j(t)\sigma_{jk}(t)dW_k(t)\right\rangle \\ &= \sum_{d=1}^D\sum_{k=1}^D \langle S_i(t)\sigma_{id}(t)dW_d(t), S_j(t)\sigma_{jk}(t)dW_k(t)\rangle = \sum_{d=1}^D S_i(t)S_j(t)\sigma_{id}(t)\sigma_{jd}(t)dt. \end{aligned} \tag{2.33}$$

In vector notation

$$\left\langle \frac{dS_t}{S_t}, \frac{dS_t}{S_t}\right\rangle = \underset{n\times n}{\sigma_t\sigma_t'}dt. \tag{2.34}$$

2.7.2 NFLVR

For pricing derivatives or searching for arbitrage opportunities we need to know when the risky asset price evolution in expression (2.30) satisfies NFLVR. By the First Fundamental Theorem 2.3 of asset pricing we have NFLVR if and only if there exists a $\mathbb{Q} \sim \mathbb{P}$ such that S is a $\mathbb{Q}$-local martingale. Using Theorem 1.7 in Chap. 1, we have NFLVR if and only if there exists a predictable process $\theta_t = \theta(t) = (\theta_1(t), \ldots, \theta_D(t))' \in \mathbb{R}^D$ with

$$\begin{aligned} &\int_0^T \|\theta_t\|^2 \, dt < \infty \qquad \text{and} \\ &E\left[e^{-\int_0^T \theta_t \cdot dW_t - \frac{1}{2}\int_0^T \|\theta_t\|^2 dt}\right] = 1 \end{aligned} \tag{2.35}$$

such that

$$\frac{d\mathbb{Q}}{d\mathbb{P}} = e^{-\int_0^T \theta_t \cdot dW_t - \frac{1}{2}\int_0^T \|\theta_t\|^2 dt} > 0, \tag{2.36}$$

and S is a $\mathbb{Q}$-local martingale.

Next, we use Girsanov's theorem (see Theorem 1.3 in Chap. 1). By Girsanov's theorem, the stochastic process $W_t^\theta = W^\theta(t) = (W_1^\theta(t), \ldots, W_D^\theta(t))' \in \mathbb{R}^D$ defined by

$$\begin{aligned} &W_i^\theta(t) = W_i(t) + \int_0^t \theta_i(u) du \text{ or} \\ &dW_i^\theta(t) = dW_i(t) + \theta_i(t) dt \end{aligned} \tag{2.37}$$

is an independent Brownian motion process under the equivalent probability measure $\mathbb{Q}$ with $W_0^\theta = (0, \ldots, 0)'$.

Substitution of expression (2.37) into expression (2.30) yields

$$dS_i(t) = S_i(t)\left((b_i(t) - r_t)dt - \sum_{d=1}^{D} \sigma_{id}(t)\theta_d(t)dt + \sum_{d=1}^{D} \sigma_{id}(t) dW_d^\theta(t)\right). \tag{2.38}$$

The risky asset price process S is a local martingale under $\mathbb{Q}$ if and only if the drift terms in these evolutions are identically zero, i.e.

$$(b_i(t) - r_t) = \sum_{d=1}^{D} \sigma_{id}(t)\theta_d(t) \text{ for } i = 1, \ldots, n \tag{2.39}$$

for almost all $t \in [0, T]$ (up to a Lebesgue measure zero set) a.s. $\mathbb{P}$ (or $\mathbb{Q}$ since equivalent). In vector notation, this is

$$b_t - r_t \mathbf{1} = \sigma_t \theta_t. \tag{2.40}$$

We have now proven the following theorem.

Theorem 2.9 (NFLVR Evolution) *The risky asset price evolution S in expression (2.30) satisfies NFLVR if and only if there exists a predictable process θ_t satisfying expression (2.35) and*

$$b_t - r_t \mathbf{1} = \sigma_t \theta_t \tag{2.41}$$

for almost all $t \in [0, T]$ (up to a Lebesgue measure zero set) a.s. $\mathbb{P}$.

The vector process $\theta_t = \theta(t) = (\theta_1(t), \ldots, \theta_D(t))' \in \mathbb{R}^D$ are the "risk premiums" associated with the risky Brownian motions.

The next theorem gives a sufficient condition on the evolution of the risky asset price process for expression (2.41) to have a solution. First, we need to define the set of predictable stochastic processes

$$K(\sigma) \equiv \left\{ \nu \in \mathscr{L}(W) : \int_0^T \|\nu_t\|^2 dt < \infty, \ \sigma_t \nu_t = 0 \quad \text{for all } t \right\}. \tag{2.42}$$

This is the set of processes that are orthogonal to the column space of the matrix σ_t for all $t \in [0, T]$ (up to a Lebesgue measure zero set) a.s. $\mathbb{P}$.

Theorem 2.10 (Sufficient Condition on S for NFLVR) *Assume that*

(1) $\operatorname{rank}(\sigma_t) = \min\{n, D\}$ *for all t a.s. $\mathbb{P}$,*
(2) $D \geq n$, *and*
(3) $\int_0^T \left\| \sigma_t' \left[\sigma_t \sigma_t'\right]^{-1} (b_t - r_t \mathbf{1}) \right\|^2 dy < \infty$. *Then, the set of solutions to expression (2.41) is nonempty and equals*

$$\{\theta + \nu : \ \nu \in K(\sigma)\},$$

where

$$\theta_t = \sigma_t' \left[\sigma_t \sigma_t'\right]^{-1} (b_t - r_t \mathbf{1}).$$

Proof Let $\mathscr{A} \equiv \left\{ \psi \in \mathscr{L}(W) : \int_0^T \|\psi_t\|^2 dt < \infty, \ b_t - r_t \mathbf{1} = \sigma_t \psi_t \right\}$ and $\mathscr{B} \equiv \left\{ \theta + \nu \in \mathscr{L}(W) : \int_0^T \|\nu_t\|^2 dt < \infty, \ \sigma_t \nu_t = 0 \right\}$.

(Step 1) Fix a $(t, \omega) \in [0, T] \times \Omega$.

Hypothesis (1) and (2) imply that $\operatorname{rank}(\sigma_t) = n$. Hence, by Theil [175, p. 11], $\operatorname{rank}\left(\sigma_t \sigma_t'\right) = n$. Hence, $\sigma_t \sigma_t'$ is invertible and θ_t is well defined.

Consider the set of solutions ν_t to the equation $\sigma_t \nu_t = 0$. Since by hypothesis (1) σ_t is of full rank, this set of equations has a solution (see Perlis [148, p. 47]).

Hence, $\{\nu : \ \sigma_t \nu_t = 0\} \neq \emptyset$. The solution set is a singleton, consisting of only $\{0\}$ if and only if $D = n$.

Note that $\nu_t \equiv 0 \in \mathscr{B}$. Hence, $\mathscr{B} \neq \emptyset$.

(Step 2) Let $\theta + \nu \in \mathscr{B}$. Then,
$\sigma_t (\theta_t + \nu_t) = \sigma_t \theta_t = \sigma_t \sigma_t' \left[\sigma_t \sigma_t'\right]^{-1} (b_t - r_t \mathbf{1}) = b_t - r_t \mathbf{1}$.
Hence, $\theta + \nu \in \mathscr{A}$.

(Step 3) Let $\psi \in \mathscr{A}$. Consider $\theta_t = \sigma_t' \left[\sigma_t \sigma_t'\right]^{-1} (b_t - r_t \mathbf{1})$. As in (Step 1), $\sigma_t \theta_t = b_t - r_t \mathbf{1}$.
Define $\nu_t = \psi_t - \theta_t$. Then, $\sigma_t \nu_t = \psi_t \theta_t - \sigma_t \theta_t = (b_t - r_t \mathbf{1}) - (b_t - r_t \mathbf{1}) = 0$.
Thus, $\psi_t = \theta_t + \nu_t \in \mathscr{B}$.
This completes the proof.

Theorem 2.10 gives sufficient conditions for the market to satisfy NFLVR. Condition (1) is that the volatility matrix must be of full rank for all t. This omits risky assets that randomly change from risky to locally riskless (finite variation) across time. Condition (2) effectively removes redundant assets from the market. Indeed, if $n > D$, then there would exist admissible s.f.t.s. that generate a locally riskless value process. Condition (2) removes these. Finally, condition (3) is a necessary integrability condition for θ_t. For subsequent use we note that conditions (1) and (2) are true if and only if $\text{rank}\,(\sigma_t) = n$ for all t a.s. $\mathbb{P}$.

Given Theorem 2.10, we can now characterize the set of local martingale measures $\mathfrak{M}_l$.

Theorem 2.11 (Characterization of $\mathfrak{M}_l$) *Assume that*

(1) $\text{rank}\,(\sigma_t) = n$ *for all* t *a.s.* $\mathbb{P}$ *and*
(2) $\int_0^T \left\| \sigma_t' \left[\sigma_t \sigma_t'\right]^{-1} (b_t - r_t \mathbf{1}) \right\|^2 dy < \infty$. *Then,*

$$\mathfrak{M}_l = \{\mathbb{Q}^\nu : \tfrac{d\mathbb{Q}^\nu}{d\mathbb{P}} = e^{-\int_0^T (\theta_t + \nu_t)\cdot dW_t - \frac{1}{2}\int_0^T \|\theta_t + \nu_t\|^2 dt} > 0,$$
$$E\left[\tfrac{d\mathbb{Q}^\nu}{d\mathbb{P}}\right] = 1, \ \nu \in K(\sigma)\} \neq \emptyset,$$

where

$$\theta_t = \sigma_t' \left[\sigma_t \sigma_t'\right]^{-1} (b_t - r_t \mathbf{1}).$$

Proof By Theorems 2.9 and 2.10, the market satisfies NFLVR and by the First Fundamental Theorem 2.3 of asset pricing $\mathfrak{M}_l \neq \emptyset$. Define

$$\mathfrak{A}_l \equiv \{\mathbb{Q}^\nu : \frac{d\mathbb{Q}^\nu}{d\mathbb{P}} = e^{-\int_0^T (\theta_t + \nu_t)\cdot dW_t - \frac{1}{2}\int_0^T \|\theta_t + \nu_t\|^2 dt} > 0, \ E\left[\frac{d\mathbb{Q}^\nu}{d\mathbb{P}}\right] = 1, \ \nu \in K(\sigma)\}.$$

(Step 1) (Show $\mathfrak{M}_l \subset \mathfrak{A}_l$)

Take $\mathbb{Q} \in \mathfrak{M}_l$. Then, by Theorem 1.7 in Chap. 1, there exists a predictable process ψ satisfying the appropriate integrability conditions such that

$$\frac{d\mathbb{Q}}{d\mathbb{P}} = e^{-\int_0^T \psi_t \cdot dW_t - \frac{1}{2}\int_0^T \|\psi_t\|^2 dt} > 0 \text{ and } E\left[\frac{d\mathbb{Q}}{d\mathbb{P}}\right] = 1.$$

By Girsanov's Theorem 1.3 in Chap. 1,

$$dW_i^{\psi} = dW_i(t) + \psi_i(t)dt$$

is a Brownian motion under $\mathbb{Q}$. Substituting into

$$dS_i(t) = S_i(t)\left((b_i(t) - r_t)dt + \sum_{d=1}^{D} \sigma_{id}(t)dW_d(t)\right)$$

gives

$$dS_i(t) = S_i(t)\left((b_i(t) - r_t)dt + \sum_{d=1}^{D} \sigma_{id}(t)\psi_{id}(t)dt + \sum_{d=1}^{D} \sigma_{id}(t)dW_d^{\psi}(t)\right).$$

But, since $\mathbb{Q} \in \mathfrak{M}_l$, this implies

$$(b_i(t) - r_t) + \sum_{d=1}^{D} \sigma_{id}(t)\psi_{id}(t) = 0.$$

Or, in vector notation, $b_t - r_t\mathbf{1} = \sigma_t\psi_t$.

By the proof of Theorem 2.10 above, this implies $\psi_t = \theta_t + \nu_t$ for some ν_t satisfying $\sigma_t\nu_t = 0$. Hence, $\mathbb{Q} \in \mathfrak{A}_l$. This completes the proof of (Step 1).

(Step 2) (Show $\mathfrak{A}_l \subset \mathfrak{M}_l$)

Take $\mathbb{Q}^{\nu} \in \mathfrak{A}_l$. Then by Girsanov's Theorem 1.3 in Chap. 1,
$dW_i^{\nu} = dW_i(t) + (\theta_i(t) + \nu_i(t))\, dt$ is a Brownian motion under $\mathbb{Q}^{\nu}$.
Substituting into

$$dS_i(t) = S_i(t)\left((b_i(t) - r_t)dt + \sum_{d=1}^{D} \sigma_{id}(t)dW_d(t)\right)$$

gives

$$dS_i(t) = S_i(t)\left((b_i(t) - r_t)dt + \sum_{d=1}^{D}\sigma_{id}(t)\left(\theta_i(t) + \nu_i(t)\right)dt + \sum_{d=1}^{D}\sigma_{id}(t)dW_d^{\nu}(t)\right).$$

By the definition of θ and ν this implies

$$dS_i(t) = S_i(t)\left(\sum_{d=1}^{D}\sigma_{id}(t)dW_d^{\nu}(t)\right),$$

which is a $\mathbb{Q}^{\nu}$-local martingale. Hence, $\mathbb{Q}^{\nu} \in \mathfrak{M}_l$.

This completes the proof of (Step 2) and the theorem.

2.7.3 Complete Markets

This subsection studies when the Brownian motion market is complete. We assume that the evolution S in expression (2.30) satisfies

(1) $\text{rank}(\sigma_t) = n$ for all t a.s. $\mathbb{P}$ and
(2) $\int_0^T \left\| \sigma_t' \left[\sigma_t \sigma_t'\right]^{-1} (b_t - r_t \mathbf{1})\right\|^2 dy < \infty$.

Then, by Theorem 2.10 above, NFLVR is satisfied.

Choose an equivalent local martingale measure $\mathbb{Q}^{\nu} \in \mathfrak{M}_l$. We have, using vector notation, that under this probability measure the risky assets evolve as

$$\frac{dS_t}{S_t} = \sigma_t dW_t^{\nu}. \tag{2.43}$$

We also know from the Second Fundamental Theorem 2.4 that a necessary condition for the market to be complete with respect to $\mathbb{Q}^{\nu}$ is that the equivalent local martingale measure must be unique. Using the previous Theorem 2.11, $\mathbb{Q}^{\nu}$ is unique if and only if the solution θ_t to the NFLVR expression $b_t - r_t\mathbf{1} = \sigma_t\theta_t$ is unique. And, this is true if and only if $K(\sigma) = \{0\}$. But, $K(\sigma) = \{0\}$ if and only if $\text{rank}(\sigma_t) = D$ for all for all t a.s. $\mathbb{P}$, see Perlis [148].

We now show that $\text{rank}(\sigma_t) = D$ for almost all $t \in [0, T]$ is also sufficient for the market to be complete. Because $\text{rank}(\sigma_t) = D$, without loss of generality we can remove redundant assets from the market (remove rows from the matrix) so that σ_t is a $D \times D$ matrix and σ_t^{-1} exists. Then, we can rewrite expression (2.43) as

$$dW_t^{\nu} = \sigma_t^{-1}\frac{dS_t}{S_t}. \tag{2.44}$$

Now, consider an arbitrary $Z_T \in L^1_+(\mathbb{Q}^\nu)$. By the Martingale Representation Theorem 1.6 in Chap. 1, there exists predictable processes H_d in $\mathscr{L}(W^\nu_d)$ with $\int_0^T (H_d(t))^2 dt < \infty$ for $d = 1, \ldots, D$ such that

$$Z_T = E^{\mathbb{Q}^\nu}[Z_T] + \sum_{d=1}^{D} \int_0^T H_d(t) dW^\nu_d(t),$$

and where the process

$$Z_t = E^{\mathbb{Q}^\nu}[Z_T] + \sum_{d=1}^{D} \int_0^t H_d(s) dW^\nu_d(s)$$

is a $\mathbb{Q}^\nu$-martingale. In vector notation, we can rewrite this as

$$dZ_t = H_t \cdot dW^\nu_t \tag{2.45}$$

with $Z_0 = E^{\mathbb{Q}^\nu}[Z_T]$ and $H_t = (H_1(t), \cdots, H_D(t))'$.

Substituting expression (2.44) into this expression gives

$$dZ_t = H_t \cdot \sigma_t^{-1} \frac{dS_t}{S_t} \quad \text{or} \tag{2.46}$$

$$Z_t = E^{\mathbb{Q}^\nu}[Z_T] + \sum_{i=1}^{D} \int_0^t \left(\sum_{d=1}^{D} \frac{H_d(t)\sigma^{-1}_{di}(t)}{S_i(t)} \right) dS_i(t). \tag{2.47}$$

Consider the trading strategy defined by $x = E^{\mathbb{Q}^\nu}[Z_T]$, $\alpha_i(t) = \sum_{d=1}^{D} \frac{H^d_s \sigma^{-1}_{di}(t)}{S^i(t)}$ for $i = 1, \cdots, D$, and $\alpha_0(t) = Z_t - \alpha_t \cdot S_t$ for all t. This trading strategy generates Z_T, it is self-financing by expression (2.47), and it is admissible because Z_t is a $\mathbb{Q}^\nu$-martingale, which implies that $Z_t = E^{\mathbb{Q}^\nu}[Z_T | \mathscr{F}_t] \geq 0$ for all t. Hence, the market is complete. We have now proven the following theorem.

Theorem 2.12 (Complete Market Evolution) *Suppose that the risky asset price evolution S in expression (2.30) satisfies*

(1) $\operatorname{rank}(\sigma_t) = n$ *for all t a.s.* $\mathbb{P}$ *and*

(2) $\int_0^T \left\| \sigma_t' \left[\sigma_t \sigma_t'\right]^{-1} (b_t - r_t \mathbf{1}) \right\|^2 dy < \infty$.

Let $\mathbb{Q} \in \mathfrak{M}_l$.

The market is complete with respect to $\mathbb{Q}$ *if and only if the volatility matrix* σ *has rank D for almost all* $t \in [0, T]$ *(up to a Lebesgue measure zero set) a.s.* $\mathbb{P}$.

When the market is complete, then without loss of generality we can remove redundant assets from the market (remove rows from the matrix) so that σ is a

$D \times D$ matrix and the NFLVR expression for the risk premium can be rewritten as

$$\theta_t = \sigma_t^{-1}(b_t - r_t\mathbf{1}). \tag{2.48}$$

2.7.4 ND

Up to this point, Theorem 2.9 provides conditions characterizing when the market satisfies NFLVR. These conditions only ensure the existence of a local martingale measure $\mathbb{Q}$. As shown when discussing the First Fundamental Theorem, NFLVR only implies that the price process S is a supermartingale under $\mathbb{Q}$. To guarantee the existence of a martingale measure, by the Third Fundamental Theorem, ND must also hold.

We now study conditions under which ND holds for the risky asset price evolution S given in expression (2.30). First, we assume that the market satisfies NFLVR. Hence, there exists a local martingale measure $\mathbb{Q}^\nu$ such that under $\mathbb{Q}^\nu$ the evolution for a risky asset can be written as

$$\begin{aligned} dS_i(t) &= S_i(t)\left(\sum_{d=1}^D \sigma_{id}(t)dW_d^\nu(t)\right) \qquad \text{or} \\ S_i(t) &= S_i(0)e^{-\frac{1}{2}\int_0^t\left(\sum_{d=1}^D \sigma_{id}^2(u)\right)du + \int_0^t \sum_{d=1}^D \sigma_{id}(u)dW_d^\nu(u)}. \end{aligned} \tag{2.49}$$

Whether or not this risky asset's price process is a martingale under $\mathbb{Q}^\nu$ is completely determined by the integrability properties of the volatility matrix σ.

A necessary and sufficient condition for $S_i(t)$ to be a $\mathbb{Q}^\nu$-martingale is that

$$E\left[e^{-\frac{1}{2}\int_0^t\left(\sum_{d=1}^D \sigma_{id}^2(u)\right)du + \int_0^t \sum_{d=1}^D \sigma_{id}(u)dW_d^\nu(u)}\right] = 1 \tag{2.50}$$

(see Protter [151, p. 138]). This is a difficult condition to verify. A stronger sufficient (an almost necessary) condition that guarantees $S_i(t)$ is a $\mathbb{Q}^\nu$-martingale is called Novikov's condition (see Protter [151], p. 140), i.e.

$$E^{\mathbb{Q}^\nu}\left[e^{\frac{1}{2}\int_0^T\left(\sum_{d=1}^D \sigma_{id}^2(t)\right)dt}\right] < \infty.$$

We have now proven the following theorem.

Theorem 2.13 (ND Evolution) *Suppose that the risky asset price evolution S in expression (2.30) satisfies NFLVR.*

The market satisfies ND if there exists an $\mathscr{F}_t$-measurable θ_t satisfying expression (2.35), where

$$b_t - r_t\mathbf{1} = \sigma_t\theta_t \tag{2.51}$$

for almost all $t \in [0, T]$ *(up to a Lebesgue measure zero set) a.s.* $\mathbb{P}$ *and*

$$E^{\mathbb{Q}^\nu}\left[e^{\frac{1}{2}\int_0^T\left(\sum_{d=1}^D \sigma_{id}^2(t)\right)dt}\right] < \infty \text{ for all } i = 1, \ldots, n.$$

2.8 Notes

This chapter assumes that markets are competitive and frictionless. When the competitive market assumption is relaxed, then strategic trading and market manipulation are possible. For models relaxing the competitive market assumption, see Jarrow [81, 82], Bank and Baum [6], and Cetin et al. [30]. In Part IV of this book, the frictionless market assumption is relaxed with the inclusion of trading constraints. Excellent references on the theory of no arbitrage for the general semimartingale market are Delbaen and Schachermayer [47], and Karatzas and Shreve [118] for the Brownian motion market.

Appendix

This appendix shows that a characterization of NFLVR implied by the original definition of NFLVR in Delbaen and Schachermayer [44, Proposition 3.6], is equivalent to Definition 2.5 of NFLVR given in the text.

Definition 2.10 (No Free Lunch with Vanishing Risk (NFLVR-D&S)) A *free lunch with vanishing risk (FLVR-D&S)* is: (i) a simple arbitrage opportunity, or (ii) a sequence of zero initial investment admissible s.f.t.s.'s $(\alpha_0, \alpha)_n \in \mathscr{A}(0)$ with value processes X_t^n, lower admissibility bounds $c_n \leq 0$, and an $\mathscr{F}_T$-measurable random variable $\chi \geq 0$ with $\mathbb{P}(\chi > 0) > 0$ such that $c_n \to 0$ and $X_T^n \to \chi$ in probability.

The market satisfies NFLVR-D&S if there exist no FLVR-D&S trading strategies.

As seen, a market satisfies NFLVR-D&S if there are no simple arbitrage opportunities and no approximate arbitrage opportunities $(\alpha_0, \alpha)_n \in \mathscr{A}(0)$, that in the limit, become simple arbitrage opportunities. Both conditions (i) and (ii) are needed in this definition. Indeed, condition (ii) does not imply condition (i). To see why, consider a (simple) arbitrage opportunity $(\alpha_0, \alpha) \in \mathscr{A}(0)$ with admissibility bound c and value process X_t that satisfies $X_0 = 0$, $X_T \geq 0$, and $\mathbb{P}(X_T > 0) > 0$. Define the (constant) sequence of admissible s.f.t.s.'s using the simple arbitrage opportunity, i.e. $(\alpha_0, \alpha)_n \equiv (\alpha_0, \alpha)$ whose time T value process $X_T^n = X_T$ approaches (equals) $\chi \equiv X_T$, where $\chi \geq 0$ with $\mathbb{P}(\chi > 0) > 0$. This constant sequence of admissible s.f.t.s.'s satisfies all of the properties of condition (ii) except that the sequence's admissibility bounds $c_n \equiv c$ do not converge to 0.

We now show that Definition 2.5 of NFLVR is equivalent to NFLVR-D&S.

Theorem 2.14 (Equivalence of NFLVR Definitions) *NFLVR is equivalent to NFLVR-D&S.*

Proof

(Step 1) Show an FLVR implies an FLVR-D&S.

Let $(\alpha_0, \alpha)_i \in \mathscr{A}(x)$ be an FLVR, i.e. $(\alpha_0, \alpha)_i \in \mathscr{A}(x)$ with initial value $x \geq 0$, value processes X_t^i, lower admissibility bounds $0 \geq c_i$ where $\exists c \leq 0$ with $c_i \geq c$ for all i, and an $\mathscr{F}_T$-measurable random variable $\chi \geq x$ with $\mathbb{P}(\chi > x) > 0$ such that $X_T^i \to \chi$ in probability.

Now, we have $0 \geq c_i \geq c$ for all i.

Thus, there exists a subsequence n such that $c_i \to c^* \in [0, c]$ with $c \leq 0$. This implies $0 \geq c^*$.

Consider $(\alpha_0, \alpha)_n \in \mathscr{A}(x)$. Define $(\tilde{\alpha}_0, \tilde{\alpha})_n = (\alpha_0 - c^*, \alpha)_n$.

This trading strategy has initial wealth $\tilde{x} = x - c^* \geq 0$.

The value process is $\tilde{X}_t = X_t - c^*$.

It is self-financing because this trading strategy just adds $-c^* \geq 0$ additional dollars to the mma in the original s.f.t.s. for all t and holds this new position for all t.

The admissibility bounds are $\tilde{c}_n = \min\{c_n - c^*, 0\}$ for all n with $\tilde{c}_n = \min\{c_n - c^*, 0\} \to 0$.

Finally, we have that

$\tilde{X}_T^n \to \tilde{\chi} = \chi - c^*$ in probability and

$\tilde{\chi} = \chi - c^* \geq x - c^* = \tilde{x}$ with $\mathbb{P}(\tilde{\chi} = \chi - c^* > x - c^* = \tilde{x}) > 0$.

Hence, this is an FLVR-D&S.

(Step 2) Show an FLVR-D&S implies an FLVR.

(Case a) Let $(\alpha_0, \alpha)_n \in \mathscr{A}(x)$ be a simple arbitrage opportunity with admissibility bound $c \leq 0$ and value process X_t that satisfies $X_0 = 0$, $X_T \geq 0$, and $\mathbb{P}(X_T > 0) > 0$. Then, define the (constant) sequence of admissible s.f.t.s.'s using the simple arbitrage opportunity, i.e. $(\alpha_0, \alpha)_n \equiv (\alpha_0, \alpha)$ whose time T value process $X_T^n = X_T$ approaches (equals) $\chi \equiv X_T$, where $\chi \geq 0$ with $\mathbb{P}(\chi > 0) > 0$, with lower admissibility bounds $c_n \equiv c \geq c$ for all n. This is an FLVR.

(Case b) Let $(\alpha_0, \alpha)_n \in \mathscr{A}(0)$ with value processes X_t^n, lower admissibility bounds $c_n \leq 0$, and an $\mathscr{F}_T$-measurable random variable $\chi \geq 0$ with $\mathbb{P}(\chi > 0) > 0$ such that $c_n \to 0$ and $X_T^n \to \chi$ in probability.

Note that since $c_n \to 0$, there exists an N and $c < 0$ such that for all $n \geq N$, $c_n \geq c$.

Consider the sequence of admissible s.f.t.s.'s $(\alpha_0, \alpha)_{n=N}^{\infty} \in \mathscr{A}(0)$. This is an FLVR. This completes the proof.

Chapter 3
Asset Price Bubbles

An important recent development in the asset pricing literature is an understanding of asset price bubbles. This chapter discusses these new insights. They are motivated by the First and Third Fundamental Theorems, which show that NFLVR only implies the existence of a local martingale measure and not a martingale measure. Asset price bubbles clarify the economic meaning of this difference. The material in this chapter is based on the papers by Jarrow et al. [109, 111].

3.1 The Set-Up

Given is a normalized market $(S, (\mathscr{F}_t), \mathbb{P})$ where the money market account $B_t \equiv 1$ and there is only *one risky asset* trading denoted S_t. The restriction to one risky asset is without loss of generality and it is imposed to simplify the notation (to avoid subscripts for the jth risky asset).

We assume that the market satisfies NFLVR, hence by the First Fundamental Theorem 2.3 of asset pricing in Chap. 2, there exists a probability measure $\mathbb{Q} \in \mathfrak{M}_l$ where $\mathfrak{M}_l = \{\mathbb{Q} \sim \mathbb{P} : S \text{ is a } \mathbb{Q}\text{-local martingale}\}$. The subsequent analysis applies to both complete and incomplete markets. Since we do not assume that the market is complete, there can exist many such equivalent local martingale measures.

To study asset price bubbles, we need to determine a unique $\mathbb{Q} \in \mathfrak{M}_l$. If the market is complete, then by the Second Fundamental Theorem 2.4 of asset pricing in Chap. 2, the local martingale measure is uniquely determined. However, we do not assume that the market is complete. If the market is incomplete, to identify a unique equivalent local martingale measure the following approach is employed.

One assumes that the market studied is embedded in a larger market that is complete and therefore the equivalent local martingale measure is again uniquely determined by the Second Fundamental Theorem of asset pricing. This extended market usually includes the trading of derivatives on the traded risky assets

R. A. Jarrow, *Continuous-Time Asset Pricing Theory*, Springer Finance,
https://doi.org/10.1007/978-3-319-77821-1_3

(e.g. call and put options with different strikes and maturities) which complete the market. See Eisenberg and Jarrow [56], Dengler and Jarrow [48], Jacod and Protter [76], and Schweizer and Wissel [167] for papers containing markets that are expanded using derivatives such that the extended market is complete, but when trading only in the risky assets themselves, the market is incomplete. For subsequent use, we say that the unique $\mathbb{Q} \in \mathfrak{M}_l$ determined in this manner by the enlarged market is the $\mathbb{Q} \in \mathfrak{M}_l$ *chosen by the market.*

3.2 The Market Price and Fundamental Value

Fix a particular equivalent local martingale measure $\mathbb{Q} \in \mathfrak{M}_l$ chosen by the market. This section characterizes the risky asset's market price in terms of its resale value on or before the risky asset's liquidation date, time T.

Theorem 3.1 (The Market Price) *Given NFLVR,*

$$S_t = \operatorname*{ess\,sup}_{\tau \in [t,T]} E^{\mathbb{Q}} [S_\tau | \mathscr{F}_t], \tag{3.1}$$

where $\tau \geq t$ is a stopping time.

Proof Since S_t is a nonnegative local martingale under $\mathbb{Q}$, it is a supermartingale (Lemma 1.3 in Chap. 1). This implies that $S_t \geq E^{\mathbb{Q}}[S_\tau | \mathscr{F}_t]$ for any stopping time τ. Hence, $S_t \geq \operatorname*{ess\,sup}_{\tau \in [t,T]} E^{\mathbb{Q}} [S_\tau | \mathscr{F}_t]$.

Conversely, consider a localizing sequence of stopping times $\tau^n \to T$. Under this sequence, S_{τ^n} is a martingale, hence $S_t = E^{\mathbb{Q}} [S_{\tau^n} | \mathscr{F}_t]$ for $\tau^n \geq t$. This implies $\operatorname*{ess\,sup}_{\tau \in [t,T]} E^{\mathbb{Q}} [S_\tau | \mathscr{F}_t] \geq S_t$, which completes the proof.

Expression (3.1) shows that the market price of the risky asset is the value computed by considering retrading and selling the asset across all possible stopping times before time T. If retrading has no value and the optimal selling time is $\tau = T$, then the stock is valued as if it is held until the liquidation date. In this case

$$S_t = E^{\mathbb{Q}} [S_T | \mathscr{F}_t], \tag{3.2}$$

i.e. the market price of the risky asset equals its expected (discounted, due to normalization by the mma) time T liquidation value. Note that this expectation adjusts for the risk of the asset's random payoff S_T by using the local martingale measure $\mathbb{Q} \in \mathfrak{M}_l$, and by not using the statistical probability measure $\mathbb{P}$. This observation justifies the subsequent definition of the risky asset's fundamental value.

Definition 3.1 (Fundamental Value) Given NFLVR, the asset's *fundamental value* is

$$E^{\mathbb{Q}} [S_T | \mathscr{F}_t].$$

3.3 The Asset Price Bubble

We can now define the asset's price bubble.

Definition 3.2 (Asset Price Bubble) Given NFLVR, the asset's *price bubble* β_t is defined by

$$\beta_t = S_t - E^{\mathbb{Q}}[S_T | \mathscr{F}_t]. \tag{3.3}$$

We have the following theorem characterizing various properties of asset price bubbles.

Theorem 3.2 (Properties of Price Bubbles) *Given NFLVR,*

1. $\beta_t \geq 0$.
2. $\beta_T = 0$.
3. *If* $\beta_s = 0$ *for some* $s \in [0, T]$, *it is zero thereafter.*

Proof

1. This follows since S_t is a supermartingale.
2. Since $S_T = E^{\mathbb{Q}}[S_T | \mathscr{F}_T] + \beta_T$, $\beta_T = 0$.
3. Suppose $\beta_s = 0$. Then $0 = \beta_s \geq E_s^{\mathbb{Q}}[\beta_t | \mathscr{F}_s] = E_s^{\mathbb{Q}}[\beta_T | \mathscr{F}_s] \geq 0$. The bubble is, thus, a nonnegative martingale after time s. Hence, it is identically zero after time s. This completes the proof.

Condition (1) states that asset price bubbles are always nonnegative, $\beta_t \geq 0$. This is because one can always choose not to retrade. Hence, the risky asset is always worth at least its fundamental value. Condition (2) states that the price bubble may burst before, but no later than time T. This is because at time T, the stock is worth its liquidation value, which also equals its fundamental value at that time. Last, condition (3) states that if the price bubble disappears before time T, it is never reborn.

Asset price bubbles can exist in a complete market, as the next examples illustrate.

Example 3.1 (A Complete Market with an Asset Price Bubble) This example gives a risky asset's market price process with an asset price bubble, where the market is complete with respect to $\mathbb{Q} \in \mathfrak{M}_l$. Consider the risky asset price that evolves under $\mathbb{P}$ as

$$dS_t = (b(S_t) - r_t)dt + \sigma(S_t)dW_t, \tag{3.4}$$

where W_t is a standard Brownian motion with $W_0 = 0$ and $(b(S_t), \sigma(S_t))$ have appropriate measurability and boundedness conditions so that this stochastic differential equation is well defined with a solution (see Protter [151, Chapter V]).

Using Theorem 2.9 in the fundamental theorems Chap. 2, we have that the market satisfies NFLVR if and only if there exists a predictable process θ_t satisfying

$$\int_0^T \theta_t^2 dt < \infty,$$
$$E\left[e^{-\int_0^T \theta_t \cdot dW_t - \frac{1}{2}\int_0^T \theta_t^2 dt}\right] = 1,$$

such that

$$\frac{b(S_t) - r_t}{\sigma(S_t)} = \theta_t.$$

We assume this condition is true so NFLVR holds and there exists an equivalent local martingale measure $\mathbb{Q} \in \mathfrak{M}_l$.

Next, using Theorem 2.12 in the fundamental theorems Chap. 2, we have the market is complete with respect to $\mathbb{Q}$ if and only if

$$\sigma(S_t) > 0.$$

We assume that this condition is also satisfied so the market is complete with respect to $\mathbb{Q}$, implying that the equivalent local martingale measure is unique by the Second Fundamental Theorem 2.4 of asset pricing.

Finally, for this asset price process, we have the following result.

$$S \text{ is a strict local martingale under } \mathbb{Q} \text{(a bubble exists)}$$
$$\text{if and only if} \quad \int_1^\infty \frac{x}{\sigma(x)^2} dx < \infty.$$

The proof is contained in Protter [152].

The above result characterizes conditions on the asset's volatility $\sigma(\cdot)$ such that there is a price bubble. The condition is that the asset's local volatility $\sigma(S_t)$ must increase quickly enough as S_t increases so that the integral $\int_1^\infty \frac{x}{\sigma(x)^2} dx$ is finite. This "explosion" in the asset's local volatility is consistent with the fact that bubbles exist only when an asset is purchased to resell, and not to hold until liquidation. For an empirical methodology that can test this condition, using historical asset price time series data, see Jarrow et al. [112].

In the fundamental theorems Chap. 2, Remark 2.15, we showed that when an asset price process is a strict local martingale in a complete market, then there exists an admissible s.f.t.s. that generates the asset's payoff more cheaply than by buying and holding the asset itself. That is, ND is violated. This completes the example.

Example 3.2 (CEV Process) A special case of Example 3.1 is a constant elasticity of variance (CEV) process. This is obtained by letting the volatility function in Example 3.1 satisfy

$$\sigma^2(x) = \beta^2 x^{2\alpha}$$

for $\alpha, \beta > 0$.

We have the following three cases.

1. $0 < \alpha < 1$. (No bubble)

$$\frac{1}{\beta^2}\int_1^\infty x^{1-2\alpha}dx = \frac{1}{\beta^2}\frac{x^{2-2\alpha}}{2-2\alpha}\Big|_1^\infty = \infty.$$

2. $\alpha = 1$ (No bubble – Geometric Brownian Motion)

$$\frac{1}{\beta^2}\int_1^\infty x^{-1}dx = \frac{1}{\beta^2}\ln(x)|_1^\infty = \infty.$$

3. $\alpha > 1$. (Bubble)

$$\frac{1}{\beta^2}\int_1^\infty x^{1-2\alpha}dx = \frac{1}{\beta^2}\frac{x^{2-2\alpha}}{2-2\alpha}\Big|_1^\infty < \infty.$$

As in the previous example, when a bubble exists under the CEV process, the risky asset price process satisfies NFLVR, but not ND. This completes the example.

The previous two examples noted that when an asset price bubble exists in a complete market, the market violates ND. This is true more generally, as the next theorem asserts.

Theorem 3.3 (Strict Local Martingales and Dominance) *Assume NFLVR and that the market is complete with respect to $\mathbb{Q} \in \mathfrak{M}_l$.*

Let S be a strict $\mathbb{Q}$-local martingale.

Then, there exists an admissible s.f.t.s. $(\alpha_0, \alpha) \in \mathscr{A}(x)$ where $x + \int_0^T \alpha_u \cdot dS_u = S_T$ and the value process $X_t \equiv x + \int_0^t \alpha_u \cdot dS_u$ satisfies $X_0 = x = E^{\mathbb{Q}}[S_T] < S_0$ with $X_t = E^{\mathbb{Q}}[S_T | \mathscr{F}_t] \leq S_t$ for all $0 < t < T$. This admissible s.f.t.s. dominates buying and holding S_t for all $t \in [0, T]$.

Proof The proof of this theorem is contained in the fundamental theorems Chap. 2, Remark 2.15. This completes the proof.

We note that when the market is incomplete, an admissible s.f.t.s. that dominates buying and holding the risky asset as characterized in this theorem may not exist.

Remark 3.1 (Discontinuous Price Processes with Bubbles) The above examples are for price processes that have continuous sample paths. For examples of discontinuous sample path price processes with bubbles, one can take the continuous sample path examples given above and add a $\mathbb{Q}$-martingale jump process (because a strict local martingale plus a martingale is again a strict local martingale). For other discontinuous sample path price processes that are strict local martingales, see Protter [153].

We note that most discontinuous sample path price processes imply that the market is incomplete. This occurs because when a jumps occurs, the distribution for the change in the price process is usually not just discrete with a finite number of jump amplitudes where the number of jump amplitudes is less than the number of traded assets (see Cont and Tankov [34, Chapter 9.2]). In this case the number of traded assets will be insufficient to hedge the different jump magnitudes possible (in mathematical terms, the martingale representation property fails for the value processes of admissible s.f.t.s.'s). This completes the remark.

We know a number of sufficient conditions for a price process not to exhibit a price bubble. The first of these is given in the next theorem.

Theorem 3.4 (Asset Price Processes with Independent Increments) *Given NFLVR, choose a* $\mathbb{Q} \in \mathfrak{M}_l$.

If $S_t = S_0 e^{L_t}$ *where* L_t *is a process with independent increments under* $\mathbb{Q}$, *then there is no price bubble.*

Proof This proof is based on Medvegyev [136, Example 1.146, p. 103]. Note that S_t a local $\mathbb{Q}$-martingale implies that $\frac{S_t}{S_0} \geq 0$ is a $\mathbb{Q}$-local martingale. By Lemma 1.3 in Chap. 1, $X_t \equiv \frac{S_t}{S_0} = e^{L_t} > 0$ is a $\mathbb{Q}$-supermartingale. Given $L_0 = 0$, $X_0 = 1 \geq E[X_t]$ for all $t \in [0, T]$. Because L_t has independent increments,

$$E[X_t] = E\left[e^{L_t - L_s} e^{L_s}\right] = E\left[e^{L_t - L_s} X_s\right] = E\left[e^{L_t - L_s}\right] E[X_s].$$

Hence,

$$E\left[\left.\frac{X_t}{E[X_t]}\right| \mathscr{F}_s\right] = E\left[\left.\frac{e^{L_t - L_s} X_s}{E[X_t]}\right| \mathscr{F}_s\right] = \frac{X_s}{E[X_t]} E\left[\left. e^{L_t - L_s}\right| \mathscr{F}_s\right] = \frac{X_s}{E[X_t]} E\left[e^{L_t - L_s}\right].$$

Using both of the above expressions yields

$$E\left[\left.\frac{X_t}{E[X_t]}\right| \mathscr{F}_s\right] = \frac{X_s}{E[X_s]}.$$

Thus, $\frac{X_t}{E[X_t]}$ is a $\mathbb{Q}$-martingale, where $0 \leq X_t \leq \frac{X_t}{E[X_t]}$ for all t.

By Lemma 1.4 in Chap. 1, $X_t = \frac{S_t}{S_0}$ is a $\mathbb{Q}$-martingale. Hence, S_t is a $\mathbb{Q}$-martingale. This completes the proof.

Remark 3.2 (Examples of No Bubble Price Processes) If L_t is a Levy process under $\mathbb{Q}$, it has independent increments, and therefore $S_t = S_0 e^{L_t}$ has no bubbles.

A special case of such a Levy process is where the normalized asset price process S_t under $\mathbb{Q}$ is a geometric Brownian motion process, i.e. $dS_t = \sigma dW_t^{\mathbb{Q}}$. This underlies the Black–Scholes–Merton option pricing model discussed in Chap. 5. Hence, by construction, the Black–Scholes–Merton market has no asset price bubbles. This completes the remark.

Another sufficient condition is provided by using Remark 1.4 following Lemma 1.4 in Chap. 1.

Theorem 3.5 (Bounded Asset Price Processes) *Given NFLVR,*
if $S_t \geq 0$ is bounded above for all $t \in [0, T]$, then there is no price bubble.

Using the previous theorem, we can prove the following collection of facts related to asset price bubbles.

Remark 3.3 (Facts About Bubbles)

(1) *(Bonds Have No Bubbles)*
Consider a default-free zero-coupon bond that pays \$1 at time T. If default-free forward interest rates are nonnegative for all $t \in [0, T]$, then the default-free zero-coupon bond price process $S_t \leq 1$ for all t (this will be proven once we study the Heath–Jarrow–Morton model in Chap. 6). By Theorem 3.5, default-free zero-coupon bond prices can have no bubbles. By extension, no default-free coupon bond, or credit risky zero- or coupon bonds can have price bubbles as well as long as ND holds, because credit risky bond prices will be less than or equal to default-free bond prices under ND. Chapter 7 studies credit risky bonds.

(2) *(Put-Call Parity Need Not Hold)*
Put-call parity is based on the identity

$$S_T - K = \max[S_T - K, 0] - \max[K - S_T, 0].$$

The fundamental value of this expression is obtained by taking expectations with respect to the equivalent local martingale measure $\mathbb{Q} \in \mathfrak{M}_l$ chosen by the market. This gives

$$E^{\mathbb{Q}}[S_T] - K = E^{\mathbb{Q}}[\max[S_T - K, 0]] - E^{\mathbb{Q}}[\max[K - S_T, 0]].$$

Let the market prices for the put and call be denoted Put_t and Call_t, respectively. Suppose that both the put and call options have no bubbles, i.e. $\text{Put}_0 = E^{\mathbb{Q}}[\max[K - S_T, 0]]$ and $\text{Call}_0 = E^{\mathbb{Q}}[\max[S_T - K, 0]]$. Suppose further that the risky asset has a price bubble, i.e. $S_0 > E^{\mathbb{Q}}[S_T]$, which is possible under NFLVR. Then,

$$S_0 - K > \text{Call}_0 - \text{Put}_0$$

and put-call parity is violated. For an example where put-call parity is violated and both the call and put options also have price bubbles, see Protter [152, Section 7, Example 2]. This completes the remark.

Remark 3.4 (Bounded at Time T) Theorem 3.5 is not true if we replace the hypothesis $S_t \geq 0$ is bounded above for all $t \in [0, T]$ with the weaker hypothesis that only $S_T \geq 0$ is bounded at time T. Indeed, an example is the following, which is based on Protter [152, Section 7, Example 2]. Consider the stock price

$$S_t = 1 + \int_0^t \frac{S_s}{T - s} dW_s,$$

where W_s is a Brownian motion initialized at $W_0 = 0$ under $\mathbb{Q}$. By construction, the process S_t is a strict $\mathbb{Q}$-local martingale. Hence, the market satisfies NFLVR, i.e. $\mathbb{Q} \in \mathfrak{M}_l$. It can be shown that $S_t \to 0$ as $t \to T$ a.s. $\mathbb{Q}$. This implies that S_T is bounded at time T because $S_T \leq 0$. This completes the remark.

3.4 Theorems Under NFLVR and ND

This section adds the additional assumption of no dominance (ND) on the market and explores the implications for the existence of asset price bubbles. An interesting theorem results.

Theorem 3.6 (No Bubbles) *Let NFLVR and ND hold.*
Suppose the market is complete with respect to $\mathbb{Q} \in \mathfrak{M}$
where $\mathfrak{M} = \{\mathbb{Q} \sim \mathbb{P} : S \text{ is a } \mathbb{Q}\text{-martingale}\}$.
Then, there is no asset price bubble.

Proof From the Third Fundamental Theorem 2.5 of asset pricing in Chap. 2, we have NFLVR and ND imply the existence of a martingale measure $\mathbb{Q} \in \mathfrak{M}$. By the Second Fundamental Theorem 2.4 of asset pricing in Chap. 2, completeness implies the martingale measure is unique. Hence, S is a $\mathbb{Q}$-martingale, and not a strict local martingale. This completes the proof.

Remark 3.5 (Incomplete Markets) In a market satisfying NFLVR and ND, when the market is incomplete, an asset price bubble can exist. Recall that in an incomplete market there are an infinite number of local martingale measures. By the Third Fundamental Theorem 2.5 of asset pricing in Chap. 2, one of these is an equivalent martingale measure. But, it may not be the $\mathbb{Q}$ chosen by the market. There are examples where both the set of martingale and local martingale measures are non-empty for a given market, see Jarrow et al. [111]. This completes the remark.

Remark 3.6 (Trading Constraints and Bubbles) It will be shown in Part IV of this book that trading constraints transform an otherwise complete market into an incomplete market. Hence, trading constraints can transform a market without any price bubbles into a market with bubbles. Unlike the bubbles under NFLVR, in a trading constrained market, bubbles do not arise because the retrade value exceeds the fundamental value of the asset. They arise due to the trading constraint itself, see the trading constraints Chap. 19, Sect. 19.3. This completes the remark.

Theorem 3.7 (Bounded Asset Price Processes) *Let NFLVR and ND hold.*
If $S_T \geq 0$ is bounded above at time T, then there is no price bubble.

Proof Let $S_T \leq K$ a.s. $\mathbb{P}$. Next, suppose $S_t > K$ for some t with positive probability. Then, buying the risky asset and selling it and investing the S_t dollars in the mma at time t in this event generates S_t dollars for sure at time T. This dynamic s.f.t.s. dominates buying and holding the risky asset until time T, contradicting ND. Hence, $S_t \leq K$ for all t. An application of Theorem 3.5 completes the proof.

Remark 3.7 (Put Options Have No Bubbles) Consider a European put option on the risky asset with time T payoff $\text{Put}_T = \max[K - S_T, 0] \leq K$. Note that the time T payoff to the put option is bounded above, hence by Theorem 3.7, if we consider a market with both the risky asset and the European put option trading, then put options can have no price bubbles. But interestingly, the underlying risky asset can! This completes the remark.

The next theorem is useful for pricing calls and puts in a market satisfying NFLVR and ND.

Theorem 3.8 (Put-Call Parity) *Let NFLVR and ND hold.*
Given European call and put options on the risky asset with identical strikes and maturity dates trade, then

$$S_0 - K = \text{Call}_0 - \text{Put}_0. \tag{3.5}$$

Proof If put-call parity is violated, then one can create a portfolio which dominates one of the component securities. This violates ND, which gives a contradiction. This is Merton's [138] original argument. This completes the proof.

Finally, one can determine the arbitrage free value of a traded call option, in a market where the underlying asset exhibits a price bubble. Using the $\mathbb{Q} \in \mathfrak{M}_l$ chosen by the market, by the definition of this selection, an enlarged market is complete with respect to $\mathbb{Q} \in \mathfrak{M}_l$. In this enlarged market, Theorem 2.8 from the fundamental theorems Chap. 2 applies to value a put option using risk neutral valuation. In conjunction with put-call parity, we get the following theorem.

Corollary 3.1 (Call Price Values with Price Bubbles) *Let NFLVR and ND hold.*
Choose a $\mathbb{Q} \in \mathfrak{M}_l$.
If S_t has a price bubble $\beta_t > 0$, then

$$\text{Call}_0 = E^{\mathbb{Q}}[\max[S_T - K, 0]] + \beta_0. \tag{3.6}$$

Proof By a math identity,

$$E^{\mathbb{Q}}[S_T] - K = E^{\mathbb{Q}}[\max[S_T - K, 0]] - E^{\mathbb{Q}}[\max[K - S_T, 0]].$$

But by Remark 3.7, $\text{Put}_0 = E^{\mathbb{Q}}[\max[K - S_T, 0]]$ and $\beta_0 = S_0 - E^{\mathbb{Q}}[S_T]$. Substitution gives $S_0 - \beta_0 - K = E^{\mathbb{Q}}[\max[S_T - K, 0]] - \text{Put}_0$. Expression (3.5) implies the result. This completes the proof.

3.5 Notes

Much more is known about the local martingale theory of bubbles. For example, the impact of asset price bubbles on derivatives, foreign currencies, and how to test for the existence of asset price bubbles is well studied, however, no textbooks exist on this topic. Recent reviews include Jarrow and Protter [102], Protter [152], and Jarrow [89]. This is a new and exciting area for future research.

Chapter 4
Spanning Portfolios, Multiple-Factor Beta Models, and Systematic Risk

This chapter studies spanning portfolios, the multiple-factor beta model, and characterizes systematic risk. This is done for an incomplete market with asset prices that can have discontinuous sample paths. Multiple-factor beta models are used for active portfolio management and the determination of positive alphas. These models can be derived using only the Third Fundamental Theorem 2.5 of asset pricing in Chap. 2. A special case of this chapter is Ross's APT, which illustrates the notion of portfolio diversification. This chapter is based on Jarrow and Protter [104].

4.1 The Set-Up

Given is a non-normalized market $((\mathbb{B}, \mathbb{S}), (\mathscr{F}_t), \mathbb{P})$. It is important to emphasize that the subsequent models are for general asset price processes, which include discontinuous sample paths (jumps). This set-up starts with the following assumption.

Assumption (Existence of an Equivalent Martingale Measure)

$$\mathfrak{M} \neq \emptyset \text{ where } \mathfrak{M} = \left\{ \mathbb{Q} \sim \mathbb{P} : \frac{\mathbb{S}}{\mathbb{B}} \text{ is a } \mathbb{Q}\text{-martingale} \right\},$$

i.e. there exists an equivalent martingale measure.

By the Third Fundamental Theorem 2.5 of asset pricing in Chap. 2, this implies both NFLVR and ND hold in the market. We do not assume that the market is complete with respect to $\mathbb{Q} \in \mathfrak{M}$, hence, there can exist many local martingale and martingale measures. Let $Y_T = \frac{d\mathbb{Q}}{d\mathbb{P}} > 0$ be this probability's density function. The subsequent analysis uses this equivalent martingale measure.

R. A. Jarrow, *Continuous-Time Asset Pricing Theory*, Springer Finance,
https://doi.org/10.1007/978-3-319-77821-1_4

Remark 4.1 (NFLVR and Not ND) The subsequent analysis can be generalized to use only the existence of an equivalent local martingale measure, i.e. $\mathfrak{M}_l \neq \emptyset$ where

$$\mathfrak{M}_l = \left\{ \mathbb{Q} \sim \mathbb{P} : \frac{\mathbb{S}}{\mathbb{B}} \text{ is a } \mathbb{Q}\text{-local martingale} \right\}$$

or equivalently assuming only NFLVR. In this case, however, asset price bubbles are introduced into the subsequent results. This extension can be found in Jarrow [93]. This completes the remark.

Recall that

$$\mathscr{A}(x) = \Big\{ (\alpha_0, \alpha) \in (\mathfrak{L}(\mathbb{B}), \mathscr{L}(\mathbb{S})) : \mathbb{X}_t = \alpha_0(t)\mathbb{B}_t + \alpha_t \cdot \mathbb{S}_t,\ \exists c \leq 0,$$

$$\mathbb{X}_t = x + \int_0^t \alpha_0(u) d\mathbb{B}_u + \int_0^t \alpha_u \cdot d\mathbb{S}_u \geq c,\ \forall t \in [0, T] \Big\}$$

is the set of admissible s.f.t.s.'s and

$$\mathscr{C}(x) = \left\{ \mathbb{X}_T \in L^0_+ : \exists (\alpha_0, \alpha) \in \mathscr{A}(x),\ x + \int_0^t \alpha_0(u) d\mathbb{B}_u + \int_0^t \alpha_u \cdot d\mathbb{S}_u = \mathbb{X}_T \right\}$$

is the set of time T payoffs generated by all admissible s.f.t.s.'s with initial value x. We are interested in the time T payoffs of all admissible s.f.t.s.'s with nonnegative initial values at time 0 whose value processes are $\mathbb{Q}$-martingales, i.e.

$$\mathscr{X} = \left\{ \mathbb{X}_T \in \underset{x \geq 0}{\cup} \mathscr{C}(x) : \frac{\mathbb{X}_t}{\mathbb{B}_t} \text{ is a } \mathbb{Q}\text{-martingale} \right\}.$$

When necessary, we will restrict ourselves to strictly positive initial values so that returns on the s.f.t.s.'s are well defined.

To understand this set of payoffs $\mathscr{X}$ note the following. If $\frac{\mathbb{S}}{\mathbb{B}}$ are $\mathscr{H}^1$-martingales under $\mathbb{Q}$, i.e. $E^{\mathbb{Q}}\left[\left[\frac{\mathbb{S}_i}{\mathbb{B}}, \frac{\mathbb{S}_i}{\mathbb{B}}\right]_T^{\frac{1}{2}}\right] < \infty$ for $i = 1, \ldots, n$ (see Protter [151, p. 193]), then $\mathscr{X}$ includes those $\mathbb{X}_T$ such that $E^{\mathbb{Q}}\left[\left[\frac{\mathbb{X}}{\mathbb{B}}, \frac{\mathbb{X}}{\mathbb{B}}\right]_T^{\frac{1}{2}}\right] < \infty$. Indeed, under this condition $\frac{\mathbb{X}}{\mathbb{B}}$ is also an $\mathscr{H}^1$-martingale under $\mathbb{Q}$ (see the proof of Theorem 2.7 in the fundamental theorems Chap. 2), hence such an $\frac{\mathbb{X}}{\mathbb{B}} \in \mathscr{X}$. In this case, however, the set $\mathscr{X}$ is larger because it also includes $\mathbb{Q}$-martingales that are not $\mathscr{H}^1$-martingales under $\mathbb{Q}$.

For this chapter, we change our terminology slightly. Rather than viewing this set of time T values as the payoffs to derivatives, we view these payoffs as the liquidation values of *traded portfolios* of the risky assets created by financial institutions such as hedge funds, mutual funds, and exchange traded funds (ETFs).

This interpretation is realistic because such financial institutions' holdings are dynamic and well represented by an element $\mathbb{X}_T \in \mathscr{X}$.

For later use, we note the properties of an arbitrary traded portfolio $\mathbb{X}_T \in \mathscr{X}$. By definition, there exists an initial investment

$$x = \mathbb{X}_0 \geq 0$$

and an admissible s.f.t.s. $(\alpha_0, \alpha) \in \mathscr{A}(x)$ such that

$$\mathbb{X}_T = x + \int_0^T \alpha_0(t) d\mathbb{B}_t + \int_0^T \alpha(t) \cdot d\mathbb{S}_t \qquad \text{where} \tag{4.1}$$

$$\mathbb{X}_t \equiv x + \int_0^t \alpha_0(s) d\mathbb{B}_s + \int_0^t \alpha(s) \cdot d\mathbb{S}_s = \alpha_0(t)\mathbb{B}_t + \alpha(t) \cdot \mathbb{S}_t \qquad \text{and} \tag{4.2}$$

$$\frac{\mathbb{X}_t}{\mathbb{B}_t} = E^{\mathbb{Q}}\left[\frac{\mathbb{X}_T}{\mathbb{B}_T}\middle|\mathscr{F}_t\right] \qquad \text{for all } t \in [0, T]. \tag{4.3}$$

In expression (4.2), the first equality is the definition of the value process and the second equality is the s.f.t.s. condition. It is important to note that since the traded portfolio's time T values in $\mathscr{X}$ are attainable securities generated by uniformly integrable martingales with the same initial values under any $\mathbb{Q} \in \mathfrak{M}$, the value processes will have identical time t values as well under any $\mathbb{Q} \in \mathfrak{M}$ (see the discussion following Theorem 2.7 in the fundamental theorems Chap. 2). This observation is important because the set of equivalent martingale measures need not be a singleton since the market need not be complete.

4.2 Spanning Portfolios

This section generates a set of spanning portfolios, a necessary step in generating a multiple-factor beta model. We claim that due to the admissibility condition, the set of portfolio values $\mathscr{X}$ is not a linear subspace, but a convex cone in the space of $\mathscr{F}_T$-measurable random variables. To prove this claim, we need to give some definitions.

Let L^0 denote the space of $\mathscr{F}_T$-measurable random variables. A subset $\mathscr{C} \subset L^0$ is a *cone* if it has the property that if $x \in \mathscr{C}$, then $rx \in \mathscr{C}$ for all $r \geq 0$, where $r \in \mathbb{R}$. A subset $\mathscr{C} \subset L^0$ is *convex* if it has the property that if $x, y \in \mathscr{C}$, then $rx + (1-r)y \in \mathscr{C}$ for all $r \in [0, 1] \subset \mathbb{R}$. Finally, a subset $\mathscr{C} \subset L^0$ is *a linear subspace* if it has the property that if $x, y \in \mathscr{C}$, then $r_1 x + r_2 y \in \mathscr{C}$ for all $r_1, r_2 \in \mathbb{R}$. Note that a linear subspace is both convex and a cone.

We can now prove the claim. To show that $\mathscr{X}$ is a convex cone, consider two random variables $\mathbb{X}_T^1, \mathbb{X}_T^2 \in \mathscr{X}$. Let $(\alpha_0^i, \alpha^i) \in \mathscr{A}(x^i)$ be the admissible s.f.t.s. for $i = 1, 2$ that generate $\mathbb{X}_T^1, \mathbb{X}_T^2 \in \mathscr{X}$. Then, $\mathscr{X}$ is a cone because $r\mathbb{X}_T^1 \in \mathscr{X}$ for any

$r \geq 0$, since $(r\alpha_0^1, r\alpha^1) \in \mathscr{A}(rx^1)$. And, $\mathscr{X}$ is convex because $r\mathbb{X}_T^1 + (1-r)\mathbb{X}_T^2 \in \mathscr{X}$ for any $r \in [0,1]$, since $(r\alpha_0^1+(1-r)\alpha^1, r\alpha_0^2+(1-r)\alpha^2) \in \mathscr{A}(rx^1+(1-r)x^2)$. The set $\mathscr{X}$ is not a linear subspace because if $\mathbb{X}_T^1 \in \mathscr{X}$ is unbounded above, then $-\mathbb{X}_T^1 \notin \mathscr{X}$ because the admissibility condition is violated, i.e. there is no uniform lower bound for the value process $\mathbb{X}_t^1$. This completes the proof of the claim.

Given this observation, consider the smallest linear subspace in L_0 containing $\mathscr{X}$, denoted $\mathscr{X}_L$. Since $\mathscr{X}$ is a convex cone containing zero, this linear subspace $\mathscr{X}_L = \mathscr{X} \oplus \{-\mathscr{X}\}$, the "positive" and "negative" traded portfolios. Next, since $\mathscr{X}_L$ is a linear space, there always exists a Hamel basis (see Friedman [64, p. 130], Simmons [170, p. 196], Taylor and Lay [174, p. 41]). A Hamel basis is a possibly uncountably infinite collection of portfolios in $\mathscr{X}_L$ such that any finite sub-collection of this set is linearly independent and the entire set spans $\mathscr{X}_L$.

In this case, the Hamel basis will also be a direct sum of "positive" traded portfolios satisfying the admissibility condition, and "negative" or short positions in traded portfolios that do not. Nonetheless, since the traded portfolios of concern in $\mathscr{X}$ are "positive," the "positive" subset of this Hamel basis is sufficient to span the traded portfolios in $\mathscr{X}$.

Let this subset of the Hamel basis be denoted by $\{\mathbb{V}_i(T) : i \in \mathscr{H}\} \subset \mathscr{X}$ where the index set $\mathscr{H}$ labels the elements in the basis. This collection can be uncountably infinite. We call the traded portfolios in $\{\mathbb{V}_i(T) : i \in \mathscr{H}\}$ the *risk factors*. This is the minimal set of linearly independent elements in the set of traded portfolios $\mathscr{X}$ from which all of $\mathscr{X}$ can be obtained. More precisely, given an arbitrary traded portfolio $\mathbb{X}_T \in \mathscr{X}$, there exists a finite set of risk factors $\Phi_X \subset \mathscr{H}$ depending upon $\mathbb{X}_T$ such that

$$\mathbb{X}_T = \sum\nolimits_{j \in \Phi_X} \eta_{Xj} \mathbb{V}_j(T), \tag{4.4}$$

where $\eta_{Xj} \neq 0$ are nonzero constants ($\mathscr{F}_0$-measurable random variables).

Different $\mathbb{X}_T \in \mathscr{X}$ will, in general, be spanned by different risk factors. But, the set of risk factors needed to span any particular traded portfolio is always finite. The number of risk factors needed to span the entire space $\mathscr{X}$ can be uncountably infinite. We note that since the value of the money market account $\mathbb{B}_T$ is random at time T, it is treated similarly to any of the other traded assets.

We first characterize the risk factors. Since $\mathbb{V}_k(T) \in \mathscr{X}$, for each $k \in \mathscr{H}$ there exists an admissible s.f.t.s. $\left(\alpha_0^k(t), \alpha^k(t)\right)_{t \in [0,T]}$ depending on k such that

$$\mathbb{V}_k(T) = \mathbb{V}_k(0) + \int_0^T \alpha_0^k(t) d\mathbb{B}_t + \int_0^T \alpha_t^k \cdot d\mathbb{S}(t), \tag{4.5}$$

$$\mathbb{V}_k(t) \equiv \mathbb{V}_k(0) + \int_0^t \alpha_0^k(s) d\mathbb{B}_s + \int_0^t \alpha_s^k \cdot d\mathbb{S}(s) = \alpha_0^k(t)\mathbb{B}_t + \alpha_t^k \cdot \mathbb{S}_t, \quad \text{and} \tag{4.6}$$

$$\frac{\mathbb{V}_k(t)}{\mathbb{B}_t} = E^{\mathbb{Q}}\left[\left.\frac{\mathbb{V}_k(T)}{\mathbb{B}_T}\right| \mathscr{F}_t\right] \quad \text{for all } t \in [0,T]. \tag{4.7}$$

Next, consider an arbitrary traded portfolio $\mathbb{X}_T \in \mathscr{X}$. As noted above, by the definition of the basis $\mathscr{H}$ there exists a finite set of risk factors $\Phi_X \subset \mathscr{H}$ such that

$$\mathbb{X}_T = \sum_{j \in \Phi_X} \eta_{Xj} \mathbb{V}_j(T), \tag{4.8}$$

where $\eta_{Xj} \neq 0$ are nonzero constants.

Dividing both sides of this expression by $\mathbb{B}_T$, and taking conditional expectations under $\mathbb{Q}$ yields a key result

$$\mathbb{X}_t = \sum_{j \in \Phi_X} \eta_{Xj} \mathbb{V}_j(t). \tag{4.9}$$

This result states that the value of an arbitrary traded portfolio at time t can be written as a linear combination of the values of a finite collection of the risk factors. And, more importantly, the same collection of risk factors apply for every $t \in [0, T]$.

Note that $\mathbb{S}_i(T) \in \mathscr{X}$ because buying and holding one unit of risky asset i is an admissible s.f.t.s. that is a $\mathbb{Q}$-martingale, after normalization by the mma. This observation implies that for the traded risky assets themselves, there exists a finite set of risk factors $\Phi_i \subset \mathscr{H}$ and constants $\eta_{ij} \neq 0$ such that

$$\mathbb{S}_i(t) = \sum_{j \in \Phi_i} \eta_{ij} \mathbb{V}_j(t) \qquad \text{for all } i = 1, \ldots, n \text{ and } t \in [0, T]. \tag{4.10}$$

This proves the following mutual fund theorem.

Theorem 4.1 (Mutual Fund Theorem) *Any investor is indifferent between*

(i) holding the portfolio $\mathbb{X}_T$, which is an admissible s.f.t.s. in the money market account and the risky assets $(\mathbb{B}, \mathbb{S})$ *and*

(ii) holding the payoff $\mathbb{X}_T$ generated by the admissible buy and hold (therefore self-financing) trading strategy constructed from the basis traded portfolios $\{\mathbb{V}_j : j \in \Phi_X\}$ with share holdings $\{\eta_{Xj} : j \in \Phi_X\}$ as in expression (4.9).

This theorem explains the existence of the plethora of traded mutual funds and ETFs. Without additional restrictions on the model's structure, an infinite number of these risk factors are needed for the validity of this mutual fund theorem.

4.3 The Multiple-Factor Beta Model

Given the preceding framework, we can now derive the multiple-factor beta model. To derive the multiple-factor model, we need to consider the returns of a traded portfolio $\mathbb{X}_T \in \mathscr{X}$ over an arbitrary time interval within the trading horizon. First, partition $[0, T]$ into a collection of subintervals of length $\Delta > 0$. Fix a time interval $[t, t + \Delta] \subset [0, T]$ where $t \geq 0$ aligns with one of these partitions.

The return on the traded portfolio $\mathbb{X}_T \in \mathscr{X}$ over this time interval is

$$R_X(t) = \frac{\mathbb{X}(t+\Delta) - \mathbb{X}(t)}{\mathbb{X}(t)},$$

where we assume that $\mathbb{X}_t \neq 0$ for all t so that these returns are well defined.

Then, identifying the portfolio with its risk factors as in expression (4.9), simple algebra yields

$$\begin{aligned} R_X(t) &= \sum_{j \in \Phi_X} \eta_{Xj} \frac{\mathbb{V}_j(t)}{\mathbb{X}(t)} \left(\frac{\mathbb{V}_j(t+\Delta) - \mathbb{V}_j(t)}{\mathbb{V}_j(t)} \right) \\ &= \sum_{j \in \Phi_X} \beta_{Xj}(t) r_j(t), \end{aligned} \tag{4.11}$$

where

$$\beta_{Xj}(t) \equiv \eta_{Xj} \frac{\mathbb{V}_j(t)}{\mathbb{X}(t)}, \quad r_j(t) \equiv \frac{\mathbb{V}_j(t+\Delta) - \mathbb{V}_j(t)}{\mathbb{V}_j(t)} \quad \text{and} \quad \sum_{j \in \Phi_X} \beta_{Xj}(t) = 1.$$

The coefficients of the risk factors $\beta_{Xj} \neq 0$ are called the portfolio's *betas*. Note that the portfolio's betas $\beta_{Xj}(t)$ are stochastic due to both $\mathbb{V}_j(t)$ and $\mathbb{X}(t)$ being stochastic. Although stochastic, the betas are $\mathscr{F}_t$-measurable, hence their values are known at time t.

To simplify the subsequent expressions, we add the following assumption.

Assumption (Traded Default-Free Zero-Coupon Bonds) *The vector of risky assets $\mathbb{S}$ contains default-free zero-coupon bonds paying \$1 at times $t = \Delta, \ldots, T$.*

Consider the default-free zero-coupon bond that matures at time $t + \Delta$, with time t price denoted $p(t, t+\Delta)$. The return on this zero-coupon bond over $[t, t+\Delta]$ is denoted $r_0(t) \equiv \frac{1}{p(t,t+\Delta)} - 1$. By its definition, $r_0(t)$ is $\mathscr{F}_t$-measurable, i.e. its return over $[t, t+\Delta]$ is known at time t. $r_0(t)$ is called the default-free *spot rate* over $[t, t+\Delta]$.

Using these traded default-free zero-coupon bonds, one can construct the following dynamic portfolio. The portfolio starts at time 0 with a dollar invested in the shortest maturity of these zero-coupon bonds (the zero-coupon bond maturing at time Δ). When this zero-coupon bond matures, the value of the portfolio is reinvested in the (then) shortest maturity zero-coupon bond (the zero-coupon bond maturing at time 2Δ). This process is continued until time $T - \Delta$. This trading strategy's value at times $t = \Delta, \ldots, T$ equals $\prod_{s=\Delta}^{t} [1 + r_0(s - \Delta)]$. This dynamic portfolio is a *discrete money market account (mma)* whose return over the time interval $[t, t+\Delta]$ is the spot rate $r_0(t)$.

Without loss of generality, we can include this discrete mma in the set of traded risk factors (see Simmons [170, Theorem A, p. 197]). Given this observation, expression (4.11) yields the next theorem. We note, however, that adding the discrete-time mma in the set of traded risk factors implies that β_0 may be equal to zero in expression (4.12) below.

Theorem 4.2 (Multiple-Factor Beta Model) *Given an arbitrary* $\mathbb{X}_T \in \mathscr{X}$*,*

$$R_X(t) = \beta_0 r_0(t) + \sum_{j \in \Phi_X} \beta_{Xj}(t) r_j(t) \qquad \textit{where} \tag{4.12}$$

$$\beta_0 + \sum_{j \in \Phi_X} \beta_{Xj}(t) = 1$$

or equivalently,

$$R_X(t) - r_0(t) = \sum_{j \in \Phi_X} \beta_{Xj}(t) \left(r_j(t) - r_0(t) \right) \tag{4.13}$$

with $\beta_{Xj}(t) \neq 0$ *all* j*.*

Remark 4.2 (Instantaneous mma Not Riskless over $[t, t + \Delta]$*)* We note that conditional on the information at time t, the *instantaneous* mma's ($\mathbb{B}_t' s$) return is only "riskless" over $[t, t + dt]$. Over the discrete time interval $[t, t + \Delta]$ its return is random and similar to any other portfolio $\mathbb{X} \in \mathscr{X}$. This is why we assumed that a term structure of default-free zero-coupon bonds trade in the market. These default-free zero-coupon bonds guarantee the existence of a default-free spot rate over $[t, t + \Delta]$ for all t (in the partition of $[0, T]$) and an admissible s.f.t.s. that earns this spot rate over this time interval. This admissible s.f.t.s. is the discrete-time mma discussed above. This completes the remark.

Remark 4.3 (Expectation Form of Multiple-Factor Model) Taking conditional expectations of expression (4.13) under $\mathbb{P}$ gives

$$E[R_X(t) \,|\mathscr{F}_t] - r_0(t) = \sum_{j \in \Phi_X} \beta_{Xj}(t) \left(E[r_j(t) \,|\mathscr{F}_t] - r_0(t) \right). \tag{4.14}$$

This is the representation of a multiple-factor beta model most often seen in the asset pricing literature. This completes the remark.

Remark 4.4 (Empirical Form of Multiple-Factor Model) To obtain an empirical form of the multiple-factor beta model, one can add an observation error to expression (4.13) to account for noise in the data and model error, i.e.

$$R_X(t) = r_0(t) + \sum_{j \in \Phi_X} \beta_{Xj}(t) \left(r_j(t) - r_0(t) \right) + \varepsilon_X(t) \qquad \text{where} \tag{4.15}$$

$\beta_{Xj}(t) \neq 0$ all j.

In empirical testing, the model is "accepted" if $\varepsilon_X(t)$ is "white noise," i.e.

1. $E[\varepsilon_X(t) \,|\mathscr{F}_t] = 0$,
2. $\varepsilon_X(t)$ is a $\mathbb{P}$-martingale with respect to $(\mathscr{F}_t)$, and
3. $\operatorname{cov}\left[\varepsilon_X(t), r_j(t) \,|\mathscr{F}_t\right] = 0$ for all $j \in \Phi_X$.

In addition, from an economic point of view, the model will be practically useful if the R^2 (goodness of fit) is large, which implies that $\varepsilon_X(t)$ is small. This completes the remark.

4.4 Positive Alphas

An important tool in portfolio evaluation and stock selection is identifying a risky asset's alpha. This section defines an asset's alpha and studies its use in portfolio management.

Recall that expression (4.13), using the Third Fundamental Theorem 2.5 of asset pricing in Chap. 2, is derived under the assumptions of NFLVR and ND. Consequently, a violation of NFLVR or ND corresponds to a violation of expression (4.13). This can be captured by adding a nonzero alpha (an $(\mathscr{F}_t)$ optional process) to the right side of expression (4.13), i.e.

$$R_X(t) - r_0(t) = \alpha_X(t) + \sum_{j \in \Phi_X} \beta_{Xj}(t) \left(r_j(t) - r_0(t) \right). \tag{4.16}$$

Definition 4.1 (A Traded Portfolio's Alpha) The traded portfolio $\mathbb{X}$'s *alpha* is defined to be the $\alpha_X(t)$ in expression (4.16).

To be consistent with the existing literature on this topic, we use the notation $\alpha_X(t)$ for a traded portfolio's alpha. This alpha, however, is unrelated to the notation for an admissible s.f.t.s. $(\alpha_0, \alpha) \in \mathscr{A}(x)$ generating $\mathbb{X}$. This double use of the notation "α" should cause no confusion in the subsequent exposition because the notation for a portfolio's alpha is only temporarily used in this section of this chapter.

The importance of a traded portfolio's alpha is due to the next theorem.

Theorem 4.3 (Positive Alphas) *A nonzero $\alpha_X(t) \neq 0$ implies there does not exist an equivalent martingale measure for the risky assets $\mathbb{S}$.*

By a nonzero alpha we mean that $\alpha_X(t)$ is an optional process that is not the zero process, i.e. there exists a (possibly and usually non-unique) stopping time τ such that $|\alpha_X(\tau) 1_{\{\tau \leq T\}}| > 0$ with positive probability.

Proof Suppose we have a nonzero and $(\mathscr{F}_t)$ optional alpha process in expression (4.16), and that there exists an equivalent martingale measure. Given the existence of an equivalent martingale measure, we have that expression (4.13) holds for all times t. This means that the process α_X must be equal to 0 under $\mathbb{Q}$, a contradiction, which completes the proof.

This result is important for active portfolio management because it implies that since arbitrage opportunities or dominated assets are rare in well-functioning markets, positive alpha trading strategies are less common than currently believed to be true by active portfolio managers, see Jarrow [90] for an elaboration.

4.5 The State Price Density

This section revisits the notion of a state price density. Given NFLVR and ND, let $Y_T = \frac{d\mathbb{Q}}{d\mathbb{P}} > 0$ be an equivalent martingale measure's density function. Recall that the state price density is given by

$$H_t = \frac{Y_t}{\mathbb{B}_t} = \frac{E\left[Y_T \,|\mathscr{F}_t\right]}{\mathbb{B}_t}. \tag{4.17}$$

Also recall that under NFLVR, we could only prove that $(\mathbb{S}_t H_t)$ is a local martingale with respect to $\mathbb{P}$ (see Lemma 2.4 in the fundamental theorems Chap. 2). Here with ND, however, $(\mathbb{S}_t H_t)$ is a martingale with respect to $\mathbb{P}$, i.e.

$$\mathbb{S}_t = E\left[\mathbb{S}_T \frac{H_T}{H_t}\middle| \mathscr{F}_t\right]. \tag{4.18}$$

Proof $\mathbb{S}_t = E^{\mathbb{Q}}\left[\frac{\mathbb{S}_T}{\mathbb{B}_T}\middle|\mathscr{F}_t\right]\mathbb{B}_t = E\left[\frac{\mathbb{S}_T}{\mathbb{B}_T}\frac{Y_T}{E[Y_T|\mathscr{F}_t]}\middle|\mathscr{F}_t\right]\mathbb{B}_t = E\left[\mathbb{S}_T\frac{H_T}{H_t}\middle|\mathscr{F}_t\right]$. This completes the proof.

In particular, at time 0 we obtain

$$\mathbb{S}_0 = E\left[\mathbb{S}_T H_T\right]. \tag{4.19}$$

This expression shows that multiplying a payoff by the state price density generates the "certainty equivalent" of the payoff. By definition, the current value of a certainty equivalent is its (discounted) expected value. We can, therefore, interpret the state price density as containing an adjustment for risk. The quantity $\frac{H_T}{H_t}$ is sometimes called a *stochastic discount factor*.

For later use, we apply expression (4.18) at time t to the traded zero-coupon bond that matures at time $t + \Delta$, i.e.

$$p(t, t+\Delta) = E\left[1 \cdot \frac{H_{t+\Delta}}{H_t}\middle| \mathscr{F}_t\right]. \tag{4.20}$$

Finally, recall that expression (4.3) for a traded portfolio is

$$\frac{\mathbb{X}_t}{\mathbb{B}_t} = E^{\mathbb{Q}}\left[\frac{\mathbb{X}_T}{\mathbb{B}_T}\middle|\mathscr{F}_t\right] = E\left[\frac{Y_T}{Y_t}\frac{\mathbb{X}_T}{\mathbb{B}_T}\middle|\mathscr{F}_t\right].$$

Using the state price density notation, this is equivalent to

$$\mathbb{X}_t = E\left[\mathbb{X}_T \frac{H_T}{H_t}\middle| \mathscr{F}_t\right], \tag{4.21}$$

which implies that $\mathbb{X}_t H_t$ is also a $\mathbb{P}$-martingale.

4.6 Arrow–Debreu Securities

This section introduces Arrow–Debreu securities. Suppose that the indicator random variable $1_A \equiv \{1 \text{ if } \omega \in A,\ 0 \text{ otherwise}\} \in \mathscr{X}$ for some $A \in \mathscr{F}_T$, then

$$p_A(t) \equiv E\left[1_A \frac{H_T}{H_t} \middle| \mathscr{F}_t \right] = E^{\mathbb{Q}}\left[\frac{1_A}{\mathbb{B}_T} \middle| \mathscr{F}_t \right] \mathbb{B}_t \tag{4.22}$$

represents the time 0 price of 1_A, i.e. the price of a security paying 1 dollar if the event $A \in \mathscr{F}_T$ occurs. $1_A \in \mathscr{X}$ is called an *Arrow–Debreu security*. Note that we have already encountered an example of an Arrow–Debreu security earlier in this section, a zero-coupon bond paying a sure dollar at time T. Indeed, here the event $A = \Omega$, so that

$$p(t,T) = p_\Omega(t) = E\left[1_\Omega \frac{H_T}{H_t} \middle| \mathscr{F}_t \right] = E\left[\frac{H_T}{H_t} \middle| \mathscr{F}_t \right]. \tag{4.23}$$

Arrow–Debreu securities enable us to obtain another economic interpretation of the state price density. Using the above expression, heuristically, the state price density is seen to represent the certainty equivalent of a security paying 1 dollar if the state $\{\omega\} \in \Omega$ occurs, i.e.

$$p_{\{\omega\}}(t) \equiv E\left[1_{\{\omega\}} \frac{H_T}{H_t} \middle| \mathscr{F}_t \right]. \tag{4.24}$$

Arrow–Debreu securities also enable us to "rank" events $A \in \mathscr{F}_T$ based on their time 0 Arrow–Debreu prices. Given two events A_1 and A_2 with $\mathbb{P}(A_1) = \mathbb{P}(A_2)$, we say that the event $A_1 \in \mathscr{F}_T$ is more valuable than event $A_2 \in \mathscr{F}_T$ if $p_{A_1} > p_{A_2}$. *An event is more valuable when its Arrow–Debreu price is larger, and it is less valuable when its Arrow–Debreu price is smaller, holding the probability of the events equal.*

4.7 Systematic Risk

We can now introduce the concept of systematic risk. To do this, we need the following theorem.

Theorem 4.4 (The Risk Return Relation) *A traded portfolio's expected return over* $[t, t+\Delta]$ *satisfies*

$$E\left[R_X(t) \,|\mathscr{F}_t\right] = r_0(t) - \text{cov}\left[R_X(t), \frac{H_{t+\Delta}}{H_t}(1 + r_0(t)) \middle| \mathscr{F}_t \right]. \tag{4.25}$$

Proof To simplify the notation in the proof, we write $E[\cdot] = E[\cdot \,|\mathscr{F}_t\,]$ and we omit the t argument in $R_X(t)$.

By expression (4.21), since $\mathbb{X}_t H_t$ is also a $\mathbb{P}$-martingale,

$$1 = E\left[\frac{\mathbb{X}_{t+\Delta}}{\mathbb{X}_t}\frac{H_{t+\Delta}}{H_t}\right]$$

$$= E\left[(1+R_X)\frac{H_{t+\Delta}}{H_t}\right] = E\left[R_X\frac{H_{t+\Delta}}{H_t}\right] + E\left[\frac{H_{t+\Delta}}{H_t}\right].$$

But, $E\left[\frac{H_{t+\Delta}}{H_t}\right] = p(t, t+\Delta)$, and

$$E\left[R_X\frac{H_{t+\Delta}}{H_t}\right] = \mathrm{cov}\left[R_X, \frac{H_{t+\Delta}}{H_t}\right] + E[R_X]E\left[\frac{H_{t+\Delta}}{H_t}\right].$$

Substitution yields

$$1 - p(t, t+\Delta) = \mathrm{cov}\left[R_X, \frac{H_{t+\Delta}}{H_t}\right] + E\,[R_X]\,p(t, t+\Delta).$$

Divide by $p(t, t+\Delta)$ and use the fact that $\frac{1-p(t,t+\Delta)}{p(t,t+\Delta)} = r_0(t)$ to get the final result. This completes the proof.

In this theorem, expression (4.25) characterizes the concept of *systematic risk*. An asset's return $R_X(t)$ contains systematic risk and requires an expected return that exceeds the default-free spot rate if and only if $\mathrm{cov}\left[R_X(t), \frac{H_{t+\Delta}}{H_t}(1+r_0(t))\,|\mathscr{F}_t\right] < 0$. Note the minus sign in front of the covariance term in expression (4.25). To understand this statement recall that

- the state price density's value $H_{t+\Delta}$ is large when the state $\omega \in \Omega$ is valuable, i.e. when its Arrow–Debreu price is large,
- if $\mathrm{corr}\left[R_X(t), \frac{H_{t+\Delta}}{H_t}\,|\mathscr{F}_t\right] > 0$, the asset $\mathbb{X}$'s returns are high when its Arrow–Debreu price is large—it is "anti -risky," and
- if $\mathrm{corr}\left[R_X(t), \frac{H_{t+\Delta}}{H_t}\,|\mathscr{F}_t\right] < 0$, the asset $\mathbb{X}$'s returns are low when its Arrow–Debreu price is large—it is "risky."

4.7.1 Risk Factors

In the multiple-factor beta model of Theorem 4.2, it is important to know which risk factors have positive risk premium (equivalently, positive excess expected returns).

To understand why this is important, as an illustration, let us consider Ross's APT (the model in Sect. 4.8 below). In this model, there are $K + 1$ risk factors identified with each primary asset. K of these risk factors are associated with all primary assets, and their risks are not diversifiable in a large portfolio. Hence, they have nonzero risk premium. One of these risk factors is idiosyncratic and unique to each primary asset. This risk factor's risk is diversifiable in a large portfolio, and has no risk premium. The implication that idiosyncratic risk has a zero risk premium is the key insight of Ross's APT.

Here, in a more general setting, we can determine which risk factors have nonzero risk premium by applying expression (4.25) to the risk factors $\{r_i(t) : i \in \mathscr{H}\}$ themselves. This yields

$$E\left[r_i(t) \,|\mathscr{F}_t\right] - r_0(t) = -\mathrm{cov}\left[r_i(t), \frac{H_{t+\Delta}}{H_t}(1 + r_0(t)) \middle| \mathscr{F}_t \right]. \tag{4.26}$$

The last expression identifies which risk factors' excess expected returns are equal to zero, i.e.

> *a risk factor has a zero risk premium if and only if its conditional covariance with the state price density is zero.*

4.7.2 The Beta Model

In general, the state price density's time T value H_T is not the value of a traded portfolio. If the state price density is a portfolio's value, then we can obtain an alternative risk return relation, the *standard beta model.* We now derive this relation.

First, if $H_T \in \mathscr{X}$, then there exists a finite set of risk factors $\Phi_H \subset \mathscr{H}$ such that the state price density's time T value is the payoff to a traded portfolio $\{\alpha_{Hj}(t) : j \in \Phi_H\}$, i.e. $H_T = \sum_{j \in \Phi_H} \alpha_{Hj} \mathbb{V}_j(T)$ with $\alpha_{Hj} \neq 0$ all $j \in \Phi_H$. The value of this traded portfolio at time t is

$$H_t = \sum_{j \in \Phi_H} \alpha_{Hj} \mathbb{V}_j(t). \tag{4.27}$$

Denote the state price density's return over $[t, t + \Delta]$ as $R_H(t) = \frac{H_{t+\Delta} - H_t}{H_t}$.

We note that expression (4.21) applied to $H_T \in \mathscr{X}$ gives

$$H_t = E\left[H_T \frac{H_T}{H_t} \middle| \mathscr{F}_t \right],$$

which implies that H_t^2 is a $\mathbb{P}$-martingale.

Theorem 4.5 (The Beta Model) *Assume that all the quantities in expression (4.28) are squared integrable with respect to* $\mathbb{P}$. *Then,*

$$E\left[R_X(t)\,|\mathscr{F}_t\right] - r_0(t) = \frac{\operatorname{cov}\left[R_H(t), R_X(t)\,|\mathscr{F}_t\right]}{\operatorname{var}\left[R_H(t)\,|\mathscr{F}_t\right]}\left(E(\left[R_H(t)\,|\mathscr{F}_t\right] - r_0(t)\right). \tag{4.28}$$

Proof To simplify the notation we write $E[\cdot] = E[\cdot\,|\mathscr{F}_t]$ and we drop the t arguments in $R_H(t), R_X(t), r_0(t)$.

$$1 + R_H = \frac{H_{t+\Delta}}{H_t}.$$

Using expression (4.25) for R_H we have

$$E\left[1 + R_H\right] = 1 + r_0 - \operatorname{cov}\left[\frac{H_{t+\Delta}}{H_t}, \frac{H_{t+\Delta}}{H_t}(1 + r_0)\right]$$

$$E\left[R_H\right] = r_0 - \operatorname{var}\left[\frac{H_{t+\Delta}}{H_t}\right](1 + r_0)$$

$$\frac{E\left[R_H\right] - r_0}{\operatorname{var}\left[\frac{H_{t+\Delta}}{H_t}\right]} = -(1 + r_0).$$

Using expression (4.25) for R_X gives

$$E[R_X] = r_0 - (1 + r_0)\operatorname{cov}\left[R_X, \frac{H_{t+\Delta}}{H_t}\right].$$

Or,

$$E[R_X] = r_0(t) + \frac{E[R_H] - r_0}{\operatorname{var}\left[\frac{H_{t+\Delta}}{H_t}\right]}\operatorname{cov}\left[R_X, R_H\right].$$

This completes the proof.

If the state price density is traded, then as shown in expression (4.28), the standard beta model implies that only a single risk factor's return, $R_H(t)$, is needed to determine a primary asset's *expected return*. This single factor consists of a finite collection of the risk factor returns $r_j(t)$ for $j \in \Phi_H$. In summary

if the state price density trades, then the same finite collection of risk factors determine any traded asset's expected return.

This is true despite the fact that an uncountable infinity may be needed to characterize realized returns.

4.8 Diversification

This section introduces the notion of diversification, i.e. "don't put all your eggs in one basket." The notion of diversification was historically introduced in finance via the use of the mean variance efficient frontier. In this context, a diversified portfolio is one that has minimum variance for a given expected return. In its most pure sense, however, the notion of diversification is an application of the law of large numbers—the cancelling of uncorrelated risks—in a large portfolio.

With respect to pricing, the intuition is that if a risk factor can be diversified away in a large portfolio, then this risk factor should not be priced, i.e. it's expected excess return should be zero. We can show this intuition is true using a modest extension of our existing arbitrage-free market with the insights from Ross [158]. The model presented is the generalization of Ross's Arbitrage Pricing Theory (APT) to our continuous-time and continuous trading market.

The modest extension needed to our market is that the primary traded risky assets are a countably infinite set $(\mathbb{S}_i(t))_{i=1}^{\infty} \geq 0$.

This extension is needed so that diversification can take place (in the limit). We consider an arbitrary time interval $[t, t + \Delta]$ (as above). All of the previous results apply unchanged to this extended market with the exception that instead of assuming NFLVR and ND, we need to directly assume the existence of a martingale measure (it is no longer implied by NFLVR and ND because the Third Fundamental Theorem 2.5 of asset pricing in Chap. 2 was only proven for a finite asset market). Recall that the existence of a martingale measure implies NFLVR and ND, even with trading in an infinite number of risky assets.

We note that by Theorem 4.2, the risky asset returns $\{R_{\mathbb{S}_i}(t)\}_{i=1}^{\infty}$ can be spanned by a countable infinite set of risk factors, denoted $\{r_j(t) : j = 0, 1, \ldots, \infty\}$. To investigate the importance of diversification on expected returns, we need to add an additional assumption that decomposes a risky asset's return into "systematic risk" and "idiosyncratic risk," idiosyncratic risk being that which can be diversified away.

Assumption (*K* Systematic Risk Factors) $\{r_j(t) : j = 0, 1, \ldots, \infty\}$ *can be written as* $\{f_1(t), \ldots, f_K(t), u_1(t), u_2(t), \ldots..\}$ *where for each risky asset expression (4.13) is written as*

$$R_{\mathbb{S}_i}(t) = r_0(t) + \sum_{k=1}^{K} \beta_{ik}(t)[f_k(t) - r_0(t)] + \beta_{iu}(t)[u_i(t) - r_0(t)] \textit{ for all } i \tag{4.29}$$

where

$$\begin{gathered} \textit{for all } t, \quad \operatorname{cov}\left[u_i(t), u_j(t) \mid \mathscr{F}_t\right] = 0 \textit{ for } i \neq j, \\ \operatorname{var}\left[\beta_{iu}(t) u_i(t) \mid \mathscr{F}_t\right] \leq \sigma_t^2 \textit{ for all } i, \\ \beta_{iu}(t) \textit{ is } \mathscr{F}_t - \textit{measurable for all } i, \qquad \textit{and} \\ \beta_{0k}(t) = \beta_{0u}(t) = 0 \textit{ for } k = 1, \ldots, K. \end{gathered}$$

This assumption states that the risky assets' risk factors can be decomposed into two sets: *(i) systematic* risk factors $\{f_1(t), \ldots, f_K(t)\}$ plus (ii) an infinite number of *idiosyncratic* risk factors $\{u_1(t), u_2(t), \ldots..\}$. The K risk factors are common across all primary assets $\mathbb{S}_i$. The risk factor $u_i(t)$ is called idiosyncratic because it is uniquely identified with a particular risky asset $\mathbb{S}_i$. In addition, the covariances of the idiosyncratic risk factors across the primary traded assets are zero and their variances over $[t, t+\Delta]$ are assumed to be bounded. The condition that $\beta_{0k}(t) = \beta_{0u}(t) = 0$ for $k = 1, \ldots, K$ just identifies the zeroth asset as the discrete mma, i.e. $R_{\mathbb{S}_0}(t) = r_0(t)$.

Under this assumption, taking conditional expectations yields

$$\begin{aligned} E[R_{\mathbb{S}_i}(t) \,|\mathscr{F}_t] = r_0(t) + \textstyle\sum_{k=1}^{K} \beta_{ik}(t)\,(E[f_k(t)\,|\mathscr{F}_t] - r_0(t)) \\ + \beta_{iu}(t)\,(E[u_i(t)\,|\mathscr{F}_t] - r_0(t)) \,. \end{aligned} \tag{4.30}$$

Of course, since the idiosyncratic risk is diversifiable, the intuition is that it should not be priced and the last term in the above expression should be approximately zero. One additional assumption gives us this result. We need to exclude arbitrage opportunities in the limit.

Assumption (No Limiting Arbitrage Opportunities) *Given a sequence of portfolios of the primary assets* $\{w_0^n, w_1^n, \ldots, w_n^n\}_{n=1}^{\infty}$ *with* $\sum_{i=0}^{n} w_i^n = 1$, *which satisfies*

$$\lim_{n\to\infty} \text{var}\left[\textstyle\sum_{i=0}^{n} w_i^n R_{\mathbb{S}_i}(t)\right] = 0 \tag{4.31}$$

then

$$\lim_{n\to\infty} E\left[\textstyle\sum_{i=0}^{n} w_i^n R_{\mathbb{S}_i}(t)\right] = r_0(t). \tag{4.32}$$

This is true for all t.

This assumption states that there is no sequence of portfolios of the primary assets that approaches a limiting arbitrage opportunity in the sense of Definition 2.3 in the fundamental theorems Chap. 2. Indeed, this limiting portfolio has no variance, hence it is riskless. Then, it must earn the default-free spot rate in the limit; or, arbitrage in the limit is possible.

It can be shown that under this assumption

$$\lim_{n\to\infty} \textstyle\sum_{i=1}^{n} (\beta_{iu}(t)\,(E[u_i(t)\,|\mathscr{F}_t] - r_0(t)))^2 < \infty. \tag{4.33}$$

Proof The proof is based on Jarrow [80]. It is a proof by contradiction.
For convenience we write $E[\cdot\,|\mathscr{F}_t] = E[\cdot]$ and we omit the t argument.
Assume that $\lim_{n\to\infty} \sum_{i=1}^{n} (\beta_{iu}\,(E[u_i] - r_0))^2 = \infty$.
Without loss of generality, we let assets $i = 1, \ldots, K$ have returns $R_{\mathbb{S}_i} = f_i$, since we can always increase the traded risky assets to include these securities.

Let $n > K + 1$ and form the portfolio

$$\begin{aligned} w_i^n &= \frac{\beta_{iu}(E[u_i]-r_0)}{\sum_{i=K+1}^{n}(\beta_{iu}(E[u_i]-r_0))^2} && \text{for } i = K+1, \ldots, \infty, \\ w_k^n &= \sum_{i=K+1}^{n} \beta_{ik} w_i^n && \text{for } k = 1, \ldots, K, \text{ and} \\ w_0^n &= -\sum_{i=1}^{n} w_i^n + 1. \end{aligned}$$

Note by the definition of w_0^n, $w_0^n + \sum_{i=1}^{n} w_i^n = 1$.

We show that this portfolio is a limiting arbitrage opportunity, yielding the contradiction.

Now,

$$\sum_{i=0}^{n} w_i^n R_{\mathbb{S}_i} = \left(-\sum_{i=1}^{n} w_i^n + 1\right) r_0 + \sum_{k=1}^{K} w_k^n f_k + \sum_{i=K+1}^{n} w_i^n R_{\mathbb{S}_i}$$

$$= r_0 + \sum_{k=1}^{K} w_k^n (f_k - r_0) + \sum_{i=K+1}^{n} w_i^n \left(R_{\mathbb{S}_i} - r_0\right).$$

Using the definition of w_k^n for $k = 1, \ldots, K$ gives

$$= r_0 + \sum_{k=1}^{K} \left(\sum_{i=K+1}^{n} \beta_{ik} w_i^n\right) (f_k - r_0) + \sum_{i=K+1}^{n} w_i^n \left(R_{\mathbb{S}_i} - r_0\right).$$

Changing the order of summation

$$= r_0 + \sum_{i=K+1}^{n} w_i^n \left[\sum_{k=1}^{K} \beta_{ik} (f_k - r_0)\right] + \sum_{i=K+1}^{n} w_i^n \left(R_{\mathbb{S}_i} - r_0\right).$$

Combining terms and using expression (4.29), we get the final result

$$\sum_{i=0}^{n} w_i^n R_{\mathbb{S}_i} = r_0 + \sum_{i=K+1}^{n} w_i^n \beta_{iu} (u_i - r_0).$$

First, taking the variance gives

$$\operatorname{var}\left[\sum_{i=0}^{n} w_i^n R_{\mathbb{S}_i}\right] = \sum_{i=K+1}^{n} \left(w_i^n\right)^2 \operatorname{var}\left[\beta_{iu} (u_i - r_0)\right].$$

But, $\text{var}(\beta_{iu} u_i) \leq \sigma_t^2$ for all i, thus

$$0 \leq \text{var}\left[\sum_{i=0}^{n} w_i^n R_{\mathbb{S}_i}\right] \leq \sigma_t^2 \sum_{i=K+1}^{n} (w_i^n)^2$$

$$= \sigma_t^2 \sum_{i=K+1}^{n} \frac{[\beta_{iu}(E[u_i] - r_0)]^2}{\left[\sum_{i=K+1}^{n} (\beta_{iu}(E[u_i] - r_0))^2\right]^2}$$

$$= \sigma_t^2 \frac{1}{\left[\sum_{i=K+1}^{n} (\beta_{iu}(E[u_i] - r_0))^2\right]} \to 0 \text{ as } n \to \infty.$$

Second, taking expectations yields

$$E\left[\sum_{i=0}^{n} w_i^n R_{\mathbb{S}_i}\right] = r_0 + \sum_{i=K+1}^{n} w_i^n \beta_{iu}(E[u_i] - r_0).$$

Substituting in the definition of w_i^n, we have that

$$\sum_{i=K+1}^{n} \left(\frac{\beta_{iu}(E[u_i] - r_0)}{\sum_{i=K+1}^{n} (\beta_{iu}(E[u_i] - r_0))^2}\right) \beta_{iu}(E[u_i] - r_0) = 1.$$

Hence, $E\left[\sum_{i=0}^{n} w_i^n R_{\mathbb{S}_i}\right] = r_0 + 1 > r_0$ for all n.

This shows that the portfolio is a limiting arbitrage opportunity, which completes the proof.

Alternatively stated, this result implies that for any $\epsilon > 0$, except for a finite number of the risky assets ($i = 1, \cdots, \infty$),

$$|\beta_{iu}(t)(E[u_i(t) | \mathscr{F}_t] - r_0(t))| < \epsilon.$$

Furthermore, if there exists a constant $\beta > 0$ such that $\beta_{iu}(t) \geq \beta$ for all i, then the expected excess returns on the idiosyncratic risk factor are approximately zero for all but a finite number of the risky assets ($i = 1, \cdots, \infty$), i.e.

$$E[u_i(t) | \mathscr{F}_t] - r_0(t) \approx 0.$$

Equivalently,

$$E[R_{\mathbb{S}_i}(t) | \mathscr{F}_t]) \approx r_0(t) + \sum_{k=1}^{K} \beta_{ik}(t)(E[f_k(t) | \mathscr{F}_t] - r_0(t)). \tag{4.34}$$

Using expression (4.26), since u_i is a risk factor for each i, this implies that

$$\operatorname{cov}\left[u_i(t), \frac{H_{t+\Delta}}{H_t} \middle| \mathscr{F}_t \right] \approx 0.$$

One can view Ross's APT as providing a set of sufficient conditions on an asset's return process under which the previous expression is valid, i.e. idiosyncratic risk is not priced.

4.9 Notes

The general multi-factor model presented in this chapter awaits empirical testing. The exception to this statement is an important special case of the multi-factor model, Ross's APT, which has already seen significant empirical testing, see Jagannathan et al. [77] for a review. Recommended books for understanding the empirical methodology for testing asset pricing models include Campbell et al. [25], Cochrane [33], and Ruppert [161].

Chapter 5
The Black–Scholes–Merton Model

This chapter presents the seminal Black–Scholes–Merton (BSM) model for pricing options. Since this chapter is a special case of the material contained in Sect. 2.7 in the fundamental theorems Chap. 2, the presentation will be brief. We are given a *normalized market* $(S, (\mathscr{F}_t), \mathbb{P})$ where there is only one risky asset trading. The money market account (mma) and risky asset's evolutions are

$$B_t \equiv 1, \tag{5.1}$$

$$\begin{aligned} dS_t &= S_t\left((b-r)dt + \sigma dW_t\right), \text{ or} \\ S_t &= S_0 e^{(b-r)t - \frac{1}{2}\sigma^2 t + \sigma W_t} \end{aligned} \tag{5.2}$$

where b, r, σ are strictly positive constants and W_t is a standard Brownian motion with $W_0 = 0$ that generates the filtration $(\mathscr{F}_t)$. This evolution is called geometric Brownian motion.

We note that this is a subcase of the evolution given in expression (2.30) in the fundamental theorems Chap. 2. Hence, we can apply the theorems from Sect. 2.7 in Chap. 2.

5.1 NFLVR, Complete Markets, and ND

First, we want to show that the risky asset price process in expression (5.2) satisfies NFLVR. In this regard, define

$$\theta = \frac{(b-r)}{\sigma} \tag{5.3}$$

R. A. Jarrow, *Continuous-Time Asset Pricing Theory*, Springer Finance,
https://doi.org/10.1007/978-3-319-77821-1_5

and the equivalent probability measure $\mathbb{Q}^\theta$ by

$$\frac{d\mathbb{Q}^\theta}{d\mathbb{P}} = e^{-\theta \cdot W_T - \frac{1}{2}\theta^2 T} > 0.$$

Note that

$$\int_0^T \|\theta\|^2 dt = \theta^2 T < \infty \qquad \text{and}$$
$$E\left[e^{-\int_0^T \theta dW_t - \frac{1}{2}\int_0^T \|\theta_t\|^2 dt}\right] = E\left[e^{-\theta W_T - \frac{1}{2}\theta^2 T}\right] = 1.$$

The second condition follows from the characteristic function of the normal random variable W_T.

By Theorem 2.9 in the fundamental theorems Chap. 2, the evolution satisfies NFLVR. This implies by the First Fundamental Theorem 2.3 of asset pricing in Chap. 2 that $\mathbb{Q}^\theta \in \mathfrak{M}_l$, where $\mathfrak{M}_l = \{\mathbb{Q} \sim \mathbb{P} : S \text{ is a } \mathbb{Q}\text{-local martingale}\}$, i.e. it is an equivalent local martingale measure.

For subsequent use, we note that the evolution of the risky asset price under $\mathbb{Q}^\theta$ is given by

$$\begin{aligned} dS_t &= S_t \sigma dW_t^\theta \quad \text{or} \\ S_t &= S_0 e^{-\frac{1}{2}\sigma^2 t + \sigma W_t^\theta}, \end{aligned} \tag{5.4}$$

where $dW^\theta(t) = dW(t) + \theta(t)dt$ is a Brownian motion under $\mathbb{Q}^\theta$. Taking natural logarithms, we get that

$$\log(S_t) = \ln(S_0) - \frac{1}{2}\sigma^2 t + \sigma W_t^\theta. \tag{5.5}$$

In this form, it is easy to see that

$$Q^\theta\left(\frac{\log(S_t/S_0) + \frac{1}{2}\sigma^2 t}{\sigma\sqrt{t}} \le x\right) = Q^\theta\left(\frac{W_t^\theta}{\sqrt{t}} \le x\right) = \int_{-\infty}^x \frac{1}{\sqrt{2\pi}} e^{-\frac{1}{2}z^2} dz, \tag{5.6}$$

where the right side of this expression is the standard cumulative normal distribution function, denoted $N(x)$.

Second, we want to show that the market is complete with respect to $\mathbb{Q}^\theta \in \mathscr{M}_l$. Since $\sigma > 0$ is a scalar, a direct application of Theorem 2.12 in the fundamental theorems Chap. 2 gives this result. This implies by the Second Fundamental Theorem 2.4 of asset pricing in Chap. 2 that $\mathbb{Q}^\theta \in \mathscr{M}_l$ is unique.

Third, we want to show that the market satisfies ND. Again, a trivial application of Theorem 2.13 in Chap. 2 gives the result since

$$e^{\frac{1}{2}\sigma^2 T} < \infty$$

for any probability measure $\mathbb{Q}^\theta \in \mathscr{M}_l$. This implies by the Third Fundamental Theorem 2.5 of asset pricing in Chap. 2 that $\mathbb{Q}^\theta \in \mathfrak{M}$, where

$$\mathfrak{M} = \{\mathbb{Q} \sim \mathbb{P} : \ S \text{ is a } \mathbb{Q}\text{-martingale}\},$$

i.e. it is an equivalent martingale measure, i.e. S_t is a $\mathbb{Q}^\theta$-martingale.

Remark 5.1 (No Asset Price Bubbles) Since under $\mathbb{Q}^\theta \in \mathscr{M}_l$, S_t is a $\mathbb{Q}^\theta$-martingale, the Black–Scholes–Merton model has no asset price bubbles as defined in the asset price bubbles Chap. 3. This is a consequence of assuming that the asset price process follows expression (5.2). This completes the remark.

5.2 The BSM Call Option Formula

This section derives the Black–Scholes–Merton (BSM) European call option formula. For this section, we need to use the non-normalized prices $(\mathbb{B}, \mathbb{S})$. A *European call option* on the risky asset with maturity T and strike price K is a financial contract that gives the owner the right, but not the obligation, to purchase the underlying asset at time T for K dollars. To maximize the payoff of the call option on the maturity date, the owner should exercise the option and buy the asset if and only if the time T asset price S_T exceeds the strike price K. The call's time T payoff is therefore

$$\mathbb{Z}_T = \max[\mathbb{S}_T - K, 0] \in L^0_+. \tag{5.7}$$

Normalizing, we get

$$Z_T = \frac{\mathbb{Z}_T}{\mathbb{B}_T} = \frac{\max[\mathbb{S}_T - K, 0]}{\mathbb{B}_T} = \max\left[S_T - Ke^{-rT}, 0\right].$$

Since $E^{\mathbb{Q}^\theta}[Z_T] = E^{\mathbb{Q}^\theta}\left[\max\left[S_T - Ke^{-rT}, 0\right]\right] \leq E^{\mathbb{Q}^\theta}[S_T] = S_0 < \infty$, we have $Z_T \in L^1_+(\mathbb{Q}^\theta)$. Using the risk neutral valuation Theorem 2.8 in Chap. 2, the option's time t value is

$$X_t = E^{\mathbb{Q}^\theta}[Z_T | \mathscr{F}_t] = E^{\mathbb{Q}^\theta}\left[\max\left[S_T - Ke^{-rT}, 0\right] \Big| \mathscr{F}_t\right].$$

By substitution

$$\frac{\mathbb{X}_t}{\mathbb{B}_t} = E^{\mathbb{Q}^\theta}\left[\max\left[\frac{\mathbb{S}_T}{\mathbb{B}_T} - Ke^{-rT}, 0\right] \Big| \mathscr{F}_t\right].$$

Using the facts that

$$\mathbb{B}_t = e^{rt}, \tag{5.8}$$

$$\begin{aligned} d\mathbb{S}_t &= \mathbb{S}_t \left(r dt + \sigma d W_t^{\theta}\right), \text{ or} \\ \mathbb{S}_t &= \mathbb{S}_0 e^{(r-\frac{1}{2}\sigma^2)t + \sigma W_t^{\theta}}, \end{aligned} \tag{5.9}$$

algebra yields

$$\begin{aligned} \mathbb{X}_t &= E^{\mathbb{Q}^{\theta}} \left[\max[\mathbb{S}_t e^{-\frac{1}{2}\sigma^2(T-t)+\sigma(W_T^{\theta}-W_t^{\theta})} - K e^{-r(T-t)}, 0] \middle| \mathscr{F}_t \right] \\ &= E^z \left[\max[\mathbb{S}_t e^{-\frac{1}{2}\sigma^2(T-t)+\sigma\sqrt{T-t}\cdot z} - K e^{-r(T-t)}, 0] \right], \end{aligned}$$

where the random variable z has the standard cumulative normal distribution $N(z)$.

Computing this expectation gives the Black–Scholes–Merton (BSM) formula for a European call option

$$\mathbb{X}_t = \mathbb{S}_t N(d_1) - K e^{-r(T-t)} N(d_2) \tag{5.10}$$

where

$$d_1 = \frac{\log(\mathbb{S}_t/K) + (r + \frac{1}{2}\sigma^2(T-t))}{\sigma\sqrt{T-t}} \quad \text{and}$$

$$d_2 = d_1 - \sigma\sqrt{T-t}.$$

Proof

$$\begin{aligned} X_t &= E^z \left[\max[\mathbb{S}_t e^{-\frac{1}{2}\sigma^2(T-t)+\sigma\sqrt{T-t}\cdot z} - K e^{-r(T-t)}, 0] \right] \\ &= E^z \left[\mathbb{S}_t e^{-\frac{1}{2}\sigma^2(T-t)+\sigma\sqrt{T-t}\cdot z} - K e^{-r(T-t)} \middle| \mathbb{S}_t e^{-\frac{1}{2}\sigma^2(T-t)+\sigma\sqrt{T-t}\cdot z} \geq K e^{-r(T-t)} \right]. \end{aligned}$$

Note that $\mathbb{S}_t e^{-\frac{1}{2}\sigma^2(T-t)+\sigma\sqrt{T-t}\cdot z} \geq K e^{-r(T-t)}$ if and only if

$$z \geq -\frac{\log\left(\frac{\mathbb{S}_t}{K}\right) + r(T-t) - \frac{1}{2}\sigma^2(T-t)}{\sigma\sqrt{T-t}} \equiv -d_2.$$

(first term in expectation)

$$\begin{aligned} &E^z \left[\mathbb{S}_t e^{-\frac{1}{2}\sigma^2(T-t)+\sigma\sqrt{T-t}\cdot z} \middle| \mathbb{S}_t e^{-\frac{1}{2}\sigma^2(T-t)+\sigma\sqrt{T-t}\cdot z} \geq K e^{-r(T-t)} \right] \\ &\quad = E^z \left[\mathbb{S}_t e^{-\frac{1}{2}\sigma^2(T-t)+\sigma\sqrt{T-t}\cdot z} \middle| z \geq -d_2 \right] \end{aligned}$$

$$= \mathbb{S}_t \int_{-d_2}^{\infty} \left(e^{-\frac{1}{2}\sigma^2(T-t)+\sigma\sqrt{T-t}\cdot z} \right) \frac{e^{-\frac{1}{2}z^2}}{\sqrt{2\pi}} dz$$

$$= \mathbb{S}_t \int_{-d_2}^{\infty} \left(e^{-\frac{1}{2}[z-\sigma\sqrt{T-t}]^2} \right) \frac{1}{\sqrt{2\pi}} dz$$

A change of variable $y = z - \sigma\sqrt{T-t}$ yields

$$= \mathbb{S}_t \int_{-d_2-\sigma\sqrt{T-t}}^{\infty} \left(e^{-\frac{1}{2}y^2} \right) \frac{1}{\sqrt{2\pi}} dy$$

$$= \mathbb{S}_t \text{Prob}\left[y \geq -d_2 - \sigma\sqrt{T-t} \right] = \mathbb{S}_t \text{Prob}\left[y \geq -d_1 \right]$$

$$= \mathbb{S}_t \text{Prob}\left[y \leq d_1 \right].$$

(second term in expectation)

$$E^z \left[K e^{-r(T-t)} \middle| \mathbb{S}_t e^{-\frac{1}{2}\sigma^2(T-t)+\sigma\sqrt{T-t}\cdot z} \geq K e^{-r(T-t)} \right]$$

$$= K e^{-r(T-t)} E^z \left[1 | z \geq -d_2 \right]$$

$$= K e^{-r(T-t)} \text{Prob}\left[z \geq -d_2 \right]$$

$$= K e^{-r(T-t)} \text{Prob}\left[z \leq d_2 \right].$$

Substitution completes the proof.

Remark 5.2 (BSM Formula Independence of the Asset's Expected Return) As seen in expression (5.10), the BSM formula does not depend on the stock's expected return

$$E\left[\frac{d\mathbb{S}_t}{\mathbb{S}_t} \middle| \mathscr{F}_t \right] = b \cdot dt.$$

This independence makes the BSM formula usable in practice because estimating the stock's expected return per unit time, b, or equivalently estimating the stock's risk premium θ is a very difficult exercise (see Jarrow and Chatterjea [95], Chapters 19 and 20 for a discussion of these statements). This completes the remark.

5.3 The Synthetic Call Option

For hedging risk, one needs to construct the synthetic call option. The synthetic call is obtained from the admissible s.f.t.s. that generates the call option's payoff. This section shows how to construct the synthetic call option in the BSM market.

Since the market is complete with respect to $\mathbb{Q}^\theta$, by the definition of a complete market, we know there exists an initial investment $x > 0$ and an admissible s.f.t.s. $(\alpha_0, \alpha) \in \mathscr{A}(x)$ with value process $X_t = x + \int_0^t \alpha_s \cdot dS_s = \alpha_0(t) + \alpha_t S_t$, a $\mathbb{Q}^\theta$-martingale, such that

$$x + \int_0^T \alpha_t \cdot dS_t = \max[S_T - Ke^{-rT}, 0]. \tag{5.11}$$

We need to determine both x and $(\alpha_0(t), \alpha_t)$.

First, taking expectations under $\mathbb{Q}^\theta$ we get that the initial investment is

$$x = X_0 = E^{\mathbb{Q}^\theta}\left[\max[S_T - Ke^{-rT}, 0]\right]. \tag{5.12}$$

This is the BSM formula's value at time 0. This observation makes sense since the initial cost of the synthetic call in an arbitrage-free market should be the arbitrage-free value of the call.

Second, we know from the self-financing condition that

$$X_t = \alpha_0(t) + \alpha_t S_t. \tag{5.13}$$

Hence, once we know α_t, then we know $\alpha_0(t)$ by solving this last expression.

Finally, to get α_t, we apply Ito's formula to the BSM formula (5.10). This gives

$$dX_t = \frac{\partial X_t}{\partial t}dt + \frac{1}{2}\frac{\partial^2 X_t}{\partial S_t^2}\langle dS_t, dS_t\rangle + \frac{\partial X_t}{\partial S_t}dS_t \qquad \text{or}$$
$$x + \int_0^T \left(\frac{\partial X_t}{\partial t} + \frac{1}{2}\frac{\partial^2 X_t}{\partial S_t^2}S_t^2\sigma^2\right)dt + \int_0^T \frac{\partial X_t}{\partial S_t}dS_t = \max[S_T - Ke^{-rT}, 0], \tag{5.14}$$

where the last expression uses the fact that

$$\langle dS_t, dS_t\rangle = S_t^2\sigma^2 dt.$$

Identifying the integrands of dS_t across Eqs. (5.11) and (5.14) gives the final result

$$\alpha_t = \frac{\partial X_t}{\partial S_t} = \frac{\partial \mathbb{X}_t}{\partial \mathbb{S}_t} = N(d_1). \tag{5.15}$$

Proof The second equality follows because $\frac{\partial\left(\frac{\mathbb{X}_t}{\mathbb{B}_t}\right)}{\partial \mathbb{S}_t} = \frac{\partial \mathbb{X}_t}{\partial \mathbb{S}_t}\frac{1}{\mathbb{B}_t}$, $\frac{\partial\left(\frac{\mathbb{S}_t}{\mathbb{B}_t}\right)}{\partial \mathbb{S}_t} = \frac{1}{\mathbb{B}_t}$, and $\frac{\partial\left(\frac{\mathbb{X}_t}{\mathbb{B}_t}\right)}{\partial \mathbb{S}_t} = \frac{\partial\left(\frac{\mathbb{X}_t}{\mathbb{B}_t}\right)}{\partial\left(\frac{\mathbb{S}_t}{\mathbb{B}_t}\right)}\frac{\partial\left(\frac{\mathbb{S}_t}{\mathbb{B}_t}\right)}{\partial \mathbb{S}_t}$.

To obtain the third equality, note that

$$\frac{\partial \mathbb{X}_t}{\partial \mathbb{S}_t} = N(d_1) + \mathbb{S}_t \frac{\partial N(d_1)}{\partial d_1} \frac{\partial d_1}{\partial \mathbb{S}_t} - Ke^{-r(T-t)} \frac{\partial N(d_2)}{\partial d_2} \frac{\partial d_2}{\partial \mathbb{S}_t}.$$

But, $\frac{\partial d_1}{\partial \mathbb{S}_t} = \frac{\partial d_2}{\partial \mathbb{S}_t}$ and

$$\begin{aligned}
\frac{\partial N(d_2)}{\partial d_2} &= \frac{1}{\sqrt{2\pi}} e^{-\frac{1}{2}(d_2)^2} = \frac{1}{\sqrt{2\pi}} e^{-\frac{1}{2}(d_1 - \sigma\sqrt{T-t})^2} \\
&= \frac{1}{\sqrt{2\pi}} e^{-\frac{1}{2}(d_1)^2 + d_1\sigma\sqrt{T-t} - \frac{1}{2}\sigma^2(T-t)} \\
&= \frac{\partial N(d_1)}{\partial d_1} e^{d_1\sigma\sqrt{T-t} - \frac{1}{2}\sigma^2(T-t)} \\
&= \frac{\partial N(d_1)}{\partial d_1} \frac{\mathbb{S}_t}{Ke^{-r(T-t)}}.
\end{aligned}$$

Substitution yields

$$\frac{\partial \mathbb{X}_t}{\partial \mathbb{S}_t} = N(d_1) + \mathbb{S}_t \frac{\partial N(d_1)}{\partial d_1} \frac{\partial d_1}{\partial \mathbb{S}_t} - \mathbb{S}_t \frac{\partial N(d_1)}{\partial d_1} \frac{\partial d_1}{\partial \mathbb{S}_t} = N(d_1).$$

This completes the proof.

The number of units held in the risky asset α_t to create a synthetic call option is called the option's *delta*.

5.4 Notes

There is a vast literature on the BSM formula and its application in practice. An introductory guide to derivative pricing is Jarrow and Chatterjea [95]. Good reference books include the classics by Cox and Rubinstein [35] and Jarrow and Rudd [105]. More recent textbooks include Back [4], Baxter and Rennie [10], Jarrow and Turnbull [108], Musiela and Rutkowski [145], and Shreve [168, 169].

Chapter 6
The Heath–Jarrow–Morton Model

This chapter presents the Heath–Jarrow–Morton (HJM) [70] model for pricing interest rate derivatives. Given frictionless and competitive markets, and assuming a complete market, this is the most general arbitrage-free pricing model possible with a stochastic term structure of interest rates. This model, with appropriate modifications, can also be used to price derivatives whose values depend on a term structure of underlying assets, examples include exotic equity derivatives where the underlyings are call and put options, commodity options where the underlyings are futures prices, and credit derivatives where the underlyings are risky zero-coupon bond prices, see Carr and Jarrow [29], Carmona [26], Carmona and Nadtochiy [27], and Kallsen and Kruhner [115].

HJM generalizes the BSM model in two important ways. One, the BSM model assumes deterministic interest rates. Two, the BSM has only one traded risky asset. The HJM model has stochastic interest rates and a continuum of traded risky assets. In some sense, the issue to be solved is that the market is "overly complete" with more risky assets trading then are needed to complete the market. This chapter is based on Jarrow [86].

6.1 The Set-Up

Given is a non-normalized market $((\mathbb{B}, \mathbb{S}), (\mathscr{F}_t), \mathbb{P})$. The money market account (mma) $\mathbb{B}_t = e^{\int_0^t r_s ds}$, $\mathbb{B}_0 = 1$, is continuous in time t, and of finite variation. r_t is the default-free spot rate of interest, adapted to $\mathscr{F}_t$ with $\int_0^t |r_s|\, ds < \infty$.

Here we modify the risky asset price vector. Instead of a finite number of risky assets $\mathbb{S}_t = (\mathbb{S}_1(t), \ldots, \mathbb{S}_n(t))'$, we let the traded risky assets correspond to the term structure of default-free zero-coupon bonds of all maturities, with time t prices $p(t, \mathscr{T})$ for a sure dollar paid at time $\mathscr{T}$ for all $0 \leq t \leq \mathscr{T} \leq T$. That is, $p(t, t) = 1$ for all t. The zero-coupon price process $p(t, \mathscr{T})$ is adapted to $\mathscr{F}_t$ for all $\mathscr{T}$.

R. A. Jarrow, *Continuous-Time Asset Pricing Theory*, Springer Finance,
https://doi.org/10.1007/978-3-319-77821-1_6

We also assume that zero-coupon bond prices are strictly positive, i.e. $p(t, \mathcal{T}) > 0$ for all $t, \mathcal{T}$. This last assumption excludes trivial arbitrage opportunities from the market. Note that there are a continuum of risky assets trading in this market.

The *forward rate* is implicitly defined by the expression

$$p(t, \mathcal{T}) = e^{-\int_t^{\mathcal{T}} f(t,u)du}. \tag{6.1}$$

Taking natural logarithms and differentiating with respect to $\mathcal{T}$ (assuming the derivative exists) yields the expression

$$f(t, \mathcal{T}) = -\frac{\partial \log p(t, \mathcal{T})}{\partial \mathcal{T}} = -\frac{\partial p(t, \mathcal{T})}{\partial \mathcal{T}} \frac{1}{p(t, \mathcal{T})}. \tag{6.2}$$

The forward rate at time t corresponds to default-free borrowing and lending implicit in the zero-coupon bonds over the future time period $[\mathcal{T}, \mathcal{T} + dt]$. To see this note that for a small time interval Δ,

$$f(t, \mathcal{T})\Delta \approx -\frac{p(t, \mathcal{T} + \Delta) - p(t, \mathcal{T})}{\Delta} \frac{\Delta}{p(t, \mathcal{T})} = \frac{p(t, \mathcal{T} + \Delta)}{p(t, \mathcal{T})} - 1.$$

The last term in this expression shows that the forward rate at time t is the implicit rate earned on $p(t, \mathcal{T} + \Delta)$ in addition to that paid by $p(t, \mathcal{T})$ over the last time period in its life $[\mathcal{T}, \mathcal{T} + \Delta]$.

The default-free *spot rate* is $r_t = f(t, t)$. This is the rate for immediate borrowing or lending over the time interval $[t, t + dt]$.

Remark 6.1 (Self-Financing Trading Strategies) Although there are a continuum of traded risky assets, an admissible s.f.t.s. in an HJM model consists of the mma and only a *finite* number of zero-coupon bonds over the trading horizon $[0, T]$. This enables us to use the fact that the existence of an equivalent martingale measure implies both NFLVR and ND, even with trading in an infinite number of risky assets. The existence of an equivalent martingale measure is no longer implied by NFLVR and ND because the Third Fundamental Theorem 2.5 of asset pricing was only proven for a finite asset market, see Chap. 2 for further elaboration. This completes the remark.

6.2 Term Structure Evolution

The HJM model starts by assuming an evolution for the default-free term structure of interest rates. Although unnecessary, for simplicity of computation, we select the forward rate curve to represent the term structure of interest rates. Alternative possibilities are the yield curve or the zero-coupon bond price curve.

In particular, the HJM model assumes a given initial forward rate curve

$$f(0, \mathcal{T}) \text{ for } 0 \leq \mathcal{T} \leq T,$$

where $\int_0^T |f(0, \mathcal{T})| \, d\mathcal{T} < \infty$, and a stochastic process for the evolution of the forward rate curve

$$f(t, \mathcal{T}) = f(0, \mathcal{T}) + \int_0^t \mu(s, \mathcal{T}) ds + \sum_{i=1}^{D} \int_0^t \sigma_i(s, \mathcal{T}) dW_i(s) \text{ for } 0 \leq t \leq \mathcal{T} \leq T, \tag{6.3}$$

where $W_i(t)$ for $i = 1, \ldots, D$ are standard independent Brownian motions with $W_i(0) = 0$ for all i that generate the filtration $\mathcal{F}_t$, $\mu(t, \mathcal{T})$ and $\sigma_i(t, \mathcal{T})$ for $i = 1, \ldots, D$ are $\mathcal{F}_t$- measurable and satisfy the regularity conditions a.s. $\mathbb{P}$

$$\int_0^T \left(\int_0^t |\mu(v, t)| \, dv \right) dt < \infty,$$

$$\int_0^t \left(\int_v^t \sigma_i(v, y) dy \right)^2 dv < \infty$$

$$\int_0^t \left(\int_t^{\mathcal{T}} \sigma_i(v, y) dy \right)^2 dv < \infty$$

for all $t \in [0, \mathcal{T}]$, $\mathcal{T} \in [0, T]$, $i = 1, \cdots, D$.

In stochastic differential equation form this is

$$df(t, \mathcal{T}) = \mu(t, \mathcal{T}) dt + \sum_{i=1}^{D} \sigma_i(t, \mathcal{T}) dW_i(t). \tag{6.4}$$

This is called a D-factor model for the forward rate evolution. The D-factors represent distinct economic forces affecting forward rates (e.g. inflation, unemployment, economic growth, monetary policy, fiscal policy).

Expression (6.3) represents a very general stochastic process. The only substantive economic restriction in expression (6.3) is that the process has continuous sample paths, however, even this restriction can be relaxed (see Jarrow and Madan [99, 100], Bjork et al. [15, 16], and Eberlein and Raible [55]). We do not pursue this relaxation here. The evolution need not be Markov (in a finite number of state variables) and it can be path dependent. In addition, given that the market does not have currency trading (only zero-coupon bonds and an mma), forward rates can be negative. To exclude negative forward rates, an additional restriction can be imposed on the evolution in expression (6.3). We do not impose such a restriction.

Under this evolution, the spot rate process is

$$r_t = r_0 + \int_0^t \frac{\partial f(0,s)}{\partial \mathcal{T}} ds + \int_0^t \mu(s,t)ds + \sum_{i=1}^{D} \int_0^t \sigma_i(s,t)dW_i(s). \tag{6.5}$$

In stochastic differential equation form this is

$$dr_t = \left(\frac{\partial f(0,t)}{\partial \mathcal{T}} + \mu(t,t) \right) dt + \sum_{i=1}^{D} \sigma_i(t,t)dW_i(t). \tag{6.6}$$

Proof Using the definition $r_t = f(t,t)$, expression (6.3) gives

$$r_t = f(0,t) + \int_0^t \mu(s,t)ds + \sum_{i=1}^{D} \int_0^t \sigma_i(s,t)dW_i(s).$$

But,

$$f(0,t) - f(0,0) = \int_0^t \frac{\partial f(0,s)}{\partial \mathcal{T}} ds.$$

Substitution, along with the fact that $r_0 = f(0,0)$, completes the proof.

Note that in this expression as time t changes, the second argument in the last two integrands also change. This makes the evolution of the spot rate more complex than that of the forward rate process itself.

From expression (6.3), we can deduce the evolution of the zero-coupon bond price curve

$$p(t,\mathcal{T}) = p(0,\mathcal{T}) e^{\int_0^t (r_s + b(s,\mathcal{T}))ds - \frac{1}{2}\sum_{i=1}^{D} \int_0^t a_i(s,\mathcal{T})^2 ds + \sum_{i=1}^{D} \int_0^t a_i(s,\mathcal{T})dW_i(s)}, \tag{6.7}$$

$$\text{where} \quad b(s,\mathcal{T}) = -\int_s^{\mathcal{T}} \mu(s,u)du + \frac{1}{2}\sum_{i=1}^{D} a_i(s,\mathcal{T})^2 \quad \text{and}$$

$$a_i(s,\mathcal{T}) = -\int_s^{\mathcal{T}} \sigma_i(s,u)du \quad \text{for all } i = 1, \ldots, D.$$

Proof First, $p(t,\mathcal{T}) = e^{-\int_t^{\mathcal{T}} f(t,u)du}$. Substitution of expression (6.3) gives

$$-\ln p(t,\mathcal{T}) = \int_t^{\mathcal{T}} f(0,u)du + \int_t^{\mathcal{T}} \int_0^t \mu(s,u)dsdu + \sum_{i=1}^{D} \int_t^{\mathcal{T}} \int_0^t \sigma_i(s,u)dW_i(s)du.$$

A stochastic Fubini's theorem (see Heath et al. [70, appendix]) gives

$$\int_t^{\mathcal{T}}\int_0^t \mu(s,u)dsdu = \int_0^t\int_t^{\mathcal{T}} \mu(s,u)duds, \quad \text{and}$$

$$\int_t^{\mathcal{T}}\int_0^t \sigma_i(s,u)dW_i(s)du = \int_0^t\left(\int_t^{\mathcal{T}} \sigma_i(s,u)du\right)dW_i(s).$$

Substitution yields

$$-\ln p(t,\mathcal{T}) = \int_t^{\mathcal{T}} f(0,u)du + \int_0^t\left(\int_t^{\mathcal{T}} \mu(s,u)du\right)ds + \sum_{i=1}^{D}\int_0^t\int_t^{\mathcal{T}} \sigma_i(s,u)dudW_i(s).$$

Rewriting yields

$$\begin{aligned}-\ln p(t,\mathcal{T}) = &\textstyle\int_0^{\mathcal{T}} f(0,u)du - \int_0^t f(0,u)du + \int_0^t\left(\int_s^{\mathcal{T}} \mu(s,u)du\right)ds\\ &\textstyle - \int_0^t\left(\int_s^t \mu(s,u)du\right)ds + \sum_{i=1}^D \int_0^t\int_s^{\mathcal{T}} \sigma_i(s,u)dudW_i(s)\\ &\textstyle - \sum_{i=1}^D \int_0^t\int_s^t \sigma_i(s,u)dudW_i(s).\end{aligned}$$

But,

$$\int_0^t\left(\int_0^u \mu(s,u)ds\right)du = \int_0^t\left(\int_s^t \mu(s,u)du\right)ds \quad \text{and}$$

$$\int_0^t\int_0^u \sigma_i(s,u)dW_i(s)du = \int_0^t\int_s^t \sigma_i(s,u)dudW_i(s).$$

Noting that $-\ln p(0,\mathcal{T}) = \int_0^{\mathcal{T}} f(0,u)du$ and using expression (6.5) yields

$$\begin{aligned}-\ln p(t,\mathcal{T}) = -\ln p(0,\mathcal{T}) + \int_0^t\left(\int_s^{\mathcal{T}} \mu(s,u)du\right)ds\\ + \sum_{i=1}^D\int_0^t\int_s^{\mathcal{T}} \sigma_i(s,u)dudW_i(s) - \int_0^t r_u du.\end{aligned}$$

Simple algebra completes the proof.

As a stochastic differential, we have that the zero-coupon bond price evolves as

$$\frac{dp(t,\mathcal{T})}{p(t,\mathcal{T})} = (r_t + b(t,\mathcal{T}))\,dt + \sum_{i=1}^D a_i(t,\mathcal{T})dW_i(t). \tag{6.8}$$

The zero-coupon bond's instantaneous return consists of a drift plus D-random shocks. Within the drift, one can interpret $b(s, \mathcal{T})$ as the risk premium on the $\mathcal{T}$-maturity zero-coupon bond in excess of the default-free spot rate of interest r_t. The bond's volatilities are $\{a_1(t, \mathcal{T}), \ldots, a_D(t, \mathcal{T})\}$. We add the following assumption.

Assumption (Nonsingular Volatility Matrix) *For any* $\mathcal{T}_1, \ldots, \mathcal{T}_D \in [0, T]$ *such that* $0 < \mathcal{T}_1 < \ldots < \mathcal{T}_D \leq T$,

$$\begin{bmatrix} a_1(t, \mathcal{T}_1) & \cdots & a_D(t, \mathcal{T}_1) \\ \vdots & & \vdots \\ a_1(t, \mathcal{T}_D) & \cdots & a_D(t, \mathcal{T}_D) \end{bmatrix}$$

is nonsingular a.s. $\mathbb{P} \times \Lambda$ *where* Λ *is Lebesgue measure.*

Later, given the existence of a probability measure $\mathbb{Q} \in \mathfrak{M}$ where

$$\mathfrak{M} = \left\{ \mathbb{Q} \sim \mathbb{P} : \frac{p(t, \mathcal{T})}{\mathbb{B}_t} \text{ for all } \mathcal{T} \in [0, T] \text{ are } \mathbb{Q}\text{-martingales} \right\},$$

this assumption will guarantee that the market is complete with respect to $\mathbb{Q}$. In fact, since this condition holds for all D subsets of the traded zero-coupon bonds, and there is an infinite collection of such D subsets, the market is "overly" complete with respect to $\mathbb{Q} \in \mathfrak{M}$.

6.3 Arbitrage-Free Conditions

This section provides the conditions that must be imposed on the evolution of the forward rate process such that there exists an equivalent martingale measure.

Theorem 6.1 (HJM Arbitrage-Free Drift Condition) *An equivalent martingale measure* $\mathbb{Q}$ *exists such that* $\left(\frac{p(t,\mathcal{T})}{\mathbb{B}_t}\right)$ *are* $\mathbb{Q}$*-martingales for all* $0 \leq \mathcal{T} \leq T$

if and only if

there exist risk premium processes

$$\phi_i(t) \text{ for } i = 1, \ldots, D$$

$\mathcal{F}_t$*-measurable with* $\int_0^T \phi_i(t)^2 dt < \infty$ *for all* i *such that*

$$\frac{d\mathbb{Q}}{d\mathbb{P}} \equiv e^{\sum_{i=1}^D \int_0^T \phi_i(s) dW_i(s) - \frac{1}{2} \sum_{i=1}^D \int_0^T \phi_i(s)^2 ds}$$

is a probability measure,

$$E\left[e^{\sum_{i=1}^{D}\int_0^{\mathcal{T}}(a_i(s,\mathcal{T})+\phi_i(s))dW_i(s)-\frac{1}{2}\sum_{i=1}^{D}\int_0^{\mathcal{T}}(a_i(s,\mathcal{T})+\phi_i(s))^2ds}\right]=1,$$

and

$$\mu(t,\mathcal{T})=-\sum_{i=1}^{D}\sigma_i(t,\mathcal{T})\left[\phi_i(t)-\int_t^{\mathcal{T}}\sigma_i(t,v)dv\right] \tag{6.9}$$

for all $0\le t\le\mathcal{T}\le T$.

Proof All equivalent martingale measures have the form

$$\frac{d\mathbb{Q}}{d\mathbb{P}}=e^{\sum_{i=1}^{n}\int_0^{T}\phi_i(s)dW_i(s)-\frac{1}{2}\sum_{i=1}^{n}\int_0^{T}\phi_i(s)^2ds}$$

for suitably measurable and integrable $\{\phi_1(t),\ldots,\phi_D(t)\}$, see Theorem 1.7 in Chap. 1.

By Girsanov's theorem (see Theorem 1.3 in Chap. 1),

$$dW_i^{\mathbb{Q}}(t)=dW_i(t)-\phi_i(t)dt\quad\text{for } i=1,\ldots,D$$

are standard independent Brownian motions under $\mathbb{Q}$.

Given expression (6.8), we have that

$$\begin{aligned}\frac{dp(t,\mathcal{T})}{p(t,\mathcal{T})}&=(r_t+b(t,\mathcal{T}))\,dt+\sum_{i=1}^{D}a_i(t,\mathcal{T})dW_i(t)\\&=(r_t+b(t,\mathcal{T}))\,dt+\sum_{i=1}^{D}a_i(t,\mathcal{T})\left(\phi_i(t)dt+dW_i^{\mathbb{Q}}(t)\right)\\&=r_tdt+\sum_{i=1}^{D}a_i(t,\mathcal{T})dW_i^{\mathbb{Q}}(t)+\left(b(t,\mathcal{T})+\sum_{i=1}^{D}\phi_i(t)a_i(t,\mathcal{T})\right)dt.\end{aligned}$$

Using the change of numeraire, expression (2.7) in the fundamental theorems Chap. 2, we have that

$$\frac{dp(t,\mathcal{T})}{p(t,\mathcal{T})}-r_tdt=\frac{d\left(\frac{p(t,\mathcal{T})}{\mathbb{B}(t)}\right)}{\frac{p(t,\mathcal{T})}{\mathbb{B}(t)}},$$

which implies that

$$\frac{d\left(\frac{p(t,\mathcal{T})}{\mathbb{B}(t)}\right)}{\frac{p(t,\mathcal{T})}{\mathbb{B}(t)}} = \sum_{i=1}^{D} a_i(t,\mathcal{T})dW_i^{\mathbb{Q}}(t) + \left(b(t,\mathcal{T}) + \sum_{i=1}^{D}\phi_i(t)a_i(t,\mathcal{T})\right)dt.$$

Thus, $\frac{p(t,\mathcal{T})}{\mathbb{B}(t)}$ is a local $\mathbb{Q}$-martingale if and only if

$$b(t,\mathcal{T}) + \sum_{i=1}^{D}\phi_i(t)a_i(t,\mathcal{T}) = 0. \text{ Or,}$$

$$-\int_s^{\mathcal{T}}\mu(s,u)du + \frac{1}{2}\sum_{i=1}^{D}\left(\int_s^{\mathcal{T}}\sigma_i(s,u)du\right)^2 - \sum_{i=1}^{D}\phi_i(t)\int_s^{\mathcal{T}}\sigma_i(s,u)du = 0.$$

Differentiate in $\mathcal{T}$, and simplify to obtain the equivalent expression

$$\mu(s,\mathcal{T}) = -\sum_{i=1}^{D}\sigma_i(s,\mathcal{T})\left[\phi_i(t) - \left(\int_s^{\mathcal{T}}\sigma_i(s,u)du\right)\right].$$

To complete the proof, we need to show that $\frac{p(t,\mathcal{T})}{\mathbb{B}(t)}$ is a $\mathbb{Q}$-martingale, and not just a local martingale. This follows from

$$E\left[e^{\sum_{i=1}^{D}\int_0^{\mathcal{T}}(a_i(s,\mathcal{T})+\phi_i(s))dW_i(s)-\frac{1}{2}\sum_{i=1}^{D}\int_0^{\mathcal{T}}(a_i(s,\mathcal{T})+\phi_i(s))^2ds}\right] = 1,$$

which is both a necessary and sufficient condition. Indeed, expression (6.7) implies that under the probability measure $\mathbb{Q}$,

$$\frac{p(\mathcal{T},\mathcal{T})}{\mathbb{B}(\mathcal{T})} = p(0,\mathcal{T})e^{\int_0^{\mathcal{T}}(b(s,\mathcal{T})+\sum_{i=1}^{D}\phi_i(s)a_i(s,\mathcal{T}))ds-\frac{1}{2}\sum_{i=1}^{D}\int_0^{\mathcal{T}}a_i(s,\mathcal{T})^2ds} \times e^{+\sum_{i=1}^{D}\int_0^{\mathcal{T}}a_i(s,\mathcal{T})dW_i^{\mathbb{Q}}(s)}.$$

Using the $\mathbb{Q}$-local martingale condition that $b(t,\mathcal{T}) + \sum_{i=1}^{D}\phi_i(t)a_i(t,\mathcal{T}) = 0$ for all t, this simplifies to

$$\frac{p(\mathcal{T},\mathcal{T})}{\mathbb{B}(\mathcal{T})} = p(0,\mathcal{T})e^{-\frac{1}{2}\sum_{i=1}^{D}\int_0^{\mathcal{T}}a_i(s,\mathcal{T})^2ds+\sum_{i=1}^{D}\int_0^{\mathcal{T}}a_i(s,\mathcal{T})dW_i^{\mathbb{Q}}(s)}.$$

$\frac{p(t,\mathcal{T})}{\mathbb{B}(t)}$ is a $\mathbb{Q}$-martingale if and only if

$$E^{\mathbb{Q}}\left[e^{-\frac{1}{2}\sum_{i=1}^{D}\int_0^{\mathcal{T}}a_i(s,\mathcal{T})^2ds+\sum_{i=1}^{D}\int_0^{\mathcal{T}}a_i(s,\mathcal{T})dW_i^{\mathbb{Q}}(s)}\right] = 1,$$

see Protter [151, p. 138]. Using $\frac{d\mathbb{Q}}{d\mathbb{P}}$, this is equivalent to

$$E\left[e^{-\frac{1}{2}\sum_{i=1}^{D}\int_0^{\mathcal{T}} a_i(s,\mathcal{T})^2 ds+\sum_{i=1}^{D}\int_0^{\mathcal{T}} a_i(s,\mathcal{T})dW_i^{\mathbb{Q}}(s)} e^{\sum_{i=1}^{D}\int_0^{\mathcal{T}} \phi_i(s)dW_i(s)-\frac{1}{2}\sum_{i=1}^{D}\int_0^{\mathcal{T}} \phi_i(s)^2 ds}\right] = 1.$$

Combining terms, and substituting $dW_i^{\mathbb{Q}}(t) = dW_i(t) - \phi_i(t)dt$, yields

$$E\left[e^{-\frac{1}{2}\sum_{i=1}^{D}\int_0^{\mathcal{T}} (a_i(s,\mathcal{T})^2+2a_i(s,\mathcal{T})\phi_i(s)+\phi_i(s)^2)ds+\sum_{i=1}^{D}\int_0^{\mathcal{T}} (a_i(s,\mathcal{T})+\phi_i(s))dW_i(s)}\right] = 1.$$

Algebra completes the proof.

We assume the existence of an equivalent martingale measure $\mathbb{Q} \in \mathfrak{M}$ for the subsequent analysis. This implies, of course, by the Third Fundamental Theorem 2.5 of asset pricing in Chap. 2, that the market satisfies NFLVR and ND. Expression (6.9) is known as the *arbitrage-free HJM drift restriction*. In the HJM drift restriction, $(\phi_1(t), \ldots, \phi_D(t))$ are *interest rate risk premiums* corresponding to the D-Brownian motions. It is important to emphasize that a characterization of these risk premiums is not determined by the arbitrage-free restrictions imposed by NFLVR and ND on the evolution of the term structure of interest rates, the theorem only assures their existence.

Theorem 6.2 (Market Completeness) *Given an equivalent martingale measure $\mathbb{Q} \in \mathfrak{M}$ exists and the non-singular volatility matrix assumption, the market is complete with respect any $\mathbb{Q} \in \mathfrak{M}$ and $\mathfrak{M}$ is the singleton set, i.e. $\mathbb{Q}$ is unique.*

As stated in this theorem, given the existence of an equivalent martingale measure $\mathbb{Q} \in \mathfrak{M}$, the non-singularity assumption for the volatility matrix of an arbitrary collection of zero-coupon bonds with maturities $\mathcal{T}_1, \ldots, \mathcal{T}_D$ satisfying $0 < \mathcal{T}_1 < \ldots < \mathcal{T}_D \leq T$, the market is complete. The proof follows by applying Theorem 2.12 in the fundamental theorems Chap. 2 for the restricted Brownian motion market consisting of just these D zero-coupon bonds and the mma proving that this restricted market is complete with respect to $\mathbb{Q} \in \mathfrak{M}$. Of course, this also implies that the larger market consisting of all the traded zero-coupon bonds is complete with respect to $\mathbb{Q} \in \mathfrak{M}$. Finally, by the Second Fundamental Theorem 2.4 of asset pricing in Chap. 2, we have that $\mathbb{Q} \in \mathfrak{M}$ is unique, completing the proof.

These results imply, of course, that the standard hedging and valuation methodology in the fundamental theorems Chap. 2 can now be applied to value and hedge interest rate derivatives.

Remark 6.2 (An Overly Complete Market with Respect to $\mathbb{Q} \in \mathfrak{M}$) As just discussed, given $\mathbb{Q} \in \mathfrak{M}$, the non-singularity assumption for the volatility matrix of an arbitrary collection of zero-coupon bonds with maturities $\mathcal{T}_1, \ldots, \mathcal{T}_D$ satisfying $0 < \mathcal{T}_1 < \ldots < \mathcal{T}_D \leq T$ implies that the market is complete with respect to $\mathbb{Q} \in \mathfrak{M}$. In fact for this term structure evolution, the market is "overly complete."

By "overly complete" we mean that the market has more risky assets trading than needed to complete the market. Indeed, any collection of zero-coupon bonds with maturities $\mathscr{T}_1, \ldots, \mathscr{T}_D$ satisfying $0 < \mathscr{T}_1 < \ldots < \mathscr{T}_D \leq T$ completes the market with respect to $\mathbb{Q} \in \mathfrak{M}$. The difficulty in proving Theorem 6.1 was to guarantee that the market is "arbitrage-free" given that it is overly complete. The key insight in the theorem is that the market is "arbitrage-free" if and only if the risk premiums are *independent of the collection of zero-coupon bonds maturing at* $\mathscr{T}_1, \ldots, \mathscr{T}_D$ *selected to complete the market,* i.e. $\phi_i(t)$ for $i = 1, \cdots, D$ are independent of $\mathscr{T}_1, \ldots, \mathscr{T}_D$. This completes the remark.

The next two corollaries formalize the implications of the previous insights for valuing zero-coupon bonds and interest rate derivatives.

Corollary 6.1 (Zero-Coupon Bond Formula) *Given an equivalent martingale measure* $\mathbb{Q} \in \mathfrak{M}$ *exists,*

$$p(t, \mathscr{T}) = E^{\mathbb{Q}}\left[e^{-\int_t^{\mathscr{T}} r_s ds}\middle| \mathscr{F}_t\right] \tag{6.10}$$

for all $0 \leq t \leq \mathscr{T} \leq T$.

Proof $\left(\frac{p(t,\mathscr{T})}{\mathbb{B}_t}\right)$ being $\mathbb{Q}$-martingales for all $0 \leq \mathscr{T} \leq T$ implies

$$\frac{p(t, \mathscr{T})}{\mathbb{B}_t} = E^{\mathbb{Q}}\left[\frac{p(\mathscr{T}, \mathscr{T})}{\mathbb{B}_{\mathscr{T}}} | \mathscr{F}_t\right].$$

Substitution of the definition for the mma and algebra gives the result. This completes the proof.

This corollary gives the intuitive result that the price of a zero-coupon bond equals its expected discounted payoff, where the discount rates correspond to the default-free spot rate of interest at each intermediate date over the life of the bond, and the expectation is computed using the equivalent martingale measure. The equivalent martingale measure includes an adjustment for the risk of the payoff since the discount rate used in expression (6.10) is the default-free spot rate of interest.

Corollary 6.2 (Risk Neutral Valuation) *Given an equivalent martingale measure* $\mathbb{Q} \in \mathfrak{M}$ *exists and the non-singular volatility matrix assumption, any derivative* $\frac{X_{\mathscr{T}}}{\mathbb{B}_{\mathscr{T}}} \in L^1_+(\mathbb{Q})$ *satisfies*

$$X_t = E^{\mathbb{Q}}\left[X_{\mathscr{T}} e^{-\int_t^{\mathscr{T}} r_s ds} | \mathscr{F}_t\right]. \tag{6.11}$$

Proof This is a direct application of Theorem 2.8 in the fundamental theorems Chap. 2, given the existence of an equivalent martingale measure $\mathbb{Q} \in \mathfrak{M}$ and the fact that the market is complete with respect to $\mathbb{Q}$. This completes the proof.

This corollary provides the justification for using risk neutral valuation to determine the arbitrage-free price of an arbitrary interest rate derivative ($\frac{X_{\mathcal{T}}}{\mathbb{B}_{\mathcal{T}}} \in L^1_+(\mathbb{Q})$). Applying the insights from the fundamental theorems Chap. 2 Sect. 2.6.3 with respect to the synthetic construction of any traded derivative, the standard hedging methodology can now also be applied.

Remark 6.3 (Assuming Only $\mathbb{Q} \in \mathfrak{M}_l$ Exists) Instead of assuming that there exists an equivalent martingale measure $\mathbb{Q} \in \mathfrak{M}$ (by the Third Fundamental Theorem 2.5 of asset pricing in Chap. 2 this implies NFLVR and ND), one can relax the previous structure slightly and only assume that an equivalent local martingale measure $\mathbb{Q} \in \mathfrak{M}_l$ exists where $\mathfrak{M}_l = \left\{\mathbb{Q} \sim \mathbb{P} : \frac{p(t,\mathcal{T})}{\mathbb{B}_t} \text{ for all } \mathcal{T} \in [0,T] \text{ are local } \mathbb{Q}\text{-martingales}\right\}$ (by the First Fundamental Theorem 2.3 of asset pricing in Chap. 2 this implies NFLVR). In this case, the market is still complete with respect to $\mathbb{Q}$ by Theorem 2.12 from the fundamental theorems Chap. 2 and the above assumption about the non-singularity of the zero-coupon bond volatility matrix.

To get the condition that $\left(\frac{p(t,\mathcal{T})}{\mathbb{B}_t}\right)$ are $\mathbb{Q}$-martingales, we need to exclude price bubbles in zero-coupon bonds. This can be accomplished by assuming that the zero-coupon bond prices are bounded above for all t (and using Theorem 3.5 from the asset price bubbles Chap. 3). This will follow, for example, if forward rates are nonnegative for all t (see expression (6.1)). This completes the remark.

6.4 Examples

To characterize an HJM model, one needs to specify the forward rate curve evolution under the equivalent martingale measure $\mathbb{Q}$. This is because in a complete market with respect to $\mathbb{Q} \in \mathfrak{M}$, the interest rate risk premiums are not restricted by the arbitrage-free conditions NFLVR and ND, and therefore the risk premiums can be arbitrarily specified. However, to uniquely specify the evolution of the forward rate curve under the statistical probability measure $\mathbb{P}$, the interest rate risk premiums need to be identified. Conceptually, interest rate risk premiums are determined by additional (above those imposed by NFLVR and ND) restrictions that an equilibrium imposes on an market (see Part III of this book).

Although the drift of the forward rate curve's evolution changes under the equivalent martingale measure $\mathbb{Q}$ (due to Girsanov's Theorem 1.3 in Chap. 1), the volatility structure remains unchanged. This is seen by noting that the evolutions of the forward rate curve, spot price, and bond price under the equivalent martingale measure are given by

$$f(t,\mathcal{T}) = f(0,\mathcal{T}) + \sum_{i=1}^{D}\left(\int_0^t \sigma_i(s,\mathcal{T})\int_s^{\mathcal{T}} \sigma_i(s,u)du\right)ds + \sum_{i=1}^{D}\int_0^t \sigma_i(s,\mathcal{T})dW_i^{\mathbb{Q}}(s), \tag{6.12}$$

$$r_t = r_0 + \int_0^t \frac{\partial f(0,s)}{\partial \mathcal{T}} ds + \sum_{i=1}^{D} \left(\int_0^t \sigma_i(s,t) \int_s^t \sigma_i(s,u) du \right) ds + \sum_{i=1}^{D} \int_0^t \sigma_i(s,t) dW_i^{\mathbb{Q}}(s), \tag{6.13}$$

$$p(t,\mathcal{T}) = p(0,\mathcal{T}) e^{\int_0^t r_s ds - \frac{1}{2}\sum_{i=1}^{D}\int_0^t a_i(s,\mathcal{T})^2 ds + \sum_{i=1}^{D}\int_0^t a_i(s,\mathcal{T}) dW_i^{\mathbb{Q}}(s)}, \tag{6.14}$$

where $W_i^{\mathbb{Q}}(t)$ for $i = 1, \ldots, D$ are independent standard Brownian motions under $\mathbb{Q} \in \mathfrak{M}$.

Given these evolutions, one can easily price and hedge any interest rate derivative using the standard techniques as presented in Sect. 2.6 in Chap. 2. For this reason, hedging and valuation will not be discussed further in this chapter with the exception of Sect. 6.6 below, which prices caps and floors using the HJM methodology. Instead, we will provide various examples for the evolution of the term structure of interest rates useful in empirical applications of this methodology. We note that for the purposes of valuation and hedging, a particular HJM model is uniquely identified by an initial forward rate curve $\{f(0,\mathcal{T}) : \mathcal{T} \in [0,T]\}$ and a collection of volatilities $\{\sigma_i(t,\mathcal{T}) : \text{ all } 0 \le t \le \mathcal{T} \le T\}_{i=1}^{D}$. Consequently, all of the subsequent examples are specified in this manner.

6.4.1 The Ho and Lee Model

The Ho and Lee model, in the HJM framework, is represented by an arbitrary initial forward rate curve $\{f(0,\mathcal{T}) : \mathcal{T} \in [0,T]\}$ and the evolution of the forward rate curve under the equivalent martingale measure $\mathbb{Q}$ as given by

$$df(t,\mathcal{T}) = \sigma^2(\mathcal{T} - t)dt + \sigma dW^{\mathbb{Q}}(t), \tag{6.15}$$

where $\sigma > 0$ is a constant. This is a single-factor model. Integration yields

$$f(t,\mathcal{T}) = f(0,\mathcal{T}) + \sigma^2 t(\mathcal{T} - \frac{t}{2}) + \sigma W^{\mathbb{Q}}(t).$$

This expression shows that the forward rate curve drifts across time in a non-parallel fashion, but with parallel random shocks. The spot rate process implied by this evolution is

$$r_t = r_0 + \int_0^t \frac{\partial f(0,s)}{\partial \mathcal{T}} ds + \sigma^2 \frac{t^2}{2} + \sigma W^{\mathbb{Q}}(t).$$

This is called an *affine model* for the spot rate because the spot rate is linear in the state variable process $W^{\mathbb{Q}}(t)$. The zero-coupon bond price evolution is

$$p(t, \mathcal{T}) = \frac{p(0, \mathcal{T})}{p(0, t)} e^{-\frac{\sigma^2}{2}\mathcal{T}t(\mathcal{T}-t)-\sigma(\mathcal{T}-t)W^{\mathbb{Q}}(t)}.$$

The natural logarithm of the zero-coupon bond's price is also seen to be linear in the state variable process $W^{\mathbb{Q}}(t)$.

6.4.2 Lognormally Distributed Forward Rates

Given an arbitrary initial forward rate curve $\{f(0, \mathcal{T}) : \mathcal{T} \in [0, T]\}$, the forward rate evolution implying lognormally distributed forward rates under the equivalent martingale measure $\mathbb{Q}$, if a solution exists to this stochastic differential equation, is given by

$$df(t, \mathcal{T}) = \left(\sigma f(t, \mathcal{T}) \int_t^{\mathcal{T}} \sigma f(t, u)du\right) dt + \sigma f(t, \mathcal{T})dW^{\mathbb{Q}}(t),$$

where $\sigma > 0$ is a constant.

Unfortunately, as shown in Heath et al. [70], the solution to this stochastic differential equation does not exist. It can be shown that under this evolution, forward rates explode (become infinite) with positive probability in finite time. When forward rates explode, zero-coupon bond prices become zero, implying the existence of arbitrage opportunities. This implies that a martingale measure $\mathbb{Q}$ does not exist for this evolution. Alternatively stated, lognormally distributed (continuously compounded) forward rates are inconsistent with a market satisfying NFLVR and ND.

6.4.3 The Vasicek Model

The Vasicek model, in the HJM framework, is represented by the evolution of the spot rate process under the equivalent martingale measure $\mathbb{Q}$ as given by

$$dr_t = k(\theta - r_t)dt + \sigma dW^{\mathbb{Q}}(t),$$

where r_0, k, θ, and σ are positive constants (see Brigo and Mercurio [24, p. 50 for the derivation]). Integrating, we get

$$r_t = r_s e^{-k(t-s)} + \theta\left(1 - e^{-k(t-s)}\right) + \sigma \int_s^t e^{-k(t-s)} dW^{\mathbb{Q}}(t).$$

Using expression (6.7), we have

$$p(t, \mathcal{T}) = A(\mathcal{T} - t)e^{-C(T-t)r_t}, \quad \text{where}$$

$$A(\mathcal{T} - t) = e^{\left(\theta - \frac{\sigma^2}{2k^2}\right)(C(\mathcal{T}-t)-(\mathcal{T}-t)) - \frac{\sigma^2}{4k}C(\mathcal{T}-t)^2} \quad \text{and}$$

$$C(\mathcal{T} - t) = \frac{1}{k}\left(1 - e^{-k(\mathcal{T}-t)}\right).$$

The forward rate curve evolution implied by this process is

$$f(t, \mathcal{T}) = \left(1 - e^{-k(\mathcal{T}-t)}\right)\left(\theta - \frac{\sigma^2}{2k^2}\left(1 - e^{-k(\mathcal{T}-t)}\right)\right) + e^{-k(\mathcal{T}-t)}r_t \text{ with}$$

$$f(0, \mathcal{T}) = \left(1 - e^{-k\mathcal{T}}\right)\left(\theta - \frac{\sigma^2}{2k^2}\left(1 - e^{-k\mathcal{T}}\right)\right) + e^{-k\mathcal{T}}r_0.$$

Not all initial forward rate curves $\{f(0, \mathcal{T}) : \mathcal{T} \in [0, T]\}$ can be fit by this model. To match an arbitrary initial forward rate curve, the Vasicek model needs to be extended with θ a function of time, and then the function θ_t needs to be calibrated to the initial forward rate curve's values.

6.4.4 The Cox–Ingersoll–Ross Model

The Cox–Ingersoll–Ross (CIR) model, in the HJM framework under the equivalent martingale measure $\mathbb{Q}$, is represented by the spot rate process

$$dr_t = k(\theta - r_t)dt + \sigma\sqrt{r_t}dW^{\mathbb{Q}}(t),$$

where r_0, θ, k, and σ are positive constants with $2k\theta > \sigma^2$ (see Brigo and Mercurio [24, p. 56 for the derivation]). Using expression (6.7), we have

$$p(t, \mathcal{T}) = A(\mathcal{T} - t)e^{-C(\mathcal{T}-t)r_t}, \quad \text{where}$$

$$A(\mathcal{T} - t) = \left[\frac{2he^{(k+h)\frac{(\mathcal{T}-t)}{2}}}{2h + (k+h)\left(e^{(\mathcal{T}-t)h} - 1\right)}\right]^{\frac{2k\theta}{\sigma^2}},$$

$$C(\mathcal{T} - t) = \frac{2\left(e^{(\mathcal{T}-t)h} - 1\right)}{2h + (k+h)\left(e^{(\mathcal{T}-t)h} - 1\right)}, \quad \text{and}$$

$$h = \sqrt{k^2 + 2\sigma^2}.$$

The forward rate curve evolution implied by this process is

$$f(t,\mathscr{T}) = -\frac{d\ln A(\mathscr{T}-t)}{d\mathscr{T}} - \frac{dC(\mathscr{T}-t)}{d\mathscr{T}} r_t \text{ with}$$

$$f(0,\mathscr{T}) = \frac{d\ln A(\mathscr{T})}{d\mathscr{T}} - \frac{dC(\mathscr{T})}{d\mathscr{T}} r_0.$$

Not all initial forward rate curves $\{f(0,\mathscr{T}) : \mathscr{T} \in [0,T]\}$ can be fit by this model. To nearly match an arbitrary initial forward rate curve, the CIR model needs to be extended with θ a function of time. However, even in this circumstance, not all forward rate curves can be attained by calibrating θ_t to market prices (see Heath et al. [70]).

6.4.5 The Affine Model

This section studies the affine model of Duffie and Kan [53] and Dai and Singleton [41]. The affine models are a multi-factor extension of the spot rate models presented above. This class of models is called affine because both the spot rate and the natural logarithm of the zero-coupon bond's price are assumed to be affine functions of a k-dimensional vector $\mathbf{X}_t$ of state variables, i.e.

$$r_t = \rho_0 + \boldsymbol{\rho}_1 \mathbf{X}_t \text{ and}$$

$$p(t,\mathscr{T}) = e^{A(\mathscr{T}-t)+\mathbf{C}(\mathscr{T}-t)\mathbf{X}_t}, \text{ where}$$

$$d\mathbf{X}_t = (\mathbf{K}_0 + \mathbf{K}_1\mathbf{X}_t)\,dt + \Sigma(t,\mathbf{X}_t)d\mathbf{W}^{\mathbb{Q}}(t) \text{ and}$$

$$\Sigma(t,\mathbf{X}_t)\Sigma(t,\mathbf{X}_t)^{\mathscr{T}} = \mathbf{H}_0 + \mathbf{H}_1\mathbf{X}_t$$

with ρ_0 a constant, $\boldsymbol{\rho}_1$ a k-vector, $\mathbb{Q}$ the equivalent martingale measure, $\mathbf{W}^{\mathbb{Q}}(t) \equiv (W_1^{\mathbb{Q}}(t), \ldots, W_n^{\mathbb{Q}}(t))$, $\Sigma(t,\mathbf{X}_t)$ is a $k \times d$ matrix, $\mathbf{K}_0$, $\mathbf{H}_0$ are k-vectors, $\mathbf{H}_1$, $\mathbf{K}_1$ are $k \times k$ matrices, $A(\mathscr{T}-t)$ is a scalar function, and $\mathbf{C}(\mathscr{T}-t)$ is a k-dimensional vector function.

Duffie and Kan [53] show that such a system exists if and only if $A(s)$, $\mathbf{C}(s)$ satisfy the following system of differential equations

$$\frac{d\mathbf{C}(s)}{ds} = \boldsymbol{\rho}_1 - \mathbf{K}_1^{\mathscr{T}}\mathbf{C}(s) - \frac{1}{2}\mathbf{C}(s)^{\mathscr{T}}\mathbf{H}_1\mathbf{C}(s)$$

$$\frac{dA(s)}{ds} = \rho_0 - \mathbf{K}_0\mathbf{C}(s) - \frac{1}{2}\mathbf{C}(s)^{\mathscr{T}}\mathbf{H}_0\mathbf{C}(s)$$

with $\mathbf{C}(\mathscr{T}) = \mathbf{0}$ and $A(\mathscr{T}) = 0$.

The forward rate evolution implied by this process is

$$f(t, \mathscr{T}) = -\frac{d \ln A(\mathscr{T} - t)}{d\mathscr{T}} - \frac{d\mathbf{C}(\mathscr{T} - t)}{d\mathscr{T}} \mathbf{X}_t \text{ with}$$

$$f(0, \mathscr{T}) = \frac{d \ln A(\mathscr{T})}{d\mathscr{T}} - \frac{d\mathbf{C}(\mathscr{T})}{d\mathscr{T}} \mathbf{X}_0.$$

6.5 Forward and Futures Contracts

A stochastic term structure of interest rate model is essential for understanding forward and futures contracts because there is no economic difference between these two contract types unless interest rates are stochastic. This observation was first proven in the classical literature by Jarrow and Oldfield [101] and Cox et al. [36]. This section studies forward and futures contracts in an HJM model.

Consider an asset with time t price $\mathbb{S}(t)$ that is adapted to the filtration $\mathscr{F}_t$. This assumption implies that the asset's randomness is generated by the Brownian motions underlying the evolution of the term structure of interest rates. For example, the asset $\mathbb{S}(t)$ could be a zero-coupon bond. Although this restriction can be easily relaxed, it is sufficient for our purpose (see Amin and Jarrow [1] for a generalization). As before, we that assume that $\mathbb{S}(t)$ is a semimartingale and that this asset has no cash flows over $[0, T)$. Formally, we view the time T payoff of this asset, $\frac{\mathbb{S}(T)}{\mathbb{B}_T} \in L^1_+(\mathbb{Q})$. In this case, by Corollary 6.2 the value process $\mathbb{S}(t)$, when normalized by the mma's value, is a $\mathbb{Q}$-martingale.

6.5.1 Forward Contracts

A *forward contract* obligates the owner (long position) to buy the asset on the delivery date $\mathscr{T}$ for a predetermined price. This predetermined price is set, by market convention, such that the value of the forward contract at initiation (time t) is zero. This market clearing price is called the *forward price* and denoted $K(t, \mathscr{T})$.

Given the underlying asset has no cash flows over the life of the forward contract, the arbitrage-free forward price must equal

$$K(t, \mathscr{T}) = \frac{\mathbb{S}(t)}{p(t, \mathscr{T})}. \tag{6.16}$$

Proof The payoff to a forward contract at time $\mathscr{T}$ is $[\mathbb{S}(\mathscr{T}) - K(t, \mathscr{T})]$. By market convention, the forward price makes the present value of this payoff equal to zero, i.e.

$$\begin{aligned} 0 &= E^{\mathbb{Q}}\left[[\mathbb{S}(\mathcal{T}) - K(t,\mathcal{T})]e^{-\int_t^{\mathcal{T}} r_s ds} |\mathcal{F}_t\right] \\ &= E^{\mathbb{Q}}\left[\mathbb{S}(\mathcal{T})e^{-\int_t^{\mathcal{T}} r_s ds} |\mathcal{F}_t\right] - K(t,\mathcal{T})E^{\mathbb{Q}}\left[e^{-\int_t^{\mathcal{T}} r_s ds} |\mathcal{F}_t\right] \\ &= \mathbb{S}(t) - K(t,\mathcal{T})p(t,\mathcal{T}). \end{aligned}$$

The second equality uses the fact that $\frac{\mathbb{S}(t)}{\mathbb{B}(t)}$ is a $\mathbb{Q}$-martingale. Algebra completes the proof.

This expression shows that the forward price is the future value of the asset's time t price. Indeed, if at time t one invests $\mathbb{S}(t)$ dollars in a zero-coupon bond maturing at time $\mathcal{T}$, the time $\mathcal{T}$ value of this investment is expression (6.16).

To facilitate understanding, it is convenient to introduce another equivalent probability measure. This equivalent probability measure makes asset payoffs at some future date $\mathcal{T}$ martingales when discounted by the $\mathcal{T}$-maturity zero-coupon bond price. This equivalent martingale measure was first discovered by Jarrow [79] and later independently again by Geman [65].

Fix $\mathcal{T} \in [0, T]$, the *forward price measure* $\mathbb{Q}^{\mathcal{T}}$ is defined by

$$\frac{d\mathbb{Q}^{\mathcal{T}}}{d\mathbb{Q}} = \frac{1}{p(0,\mathcal{T})\mathbb{B}_{\mathcal{T}}} > 0. \tag{6.17}$$

Proof Note that $\frac{d\mathbb{Q}^{\mathcal{T}}}{d\mathbb{Q}} = \frac{1}{p(0,\mathcal{T})\mathbb{B}_{\mathcal{T}}} > 0$ with

$$\begin{aligned} E^{\mathbb{Q}}\left[\frac{d\mathbb{Q}^{\mathcal{T}}}{d\mathbb{Q}}\right] &= E^{\mathbb{Q}}\left[\frac{1}{p(0,\mathcal{T})\mathbb{B}_{\mathcal{T}}}\right] \\ &= \frac{1}{p(0,\mathcal{T})}E^{\mathbb{Q}}\left[\frac{1}{\mathbb{B}_{\mathcal{T}}}\right] = 1. \end{aligned}$$

So, it is an equivalent probability measure. This completes the proof.

Remark 6.4 (Alternative Characterization of $\mathbb{Q} \in \mathfrak{M}$*)* This remark gives an alternative characterization of $\mathbb{Q} \in \mathfrak{M}$ in an HJM model. Using the forward price measure $\frac{d\mathbb{Q}^T}{d\mathbb{Q}} = \frac{1}{p(0,T)\mathbb{B}_T} > 0$, we can prove the following characterization.

Theorem (Characterizaton of NFLVR and ND) *There exists a* $\mathbb{Q} \sim \mathbb{P}$ *such that* $\frac{p(t,\mathcal{T})}{\mathbb{B}_t}$ *are* $\mathbb{Q}$*-martingales for all* $0 \le \mathcal{T} \le T$
if and only if
there exists a $\mathbb{Q}^T \sim \mathbb{P}$ *such that* $\frac{p(t,\mathcal{T})}{p(t,T)}$ *are* $\mathbb{Q}^T$*-martingales for all* $0 \le \mathcal{T} \le T$.

Proof Before proving the theorem, we need the following facts.

(1) At time $\mathscr{T}$, $p(\mathscr{T},\mathscr{T})=1$. Investing this dollar into the mma at $\mathscr{T}$ results in the value $\frac{\mathbb{B}_T}{\mathbb{B}_{\mathscr{T}}}$ at time T. For clarity, denote this value as $p(T,\mathscr{T})=\frac{\mathbb{B}_T}{\mathbb{B}_{\mathscr{T}}}$ when $T>\mathscr{T}$.

(2) Define

$$Z_t \equiv E^{\mathbb{Q}}\left[\frac{d\mathbb{Q}^T}{d\mathbb{Q}}\,|\mathscr{F}_t\right]=E^{\mathbb{Q}}\left[\frac{1}{p(0,T)\mathbb{B}_T}\,|\mathscr{F}_t\right]=\frac{1}{p(0,T)}E^{\mathbb{Q}}\left[\frac{1}{\mathbb{B}_T}\,|\mathscr{F}_t\right]$$
$$=\frac{p(t,T)}{p(0,T)\mathbb{B}_t}.$$

Given a suitably integrable $\mathscr{F}_T$- measurable X, then using Shreve [169, Lemma 5.2.2, p. 212],

$$E^{\mathbb{Q}^T}[X\,|\mathscr{F}_t]=\frac{1}{Z_t}E^{\mathbb{Q}}\left[X\frac{d\mathbb{Q}^T}{d\mathbb{Q}}\,|\mathscr{F}_t\right]=\frac{\mathbb{B}_\mathbf{t}}{p(t,T)}E^{\mathbb{Q}}\left[X\frac{1}{\mathbb{B}_\mathbf{T}}\,|\mathscr{F}_t\right],\quad\text{and}$$
$$E^{\mathbb{Q}}[X\,|\mathscr{F}_t]=Z_tE^{\mathbb{Q}^T}\left[X\frac{d\mathbb{Q}}{d\mathbb{Q}^T}\,|\mathscr{F}_t\right]=\frac{p(t,T)}{\mathbb{B}_t}E^{\mathbb{Q}^T}[X\mathbb{B}_T\,|\mathscr{F}_t].$$

($\Longrightarrow$) Given $\mathbb{Q}$ satisfying the hypothesis, show $E^{\mathbb{Q}^T}\left[\frac{p(\mathscr{T},\mathscr{T})}{p(\mathscr{T},T)}\,|\mathscr{F}_t\right]=\frac{p(t,\mathscr{T})}{p(t,T)}$.

Now, by fact (2) we have

$$E^{\mathbb{Q}^T}\left[\frac{p(\mathscr{T},\mathscr{T})}{p(\mathscr{T},T)}\,|\mathscr{F}_t\right]=E^{\mathbb{Q}^T}\left[\frac{1}{p(\mathscr{T},T)}\,|\mathscr{F}_t\right]$$
$$=\frac{\mathbb{B}_\mathbf{t}}{p(t,T)}E^{\mathbb{Q}}\left[\frac{1}{p(\mathscr{T},T)}\frac{1}{\mathbb{B}_\mathbf{T}}\,|\mathscr{F}_t\right]$$
$$=\frac{\mathbb{B}_\mathbf{t}}{p(t,T)}E^{\mathbb{Q}}\left[\frac{1}{p(\mathscr{T},T)}E^{\mathbb{Q}}\left[\frac{1}{\mathbb{B}_\mathbf{T}}\,|\mathscr{F}_{\mathscr{T}}\right]|\mathscr{F}_t\right]$$
$$=\frac{\mathbb{B}_\mathbf{t}}{p(t,T)}E^{\mathbb{Q}}\left[\frac{1}{p(\mathscr{T},T)}\frac{p(\mathscr{T},T)}{\mathbb{B}_{\mathscr{T}}}\,|\mathscr{F}_t\right]$$
$$=\frac{\mathbb{B}_\mathbf{t}}{p(t,T)}E^{\mathbb{Q}}\left[\frac{1}{\mathbb{B}_{\mathscr{T}}}\,|\mathscr{F}_t\right]=\frac{p(t,\mathscr{T})}{p(t,T)}.$$

This completes the proof.

($\Longleftarrow$) Given $\mathbb{Q}^T$ satisfying the hypothesis, show $E^{\mathbb{Q}}\left[\frac{p(\mathscr{T},\mathscr{T})}{\mathbb{B}_\mathbf{T}}\,|\mathscr{F}_t\right]=\frac{p(t,\mathscr{T})}{\mathbb{B}_\mathbf{t}}$.

Now, by facts (1) and (2) we have

$$
\begin{aligned}
E^{\mathbb{Q}}\left[\frac{p(\mathcal{T},\mathcal{T})}{\mathbb{B}_{\mathcal{T}}}|\mathcal{F}_t\right] &= E^{\mathbb{Q}}\left[\frac{1}{\mathbb{B}_{\mathcal{T}}}|\mathcal{F}_t\right] = \frac{p(t,T)}{\mathbb{B}_t}E^{\mathbb{Q}^T}\left[\frac{1}{\mathbb{B}_{\mathcal{T}}}\mathbb{B}_T|\mathcal{F}_t\right] \\
&= \frac{p(t,T)}{\mathbb{B}_t}E^{\mathbb{Q}^T}\left[E^{\mathbb{Q}^T}\left[\frac{1}{\mathbb{B}_{\mathcal{T}}}\mathbb{B}_T|\mathcal{F}_{\mathcal{T}}\right]|\mathcal{F}_t\right] \\
&= \frac{p(t,T)}{\mathbb{B}_t}E^{\mathbb{Q}^T}\left[E^{\mathbb{Q}^T}\left[p(T,\mathcal{T})|\mathcal{F}_{\mathcal{T}}\right]|\mathcal{F}_t\right] \\
&= \frac{p(t,T)}{\mathbb{B}_t}E^{\mathbb{Q}^T}\left[E^{\mathbb{Q}^T}\left[\frac{p(T,\mathcal{T})}{p(T,T)}|\mathcal{F}_{\mathcal{T}}\right]|\mathcal{F}_t\right] \\
&= \frac{p(t,T)}{\mathbb{B}_t}E^{\mathbb{Q}^T}\left[\frac{p(\mathcal{T},\mathcal{T})}{p(\mathcal{T},T)}|\mathcal{F}_t\right] \\
&= \frac{p(t,T)}{\mathbb{B}_t}\frac{p(t,\mathcal{T})}{p(t,T)} = \frac{p(t,\mathcal{T})}{\mathbb{B}_t}.
\end{aligned}
$$

This completes the proof.

This completes the remark.

Using this forward price measure, for any suitably measurable and integrable random payoff $\frac{X_{\mathcal{T}}}{\mathbb{B}_{\mathcal{T}}} \in L^1_+(\mathbb{Q})$ received at time, $\frac{X_t}{p(t,\mathcal{T})}$ is a $\mathbb{Q}^{\mathcal{T}}$-martingale, i.e.

$$X_t = p(t,\mathcal{T})E^{\mathbb{Q}^{\mathcal{T}}}\left[X_{\mathcal{T}}|\mathcal{F}_t\right]. \tag{6.18}$$

Proof First, note that

$$
\begin{aligned}
E^{\mathbb{Q}}\left[\frac{d\mathbb{Q}^{\mathcal{T}}}{d\mathbb{Q}}|\mathcal{F}_t\right] &= E^{\mathbb{Q}}\left[\frac{1}{p(0,\mathcal{T})\mathbb{B}_{\mathcal{T}}}|\mathcal{F}_t\right] \\
&= \frac{1}{p(0,\mathcal{T})}E^{\mathbb{Q}}\left[\frac{1}{\mathbb{B}_{\mathcal{T}}}|\mathcal{F}_t\right] = \frac{p(t,\mathcal{T})}{p(0,\mathcal{T})\mathbb{B}_t}.
\end{aligned}
$$

Then, given $X_t = E^{\mathbb{Q}}\left[X_{\mathcal{T}}e^{-\int_t^{\mathcal{T}} r_s ds}|\mathcal{F}_t\right] = E^{\mathbb{Q}}\left[X_{\mathcal{T}}\frac{\mathbb{B}_t}{\mathbb{B}_{\mathcal{T}}}|\mathcal{F}_t\right]$, we have using Shreve [169, Lemma 5.2.2, p. 212], that

$$
\begin{aligned}
p(t,\mathcal{T})E^{\mathbb{Q}^{\mathcal{T}}}\left[X_{\mathcal{T}}|\mathcal{F}_t\right] &= p(t,\mathcal{T})\frac{1}{E^{\mathbb{Q}}\left[\frac{d\mathbb{Q}^{\mathcal{T}}}{d\mathbb{Q}}|\mathcal{F}_t\right]}E^{\mathbb{Q}}\left[X_{\mathcal{T}}\frac{d\mathbb{Q}^{\mathcal{T}}}{d\mathbb{Q}}|\mathcal{F}_t\right] \\
&= p(t,\mathcal{T})\frac{p(0,\mathcal{T})\mathbb{B}_t}{p(t,\mathcal{T})}E^{\mathbb{Q}}\left[X_{\mathcal{T}}\frac{1}{p(0,\mathcal{T})\mathbb{B}_{\mathcal{T}}}|\mathcal{F}_t\right] \\
&= E^{\mathbb{Q}}\left[X_{\mathcal{T}}\frac{\mathbb{B}_t}{\mathbb{B}_{\mathcal{T}}}|\mathcal{F}_t\right] = X_t.
\end{aligned}
$$

This completes the proof.

Expression (6.18) gives a useful alternative procedure for computing present values. This procedure is to take the payoff's expectation under the forward price measure $\mathbb{Q}^{\mathcal{T}}$ and discount the expectation to time t using the appropriate zero-coupon bond's price. Note that the default-free spot rate of interest process does not explicitly appear in this expression. It is implicit, however, in the forward price measure $\mathbb{Q}^{\mathcal{T}}$. It is important to emphasize that unlike the martingale probability $\mathbb{Q}$, the forward price measure depends on a particular future date $\mathcal{T}$ corresponding to the maturity of the zero-coupon bond used in its definition.

Applying this result to the time $\mathcal{T}$ price of the asset $\mathbb{S}(\mathcal{T})$ yields an equivalent expression for the forward price

$$K(t, \mathcal{T}) = E^{\mathbb{Q}^{\mathcal{T}}}\left[\mathbb{S}(\mathcal{T})\,|\mathcal{F}_t\right], \tag{6.19}$$

which explains the name for this probability measure.

Proof $\mathbb{S}(t) = p(t, \mathcal{T})E^{\mathbb{Q}^{\mathcal{T}}}\left[\mathbb{S}(\mathcal{T})\,|\mathcal{F}_t\right]$. Equating this to expression (6.16) completes the proof.

For subsequent usage, it is easy to show that the forward rate equals the expected time $\mathcal{T}$ spot rate under the forward price measure, i.e.

$$f(t, \mathcal{T}) = E^{\mathbb{Q}^{\mathcal{T}}}\left[r_{\mathcal{T}}\,|\mathcal{F}_t\right]. \tag{6.20}$$

Proof Given expression (6.10), $p(t, \mathcal{T}) = E^{\mathbb{Q}}\left[e^{-\int_t^{\mathcal{T}} r_s ds}\,|\mathcal{F}_t\right]$. Differentiating yields

$$-\frac{\partial p(t, \mathcal{T})}{\partial \mathcal{T}} = E^{\mathbb{Q}}\left[r_{\mathcal{T}}e^{-\int_t^{\mathcal{T}} r_s ds}\,|\mathcal{F}_t\right] = p(t, \mathcal{T})E^{\mathbb{Q}^{\mathcal{T}}}\left[r_{\mathcal{T}}\,|\mathcal{F}_t\right],$$

where the second equality follows from expression (6.18). Dividing by $p(t, \mathcal{T})$ completes the proof.

This implies, of course, that the forward price is not an unbiased estimate of the future spot price (under the statistical probability $\mathbb{P}$).

6.5.2 Futures Contracts

A futures contract is similar to a forward contract. It is a financial contract, written on the asset $\mathbb{S}(t)$, with a fixed maturity $\mathcal{T}$. It represents the purchase of the underlying asset at time $\mathcal{T}$ via a prearranged payment procedure. The prearranged payment procedure is called *marking-to-market*. Marking-to-market obligates the purchaser (long position) to accept a cash flow stream equal to the continuous changes in the futures prices for this contract.

The time t *futures prices*, denoted $k(t, \mathcal{T})$, are set (by market convention) such that newly issued futures contracts (at time t) on the same underlying asset with the same maturity date $\mathcal{T}$ have zero market value. Hence, futures contracts (by construction) have zero market value at all times, and a continuous cash flow stream equal to $dk(t, \mathcal{T})$. At maturity, the last futures price must equal the underlying asset's price $k(\mathcal{T}, \mathcal{T}) = \mathbb{S}(\mathcal{T})$.

It can be shown that the arbitrage-free futures price must satisfy

$$k(t, \mathcal{T}) = E^{\mathbb{Q}}\left[\mathbb{S}(\mathcal{T})\,|\mathcal{F}_t\right]. \tag{6.21}$$

Proof The accumulated value from buying and holding a futures contract over $[t, \mathcal{T}]$ and investing all proceeds into the mma is

$$\mathbb{B}_{\mathcal{T}} \int_t^{\mathcal{T}} \frac{1}{\mathbb{B}_s} dk(s, \mathcal{T}).$$

Hence, the futures price $k(t, \mathcal{T})$ solves

$$E^{\mathbb{Q}}\left[\frac{\mathbb{B}_{\mathcal{T}} \int_t^{\mathcal{T}} \frac{1}{\mathbb{B}_s} dk(s, \mathcal{T})}{\mathbb{B}_{\mathcal{T}}}\middle|\, \mathcal{F}_t\right] \mathbb{B}_t = 0 \text{ for all } t \text{ and } k(\mathcal{T}, \mathcal{T}) = \mathbb{S}(\mathcal{T}).$$

This implies $E^{\mathbb{Q}}\left[\int_t^{\mathcal{T}} \frac{1}{\mathbb{B}_s} dk(s, \mathcal{T})\,|\mathcal{F}_t\right] = 0$, i.e. $M_t = \int_0^t \frac{1}{\mathbb{B}_s} dk(s, \mathcal{T})$ is a $\mathbb{Q}$-martingale. Using associativity of the stochastic integral (Protter [151, p. 165]), we have that $\int_0^t \mathbb{B}_s dM_s = \int_0^t \mathbb{B}_s \frac{1}{\mathbb{B}_s} dk(s, \mathcal{T}) = k(t, \mathcal{T}) - k(0, \mathcal{T})$ is a $\mathbb{Q}$-martingale, or $k(t, \mathcal{T}) = E^{\mathbb{Q}}\left[k(\mathcal{T}, \mathcal{T})\,|\mathcal{F}_t\right] = E^{\mathbb{Q}}\left[\mathbb{S}(\mathcal{T})\,|\mathcal{F}_t\right]$. This completes the proof.

This expression shows that the futures price is not the expected value of the asset's time T payoff under the statistical probability measure $\mathbb{P}$, unless the equivalent martingale measure equals the statistical probability, i.e. $\mathbb{Q} = \mathbb{P}$. Recall that these two probability measures are equal, if and only if, interest rate risk premia are identically zero. Hence, in general, the futures price will not be an unbiased predictor of the future spot price of an asset.

The relation between forward and futures prices follows directly from these expressions

$$K(t, \mathcal{T}) = k(t, \mathcal{T}) + \mathrm{cov}^{\mathbb{Q}}\left[\mathbb{S}(\mathcal{T}), \frac{1}{\mathbb{B}_{\mathcal{T}}}\middle|\, \mathcal{F}_t\right] \frac{\mathbb{B}_t}{p(t, \mathcal{T})}. \tag{6.22}$$

Proof Using expression (6.16), we get

$$\begin{aligned} K(t,\mathcal{T})p(t,\mathcal{T}) = S_t &= E^{\mathbb{Q}}\left[\frac{\mathbb{S}_{\mathcal{T}}}{\mathbb{B}_{\mathcal{T}}}\,|\mathscr{F}_t\right]\mathbb{B}_t \\ &= E^{\mathbb{Q}}\left[\mathbb{S}_{\mathcal{T}}\,|\mathscr{F}_t\right]E^{\mathbb{Q}}\left[\frac{1}{\mathbb{B}_{\mathcal{T}}}\,|\mathscr{F}_t\right]\mathbb{B}_t + \mathrm{cov}^{\mathbb{Q}}\left[\mathbb{S}(\mathcal{T}),\,\frac{1}{\mathbb{B}_{\mathcal{T}}}\,|\mathscr{F}_t\right]\mathbb{B}_t. \end{aligned}$$

Using expression (6.21) and algebra completes the proof.

Expression (6.22) shows that forward and futures prices are equal if and only if, under the martingale measure $\mathbb{Q}$, the covariance between the asset's spot price and the reciprocal of the mma's value is zero. A sufficient condition for this covariance to equal zero, and hence for the equivalence of forward and futures prices, is that interest rates are deterministic (the classical result).

6.6 The Libor Model

This section presents the Libor model for pricing caps and floors. Caps and floors are interest rate derivatives written on Eurodollar deposit rates. Caps are a portfolio of European caplets, where a caplet is a European call option on an interest rate, and floors are a portfolio of floorlets, where a floorlet is a European put option on an interest rate. These financial instruments are usually based on LIBOR (London InterBank Offer Rates), an index of Eurodollar borrowing rates (see Jarrow and Chatterjea [95, Chapter 2 for more explanation]). In our context, we assume that the LIBOR rate corresponds to the default-free interest rate that one can earn from investing dollars for a fixed time period, say δ units of a year (e.g. $\frac{1}{4}$ of a year), quoted as a *discrete interest rate*, and not as the continuously compounded and instantaneous default-free spot rate r_t.

Just as with the continuously compounded and instantaneous forward rates, there is an analogous discrete forward rate. Consider the future time interval $[\mathcal{T}, \mathcal{T}+\delta]$ where δ corresponds to the "earning" interval. The *discrete forward rate* at time t for the time interval $[\mathcal{T}, \mathcal{T}+\delta]$ is defined by

$$1+\delta L(t,\mathcal{T}) = \frac{p(t,\mathcal{T})}{p(t,\mathcal{T}+\delta)}. \tag{6.23}$$

The right side of this expression isolates the implicit interest embedded in the zero-coupon bonds over $[\mathcal{T}, \mathcal{T}+\delta]$. The *discrete spot rate* is $L(t,t)$ for $[t, t+\delta]$.

Given the evolution for a 1-factor (continuous) forward rate process as in expression (6.3) and the definition (6.23) of the discrete forward rate involving zero-coupon bond prices, the evolution of the discrete forward rate is

$$dL(t,\mathcal{T}) = \left(\frac{1+\delta L(t,\mathcal{T})}{\delta}\right)(a_1(t,\mathcal{T}) - a_1(t,\mathcal{T}+\delta))\,dW_1^{\mathbb{Q}^{\mathcal{T}+\delta}}(t), \qquad (6.24)$$

where $W_1^{\mathbb{Q}^{\mathcal{T}+\delta}}(t)$ is a standard independent Brownian motion under the forward price measure $\mathbb{Q}^{\mathcal{T}+\delta}$.

Proof For ease of notation we drop the subscript 1. From expression (6.14), we have that

$$p(t,\mathcal{T}) = p(0,\mathcal{T})e^{\int_0^t r_s ds - \frac{1}{2}\int_0^t a(s,\mathcal{T})^2 ds + \int_0^t a(s,\mathcal{T})dW^{\mathbb{Q}}(s)} \text{ and}$$

$$p(t,\mathcal{T}+\delta) = p(0,\mathcal{T}+\delta)e^{\int_0^t r_s ds - \frac{1}{2}\int_0^t a(s,\mathcal{T}+\delta)^2 ds + \int_0^t a(s,\mathcal{T}+\delta)dW^{\mathbb{Q}}(s)}.$$

Thus,

$$\frac{p(t,\mathcal{T})}{p(t,\mathcal{T}+\delta)} = \frac{p(0,\mathcal{T})}{p(0,\mathcal{T}+\delta)}e^{-\frac{1}{2}\int_0^t a(s,\mathcal{T})^2 ds + \frac{1}{2}\int_0^t a(s,\mathcal{T}+\delta)^2 ds + \int_0^t [a(s,\mathcal{T}) - a(s,\mathcal{T}+\delta)]dW^{\mathbb{Q}}(s)}.$$

Define an equivalent change of measure

$$\frac{d\mathbb{Q}^x}{d\mathbb{Q}} = e^{-\int_0^{\mathcal{T}} \frac{1}{2}x_s^2 ds + \int_0^{\mathcal{T}} x_s dW^{\mathbb{Q}}(s)} > 0.$$

By Girsanov's Theorem 1.3 in Chap. 1,

$$dW^{\mathbb{Q}^x}(s) = -x_s ds + dW^{\mathbb{Q}}(s)$$

is a Brownian motion under $\mathbb{Q}^x$.

We want to determine x_s such that $\frac{p(t,\mathcal{T})}{p(t,\mathcal{T}+\delta)}$ is a $\mathbb{Q}^x$-martingale for all $\mathcal{T}$.

Given the market is complete with respect to $\mathbb{Q}$, this implies the market is complete with respect to $\mathbb{Q}^{\mathcal{T}+\delta}$ when using $p(t,\mathcal{T}+\delta)$ as the numeraire. Hence, by the definition of the forward price measure $\mathbb{Q}^{\mathcal{T}+\delta}$, this implies that $\mathbb{Q}^x = \mathbb{Q}^{\mathcal{T}+\delta}$ by uniqueness of the equivalent martingale measure (using the numeraire $p(t,\mathcal{T}+\delta)$) due to the Second Fundamental Theorem 2.4 of asset pricing in Chap. 2. We use this observation in the last step of the proof.

By substitution,

$$\frac{p(t,\mathcal{T})}{p(t,\mathcal{T}+\delta)} = \frac{p(0,\mathcal{T})}{p(0,\mathcal{T}+\delta)}e^{-\frac{1}{2}\int_0^t a(s,\mathcal{T})^2 ds + \frac{1}{2}\int_0^t a(s,\mathcal{T}+\delta)^2 ds + \int_0^t [a(s,\mathcal{T}) - a(s,\mathcal{T}+\delta)]x_s ds} \cdot e^{+\int_0^t [a(s,\mathcal{T}) - a(s,\mathcal{T}+\delta)]dW^{\mathbb{Q}^x}(s)}.$$

Using the Dolean–Dades exponential (see Medvegyev [136, p. 412]), Z is a $\mathbb{Q}^x$-local martingale if and only if $e^{Z-\frac{1}{2}[Z,Z]}$ is a $\mathbb{Q}^x$-local martingale.

Hence, adding and subtracting $+\frac{1}{2}\int_0^t [a(s,\mathcal{T})-a(s,\mathcal{T}+\delta)]^2\,ds$ in the exponent yields

$$-\frac{1}{2}\int_0^t a(s,\mathcal{T})^2 ds+\frac{1}{2}\int_0^t a(s,\mathcal{T}+\delta)^2 ds+\int_0^t [a(s,\mathcal{T})-a(s,\mathcal{T}+\delta)]\,x_s ds$$
$$+\frac{1}{2}\int_0^t [a(s,\mathcal{T})-a(s,\mathcal{T}+\delta)]^2$$
$$+\int_0^t [a(s,\mathcal{T})-a(s,\mathcal{T}+\delta)]\,dW^{\mathbb{Q}^x}(s)-\frac{1}{2}\int_0^t [a(s,\mathcal{T})-a(s,\mathcal{T}+\delta)]^2\,ds$$

is a $\mathbb{Q}^x$-local martingale
if and only if

$$-\frac{1}{2}\int_0^t a(s,\mathcal{T})^2 ds+\frac{1}{2}\int_0^t a(s,\mathcal{T}+\delta)^2 ds+\int_0^t [a(s,\mathcal{T})-a(s,\mathcal{T}+\delta)]\,x_s ds$$
$$+\frac{1}{2}\int_0^t [a(s,\mathcal{T})-a(s,\mathcal{T}+\delta)]^2\,ds=0 \quad \text{for all } t$$

if and only if

$$-\frac{1}{2}a(s,\mathcal{T})^2+\frac{1}{2}a(s,\mathcal{T}+\delta)^2+[a(s,\mathcal{T})-a(s,\mathcal{T}+\delta)]\,x_s$$
$$+\frac{1}{2}[a(s,\mathcal{T})-a(s,\mathcal{T}+\delta)]^2=0 \quad \text{for all } s$$

if and only if

$$-\frac{1}{2}a(s,\mathcal{T})^2+\frac{1}{2}a(s,\mathcal{T}+\delta)^2+[a(s,\mathcal{T})-a(s,\mathcal{T}+\delta)]\,x_s$$
$$+\frac{1}{2}a(s,\mathcal{T})^2-a(s,\mathcal{T})a(s,\mathcal{T}+\delta)+\frac{1}{2}a(s,\mathcal{T}+\delta)^2=0 \quad \text{for all } s$$

if and only if

$$a(s,\mathcal{T}+\delta)^2-a(s,\mathcal{T})a(s,\mathcal{T}+\delta)=-[a(s,\mathcal{T})-a(s,\mathcal{T}+\delta)]\,x_s \quad \text{for all } s$$

if and only if

$$-[a(s,\mathcal{T})-a(s,\mathcal{T}+\delta)]\,a(s,\mathcal{T}+\delta)=-[a(s,\mathcal{T})-a(s,\mathcal{T}+\delta)]\,x_s \quad \text{for all } s.$$

Hence, $x_s=a(s,\mathcal{T}+\delta)$ for all s.

By substitution, we get using $W^{\mathbb{Q}^x}(s)$ that

$$\frac{p(t,\mathcal{T})}{p(t,\mathcal{T}+\delta)} = \frac{p(0,\mathcal{T})}{p(0,\mathcal{T}+\delta)} e^{\int_0^t [a(s,\mathcal{T})-a(s,\mathcal{T}+\delta)]dW^{\mathbb{Q}^x}(s) - \frac{1}{2}\int_0^t [a(s,\mathcal{T})-a(s,\mathcal{T}+\delta)]^2 ds}.$$

Using the definition of $L(t,\mathcal{T})$ yields

$$1+\delta L(t,\mathcal{T}) = [1+\delta L(0,\mathcal{T})]\, e^{\int_0^t [a(s,\mathcal{T})-a(s,\mathcal{T}+\delta)]dW^{\mathbb{Q}^x}(s) - \frac{1}{2}\int_0^t [a(s,\mathcal{T})-a(s,\mathcal{T}+\delta)]^2 ds}.$$

This is the unique solution to (see Medvegyev [136, p. 412]),

$$d[1+\delta L(s,\mathcal{T})] = [1+\delta L(s,\mathcal{T})]\,[a(s,\mathcal{T}) - a(s,\mathcal{T}+\delta)]\, dW^{\mathbb{Q}^x}(s).$$

Algebra gives

$$dL(s,\mathcal{T}) = \frac{[1+\delta L(s,\mathcal{T})]}{\delta}[a(s,\mathcal{T}) - a(s,\mathcal{T}+\delta)]\, dW^{\mathbb{Q}^x}(s).$$

Using the fact that $\mathbb{Q}^x = \mathbb{Q}^{\mathcal{T}+\delta}$ completes the proof.

This discrete forward rate evolution is consistent with no-arbitrage (by construction), since it is derived from the arbitrage free (continuously compounded) forward rate evolution in expression (6.12), after a change of probability measure from $\mathbb{Q}$ to $\mathbb{Q}^{\mathcal{T}+\delta}$.

To facilitate an analytic formula for a cap's value, the LIBOR model *assumes* that

$$\gamma(t,\mathcal{T}) \equiv \left(\frac{1+\delta L(t,\mathcal{T})}{\delta L(t,\mathcal{T})}\right)(a_1(t,\mathcal{T}) - a_1(t,\mathcal{T}+\delta)) \tag{6.25}$$

is a deterministic function of time. Hence, under this assumption, the discrete forward rate evolves as

$$dL(t,\mathcal{T}) = \gamma(t,\mathcal{T})L(t,\mathcal{T})d\bar{W}_1^{\mathbb{Q}^{\mathcal{T}+\delta}}(t). \tag{6.26}$$

This is a (generalized) geometric Brownian motion process. Note the analogy to the evolution of the risky asset price process in the BSM model in Chap. 5.

We now value caps and floors under this evolution. But first, some definitions are needed. A *European caplet* is a European call option with maturity $\mathcal{T}+\delta$ and strike k on the discrete spot rate at time $\mathcal{T}$. *Caps* are a portfolio of European caplets, with the same strike but with different and increasing maturity dates. The maturity dates of the included caplets are evenly spaced at fixed intervals (e.g. every 3 months) up to the maturity of the cap (e.g. 5 years). Hence, to value a cap, one only needs to value the constituent caplets.

Similarly, a *floorlet* is a European put option on the forward interest rate. *Floors* are a portfolio of floorlets with the same strike and with increasing maturity dates. The sequence of maturity dates is similar to that of a cap. To value a floor, one need only value the constituent floorlets. Using put-call parity (see Jarrow and Chatterjea [95, Chapter 16 for a discussion of put-call parity]), if one prices European caplets, then the formula for European floorlets immediately follows. For this reason, this section only concentrates on pricing caplets.

Now, consider a caplet on the discrete forward rate with maturity $\mathscr{T} + \delta$ and strike k. The standard caplet pays off at time $\mathscr{T} + \delta$, based on the discrete forward rate from time $\mathscr{T}$, i.e. the payoff is defined by

$$X_{\mathscr{T}+\delta} \equiv \max[(L(\mathscr{T}, \mathscr{T}) - k)\,\delta, 0]$$

at time $\mathscr{T} + \delta$. This can be interpreted as the differential interest $(L(\mathscr{T}, \mathscr{T}) - k)\,\delta$ earned on the principal of 1 dollar over the time period $[\mathscr{T}, \mathscr{T} + \delta]$, but only if the differential is positive. Note that the interest paid is prorated by the time period δ over which it is earned.

Using expression (6.18), the caplet's time 0 value is

$$X_0 = p(0, \mathscr{T} + \delta) E^{\mathbb{Q}^{\mathscr{T}+\delta}} \left[\max[(L(\mathscr{T}, \mathscr{T}) - k)\,\delta, 0]\right]. \tag{6.27}$$

Given that $L(\mathscr{T}, \mathscr{T})$ follows a lognormal distribution under expression (6.26), it is straightforward to show that the caplet's value satisfies the subsequent formula, which appears remarkably similar to the BSM formula (5.10) in Chap. 5.

$$X_0 = p(0, \mathscr{T} + \delta)\delta[L(0, \mathscr{T})N(d_1) - kN(d_2)] \quad \text{where} \tag{6.28}$$

$$d_1 = \left(\frac{\log\left(\frac{L(0,\mathscr{T})}{k}\right) + \frac{1}{2}\int_0^{\mathscr{T}} \gamma(t, \mathscr{T})^2 dt}{\sqrt{\int_0^{\mathscr{T}} \gamma(t, \mathscr{T})^2 dt}} \right) \quad \text{and}$$

$$d_2 = d_1 - \sqrt{\int_0^{\mathscr{T}} \gamma(t, \mathscr{T})^2 dt}.$$

Proof Using expression (6.27), we have

$$X_0 = p(0, \mathscr{T} + \delta)\delta E^{\mathbb{Q}^{\mathscr{T}+\delta}} \left[\max[(L(\mathscr{T}, \mathscr{T}) - k), 0]\right].$$

The same proof as used to prove the BSM formula (5.10) in Chap. 5 applies with the following identifications: $\mathbb{S}(t) = L(t, \mathscr{T})$, $T = \mathscr{T}$, $K = k$, $r = 0$, and $\sigma^2 = \frac{\int_0^{\mathscr{T}} \gamma(t,\mathscr{T})^2 dt}{\mathscr{T}}$. This completes the proof.

It is easy to generalize this 1-factor to a D-factor model. The lognormal evolution for forward Libor rates has also been extended to jump diffusions, Levy processes, and stochastic volatilities (see Jarrow et al. [110] and references therein). For textbooks expanding on the LIBOR model, as used in practice, see Rebanato [156] and Schoenmakers [165].

Remark 6.5 (Geometric Brownian Motion (Continuously Compounded) Forward Rates) It was shown in Sect. 6.4 above that (continuously compounded) forward rates $f(t, \mathscr{T})$ following a geometric Brownian motion process are inconsistent with a market satisfying NFLVR and ND. This fact implies that a formula for caplets analogous to the Black–Scholes–Merton formula, using continuously compounded forward rates, is also inconsistent with NFLVR and ND. This inconsistency was the motivation for obtaining the LIBOR model (see Sandmann et al. [163], Miltersen et al. [142], and Brace et al. [20]), and it is the reason *discrete* forward rates $L(t, \mathscr{T})$ are used in the derivation of expression (6.28). This completes the remark.

6.7 Notes

There is much more known about implementing the HJM model and its empirical testing, especially as related to fixed incomes securities. Excellent books in this regard include Brigo and Mercurio [24], Carmona and Tehranchi [28], Filipovic [60], Jarrow [83], Rebonato [156], Schoenmakers [165], and Zagst [177].

Chapter 7
Reduced Form Credit Risk Models

Credit risk arises whenever two counter parties engage in borrowing and lending. Borrowing can be in cash, which is the standard case, or it can be through the "shorting" of securities. *Shorting* a security is selling a security one does not own. To do this, the security must first be borrowed from an intermediate counter party, with an obligation to return the borrowed security at a later date. The borrowing part of this shorting transaction involves credit risk. Since the majority of transactions in financial and commodity markets involve some sort of borrowing, understanding the economics of credit risk is fundamental to the broader understanding of economics itself.

There are two models for studying credit risk. The first is called the *structural approach*, which was introduced by Merton [139]. This model assumes that all of the assets of the firm trade, an unrealistic assumption. Consequently, this model is best used for conceptual understanding (see Jarrow [88] for a detailed discussion). The second is called the *reduced form model*, which was introduced by Jarrow and Turnbull [106, 107]. This model assumes that only a subset of the firm's liabilities trade, those that need to be priced and hedged. This is the model studied in this chapter. This chapter is based on Jarrow [85].

7.1 The Set-Up

This chapter builds on the HJM model in Chap. 6 by adding trading in risky zero-coupon bonds. Given is a non-normalized market $((\mathbb{B}, \mathbb{S}), (\mathscr{F}_t), \mathbb{P})$. The money market account (mma) $\mathbb{B}_t = e^{\int_0^t r_s ds}$, $\mathbb{B}_0 = 1$, is continuous in time t, and of finite variation. r_t is the default-free spot rate of interest, adapted to $\mathscr{F}_t$ with $\int_0^t |r_s| ds < \infty$.

R. A. Jarrow, *Continuous-Time Asset Pricing Theory*, Springer Finance,
https://doi.org/10.1007/978-3-319-77821-1_7

As in the HJM model in Chap. 6 we modify the risky asset price vector. Instead of a finite number of risky assets $\mathbb{S}_t = (\mathbb{S}_1(t), \ldots, \mathbb{S}_n(t))'$, we let the traded risky assets include the collection of default-free zero-coupon bonds of all maturities, with time t price $p(t, \mathcal{T}) > 0$ for a sure dollar paid at time $\mathcal{T}$ for all $0 \leq t \leq \mathcal{T} \leq T$. The default-free zero-coupon price process is assumed to be a strictly positive continuous semimartingale adapted to $\mathcal{F}_t$ for all $\mathcal{T}$.

In addition, also traded is a risky zero-coupon bond promising to pay a dollar at time $\mathcal{T}$ with time t price denoted $D(t, \mathcal{T}) \geq 0$. This bond may default, and the promised dollar not paid. The value of the risky zero-coupon bond at time $\mathcal{T}$, $D(\mathcal{T}, \mathcal{T})$, may be strictly less than one if default occurs. The risky zero-coupon price process is assumed to be a nonnegative semimartingale adapted to $\mathcal{F}_t$. This price process, in general, is not continuous. More risky securities could be traded, but they are not needed for the subsequent presentation.

Remark 7.1 (Self-Financing-Trading Strategies) As in the HJM model in Chap. 6, although there are a continuum of traded risky assets, an admissible s.f.t.s. consists of the mma and only a finite number of default-free and risky zero-coupon bonds over the trading horizon $[0, T]$. This enables us to use the fact that the existence of an equivalent martingale measure $\mathbb{Q} \in \mathfrak{M}$ implies NFLVR and ND by the Third Fundamental Theorem 2.5 of asset pricing in Chap. 2, even with trading in an infinite number of risky assets. The existence of an equivalent martingale measure is no longer implied by NFLVR and ND because the First and Third Fundamental theorems were only proven for a finite asset market (see the discussion following the First Fundamental Theorem 2.3 in Chap. 2). This completes the remark.

7.2 The Risky Firm

To provide an increased understanding of the pricing of the risky zero-coupon bond, we add some additional structure related to default and the recovery rate process of the risky zero-coupon bond. These are best characterized by considering the firm (more generally a credit entity) that issued this risky zero-coupon bond. The firm can have many outstanding liabilities in addition to this zero-coupon bond. All of these liabilities contain contractual payments to be made by the firm at various times over their lives. The first time the firm does not make a contractual payment on any of its liabilities, the firm is in "default" on that liability's payments. Due to cross-defaulting provisions, this means that all of the firm's liabilities are in default too (usually all contractual payments—interest owed and principal outstanding—become due).

We want to characterize default and the recovery rate on the risky zero-coupon bond if default occurs. Let $\Gamma_t = (\Gamma_1(t), \ldots, \Gamma_m(t))' \in \mathbb{R}^m$ be a collection of stochastic processes characterizing the state of the market at time t with $\mathcal{F}_t^{\Gamma}$ representing the filtration generated by the state variables Γ_t up to and including time $t \geq 0$. These *state variables* are assumed to be nonnegative semimartingales adapted to $\mathscr{F}_t$, which implies that $\mathscr{F}_t^{\Gamma} \subset \mathscr{F}_t$. We include the default-free spot rate of interest r_t in this set of state variables, which implies that r_t is $\mathscr{F}_t^{\Gamma}$-measurable. Examples of additional state variables that could be included in this set are the remaining default-free forward rates, unemployment rates, and inflation rates.

Let $\lambda : [0, T] \times \mathbb{R}^m \longrightarrow [0, \infty)$, denoted $\lambda_t = \lambda_t(\Gamma_t) \geq 0$, be jointly Borel measurable with $\int_0^T \lambda_t(\Gamma_t)dt < \infty$ a.s. $\mathbb{P}$.

Let $N_t \in \{0, 1, 2, \cdots\}$ with $N_0 = 0$ be a Cox process conditioned on $\mathscr{F}_T^{\Gamma}$ with $\lambda_t(\Gamma_t)$ its intensity process, i.e. for all $0 \leq s < t$ and $n = 0, 1, 2, \cdots$

$$\mathbb{P}\left(N_t - N_s = n \mid \mathscr{F}_T^{\Gamma} \vee \mathscr{F}_s\right) = \frac{e^{-\int_s^t \lambda_u(\Gamma_u)du} \left(\int_s^t \lambda_u(\Gamma_u)du\right)^n}{n!} \tag{7.1}$$

where $\mathscr{F}_T^{\Gamma} \vee \mathscr{F}_s$ is the smallest σ-algebra containing both $\mathscr{F}_T^{\Gamma}$ and $\mathscr{F}_s$ (see Definition 1.13 in Chap. 1).

Finally, let $\tau \in [0, T]$ be the stopping time adapted to the filtration $\mathscr{F}_t$ defined by

$$\tau \equiv \inf\{t > 0 : N_t = 1\}.$$

We let τ represent the firm's *default time*. The function $\lambda_t(\Gamma_t)$ is the firm's default intensity, which can be interpreted as the probability of default over a small time interval $[t, t+\Delta]$ conditional upon no default prior to time t. We note that the default time is *totally inaccessible*, i.e. it cannot be written as a predictable stopping time. A stopping time is predictable if there exists an increasing sequence of stopping times that approach τ from below (see Protter [151] for these definitions). For simplicity of notation, we often write $\lambda_t(\Gamma_t)$ as λ_t below.

Intuitively, a Cox process is a point process which, conditional upon the information set generated by the state variables process Γ_t over the entire trading horizon $[0, T]$, behaves like a standard Poisson process. For subsequent use, we note the next lemma (for a proof, see Lando [130]).

Lemma 7.1 (Cox Process Probabilities) *For* $\tau > t$,

$$\mathbb{P}(\tau > \mathcal{T} \mid \mathscr{F}_T^{\Gamma} \vee \mathscr{F}_t) = E\left[1_{\tau > \mathcal{T}} \mid \mathscr{F}_T^{\Gamma} \vee \mathscr{F}_t\right] = e^{-\int_t^{\mathcal{T}} \lambda_u du}, \tag{7.2}$$

$$\mathbb{P}(\tau > \mathcal{T} \mid \mathscr{F}_t) = E\left[E\left[1_{\tau > \mathcal{T}} \mid \mathscr{F}_T^{\Gamma} \vee \mathscr{F}_t\right] \mid \mathscr{F}_t\right] = E\left[e^{-\int_t^{\mathcal{T}} \lambda_u du} \mid \mathscr{F}_t\right], \tag{7.3}$$

$$\mathbb{P}(t < \tau \leq \mathcal{T} \mid \mathscr{F}_t) = 1 - E\left[e^{-\int_t^{\mathcal{T}} \lambda_u du} \mid \mathscr{F}_t\right], \tag{7.4}$$

$$\mathbb{P}(\tau \in [s, s+dt) \mid \mathscr{F}_T^{\Gamma} \vee \mathscr{F}_t) = \frac{d\mathbb{P}(\tau \leq s \mid \mathscr{F}_T^{\Gamma} \vee \mathscr{F}_t)}{ds} = \lambda_s e^{-\int_t^s \lambda_u du}. \tag{7.5}$$

If the firm defaults prior to the zero-coupon bond's maturity date, we assume that at time $\mathscr{T}$ the bond receives a *recovery payment* less than or equal to the promised dollar, i.e.

$$D(\mathscr{T}, \mathscr{T}) = \begin{cases} \frac{R(\tau, \mathscr{T})\mathbb{B}(\mathscr{T})}{\mathbb{B}(\tau)} \leq 1 & \text{if } 0 < \tau \leq \mathscr{T}, \\ 1 & \text{if } 0 \leq \mathscr{T} < \tau, \end{cases} \tag{7.6}$$

where $1 \geq R(t, \mathscr{T}) \geq 0$ is $\mathscr{F}_t^{\Gamma}$-measurable.

Note that the recovery rate is assumed to be measurable with respect to the information set generated by the state variables process. In the literature, three recovery rate processes have been frequently used.

1. *Recovery of Face Value.*

$$R(\tau, \mathscr{T}) \equiv \delta \text{ where } \delta \in [0, 1].$$

This states that on the default date, the debt is worth some constant recovery rate between 0 and 1.

2. *Recovery of Treasury.*

$$R(\tau, \mathscr{T}) \equiv \delta p(\tau, \mathscr{T}) \text{ where } \delta \in [0, 1].$$

This states that on the default date, the debt is worth some constant percentage of an otherwise equivalent, but default-free zero-coupon bond.

3. *Recovery of Market Value*

$$R(\tau, \mathscr{T}) \equiv \delta D(\tau-, \mathscr{T}) \text{ where } \delta \in [0, 1]$$

and $D(\tau-, \mathscr{T}) \equiv \lim_{t \to \tau, t \leq \tau} D(t, \mathscr{T})$ is the value of the debt issue an instant before default. This states that on the default date, the debt is worth some constant fraction of its value an instant before default occurred, at time $\tau-$.

7.3 Existence of an Equivalent Martingale Measure

For pricing and hedging credit derivatives we need to assume that there exists an equivalent probability measure $\mathbb{Q} \in \mathfrak{M}$ where

$$\mathfrak{M} = \left\{ \mathbb{Q} \sim \mathbb{P} : \frac{p(t, \mathscr{T})}{\mathbb{B}_t} \text{ for all } \mathscr{T} \in [0, T] \text{ and } \frac{D(t, \mathscr{T})}{\mathbb{B}_t} \text{ are } \mathbb{Q}\text{-martingales} \right\}.$$

Assumption (Existence of an Equivalent Martingale Measure) *There exists an equivalent probability measure* $\mathbb{Q} \in \mathfrak{M}$ *such that*

$$\frac{p(t,\mathscr{T})}{\mathbb{B}_t} \quad \textit{for all } \mathscr{T} \in [0,T] \quad \textit{and} \quad \frac{D(t,\mathscr{T})}{\mathbb{B}_t}$$

are $\mathbb{Q}$*-martingales.*

This assumption implies that the market satisfies NFLVR and ND by the Third Fundamental Theorem 2.5 of asset pricing in Chap. 2. Of course, as shown in Chap. 2, given there are an infinite number of traded risky assets, the converse of this theorem is not necessarily true. For the purposes of this chapter, we do not need to specify a particular evolution for the term structure of interest rates as done in the HJM model in Chap. 6.

The previous assumption immediately implies the next corollary.

Corollary 7.1 (Zero-Coupon Bond Formulas)

$$p(t,\mathscr{T}) = E^{\mathbb{Q}}\left[e^{-\int_t^{\mathscr{T}} r_s ds} \,|\mathscr{F}_t\right] \tag{7.7}$$

for all $0 \le t \le \mathscr{T} \le T$ *and*

$$D(t,\mathscr{T}) = E^{\mathbb{Q}}\left[D(\mathscr{T},\mathscr{T})e^{-\int_t^{\mathscr{T}} r_s ds} \,|\mathscr{F}_t\right]. \tag{7.8}$$

The zero-coupon bond prices are given by the expected discounted cash flows using the equivalent martingale measure $\mathbb{Q} \in \mathfrak{M}$. As before, the risk of the payoffs is incorporated into these present value formulas because expectations are taken with respect to the equivalent martingale measure $\mathbb{Q}$ and not the statistical probability measure $\mathbb{P}$.

Expression (7.8) can be written as

$$D(t,\mathscr{T}) = E^{\mathbb{Q}}\left[R(\tau,\mathscr{T})e^{-\int_t^{\tau} r_s ds} 1_{\{\tau \le \mathscr{T}\}} + 1 \cdot e^{-\int_t^{\mathscr{T}} r_s ds} 1_{\{\mathscr{T} < \tau\}} \,|\mathscr{F}_t\right]. \tag{7.9}$$

Expression (7.9) states that the risky zero-coupon bond's value is equal to its expected discounted payoff across the two possibilities regarding default: (i) default occurs prior to maturity in which case the debt receives its recovery value $R(\tau,\mathscr{T})$, or (ii) the debt does not default before maturity in which case it receives its face value 1.

Definition 7.1 (Martingale Measure Intensity) Given an equivalent martingale measure $\mathbb{Q} \in \mathfrak{M}$ define $\widetilde{\lambda}_t \equiv \lambda_t \kappa_t$ to be the *intensity process* of the Cox process under $\mathbb{Q}$ where $\kappa_t(\omega) \ge 0$ is a predictable process with $\int_0^T \lambda_t(\Gamma_t)\kappa_t dt < \infty$ a.s. $\mathbb{P}$ (see Bremaud [23, p. 167]).

Under the Cox process assumption, and using the intensity under the martingale measure, this expression simplifies to

$$\begin{aligned} D(t, \mathcal{T}) = \int_t^{\mathcal{T}} E^{\mathbb{Q}}\left[R(s, \mathcal{T})\widetilde{\lambda}_s e^{-\int_t^s (r_u+\widetilde{\lambda}_u)du} | \mathcal{F}_t\right] ds \\ +1 \cdot E^{\mathbb{Q}}\left[e^{-\int_t^{\mathcal{T}} (r_u+\widetilde{\lambda}_u)du} | \mathcal{F}_t\right]. \end{aligned} \tag{7.10}$$

Proof This follows from expressions (7.13) and (7.15) below. This completes the proof.

Note that the omission of the default time τ from expression (7.10) implies that the computation of the debt's value is analogous to computing the value of an interest rate derivative under an HJM model where the default-free spot rate process r_u is replaced by a *pseudo* default-free spot rate process $(r_u + \widetilde{\lambda}_u)$.

For the different recovery rate processes, expression (7.10) simplifies further

$$D(t, \mathcal{T}) = \begin{cases} \int_t^{\mathcal{T}} \delta E^{\mathbb{Q}}\left[\widetilde{\lambda}_s e^{-\int_t^s (r_u+\widetilde{\lambda}_u)du} | \mathcal{F}_t\right] ds \\ \quad + E^{\mathbb{Q}}\left[e^{-\int_t^{\mathcal{T}} (r_u+\widetilde{\lambda}_u)du} | \mathcal{F}_t\right] & \text{recovery of face value} \\ \delta p(t, \mathcal{T}) + (1-\delta)E^{\mathbb{Q}}\left[e^{-\int_t^{\mathcal{T}} (r_s+\widetilde{\lambda}_s)ds} | \mathcal{F}_t\right] & \text{recovery of Treasury} \\ E^{\mathbb{Q}}\left[e^{-\int_t^{\mathcal{T}} (r_u+\widetilde{\lambda}_u(1-\delta))du} | \mathcal{F}_t\right] & \text{recovery of market value} \end{cases} \tag{7.11}$$

Proof The proof for the recovery of face value case is just simple algebra and expression (7.10).

The proof for the recovery of market value is given by expression (7.16).

The proof for the recovery of Treasury is as follows

$$\begin{aligned} D(t, \mathcal{T}) &= E^{\mathbb{Q}}\left[\delta p(\tau, \mathcal{T}) e^{-\int_t^{\tau} r_s ds} 1_{\{\tau \le \mathcal{T}\}} + e^{-\int_t^{\mathcal{T}} r_s ds} 1_{\{\mathcal{T} < \tau\}} | \mathcal{F}_t\right] \\ &= E^{\mathbb{Q}}\left[\delta E^{\mathbb{Q}}\left[e^{-\int_\tau^{\mathcal{T}} r_s ds} | \mathcal{F}_\tau\right] e^{-\int_t^{\tau} r_s ds} 1_{\{\tau \le \mathcal{T}\}} + e^{-\int_t^{\mathcal{T}} r_s ds} 1_{\{\mathcal{T} < \tau\}} | \mathcal{F}_t\right] \\ &= E^{\mathbb{Q}}\left[\delta e^{-\int_t^{\mathcal{T}} r_s ds} 1_{\{\tau \le \mathcal{T}\}} + e^{-\int_t^{\mathcal{T}} r_s ds} 1_{\{\mathcal{T} < \tau\}} | \mathcal{F}_t\right] \\ &= E^{\mathbb{Q}}\left[\delta e^{-\int_t^{\mathcal{T}} r_s ds} + (1-\delta) e^{-\int_t^{\mathcal{T}} r_s ds} 1_{\{\mathcal{T} < \tau\}} | \mathcal{F}_t\right] \\ &= \delta p(t, \mathcal{T}) + (1-\delta) E^{\mathbb{Q}}\left[e^{-\int_t^{\mathcal{T}} r_s ds} 1_{\{\mathcal{T} < \tau\}} | \mathcal{F}_t\right]. \end{aligned}$$

Using expression (7.14) gives

$$= \delta p(t, \mathcal{T}) + (1-\delta) E^{\mathbb{Q}}\left[e^{-\int_t^{\mathcal{T}} (r_s+\widetilde{\lambda}_s) ds} | \mathcal{F}_t\right].$$

This completes the proof.

7.4 Risk Neutral Valuation

This section studies the pricing of various hypothetical credit derivatives using risk neutral valuation. If the market is complete with respect to $\mathbb{Q} \in \mathfrak{M}$, then the martingale probability is unique by the Second Fundamental Theorem 2.6 of asset pricing in Chap. 2. In this case, the pricing and hedging follows the standard procedures discussed therein. However, considering trading in only the mma, the default-free zero-coupon bonds, and the risky zero-coupon bond, the market is most likely incomplete. This follows because the added randomness due to the default time τ being an inaccessible stopping time implies that $D(t, \mathscr{T})$ is a discontinuous sample path process (with jumps). Discontinuous sample path processes exclude market completeness when the amplitude of the jumps has a distribution with more than a finite number of values possible, and the number of these possible values exceeds the number of traded risky assets (see Cont and Tankov [34, Chapter 9.2]). The reason for this is that the number of traded assets are typically insufficient to hedge the quantity of different jump magnitudes possible (in mathematical terms, the martingale representation property fails for the value processes of admissible s.f.t.s.'s).

If the market is incomplete, to price credit derivatives, we select a unique equivalent martingale probability $\mathbb{Q}$ from the set of martingale probabilities by *assuming that enough derivatives trade so that the extended market is complete* and by the Second Fundamental Theorem 2.4 of asset pricing in Chap. 2, $\mathbb{Q} \in \mathfrak{M}$ is unique. We say that this unique martingale measure is *chosen by the market* (see Sect. 3.1 in the asset price bubbles Chap. 3 for a related discussion). See Eisenberg and Jarrow [56], Dengler and Jarrow [48], Jacod and Protter [76], and Schweizer and Wissel [167] for papers containing markets that are enlarged using derivatives such that the extended market is complete, but when trading only in the risky assets themselves, the market is incomplete.

Assuming that the enlarged market that includes trading in derivatives is complete implies that any derivative security $Z_T \in L^1_+(\mathbb{Q})$ is attainable and its value process is a $\mathbb{Q}$-martingale. This, in turn, implies by Theorem 2.8 in the fundamental theorems Chap. 2 that one can use risk neutral valuation to price any derivative security $Z_T \in L^1_+(\mathbb{Q})$.

In the incomplete market case, however, exact hedging is in general not possible using the mma, default-free zero-coupon bonds, and the risky zero-coupon bond alone. As an alternative to exact hedging, Chap. 8 provides a discussion of super- and sub-replication, which can be used in this circumstance.

For the purposes of the subsequent valuation, we fix the equivalent martingale measure $\mathbb{Q} \in \mathfrak{M}$ chosen by the market. We now apply these insights to price various hypothetical credit derivatives that can be used as building blocks for a larger class of traded credit derivatives.

Consider a random cash flow received at time $\mathscr{T}$, denoted $Y_{\mathscr{T}}$, which is $\mathscr{F}^{\Gamma}_{\mathscr{T}}$-measurable with $E^{\mathbb{Q}}[|Y_{\mathscr{T}}|] < \infty$. Note that the random cash flow is measurable with respect to the filtration generated by the state variables process. Then, the time

t value of this cash flow, Y_t, is given by

$$Y_t = E^{\mathbb{Q}}\left[Y_{\mathcal{T}} e^{-\int_t^{\mathcal{T}} r_u du} \middle| \mathcal{F}_t \right]. \tag{7.12}$$

7.4.1 Cash Flow 1

Random cash flow $Y_{\mathcal{T}}$ which is $\mathcal{F}_{\mathcal{T}}^{\Gamma}$-measurable at time $\mathcal{T}$, but the cash flow occurs only if no default ($\tau > \mathcal{T}$). Then,

$$E^{\mathbb{Q}}\left[Y_{\mathcal{T}} 1_{\tau > \mathcal{T}} e^{-\int_t^{\mathcal{T}} r_u du} \middle| \mathcal{F}_t \right] = E^{\mathbb{Q}}\left[Y_{\mathcal{T}} e^{-\int_t^{\mathcal{T}} (r_u + \widetilde{\lambda}_u) du} \middle| \mathcal{F}_t \right]. \tag{7.13}$$

Proof

$$E^{\mathbb{Q}}\left[Y_{\mathcal{T}} 1_{\tau > \mathcal{T}} e^{-\int_t^{\mathcal{T}} r_u du} | \mathcal{F}_t \right]$$
$$= E^{\mathbb{Q}}\left[E^{\mathbb{Q}}\left[Y_{\mathcal{T}} 1_{\tau > \mathcal{T}} e^{-\int_t^{\mathcal{T}} r_u du} \middle| \mathcal{F}_T^{\Gamma} \vee \mathcal{F}_t \right] | \mathcal{F}_t \right]$$

Noting that r_t is $\mathcal{F}_t^{\Gamma}$-adapted yields

$$= E^{\mathbb{Q}}\left[Y_{\mathcal{T}} e^{-\int_t^{\mathcal{T}} r_u du} E^{\mathbb{Q}}\left[1_{\tau > \mathcal{T}} | \mathcal{F}_T^{\Gamma} \vee \mathcal{F}_t \right] | \mathcal{F}_t \right]$$
$$= E^{\mathbb{Q}}\left[Y_{\mathcal{T}} e^{-\int_t^{\mathcal{T}} r_u du} e^{-\int_t^{\mathcal{T}} \widetilde{\lambda}_u du} | \mathcal{F}_t \right].$$

This completes the proof.

7.4.2 Cash Flow 2

Random payment rate of $y_s ds$ at time s over $[0, \mathcal{T}]$ that is $\mathcal{F}_s^{\Gamma}$-measurable, but the cash flow occurs only if no default ($\tau > s$). Then,

$$E^{\mathbb{Q}}\left[\int_t^{\mathcal{T}} y_s 1_{\tau > s \geq t} e^{-\int_t^s r_u du} ds \,| \mathcal{F}_t \right] = E^{\mathbb{Q}}\left[\int_t^{\mathcal{T}} y_s e^{-\int_t^s (r_u + \widetilde{\lambda}_u) du} ds \middle| \mathcal{F}_t \right]. \tag{7.14}$$

Proof

$$E^{\mathbb{Q}}\left[\int_t^{\mathscr{T}} y_s 1_{\tau>s} e^{-\int_t^s r_u du} ds \,|\mathscr{F}_t\right]$$

$$= E^{\mathbb{Q}}\left[E^{\mathbb{Q}}\left[\int_t^{\mathscr{T}} y_s 1_{\tau>s} e^{-\int_t^s r_u du} ds \,|\mathscr{F}_T^{\Gamma} \vee \mathscr{F}_t\right] |\mathscr{F}_t\right]$$

Noting that r_t is $\mathscr{F}_t^{\Gamma}$-adapted yields

$$= E^{\mathbb{Q}}\left[\int_t^{\mathscr{T}} y_s e^{-\int_t^s r_u du} E^{\mathbb{Q}}\left[1_{\tau>s}| \mathscr{F}_T^{\Gamma} \vee \mathscr{F}_t\right] ds \,|\mathscr{F}_t\right]$$

$$= E^{\mathbb{Q}}\left[\int_t^{\mathscr{T}} y_s e^{-\int_t^s r_u du} e^{-\int_t^s \widetilde{\lambda}_u du} ds \,|\mathscr{F}_t\right].$$

This completes the proof.

7.4.3 Cash Flow 3

Random cash flow Y_τ which is $\mathscr{F}_\tau^{\Gamma}$-measurable at time τ, but the cash flow occurs only if default has occurred within $[0, \mathscr{T}]$, i.e. $(\tau \leq \mathscr{T})$. Then,

$$E^{\mathbb{Q}}\left[Y_\tau 1_{t<\tau\leq\mathscr{T}} e^{-\int_t^\tau r_u du} \,|\mathscr{F}_t\right] = E^{\mathbb{Q}}\left[\int_t^{\mathscr{T}} Y_s \widetilde{\lambda}_s e^{-\int_t^s (r_u+\widetilde{\lambda}_u) du} \middle| \mathscr{F}_t\right]. \tag{7.15}$$

Proof

$$E^{\mathbb{Q}}\left[Y_\tau 1_{\tau\leq\mathscr{T}} e^{-\int_t^\tau r_u du} \,|\mathscr{F}_t\right]$$

$$= E^{\mathbb{Q}}\left[\int_t^{\mathscr{T}} E^{\mathbb{Q}}\left[Y_s e^{-\int_t^s r_u du} 1_{\tau=s} \,|\mathscr{F}_T^{\Gamma} \vee \mathscr{F}_t\right] ds \,|\mathscr{F}_t\right]$$

Noting that r_t is $\mathscr{F}_t^{\Gamma}$-adapted yields

$$= E^{\mathbb{Q}}\left[\int_t^{\mathscr{T}} Y_s e^{-\int_t^s r_u du} E^{\mathbb{Q}}\left[1_{\tau=s} \,|\mathscr{F}_T^{\Gamma} \vee \mathscr{F}_t\right] ds \,|\mathscr{F}_t\right]$$

$$= E^{\mathbb{Q}}\left[\int_t^{\mathscr{T}} Y_s e^{-\int_t^s r_u du} \widetilde{\lambda}_s e^{-\int_t^s \widetilde{\lambda}_u du} ds \,|\mathscr{F}_t\right].$$

This completes the proof.

7.4.4 Cash Flow 4

Random cash flow $Y_{\mathscr{T}}$ which is $\mathscr{F}^{\Gamma}_{\mathscr{T}}$-measurable at time $\mathscr{T}$, but the cash flow occurs only if no default ($\tau > \mathscr{T}$). If default occurs before $\mathscr{T}$, i.e. ($\tau \leq \mathscr{T}$), then the payment is $\delta_\tau V_{\tau-}$, where δ_τ is $\mathscr{F}^{\Gamma}_{\tau}$-measurable and V_t is the implicit solution to the following expression.

$$V_t = E^{\mathbb{Q}}\left[Y_{\mathscr{T}} 1_{\tau > \mathscr{T}} e^{-\int_t^{\mathscr{T}} r_u du} + \delta_\tau V_{\tau-} 1_{\tau \leq \mathscr{T}} e^{-\int_t^{\tau} r_u du} | \mathscr{F}_t\right].$$

The solution is

$$V_t = E^{\mathbb{Q}}\left[Y_{\mathscr{T}} e^{-\int_t^{\mathscr{T}} r_u du} e^{-\int_t^{\mathscr{T}} \widetilde{\lambda}_u (1-\delta_u) du} | \mathscr{F}_t\right]. \tag{7.16}$$

Proof This proof is due to Lane Hughston and Stuart Turnbull. For the proof define $E^{\mathbb{Q}}[\cdot | \mathscr{F}_t] \equiv E^{\mathbb{Q}}_t$.

Note $\mathbb{B}_t = e^{\int_0^t r_u du}$. Then, $\frac{\mathbb{B}_{\mathscr{T}}}{\mathbb{B}_t} = e^{\int_t^{\mathscr{T}} r_u du}$.

Define $\Lambda_t = e^{-\int_0^t \widetilde{\lambda}_u du}$. Then, $\frac{\Lambda_{\mathscr{T}}}{\Lambda_t} = e^{-\int_t^{\mathscr{T}} \widetilde{\lambda}_u du}$.

$$\begin{aligned}
\frac{V_t 1_{\tau > t}}{\mathbb{B}_t} &= E^{\mathbb{Q}}_t\left[\frac{Y_{\mathscr{T}} 1_{\tau > \mathscr{T}}}{\mathbb{B}_t} + \delta_\tau \frac{V_{\tau-}}{\mathbb{B}_\tau} 1_{t < \tau \leq \mathscr{T}}\right] \\
&= E^{\mathbb{Q}}_t\left(\frac{Y_{\mathscr{T}} 1_{\tau > \mathscr{T}}}{\mathbb{B}_t} + \int_t^{\mathscr{T}} \delta_s \frac{V_{s-}}{\mathbb{B}_s} 1_{\tau = s} ds\right).
\end{aligned}$$

But, $V_{s-} = 1_{\tau > s-} V_{s-}$ on $\{\tau = s\}$. Thus,

$$= E^{\mathbb{Q}}_t\left(\frac{Y_{\mathscr{T}} 1_{\tau > \mathscr{T}}}{\mathbb{B}_t} + \int_t^{\mathscr{T}} \delta_s \frac{V_{s-} 1_{\tau > s-}}{\mathbb{B}_s} 1_{\tau = s} ds\right).$$

Using expression (7.15), we can write this as

$$E^{\mathbb{Q}}_t\left(\frac{Y_{\mathscr{T}}}{\mathbb{B}_t}\frac{\Lambda_{\mathscr{T}}}{\Lambda_t} + \int_t^{\mathscr{T}} \delta_s \frac{V_{s-} 1_{\tau > s-}}{\mathbb{B}_s}\frac{\Lambda_s}{\Lambda_t} \widetilde{\lambda}_s ds\right).$$

Define $\widetilde{V}_t = \frac{V_{s-} 1_{\tau > s-} \Lambda_s}{\mathbb{B}_s}$. Note that $V_{\mathscr{T}} 1_{\tau > \mathscr{T}} = Y_{\mathscr{T}}$.
Also $\Lambda_s = \Lambda_{s-}$ and $\mathbb{B}_s = \mathbb{B}_{s-}$.
We rewrite the last equation as

$$\widetilde{V}_t + \int_0^t \delta_s \widetilde{V}_{s-} \widetilde{\lambda}_s ds = E^{\mathbb{Q}}_t\left[\widetilde{V}_{\mathscr{T}} + \int_0^{\mathscr{T}} \delta_s \widetilde{V}_{s-} \widetilde{\lambda}_s ds\right].$$

Define $M_t = \widetilde{V}_t + \int_0^t \delta_s \widetilde{V}_{s-} \widetilde{\lambda}_s ds$.

The last equation shows that M_t is a $\mathbb{Q}$-martingale, i.e. $M_t = E_t^{\mathbb{Q}}[M_{\mathcal{T}}]$. Rewriting

$$\widetilde{V}_t = M_t - \int_0^t \delta_s \widetilde{V}_{s-} \widetilde{\lambda}_s ds.$$

Or,

$$d\widetilde{V}_t = dM_t - \delta_t \widetilde{V}_{t-} \widetilde{\lambda}_t dt.$$

Define $\frac{d\widetilde{M}_t}{\widetilde{M}_t} = \frac{dM_t}{\widetilde{V}_{t-}}$. Note that $\widetilde{M}_t$ is also a $\mathbb{Q}$-martingale.

We can rewrite this as

$$\frac{d\widetilde{V}_t}{\widetilde{V}_t} = \frac{d\widetilde{M}_t}{\widetilde{M}_t} - \delta_t \widetilde{\lambda}_t dt.$$

Using Ito's formula, we get

$$d \log \widetilde{V}_t = d \log \widetilde{M}_t - \delta_t \widetilde{\lambda}_t dt.$$

The solution is

$$\widetilde{V}_t = \widetilde{M}_t e^{-\int_0^t \delta_s \widetilde{\lambda}_s ds}$$

since $\widetilde{V}_0 = \widetilde{M}_0$.

Substituting back, we get

$$\widetilde{M}_t = V_t e^{-\int_0^t [r_s + (1-\delta_s)\widetilde{\lambda}_s] ds}.$$

As a $\mathbb{Q}$-martingale,

$$\widetilde{M}_t = E_t^{\mathbb{Q}}[\widetilde{M}_T],$$

or

$$V_t 1_{\tau > t} e^{-\int_0^t [r_s + (1-\delta_s)\widetilde{\lambda}_s] ds} = E_t^{\mathbb{Q}}\left[V_{\mathcal{T}} 1_{\tau > \mathcal{T}} e^{-\int_0^{\mathcal{T}} [r_s + (1-\delta_s)\widetilde{\lambda}_s] ds}\right].$$

Or,

$$V_t 1_{\tau > t} = E_t^{\mathbb{Q}}\left[Y_{\mathcal{T}} e^{-\int_t^{\mathcal{T}} [r_s + (1-\delta_s)\widetilde{\lambda}_s] ds}\right].$$

This completes the proof.

Remark 7.2 (Assuming Only $\mathbb{Q} \in \mathfrak{M}_l$ Exists) Instead of assuming that there exists an equivalent martingale measure $\mathbb{Q} \in \mathfrak{M}$ as done above (by the Third Fundamental Theorem 2.5 of asset pricing in Chap. 2 this implies NFLVR and ND), one can relax this structure and only assume that an equivalent local martingale measure $\mathbb{Q} \in \mathfrak{M}_l$ exists where $\mathfrak{M}_l = \left\{\mathbb{Q} \sim \mathbb{P} : \frac{p(t,\mathcal{T})}{\mathbb{B}_t} \text{ for all } \mathcal{T} \in [0,T] \text{ and } \frac{D(t,\mathcal{T})}{\mathbb{B}_t} \text{ are local } \mathbb{Q}\text{-martingales}\right\}$, which by the First Fundamental Theorem 2.3 of asset pricing in Chap. 2 implies just NFLVR.

To get the condition that $\left(\frac{p(t,\mathcal{T})}{\mathbb{B}_t}\right)$ and $\frac{D(t,\mathcal{T})}{\mathbb{B}_t}$ are $\mathbb{Q}$-martingales, we need to exclude price bubbles in these zero-coupon bonds. This can be accomplished, for example, by assuming that these zero-coupon bond prices are bounded above for all t (and using Theorem 3.5 from the asset price bubbles Chap. 3). This follows for default-free zero-coupon bonds if forward rates are nonnegative for all t (see expression (6.1) in the HJM model Chap. 6). Defining forward rates for risky zero-coupon bonds in an analogous fashion, if these are nonnegative for all t, then the same result applies to them as well.

To price any derivative using risk neutral valuation under the equivalent local martingale measure $\mathbb{Q} \in \mathfrak{M}_l$ chosen by the market, one needs to assume

1. that the derivative's payoffs $\frac{X_{\mathcal{T}}}{\mathbb{B}_{\mathcal{T}}} \in L^0_+$ are attainable in an enlarged market that includes trading in derivatives, and
2. that $\frac{X_{\mathcal{T}}}{\mathbb{B}_{\mathcal{T}}} \in L^0_+$ satisfies suitable integrability conditions as in Theorem 2.7 in the fundamental theorems Chap. 2 so that $\frac{X_{\mathcal{T}}}{\mathbb{B}_{\mathcal{T}}} \in L^0_+$ is a $\mathbb{Q}$-martingale.
This completes the remark.

7.5 Examples

This section illustrates the use of the previous formulas for various more complex credit derivatives.

7.5.1 Coupon Bonds

This section studies a coupon bond issued by the firm. A coupon bond has a coupon rate $C \in [0, 1]$, a face value which we assume to be one dollar, and a maturity date T. The bond pays the coupon rate C times the notional at intermediate dates $\{t_1, \ldots, t_m = T\}$, but only up to the default time τ. If default happens prior to the maturity date, we assume the bond pays the recovery rate $R_\tau \in [0, 1]$, which is $\mathscr{F}^{\Gamma}_{\tau}$-measurable. If default does not happen, the face value is repaid at time T.

Denote the time $t \leq t_1$ value of the coupon bond as v_t. Then,

$$v_t = E^{\mathbb{Q}}\left[\sum_{k=1}^{m} C 1_{\{\tau > t_k\}} e^{-\int_t^{t_k} r_u du} + 1_{\{\tau > T\}} e^{-\int_t^T r_u du} | \mathscr{F}_t\right]$$
$$+E^{\mathbb{Q}}\left[R_\tau 1_{\{\tau \leq T\}} e^{-\int_t^\tau r_u du} | \mathscr{F}_t\right]$$
$$= \sum_{k=1}^{m} C E^{\mathbb{Q}}\left[e^{-\int_t^{t_k}(r_u+\widetilde{\lambda}_u)du} | \mathscr{F}_t\right] + E^{\mathbb{Q}}\left[1 e^{-\int_t^T (r_u+\widetilde{\lambda}_u)du} | \mathscr{F}_t\right]$$
$$+ \int_t^T E^{\mathbb{Q}}\left[R_s \widetilde{\lambda}_s e^{-\int_t^s (r_u+\widetilde{\lambda}_u)du} | \mathscr{F}_t\right] ds. \tag{7.17}$$

If the bond is priced at par, then the coupon rate C is such that this expression satisfies $v_t = 1$.

In general, by comparing expression (7.10) with expression (7.17), we see that the bond's price is not equal to the sum of the promised cash payments multiplied by the risky zero-coupon bond prices, i.e.

$$v_t \neq \sum_{k=1}^{m} C D(t, t_k) + D(t, T). \tag{7.18}$$

We note, however, that if either

1. $R(\tau, \mathscr{T}) = \delta$ for all $\mathscr{T}$, $R_\tau = \delta\left(\sum_{k=1}^{m} C + 1\right)$ where δ is a constant (recovery of face value), or
2. $R(\tau, \mathscr{T}) = \delta p(\tau, \mathscr{T})$ for all $\mathscr{T}$, $R_\tau = \delta\left(\sum_{k=1}^{m} C p(\tau, t_k) + p(\tau, T)\right)$ where δ is a constant (recovery of Treasury), or
3. $R(\tau, \mathscr{T}) = \delta p(\tau-, \mathscr{T})$ for all $\mathscr{T}$, $R_\tau = \delta v_{\tau-}$ where δ is a constant (recovery of market value), then

$$v_t = \sum_{k=1}^{m} C D(t, t_k) + D(t, T). \tag{7.19}$$

The proof of this statement follows directly from expressions (7.11), (7.17), and algebraic manipulation. General necessary and sufficient conditions for the satisfaction of expression (7.19) are contained in Jarrow [84].

Expression (7.19) is an important result. Similar to default-free zero-coupon bonds, it facilitates the estimation of credit risk yield curves using coupon bonds and smoothing procedures.

7.5.2 *Credit Default Swaps (CDS)*

This section studies a credit default swap (CDS) on corporate debt. A CDS has a maturity date T and a notional value, which we assume to be one dollar. The protection seller agrees to pay the protection buyer the difference between the face value of the debt and the recovery value at default, if the firm defaults before the swap's maturity date. In return, the protection buyer pays a constant dollar spread times the notional at fixed intermediate dates until the swap's maturity or the default date, whichever comes first.

Consider a coupon bearing bond, as in the previous section, with recovery rate $R_\tau \geq 0$ which is $\mathscr{F}_\tau^\Gamma$-measurable. The protection seller pays at default:

$$L_\tau = \begin{cases} 1 - R_\tau \text{ if } & \tau \leq T, \\ 0 & \text{otherwise},. \end{cases} \tag{7.20}$$

The protection buyer pays a constant dollar spread c times the notional at the intermediate dates $\{t_1, \ldots, t_m = T\}$, but only up to the default time τ. The time $t \leq t_1$ value of the CDS to the protection seller is therefore

$$E^{\mathbb{Q}}\left[\sum_{k=1}^{m} c 1_{\{\tau > t_k\}} e^{-\int_t^{t_k} r_u du} - L_\tau 1_{\{\tau \leq T\}} e^{-\int_t^{\tau} r_u du} \mid \mathscr{F}_t\right] \tag{7.21}$$

$$= c \sum_{k=1}^{m} E^{\mathbb{Q}}\left[e^{-\int_t^{t_k}(r_u + \widetilde{\lambda}_u)du} \mid \mathscr{F}_t\right] - \int_t^T E^{\mathbb{Q}}\left[L_s \widetilde{\lambda}_s e^{-\int_t^s (r_u + \widetilde{\lambda}_u)du} \mid \mathscr{F}_t\right] ds.$$

The proof uses expressions (7.13) and (7.15).

Note that the protection seller's position is similar to that of being long a coupon bond, except that: (i) the value of the bond is not exchanged at time t (the start date), and (ii) the notional (face) value's cash flow is not included if the bond does not default. This difference helps explain why a swap has zero value at initiation, but a coupon bond's value if priced at par is unity.

The *market clearing CDS rate* c_{mkt} makes this value zero, i.e.

$$c_{mkt} = \frac{\int_t^T E^{\mathbb{Q}}\left[L_s \widetilde{\lambda}_s e^{-\int_t^s (r_u + \widetilde{\lambda}_u)du} \mid \mathscr{F}_t\right] ds}{\sum_{k=1}^{m} E^{\mathbb{Q}}\left[e^{-\int_t^{t_k}(r_u + \widetilde{\lambda}_u)du} \mid \mathscr{F}_t\right]}. \tag{7.22}$$

Example 7.1 (CDS Rate Rule of Thumb) If $L_s = L$ and $\widetilde{\lambda}_s = \widetilde{\lambda}$ are constants, then this expression simplifies to

$$c_{mkt} = \frac{L\widetilde{\lambda} \int_t^T E^{\mathbb{Q}}\left[e^{-\int_t^s (r_u + \widetilde{\lambda})du} \mid \mathscr{F}_t\right] ds}{\sum_{k=1}^{m} E^{\mathbb{Q}}\left[e^{-\int_t^{t_k}(r_u + \widetilde{\lambda})du} \mid \mathscr{F}_t\right]} \approx L\widetilde{\lambda}. \tag{7.23}$$

Here, we see that the CDS rate is equal to the expected loss rate, per year. This CDS rate rule of thumb is often used in the financial press. This completes the example.

If the loss rate L_s is a constant and both $\widetilde{\lambda}_s, r_s$ follow an affine process, then an analytic solution for the CDS rate can be obtained (see Lando [131]). Otherwise, numerical methods need to be employed to evaluate this expression.

7.5.3 First-to-Default Swaps

This section studies a simple basket credit derivative on n firms called a *first-to-default* swap. This swap has a maturity date T and a notional value, which we assume to be one dollar. The protection seller agrees to pay the protection buyer a cash flow if any of the firms in the basket default before the swap's maturity date. But, only one payment. This payment can depend on the firm defaulting. In return, the protection buyer pays a constant dollar spread times the notional at fixed intermediate dates until the swap's maturity or the first default date, whichever comes first.

We consider a collection of n firms. We let τ_i denote the default time for firm i generated by the Cox process $1_{\tau_i \geq t}$ with intensity $\lambda_i(t) = \lambda_i(t, \Gamma_t) \geq 0$. We assume that conditioned upon $\mathscr{F}_T^{\Gamma}$ the point processes $1_{\tau_i \geq t}$ are independent across firms.

Let $D_i(t, T)$ denote the time t value of firm i's zero-coupon bond with maturity T and a face value of one dollar.

We assume that all of these risky zero-coupon bonds trade in a market with an equivalent martingale measure $\mathbb{Q} \in \mathfrak{M}$ that is chosen by the market.

Let $\widehat{\tau} \equiv \min\{\tau_1, \ldots, \tau_n\}$ denote the time of the first firm's default, and $\widehat{i} \equiv \arg\min\{\tau_1, \ldots, \tau_n\}$ denote the first defaulting firm. The protection seller's payoff is

$$1_{\{\widehat{\tau} \leq T\}} L_{\widehat{i}}(\widehat{\tau}) = \begin{cases} L_{\widehat{i}}(\widehat{\tau}) \text{ if } & \widehat{\tau} \leq T \\ 0 & \text{otherwise} \end{cases} \tag{7.24}$$

where $L_i(\widehat{\tau})$ represents the payment made on the swap if firm i is the first to default at time $\widehat{\tau}$. $L_i(\widehat{\tau})$ is assumed to be $\mathscr{F}_{\widehat{\tau}}^{\Gamma}$-measurable.

The protection buyer pays a constant dollar spread c times the notional at the intermediate dates $\{t_1, \ldots, t_m = T\}$, but only up to the $\min\{\widehat{\tau}, T\}$. The time $t \leq t_1$ value of the first-to-default swap to the protection seller is therefore

$$E^{\mathbb{Q}}\left[\sum_{k=1}^{m} c 1_{\{\widehat{\tau} > t_k\}} e^{-\int_t^{t_k} r_u du} - L_{\widehat{i}}(\widehat{\tau}) 1_{\{\widehat{\tau} \leq T\}} e^{-\int_t^{\widehat{\tau}} r_u du} \middle| \mathscr{F}_t\right]$$

$$= c \sum_{k=1}^{m} E^{\mathbb{Q}}\left[e^{-\int_t^{t_k}\left(r_u + \sum_{j=1}^{n} \widetilde{\lambda}_j(u)\right)du} \middle| \mathscr{F}_t\right]$$

$$- \sum_{i=1}^{n} \int_t^T E^{\mathbb{Q}}\left[L_i(s) \widetilde{\lambda}_i(s) e^{-\int_t^s \left(r_u + \sum_{j=1}^{n} \widetilde{\lambda}_j(u)\right)du} \middle| \mathscr{F}_t\right] ds. \tag{7.25}$$

Proof For the proof define $E^{\mathbb{Q}}[\cdot\,|\mathscr{F}_t] \equiv E_t^{\mathbb{Q}}$.

The protection buyer pays

$$E_t^{\mathbb{Q}}\left[E^{\mathbb{Q}}\left[1_{\{\widehat{\tau}>T\}}e^{-\int_t^T r_u du}\,|\mathscr{F}_T^{\Gamma}\right]\right]$$

$$= E_t^{\mathbb{Q}}\left[E^{\mathbb{Q}}\left[1_{\{\tau_1>T\}}\cdots 1_{\{\tau_n>T\}}e^{-\int_t^T r_u du}\,|\mathscr{F}_T^{\Gamma}\right]\right]$$

and by conditional independence

$$= E_t^{\mathbb{Q}}\left[E^{\mathbb{Q}}\left[1_{\{\tau_1>T\}}\,|\mathscr{F}_T^{\Gamma}\right]\cdots E^{\mathbb{Q}}\left[1_{\{\tau_n>T\}}\,|\mathscr{F}_T^{\Gamma}\right]e^{-\int_t^T r_u du}\right]$$

using (7.3)

$$= E_t^{\mathbb{Q}}\left[e^{-\int_t^T \widetilde{\lambda}_1(u)du}\cdots e^{-\int_t^T \widetilde{\lambda}_n(u)du}e^{-\int_t^T r_u du}\right]$$

$$= E_t^{\mathbb{Q}}\left[e^{-\int_t^T r_u du}e^{-\sum_{i=1}^n \int_t^T \widetilde{\lambda}_i(u)du}\right].$$

To compute the protection seller's payment, note that

$$\mathbb{Q}_t(i=\widehat{i},\tau_i\in[s,s+dt)\big|\,\mathscr{F}_T^{\Gamma})$$

$$= \mathbb{Q}_t(\tau_i\in[s,s+dt),\tau_j>s \text{ for } i\neq j\big|\,\mathscr{F}_T^{\Gamma})$$

and by conditional independence

$$= \mathbb{Q}_t(\tau_i\in[s,s+dt)|\,\mathscr{F}_T^{\Gamma})\mathbb{Q}_t(\tau_j>s \text{ for } i\neq j\big|\,\mathscr{F}_T^{\Gamma})$$

$$= \mathbb{Q}_t(\tau_i\in[s,s+dt)|\,\mathscr{F}_T^{\Gamma})\prod_{i\neq j}e^{-\int_t^s \widetilde{\lambda}_j(u)du}$$

$$= \widetilde{\lambda}_i(s)e^{-\int_t^s \widetilde{\lambda}_i(u)du}\prod_{i\neq j}e^{-\int_t^s \widetilde{\lambda}_j(u)du} = \widetilde{\lambda}_i(s)\prod_{j=1}^n e^{-\int_t^s \widetilde{\lambda}_j(u)du}.$$

Finally,

$$E_t^{\mathbb{Q}}\left[L_{\widehat{i}}(\widehat{\tau})1_{\{\widehat{\tau}\leq T\}}e^{-\int_t^{\widehat{\tau}} r_u du}\,|\mathscr{F}_T^{\Gamma}\right]$$

$$= \sum_{i=1}^n E_t^{\mathbb{Q}}\left[L_i(\tau_i)1_{\{\tau_i\leq T\}}e^{-\int_t^{\tau_t} r_u du}\,|i=\widehat{i}\vee\mathscr{F}_T^{\Gamma}\right]\mathbb{Q}_t(i=\widehat{i}\,|\mathscr{F}_T^{\Gamma})$$

$$= \sum_{i=1}^n \int_t^T E_t^{\mathbb{Q}}\left[L_i(s)e^{-\int_t^s r_u du}\,|i=\widehat{i},\tau_i\in[s,s+dt)\vee\mathscr{F}_T^{\Gamma}\right]$$

$$\times\,\mathbb{Q}_t(i=\widehat{i},\tau_i\in[s,s+dt)\,|\mathscr{F}_T^{\Gamma})ds.$$

By the measurability of $L_{\widehat{i}}(\widehat{\tau})$ and $e^{-\int_t^{\widehat{\tau}} r_u du}$ we have

$$= \int_t^T L_i(s) e^{-\int_t^s r_u du} \sum_{i=1}^n \mathbb{Q}(i = \widehat{i}, \tau_i \in [s, s+dt) \,|\mathscr{F}_T^r) ds$$

$$= \sum_{i=1}^n \int_t^T L_i(s) \widetilde{\lambda}_i(s) e^{-\int_t^s r_u du} \prod_{j=1}^n e^{-\int_t^s \widetilde{\lambda}_j(u) du} ds$$

$$= \sum_{i=1}^n \int_t^T L_i(s) \widetilde{\lambda}_i(s) e^{-\int_t^s \left(r_u + \sum_{j=1}^n \widetilde{\lambda}_j(u)\right) du} ds.$$

Taking expectations $E^{\mathbb{Q}}[\cdot\,|\mathscr{F}_t] \equiv E_t^{\mathbb{Q}}$ completes the proof.

The *market clearing swap rate c* makes this value zero, i.e.

$$c_{mkt} = \frac{\sum_{i=1}^n \int_t^T E^{\mathbb{Q}}\left[L_i(s) \widetilde{\lambda}_i(s) e^{-\int_t^s \left(r_u + \sum_{j=1}^n \widetilde{\lambda}_j(u)\right) du} \middle| \mathscr{F}_t \right] ds}{\sum_{k=1}^m E^{\mathbb{Q}}\left[e^{-\int_t^{t_k} \left(r_u + \sum_{j=1}^n \widetilde{\lambda}_j(u)\right) du} \middle| \mathscr{F}_t \right]}. \tag{7.26}$$

Note that if the payment rate $L_i(s)$ is a constant and both $\widetilde{\lambda}_s, r_s$ follow an affine process, then an analytic solution for this swap rate can be obtained. Otherwise, as with CDS, numerical methods need to be employed.

7.6 Notes

Much more is known about credit risk and credit derivatives. Good textbooks in this regard are Bielecki and Rutkowski [12], Bluhm and Overbeck [17], Bluhm et al. [18], Duffie and Singleton [54], Lando [131], and Schonbucher [166].

Chapter 8
Incomplete Markets

This chapter studies the arbitrage-free pricing of derivatives in an incomplete market satisfying NFLVR. This chapter is a modest generalization of the presentation contained in Pham [149] to discontinuous risky asset price processes.

8.1 The Set-Up

Given is a normalized market $(S, (\mathscr{F}_t), \mathbb{P})$ where the value of a money market account $B_t \equiv 1$ for all t. We assume that there are no arbitrage opportunities in the market, i.e.

Assumption (NFLVR)

$$\mathfrak{M}_l \neq \emptyset$$

where $\mathfrak{M}_l = \{\mathbb{Q} \sim \mathbb{P} : S \text{ is a } \mathbb{Q}\text{-local martingale}\}$.

In this chapter the market is incomplete. In an incomplete market, for a given $\mathbb{Q} \in \mathfrak{M}_l$, there exist integrable derivatives $X_T \in L^1_+(\mathbb{Q})$ that may not be attainable or if attainable, the value process implied by the admissible s.f.t.s. may not be a $\mathbb{Q}$-martingale. If the set of equivalent local martingale measures is not a singleton, then the cardinality of $|\mathfrak{M}_l| = \infty$, i.e. the set contains an infinite number of elements. Thus, there usually is no unique arbitrage free (NFLVR) price for a derivative. In this case, the best we can do is to determine upper and lower bounds for these derivative prices, based on super- and sub-replication. This is the task to which we now turn.

R. A. Jarrow, *Continuous-Time Asset Pricing Theory*, Springer Finance,
https://doi.org/10.1007/978-3-319-77821-1_8

8.2 The Super-Replication Cost

This section derives the super-replication cost of an arbitrary derivative on the risky assets. The *super-replication cost* $\bar{c}_0$ is defined to be the minimum investment required if one sells the derivative $Z_T \in L_+^0$ and forms an admissible s.f.t.s. to cover the derivative's payoffs at time T for all $\omega \in \Omega$ a.s. $\mathbb{P}$, i.e.

$$\bar{c}_0 = \inf\{x \in \mathbb{R} : \exists(\alpha_0, \alpha) \in \mathscr{A}(x),\ x + \int_0^T \alpha_t \cdot dS_t \geq Z_T\}. \tag{8.1}$$

We now characterize the super-replication cost of an arbitrary derivative security using the set of equivalent local martingale measures $\mathbb{Q} \in \mathfrak{M}_l$.

Theorem 8.1 (Super-Replication Cost) *Let $Z_T \in L_+^1(\mathbb{Q})$ for all $\mathbb{Q} \in \mathfrak{M}_l$.*
The super-replication cost is

$$\bar{c}_0 = \sup_{\mathbb{Q} \in \mathfrak{M}_l} E^{\mathbb{Q}}[Z_T] \geq 0. \tag{8.2}$$

At time t, the super-replication cost is

$$\bar{c}_t = \operatorname{ess\,sup}_{\mathbb{Q} \in \mathfrak{M}_l} E^{\mathbb{Q}}[Z_T \,|\mathscr{F}_t] \geq 0. \tag{8.3}$$

Furthermore, there exists an admissible s.f.t.s. $(\alpha_0, \alpha) \in \mathscr{A}(\bar{c}_0)$ with cash flows, an adapted right continuous nondecreasing process A_t with $A_0 = 0$, and a value process $X_t \equiv \alpha_0(t) + \alpha_t \cdot S_t$ such that

$$X_t = \bar{c}_0 + \int_0^t \alpha_s \cdot dS_s - A_t = \bar{c}_t$$

for all $t \in [0, T]$.

The admissible s.f.t.s. $(\alpha_0, \alpha) \in \mathscr{A}(\bar{c}_0)$ with cash flows is the super-replication trading strategy.

The self-financing condition with cash flows implies that $\bar{c}_0 + \int_0^t \alpha_s \cdot dS_s = \alpha_0(t) + \alpha_t \cdot S_t + A_t$ for all t.

The adapted process A corresponds to the cash one can withdraw from the trading strategy across time and still obtain the derivative's payoff at time T.

Proof Define

$$\mathscr{C}_1(x) = \{Z_T \in L_+^0 : \sup_{\mathbb{Q} \in \mathfrak{M}_l} E^{\mathbb{Q}}[Z_T] \leq x\}.$$

$$\mathscr{C}_2(x) = \{Z_T \in L_+^0 : \exists(\alpha_0, \alpha) \in \mathscr{A}(x),\ x + \int_0^T \alpha_t \cdot dS_t \geq Z_T\}.$$

(Step 1) Show $\mathscr{C}_2(x) = \mathscr{C}_1(x)$.

(Step 1, Part 1) Show $(\mathscr{C}_2(x) \subset \mathscr{C}_1(x))$.

Take $Z_T \in \mathscr{C}_2(x)$, then $\exists(\alpha_0, \alpha) \in \mathscr{A}(x)$ such that $x + \int_0^T \alpha_t \cdot dS_t \geq Z_T$.

Note that $\int_0^t \alpha_u \cdot dS_u$ is a local martingale under $\mathbb{Q}$ that is bounded below. Hence a supermartingale (Lemma 1.3 in Chap. 1), i.e. $E^{\mathbb{Q}}[\int_0^T \alpha_t \cdot dS_t] \leq 0$.

This implies $E^{\mathbb{Q}}[Z_T] \leq x$. Since this is true for all $\mathbb{Q} \in \mathfrak{M}_l$, $\sup_{\mathbb{Q}\in\mathfrak{M}_l} E^{\mathbb{Q}}[Z_T] \leq x$, i.e. $Z_T \in \mathscr{C}_1(x)$.

(Step 1, Part 2) Show $(\mathscr{C}_1(x) \subset \mathscr{C}_2(x))$.

Take $Z_T \in \mathscr{C}_1(x)$. Define $Z_t = \operatorname{ess\,sup}_{\mathbb{Q}\in\mathfrak{M}_l} E^{\mathbb{Q}}[Z_T | \mathscr{F}_t] \geq 0$.

(Step 1, Part 2a) From Theorem 1.4 in Chap. 1 we know that Z_t is a supermartingale for all $\mathbb{Q} \in \mathfrak{M}_l$.

(Step 1, Part 2b) The optional decomposition Theorem 1.5 in Chap. 1 implies that $\exists\alpha \in \mathscr{L}(S)$ and an adapted right continuous nondecreasing process A_t with $A_0 = 0$ such that $X_t \equiv Z_0 + \int_0^t \alpha_u \cdot dS_u - A_t = Z_t$ for all t.

Let α_t be the position in the risky assets. Define the position in the mma $\alpha_0(t)$ by the expression $X_t = \alpha_0(t) + \alpha_t \cdot S_t$ for all t. By the definition of X_t we get that $Z_0 + \int_0^t \alpha_u \cdot dS_u = \alpha_0(t) + \alpha_t \cdot S_t + A_t$. Hence, the trading strategy is self-financing with cash flow A_t and initial value $x = Z_0$.

We have $Z_0 + \int_0^t \alpha_u \cdot dS_u = X_t + A_t \geq Z_t \geq 0$ since $A_t \geq 0$ and $Z_t \geq 0$. Hence, $Z_0 + \int_0^t \alpha_u \cdot dS_u$ is bounded below by zero, and is therefore admissible, i.e. $(\alpha_0, \alpha) \in \mathscr{A}(x)$ with $x = Z_0$ (this argument extends to $Z_t \geq k \in \mathbb{R}$). Thus, $Z_T \in \mathscr{C}_2(x)$.

(Step 2)

$$
\begin{aligned}
\bar{c}_0 &= \inf\{x \in \mathbb{R} : \exists(\alpha_0, \alpha) \in \mathscr{A}(x),\ x + \int_0^T \alpha_t \cdot dS_t \geq Z_T\} \\
&= \inf\{x \in \mathbb{R} : Z_T \in \mathscr{C}_2(x)\} = \inf\{x \in \mathbb{R} : Z_T \in \mathscr{C}_1(x)\} \\
&= \inf\{x \in \mathbb{R} : \sup_{\mathbb{Q}\in\mathfrak{M}_l} E^{\mathbb{Q}}[Z_T] \leq x\} = \sup_{\mathbb{Q}\in\mathfrak{M}_l} E^{\mathbb{Q}}[Z_T].
\end{aligned}
$$

This completes the proof.

The admissible s.f.t.s. $(\alpha_0, \alpha) \in \mathscr{A}(\bar{c}_0)$ with cash flows satisfying the conclusions of the theorem needs some explanation. The terminology "self-financing" is somewhat misleading in this case. It is only self-financing to the extent that the trading strategy generates both the sum of the value process $X_t = \alpha_0(t) + \alpha_t \cdot S_t$ and the cash flow process A_t, i.e.

$$
\bar{c}_0 + \int_0^t \alpha_s \cdot dS_s = \alpha_0(t) + \alpha_t \cdot S_t + A_t.
$$

Remark 8.1 (Z_T Bounded Below) Important in this proof is that the derivative's time T payoff is nonnegative, i.e. $Z_T \geq 0$. The proof can be extended to any $Z_T \in L^0$ bounded below by a constant $k \in \mathbb{R}$. This completes the remark.

8.3 The Super-Replication Trading Strategy

Theorem 8.1 from the previous section gives the super-replication cost and guarantees the existence of an admissible s.f.t.s. that attains this super-replication cost. To determine the super-replication trading strategy, one uses the insights from the synthetic construction of a derivative's payoffs discussed in the fundamental theorems Chap. 2 Sect. 2.6.3. We illustrate this construction.

Choose a derivative $Z_T \in L^1_+(\mathbb{Q})$ for all $\mathbb{Q} \in \mathfrak{M}_l$. By Theorem 8.1, we know there exists an admissible s.f.t.s $(\alpha_0, \alpha) \in \mathscr{A}(\bar{c}_0)$, an adapted right continuous nondecreasing process A_t with $A_0 = 0$, and a value process $X_t = \alpha_0(t) + \alpha_t \cdot S_t$ such that

$$X_t = \bar{c}_0 + \int_0^t \alpha_s \cdot dS_s - A_t = \bar{c}_t$$

for all t. The stochastic process A_t corresponds to cash flows that can be withdrawn from the trading strategy and still match the derivative's payoffs at time T.

The initial investment in the strategy is $X_0 = \bar{c}_0$. Given the holdings in the risky assets α_t and the cash flow A_t, the position in the mma $\alpha_0(t)$ is determined from the self-financing condition $\bar{c}_0 + \int_0^t \alpha_s \cdot dS_s = \alpha_0(t) + \alpha_t \cdot S_t + A_t$ for all t. To determine the holdings in the risky assets α_t and the cash flow A_t, we suppose that S is a Markov diffusion process. In this case, by the diffusion assumption, S is a continuous process. Then, we can write the cost of super-replication,

$$\operatorname*{ess\ sup}_{\mathbb{Q} \in \mathfrak{M}_l} E^{\mathbb{Q}}[Z_T | \mathscr{F}_t] = \bar{c}_t(S_t),$$

as a deterministic function of t and S_t. Using Ito's formula on this expression (assuming the appropriate differentiability conditions are satisfied by $\bar{c}_t(S_t)$) yields

$$\bar{c}_t(S_t) - \bar{c}_0 = \int_0^T \frac{\partial \bar{c}_t}{\partial t} dt + \frac{1}{2} \int_0^T \sum_{i=1}^n \sum_{j=1}^n \frac{\partial^2 \bar{c}_t}{\partial S_{ij}^2} d \langle S_i(t), S_j(t) \rangle + \int_0^T \frac{\partial \bar{c}_t}{\partial S} \cdot dS_t,$$

where $\frac{\partial \bar{c}}{\partial S} = (\frac{\partial \bar{c}}{\partial S_1}, \ldots, \frac{\partial \bar{c}}{\partial S_n})' \in \mathbb{R}^n$. Matching the integrands of the stochastic integral involving dS_t implies that the holdings in the risky assets are given by

$$\frac{\partial \bar{c}_t(S_t)}{\partial S_t} = \alpha_t.$$

Finally, the cash flows are given by

$$A_t = -\left(\int_0^T \frac{\partial \bar{c}_t}{\partial t} dt + \frac{1}{2} \int_0^T \sum_{i=1}^n \sum_{j=1}^n \frac{\partial^2 \bar{c}_t}{\partial S_{ij}^2} d\left\langle S_i(t), S_j(t) \right\rangle \right).$$

If the underlying asset price process is not Markov, then for a large class of processes Malliavin calculus can be used to determine α_t (see Nunno et al. [51]).

8.4 The Sub-Replication Cost

This section derives the sub-replication cost of an arbitrary derivative on the risky assets. Similar to the section on super-replication, the *sub-replication cost* $\underline{c}_0$ is the maximum amount one can borrow to buy the derivative $Z_T \in L_+^0$ and form an admissible s.f.t.s. that covers the borrowing for all $\omega \in \Omega$ a.s. $\mathbb{P}$, i.e.

$$\underline{c}_0 = \sup\{x \in \mathbb{R} : \exists(\alpha_0, \alpha) \in \mathscr{A}(x),\ x + \int_0^T \alpha_t \cdot dS_t \leq Z_T\}. \tag{8.4}$$

As before, we can characterize the sub-replication cost using the set of equivalent local martingale measures $\mathbb{Q} \in \mathfrak{M}_l$.

Theorem 8.2 (Sub-Replication Cost) *Let $Z_T \in L_+^1(\mathbb{Q})$ for all $\mathbb{Q} \in \mathfrak{M}_l$ be bounded from above.*

The sub-replication cost is

$$\underline{c}_0 = \inf_{\mathbb{Q} \in \mathfrak{M}_l} E^{\mathbb{Q}}[Z_T]. \tag{8.5}$$

At time t, the sub-replication cost is

$$\underline{c}_t = \operatorname{ess\,inf}_{\mathbb{Q} \in \mathfrak{M}_l} E^{\mathbb{Q}}[Z_T | \mathscr{F}_t] \geq 0. \tag{8.6}$$

Furthermore, there exists an admissible s.f.t.s. $(\alpha_0, \alpha) \in \mathscr{A}(\underline{c}_0)$ with cash flows, an adapted right continuous nondecreasing process A_t with $A_0 = 0$, and a value process $X_t \equiv \alpha_0(t) + \alpha_t \cdot S_t$ such that

$$X_t = \underline{c}_0 + \int_0^t \alpha_s dS_s + A_t = \underline{c}_t$$

for all $t \in [0, T]$.

The admissible s.f.t.s. $(\alpha_0, \alpha) \in \mathscr{A}(\underline{c}_0)$ with cash flows is the sub-replication trading strategy.

The self-financing condition with cash flows implies that $\underline{c}_0 + \int_0^t \alpha_s dS_s = \alpha_0(t) + \alpha_t \cdot S_t - A_t$ *for all* t.

The adapted process A *corresponds to the additional cash one must add to the trading strategy across time to obtain the derivative's payoff at time* T.

Proof Consider $0 \geq -Z_T \geq -k$ for some constant $k > 0$.

Note that

$$\underline{c}_0 = \sup\{x \in \mathbb{R} : \exists(\alpha_0, \alpha) \in \mathscr{A}(x),\ x + \int_0^T \alpha_t \cdot dS_t \leq Z_T\}$$

$$= -\inf\{-x \in \mathbb{R} : \exists(\alpha_0, \alpha) \in \mathscr{A}(x),\ -\left(x + \int_0^T \alpha_t \cdot dS_t\right) \geq -Z_T\}.$$

An application of Theorem 8.1 to $-Z_T$ using the fact that

$$-\text{ess}\sup_{\mathbb{Q} \in \mathfrak{M}_l} E^{\mathbb{Q}}[-Z_T \,|\mathscr{F}_t] = \text{ess}\inf_{\mathbb{Q} \in \mathfrak{M}_l} E^{\mathbb{Q}}[Z_T \,|\mathscr{F}_t]$$

completes the proof.

Note the condition that the derivative's time T payoff is bounded above in the hypothesis of this theorem. This condition is needed in the proof to maintain the admissibility condition when shorting the risky assets. The determination of the sub-replication trading strategy is done in the same manner as for a super-replication trading strategy and, therefore, no additional discussion is needed.

Corollary 8.1 (Arbitrage-Free Price Range) *Given* $Z_T \in L^1_+(\mathbb{Q})$ *for all* $\mathbb{Q} \in \mathfrak{M}_l$ *which is bounded above, the arbitrage-free price range is*

$$\left[\underline{c}_0 = \inf_{\mathbb{Q} \in \mathfrak{M}_l} E^{\mathbb{Q}}[Z_T], \quad \sup_{\mathbb{Q} \in \mathfrak{M}_l} E^{\mathbb{Q}}[Z_T] = \bar{c}_0\right].$$

Remark 8.2 (Bubbles) This corollary is very general. We have not assumed the existence of a martingale measure, hence, the market could exhibit bubbles. Alternatively stated, the market may have dominated assets (by the Third Fundamental Theorem 2.5 of asset pricing in Chap. 2). This completes the remark.

8.5 Notes

Additional readings on super- and sub-replication can be found in the textbooks by Dana and Jeanblanc [42], Karatzas and Shreve [118], and Pham [149]. For alternative pricing methods in incomplete markets, see Cont and Tankov [34].

Part II
Portfolio Optimization

Overview

Beyond the insights obtained from the three fundamental theorems of asset pricing, portfolio optimization generates results related to optimal trading strategies and consumption flows. In addition, portfolio optimization also characterizes the equivalent local martingale measure with respect to an investor's beliefs, preferences, and endowment.

We emphasize the martingale duality approach (with convex optimization) in these lectures because it is the natural extension of the martingale approach for pricing and hedging (the synthetic construction) of derivatives. Indeed, this approach to portfolio optimization transforms the dynamic optimization problem into two steps.

(Step 1) A "static" one period optimization problem for the optimal wealth at time T or equivalently, the optimal "derivative". This static optimization problem can be solved using the methods of convex optimization.

(Step 2) After determining the optimal "derivative," the (optimal) trading strategy that constructs this "derivative" can be determined using the standard approach for constructing a synthetic derivative in a complete market (using martingale representation). In an incomplete market, the only change is that one restricts the analysis to those time T wealths (or, equivalently "derivatives") that are attainable (and the complete market argument applies again).

Although the traditional techniques of dynamic programming can be employed, these are typically more restrictive because they impose a Markov process assumption on the evolution of the traded securities. In addition, the solution is usually only characterizable implicitly via a partial differential equation subject to boundary conditions. Last, the connection to derivative pricing and the synthetic construction of a derivative is hidden in the traditional approach.

Chapter 9
Utility Functions

This chapter studies an investor's utility function. We start with a normalized market $(S, (\mathscr{F}_t), \mathbb{P})$ where the money market account's (mma's) value is $B_t \equiv 1$. For the chapters in Part I of this book, although unstated, we implicitly assumed that the *trader's beliefs* were equivalent to the statistical probability measure $\mathbb{P}$, i.e. the trader's beliefs and the statistical probability measure agree on zero probability events. For this part of the book, Part II, we let the probability measure $\mathbb{P}$ correspond to the trader's beliefs. This should cause no confusion since we do not need additional notation for the statistical probability measure in this part of the book. We discuss differential beliefs in Sect. 9.7 below. In addition, consistent with this interpretation, we let the information filtration $\mathscr{F}_t$ given above correspond to the trader's information set. When we study the notion of an equilibrium in Part III of this book, we will introduce a distinction between the trader's beliefs and the statistical probability measure, and a distinction between the trader's information set and the market's information set.

9.1 Preference Relations

Given is a measurable space $(\Omega, \mathscr{F})$. The choice set for our trader is the set of nonnegative $\mathscr{F}$-measurable random variables on this measurable space $X \in L^0_+ \equiv L^0_+(\Omega, \mathscr{F}, \mathbb{P})$. These random variables are interpreted as the possible terminal wealths at time T, denominated in units of the mma. We assume that traders are endowed with a preference relation over the elements in the choice set, denoted by $\overset{\rho}{\succeq}$, where $X_1 \overset{\rho}{\succeq} X_2$ means that X_1 is preferred or indifferent to X_2. We define indifference $\overset{\rho}{\sim}$ as $X_1 \overset{\rho}{\sim} X_2$ if and only $X_1 \overset{\rho}{\succeq} X_2$ and $X_2 \overset{\rho}{\succeq} X_1$. We define strict preference $\overset{\rho}{\succ}$ as $X_1 \overset{\rho}{\succ} X_2$ if and only if $X_1 \overset{\rho}{\succeq} X_2$ and not $X_1 \overset{\rho}{\sim} X_2$.

R. A. Jarrow, *Continuous-Time Asset Pricing Theory*, Springer Finance,
https://doi.org/10.1007/978-3-319-77821-1_9

The preference relation $\overset{\rho}{\succeq}$ on L_0^+ may satisfy certain properties. The subsequent six properties are called rationality axioms.

1. *Reflexivity*

   For any $X \in L_+^0$, $X \overset{\rho}{\succeq} X$.
2. *Comparability*

   For any $X_1, X_2 \in L_+^0$, either $X_1 \overset{\rho}{\succeq} X_2$ or $X_2 \overset{\rho}{\succeq} X_1$.
3. *Transitivity*

   For any $X_1, X_2, X_3 \in L_+^0$, if $X_1 \overset{\rho}{\succeq} X_2$ and $X_2 \overset{\rho}{\succeq} X_3$, then $X_1 \overset{\rho}{\succeq} X_3$.
4. *Order Preserving*

   For any $X_1, X_2 \in L_+^0$ where $X_1 \overset{\rho}{\succ} X_2$ and $\alpha_1, \alpha_2 \in (0, 1)$, $[\alpha_1 X_1 + (1 - \alpha_1) X_2] \overset{\rho}{\succ} [\alpha_2 X_1 + (1 - \alpha_2) X_2]$ if and only if $\alpha_1 > \alpha_2$.
5. *Intermediate Value*

   For any $X_1, X_2, X_3 \in L_+^0$, if $X_1 \overset{\rho}{\succ} X_2 \overset{\rho}{\succ} X_3$ then there exists a unique $\alpha \in (0, 1)$ such that $X_2 \overset{\rho}{\sim} \alpha X_1 + (1 - \alpha) X_3$.
6. *Strong Independence*

   For any $X_1, X_2 \in L_+^0$, if $X_1 \overset{\rho}{\succ} X_2$ then for any $\alpha \in (0, 1)$ and any $X \in L_+^0$, $\alpha X_1 + (1 - \alpha) X \overset{\rho}{\succ} \alpha X_2 + (1 - \alpha) X$.

   These properties are called rationality axioms because they reflect rationality on the trader's choices. Properties (1)–(3) make $\overset{\rho}{\succeq}$ an equivalence relation. Essentially, as an equivalent relation, the trader can decide likes and dislikes on all elements in the choice set L_+^0 and his choices are transitive. Property (4), order preserving, states that if X_1 is preferred to X_2 then combinations of X_1 and X_2 with more X_1 are preferred to those with more X_2. Property (5), intermediate value, is a continuity restriction on preferences. It states that if X_1 is preferred to X_2 which is preferred to X_3 then there is a unique combination of X_1 and X_3 that is indifferent to X_2. Last, property (5) is the strong independence axiom. It states that if X_1 is preferred to X_2, then given any third choice X, any combination of X_1 and X is preferred to a combination of X_2 and X.

   Another important set of properties depend on specifying a probability measure $\mathbb{P} : \mathscr{F} \to [0, 1]$, interpreted as the trader's beliefs. These properties are defined below.
7. *Strictly Monotone*

   For any $X_1, X_2 \in L_+^0$, if $X_1 \geq X_2$ a.s. $\mathbb{P}$ and $X_1 \neq X_2$, i.e. $\mathbb{P}(X_1 > X_2) > 0$ then $X_1 \overset{\rho}{\succ} X_2$.
8. *Risk Aversion Characteristics*

$$\begin{aligned}
&\overset{\rho}{\succeq} \textit{ is risk averse} \quad \text{if} \quad && 1_\Omega(\omega) E[X] \overset{\rho}{\succ} X, \\
&\overset{\rho}{\succeq} \textit{ is risk neutral} \quad \text{if} \quad && 1_\Omega(\omega) E[X] \overset{\rho}{\sim} X, \\
&\overset{\rho}{\succeq} \textit{ is risk loving} \quad \text{if} \quad && 1_\Omega(\omega) E[X] \overset{\rho}{\prec} X
\end{aligned}$$

   for any $X \in L_+^0$ with $E[X] < \infty$.

If a trader's preferences are strictly monotone as in property (7), then the trader prefers more wealth to less. A trader's aversion to risk is characterized in property (8). A trader is risk averse with respect to X if she prefers receiving the expected value of the terminal wealth $E[X]$ instead of receiving the random wealth X itself. A trader is risk neutral if she is indifferent between the two. And, a trader is risk loving if she prefers the random wealth to the expected value of the wealth.

For the subsequent theory, the following property of a preference relation is crucial.

9. *State Dependent Expected Utility Representation*

Given a probability space $(\Omega, \mathscr{F}, \mathbb{P})$ and a function $U : (0, \infty) \times \Omega \to \mathbb{R}$ such that $U(x, \omega)$ is $\mathscr{B}(0, \infty) \otimes \mathscr{F}_T$-measurable, where $\mathscr{B}(0, \infty)$ is the Borel σ-algebra on $(0, \infty)$, we say that $\overset{\rho}{\succeq}$ has a state dependent $E[U]$ representation if

$$X_1 \overset{\rho}{\succ} X_2 \iff E[U(X_1, \omega)] > E[U(X_2, \omega)]$$

for all $X_1, X_2 \in L^0_+$ with $E[U(X_1, \omega)] < \infty$ and $E[U(X_2, \omega)] < \infty$.

In this representation of a trader's preferences, the probability measure $\mathbb{P}$ is interpreted as the trader's beliefs and U as the trader's utility function. The first important observation about this property is that the utility function representing the preference relation depends on the realized state $\omega \in \Omega$ in addition to the terminal wealth. Hence, the "past" affects the trader's "current" preferences. An important special case of this representation are those utility functions that are state independent, where $U(x, \omega) = U(x)$ does not depend on $\omega \in \Omega$.

Remark 9.1 (Preferences over Lotteries) An alternative choice set is the set of all probability measures on $\mathscr{B}(0, \infty)$, called *lotteries*. For the subsequent discussion, denote this set of lotteries as $\mathfrak{U}$. $\mathfrak{U}$ can be interpreted as the set of all possible probability measures over the realizations of terminal wealth X at time T. Preferences are defined over these probability measures $\mathbb{P} \in \mathfrak{U}$, and the rationality axioms are imposed on the these preferences, e.g. see Follmer and Schied [63, Chapter 2 for such a formulation].

The choice set $\mathfrak{U}$ is richer and more complex than L^0_+. To understand the relation between the two choice sets, fix a probability measure $\mathbb{P} : \Omega \to [0, 1]$. We can interpret $\mathbb{P}$ as the trader's beliefs. Then, given an $X \in L^0_+$ and an $A \in \mathscr{B}(0, \infty)$, $\mathbb{P}(X \in A)$ determines a probability measure over $(0, \infty)$, which is an element in the choice set $\mathfrak{U}$. A different random wealth $X^* \in L^0_+$ determines another probability measure $\mathbb{P}(X^* \in A)$, which is also in $\mathfrak{U}$. The set of all probability measures in $\mathfrak{U}$ determined by varying $X \in L^0_+$ in this fashion generates a subset of $\mathfrak{U}$, call it $\mathfrak{U}_{\mathbb{P}}$. $\mathfrak{U}_{\mathbb{P}}$ is a strict subset because a different probability measure $\mathbb{Q} : \Omega \to [0, 1]$ gives another subset $\mathfrak{U}_{\mathbb{Q}}$. Taking the union of these subsets across all probability measures $\mathbb{P}$ generates the set of lotteries, i.e. $\cup_{\mathbb{P}} \mathfrak{U}_{\mathbb{P}} = \mathfrak{U}$.

Defining a preference relation and rationality axioms on the set of lotteries $\mathfrak{U}$ requires more sophistication on the part of traders because the choice set $\mathfrak{U}$ is larger and more complex. For the subsequent theory, we do not need to use this more complex setting. One can view the subsequent theory as restricting the choice set of the trader to the subset $\mathfrak{U}_{\mathbb{P}}$ as constructed above. This completes the remark.

9.2 State Dependent EU Representation

This section studies the implications of the existence of a state dependent EU representation. We first study the rationality axioms implied by such a state dependent EU representation.

Lemma 9.1 (Necessary Rationality Axioms) *Let $\overset{\rho}{\succeq}$ have a state dependent EU representation.*

Then, $\overset{\rho}{\succeq}$ satisfies reflexivity, comparability, and transitivity.

Proof (reflexivity) Let $X \in L_+^0$, then $E[U(X, \omega)] \geq E[U(X, \omega)]$.

Or, $X \overset{\rho}{\succeq} X$. This completes the proof of reflexivity.
(comparability) Let $X_1, X_2 \in L_+^0$, then by the properties of $\mathbb{R}$ either

$$E[U(X_1, \omega)] \geq E[U(X_2, \omega)] \text{ or } E[U(X_2, \omega)] \geq E[U(X_1, \omega)].$$

Or, $X_1 \overset{\rho}{\succeq} X_2$ or $X_2 \overset{\rho}{\succeq} X_1$. This completes the proof of comparability.
(transitivity) Let $X_1, X_2, X_3 \in L_+^0$, if $X_1 \overset{\rho}{\succeq} X_2$ and $X_2 \overset{\rho}{\succeq} X_3$, then

$$E[U(X_1, \omega)] \geq E[U(X_2, \omega)] \text{ and } E[U(X_2, \omega)] \geq E[U(X_3, \omega)].$$

This implies $E[U(X_1, \omega)] \geq E[U(X_3, \omega)]$.

Or, $X_1 \overset{\rho}{\succeq} X_3$. This completes the proof of transitivity.
This completes the proof.

A state dependent EU representation does not imply preferences $\overset{\rho}{\succeq}$ satisfy the order preserving, the intermediate value, or the strong independence properties. The next example illustrates this observation.

Example 9.1 (EU Representation But Not Order Preserving, Not Intermediate Value, or Not Strong Independence Preferences) Assume that a state dependent EU representation exists for $\overset{\rho}{\succeq}$. Consider the following examples. They show that $\overset{\rho}{\succeq}$ need not satisfy the order preserving, the intermediate value, nor the strong independence properties.
(Not Order Preserving)

Let $U(x, \omega) = x1_{x<1}$ for all $\omega \in \Omega$.
Consider $X_1(\omega) = \frac{1}{2}$ and $X_2(\omega) = 2$.
We have $E[U(X_1, \omega)] = \frac{1}{2} > E[U(X_2, \omega)] = 0$.
Consider $\alpha_1 = \frac{1}{2}$ and $\alpha_2 = \frac{1}{3}$.

$\alpha_1 X_1 + (1 - \alpha_1) X_2 = \frac{5}{4} > 1$, which implies $E[U(\alpha_1 X_1 + (1 - \alpha_1) X_2, \omega)] = 0$, and

$\alpha_2 X_1 + (1 - \alpha_2) X_2 = \frac{3}{2} > 1$, which implies $E[U(\alpha_2 X_1 + (1 - \alpha_2) X_2, \omega)] = 0$.

In conjunction, this gives an example where $X_1 \overset{\rho}{\succ} X_2$ and there exist $\alpha_1, \alpha_2 \in (0,1)$ with $\alpha_1 > \alpha_2$ such that $[\alpha_1 X_1 + (1-\alpha_1)X_2] \overset{\rho}{\sim} [\alpha_2 X_1 + (1-\alpha_2)X_2]$, which violates the order preserving property.

(Not Intermediate Value)

Let $U(x,\omega) = 1_{x\leq 1} + 2 \cdot 1_{x>1}$ for all $\omega \in \Omega$.
Consider $X_1(\omega) = 2$, $X_2(\omega) = \frac{1}{2}1_B(\omega) + 2 \cdot 1_{B^c}(\omega)$ where $B \in \mathscr{F}$ with $\mathbb{P}(B_c) = \frac{3}{4}$, and $X_3(\omega) = 1$.
We have $E[U(X_1,\omega)] = 2 > E[U(X_2,\omega)] = \mathbb{P}(B) + 2\mathbb{P}(B_c) = \frac{7}{4} > E[U(X_3,\omega)] = 1$.
Note that $E[U(\alpha X_1 + (1-\alpha)X_3, \omega)] = 2$ for all $\alpha \in (0,1)$,
hence, $E[U(\alpha X_1 + (1-\alpha)X_3, \omega)] \neq E[U(X_2,\omega)]$.

In conjunction, this gives an example where $X_1 \overset{\rho}{\succ} X_2 \overset{\rho}{\succ} X_3$, but there exists no $\alpha \in (0,1)$ such that $X_2 \overset{\rho}{\sim} \alpha X_1 + (1-\alpha)X_3$, which violates the intermediate value property.

(Not Strong Independence)

Let $U(x,\omega) = x 1_{x<1}$ for all $\omega \in \Omega$.
Consider $X_1(\omega) = \frac{1}{2}$, $X_2(\omega) = \frac{1}{3}$, and $X(\omega) = 2$.
We have $E[U(X_1,\omega)] = \frac{1}{2} > E[U(X_2,\omega)] = \frac{1}{3}$.
Letting $\alpha = \frac{1}{2}$,
$\alpha X_1 + (1-\alpha)X = \frac{5}{4} > 1$, which implies $E[U(\alpha X_1 + (1-\alpha)X, \omega)] = 0$, and
$\alpha X_2 + (1-\alpha)X = \frac{3}{2} > 1$, which implies $E[U(\alpha X_2 + (1-\alpha)X, \omega)] = 0$.

In conjunction, this gives an example where $X_1 \overset{\rho}{\succ} X_2$ and not for any $\alpha \in (0,1)$, $\alpha X_1 + (1-\alpha)X \overset{\rho}{\succ} \alpha X_2 + (1-\alpha)X$, which violates the strong independence property.

This completes the example.

Remark 9.2 (Behavioral Finance) Behavioral finance studies financial markets where traders violate rationality. The definition of rationality in this literature is often taken to be very restrictive. Under a restrictive definition, a trader is defined to be *rational* if their utility function $U(x) : (0,\infty) \to \mathbb{R}$ is state independent and has an *EU* representation. For a survey of this literature, see Barberis and Thaler [7].

Using a more relaxed definition of rationality that includes state dependent utility functions with beliefs that can differ from statistical probabilities (as above), it can be shown that many of the "observed" behavioral biases are no longer irrational (see Kreps [128, Chapter 3]). Violations of the laws of probability (e.g. Bayes' law), however, remain inexplicable by the general utility framework used here. This completes the remark.

Remark 9.3 (Sufficient Conditions for a EU Representation) Sufficient conditions on preferences $\overset{\rho}{\succeq}$ to obtain either a state dependent or a state independent *EU* representation are known. See Jarrow [78] and Follmer and Schied [63] for the sufficient conditions for a state independent EU representation where the choice set is over lotteries; see Kreps [128] and Mas-Colell et al. [135] for a discussion of sufficient conditions for both state independent and state dependent EU representations over lotteries and realizations; and see Wakker and Zank [176] for sufficient conditions for state dependent EU representations. This completes the remark.

The next lemma explores some additional implications of the existence of a state dependent EU representation.

Lemma 9.2 (Utility Functions Are Non-Unique) *Let $\overset{\rho}{\succeq}$ have a state dependent EU representation. Then, for any $\mathscr{F}$-measurable random variable $a(\omega)$ and constant $b > 0$,*

$$U^*(x, \omega) = a(\omega) + bU(x, \omega) \tag{9.1}$$

is also an equivalent state dependent EU representation.

Proof

(Step 1) Let EU represent $\overset{\rho}{\succeq}$ and assume that there exists an $\mathscr{F}$-measurable random variable $a(\omega)$ and a constant $b > 0$ such that $U^*(x, \omega) = a(\omega) + bU(x, \omega)$. Then,

$$\begin{aligned}
X_1 \overset{\rho}{\succ} X_2 &\iff E\left[U(X_1, \omega)\right] > E\left[U(X_2, \omega)\right] \\
&\iff E\left[a\right] + bE\left[U(X_1, \omega)\right] > E\left[a\right] + bE\left[U(X_2, \omega)\right] \\
&\iff E\left[a + U(X_1, \omega)\right] > E\left[a + U(X_2, \omega)\right] \\
&\iff E\left[U^*(X_1, \omega)\right] > E\left[U^*(X_2, \omega)\right].
\end{aligned}$$

Thus, EU^* represents $\overset{\rho}{\succeq}$. This completes the proof.

Remark 9.4 (Converse of Lemma 9.2) The converse of Lemma 9.2 is true for any state independent EU representation of $\overset{\rho}{\succeq}$, where $\overset{\rho}{\succeq}$ is defined over the lottery choice set in Remark 9.1 and $a(\omega)$ is also a constant, i.e. if U and U^* represent $\overset{\rho}{\succeq}$ defined over the set of lotteries, then $U^*(x) = a + bU(x)$ for some constants $a \in \mathbb{R}$ and $b > 0$ gives an equivalent EU representation (see Follmer and Schied [63, Chapter 2]). This completes the remark.

The importance of Lemma 9.2 is that it implies for a given preference relation $\overset{\rho}{\succeq}$ one cannot compare utility levels across investors. That is, a utility level of 200 for trader A and 100 for trader B does not imply that trader A is twice as well off as trader B. Despite the non-uniqueness of the utility function in an EU representation, the optimal choice of a trader is invariant to the equivalent utility function used in the representation. That is, the solution $\hat{X}$ to $\sup_{X \in L_+^0} E[U(X, \omega)]$ subject to constraints is the same as the solution $\hat{X} = \arg\max \{ \overset{\rho}{\succeq} : X \in L_+^0 \}$ subject to the same constraints for all U satisfying the EU representation for $\overset{\rho}{\succeq}$. Hence, for the purposes of characterizing a trader's optimal portfolio decision, therefore, the non-uniqueness is unimportant.

Given the existence of a state dependent EU representation, we now explore sufficient conditions on the utility function U such that the underlying preferences $\overset{\rho}{\succeq}$ are strictly monotone.

Lemma 9.3 (Strictly Monotone) *Suppose $\overset{\rho}{\succeq}$ has a state dependent EU representation. Then,*

$$U \textit{ strictly increasing in } x \textit{ implies } \overset{\rho}{\succeq} \textit{ is strictly monotone.}$$

Proof Let $X_1, X_2 \in L_+^0$ be such that $X_1 \geq X_2$ a.s. $\mathbb{P}$ and $X_1 \neq X_2$. Then, U strictly increasing in x implies that

$$U(X_1, \omega) \geq U(X_2, \omega) \text{ and } U(X_1, \omega) \neq U(X_2, \omega).$$

Hence, $E[U(X_1, \omega)] > E[U(X_2, \omega)]$, which implies $X_1 \overset{\rho}{\succ} X_2$. This completes the proof.

Lemma 9.4 (Risk Aversion) *Let $\overset{\rho}{\succeq}$ have a state dependent EU representation. The preference relation $\overset{\rho}{\succ}$ is*

$$\begin{aligned} &\textit{risk averse if and only if} \quad E[U(E[X], \omega)] > E[U(X, \omega)], \\ &\textit{risk neutral if and only if} \ E[U(E[X], \omega)] = E[U(X, \omega)], \textit{ and} \\ &\textit{risk loving if and only if} \quad E[U(E[X], \omega)] < E[U(X, \omega)] \end{aligned}$$

for all $X \in L_+^0$ with $E[X] < \infty$.

Proof Given any $X \in L_+^0$ with $E[X] < \infty$, $\overset{\rho}{\succeq}$ is risk averse if and only if $1_\Omega(\omega) E[X] \overset{\rho}{\succ} X$. Using the fact that $\overset{\rho}{\succeq}$ has a state dependent EU representation, this is equivalent to $E[U(E[X], \omega)] > E[U(X, \omega)]$. The arguments for the risk neutral and risk loving cases follows similarly. This completes the proof.

Next, we study the relation between risk averse preferences $\overset{\rho}{\succeq}$ with an EU representation and strictly concave utility functions. The next example shows that the state dependent utility function U being strictly increasing and strictly concave in x does not imply that preferences $\overset{\rho}{\succeq}$ are risk averse.

Example 9.2 ($U(\cdot)$ Strictly Increasing and Strictly Concave, But Not Risk Averse) The following is an example of a state dependent utility function U that is strictly increasing and strictly concave in x, but where preferences $\overset{\rho}{\succeq}$ are not risk averse.

Let $\omega \in \Omega = [0, 1]$, $\mathscr{F} = \mathscr{B}[0, 1]$, and $d\mathbb{P}(\omega) = d\omega$, which is the uniform distribution. The mean $E(\omega) = \frac{1}{2}$ is equal to the median, i.e. $P(\omega \leq E(\omega)) = P(\omega > E(\omega)) = \frac{1}{2}$.

Let $U(x, \omega) = \varepsilon(2\sqrt{x})1_{\omega \leq E(\omega)} + (2\sqrt{x})1_{\omega > E(\omega)}$ for $\varepsilon > 0$ a constant. The function $2\sqrt{x}$ is a power utility function with $\rho = \frac{1}{2}$. For all $\omega \in \Omega$, $U(x, \omega)$ is strictly increasing and strictly concave in x. We have

$$\begin{aligned} E(U(X, \omega)) &= 2\varepsilon E\left(\sqrt{X} \,\middle|\, \omega \leq E(\omega)\right) P(\omega \leq E(\omega)) \\ &\quad + 2E\left(\sqrt{X} \,\middle|\, \omega > E(\omega)\right) P(\omega > E(\omega)) \\ &= \varepsilon E\left(\sqrt{X} \,\middle|\, \omega \leq E(\omega)\right) + E\left(\sqrt{X} \,\middle|\, \omega > E(\omega)\right). \end{aligned}$$

And, $E(U(\mu, \omega)) = (\varepsilon + 1)\sqrt{\mu}$ for μ a constant.

Consider $X(\omega) = \omega \in L^0_+$. Then,

$$E(U(X, \omega)) = \varepsilon E\left(\sqrt{\omega} \mid \omega \leq E(\omega)\right) + E\left(\sqrt{\omega} \mid \omega > E(\omega)\right)$$

and

$$E(U(E(X), \omega)) = (\varepsilon + 1)\sqrt{E(\omega)}.$$

As $\varepsilon \to 0$,

$$E(U(X, \omega)) \to E\left(\sqrt{\omega} \mid \omega > E(\omega)\right)$$

and

$$E(U(E(X), \omega)) \to \sqrt{E(\omega)}.$$

Since $E\left(\sqrt{\omega} \mid \omega > E(\omega)\right) > \sqrt{E(\omega)}$, there exists an $\varepsilon > 0$ such that

$$E(U(X, \omega)) > E(U(E(X), \omega)).$$

Hence, $\overset{\rho}{\succeq}$ is not risk averse. This completes the example.

The next theorem gives sufficient conditions on a state dependent utility function U such that the underlying preference relation $\overset{\rho}{\succeq}$ is risk averse. We write the utility function's derivative in $x \in (0, \infty)$ for a given $\omega \in \Omega$ as $U'(x, \omega)$.

Theorem 9.1 (Sufficient Conditions for Risk Averse Preferences) *Suppose* $\overset{\rho}{\succeq}$ *has a state dependent EU representation where U is*

(1) continuously differentiable in x,
(2) strictly increasing in x,
(3) strictly concave in x, and
(4) $\operatorname{cov}\left[X, \frac{d\mathbb{Q}_X}{d\mathbb{P}}\right] = 0$ *is satisfied for all* $X \in L_+^0$ *with* $E[X] < \infty$ *and*

$$E\left[U'(E[X], \omega)\right] < \infty,$$

where

$$\frac{d\mathbb{Q}_X}{d\mathbb{P}} = \frac{U'(E[X], \omega)}{E[U'(E[X], \omega)]} > 0.$$

Then, $\overset{\rho}{\succeq}$ *is risk averse.*

Proof By strict concavity (see Guler [66, p. 98])

$$U(X, \omega) - U(E[X], \omega) < U'(E[X], \omega)(X - E[X])$$

Since $U' > 0$ is strictly increasing,

$$\frac{U(X, \omega) - U(E[X], \omega)}{E[U'(E[X], \omega)]} < \frac{U'(E[X], \omega)}{E[U'(E[X], \omega)]}(X - E[X]).$$

Taking expectations yields

$$\frac{E[U(X, \omega)] - E[U(E[X], \omega)]}{E[U'(E[X], \omega)]} < E\left[\frac{U'(E[X], \omega)}{E[U'(E[X], \omega)]}(X - E[X])\right]$$
$$= E[X]\, E^{\mathbb{Q}_X}[X] + \operatorname{cov}\left[X, \frac{d\mathbb{Q}_X}{d\mathbb{P}}\right] - E[X] = 0$$

since $E\left[\frac{d\mathbb{Q}_X}{d\mathbb{P}}\right] = 1$ and by hypothesis (4).

This completes the proof.

In the hypotheses of Theorem 9.1, condition (1) is a smoothness condition. Condition (2) guarantees that the preference relation $\overset{\rho}{\succeq}$ is strictly monotone by Lemma 9.3. Condition (3) was shown not to be sufficient for risk aversion. To guarantee the preference relation $\overset{\rho}{\succeq}$ is risk averse, condition (4) is needed. Condition (4) effectively states that the state dependent utility function's valuation density ($\frac{d\mathbb{Q}_X}{d\mathbb{P}}$) is not correlated with X for all possible random wealths X. It is a weaker condition than state independence of the utility function, but which yields a similar conclusion when evaluating the risk of any possible wealth X.

Corollary 9.1 (State Independent Utility) *Suppose* $\overset{\rho}{\succeq}$ *has a state independent* EU *representation* ($U(x,\omega) \equiv U(x)$ *does not depend on* $\omega \in \Omega$) *where* U *is*

(1) continuously differentiable in x,
(2) strictly increasing in x,
(3) strictly concave in x.

Then, $\overset{\rho}{\succeq}$ *is risk averse.*

Proof Note that $\frac{d\mathbb{Q}_X}{d\mathbb{P}} = \frac{U'(E[X])}{E[U'(E[X])]} \equiv 1$ for all $\omega \in \Omega$. Hence, $\text{cov}\left[X, \frac{d\mathbb{Q}_X}{d\mathbb{P}}\right] = 0$. This completes the proof.

For the subsequent chapters, we will be assuming that the trader's preference relation $\overset{\rho}{\succeq}$ has a state dependent EU representation where U satisfies only hypotheses (1)–(3) in this Corollary. Consequently, the trader's preference structure utilized below is weaker than assuming that the trader's preferences are risk averse.

9.3 Measures of Risk Aversion

This section studies measures of risk aversion for preferences $\overset{\rho}{\succeq}$ with a state dependent EU representation. We start with a definition.

Definition 9.1 (Risk Aversion Measure) Let $\overset{\rho}{\succeq}$ have a state dependent EU representation with U continuously differentiable in x. For a given $X \in L^0_+$ with $E[X] < \infty$ and $E\left[U'(E[X], \omega)\right] < \infty$, a *measure of risk aversion* for U given X is

$$\mathscr{R}(U, X) = \frac{E[U(E[X], \omega)] - E[U(X, \omega)]}{E[U'(E[X], \omega)]}. \tag{9.2}$$

The difference $E[U(E[X], \omega)] - E[U(X, \omega)]$ in the numerator measures, for a given terminal wealth X, the difference in utility due to taking $E[X]$ for sure versus taking the risky wealth X. The normalization by $E\left[U'(E[X], \omega)\right]$ in the denominator makes this measure of risk aversion invariant with respect to a positive affine transformation, i.e. $\mathscr{R}(U, X) = \mathscr{R}(U^*, X)$ for $U(x, \omega) = a(\omega) + bU^*(x, \omega)$ with $a(\omega)$ $\mathscr{F}$-measurable and $b > 0$. It is important to note that $\mathscr{R}(U, X)$ can be positive for some wealths X and negative for others, hence this measure of risk aversion depends on both U and X.

This measure of risk aversion can be used to compare which traders are more risk averse. Indeed, $\mathscr{R}(U_1, X) > \mathscr{R}(U_2, X)$ for all $X \in L^0_+$ implies that trader U_1 is uniformly more risk averse than trader U_2. We can relate this measure of risk aversion to others in the literature. We start with the absolute risk aversion coefficient.

Lemma 9.5 (Absolute Risk Aversion) *Suppose* $\overset{\rho}{\succeq}$ *has a state dependent EU representation where U is*

(1) twice continuously differentiable in x,
(2) strictly increasing in x,
(3) strictly concave in x.

Then,

$$ARA(U, X, \omega) = -\frac{U''(X, \omega)}{U'(X, \omega)} > 0$$

is a measure of risk aversion *given* U, X, *and* ω.

Proof First note that $ARA(U, X)$ is invariant to a positive affine transformation.
Using a Taylor series expansion,

$$U(X, \omega) = U(E[X], \omega) + U'(E[X], \omega)(X - E[X]) + U''(\xi, \omega)\frac{(X - E[X])^2}{2} \text{ for } \xi \in [X, E[X]].$$

Rearranging terms and taking expectations yields

$$E[U(E[X], \omega)] - E[U(X, \omega)] = E\left[U'(E[X], \omega)(E[X] - X)\right] + E\left[-U''(\xi, \omega)\frac{(X - E[X])^2}{2}\right].$$

Divide both sides by $E\left[U'(E[X], \omega)\right]$.
Then, changing measures using $\frac{d\mathbb{Q}_x}{d\mathbb{P}} = \frac{U'(E[X],\omega)}{E[U'(E[X],\omega)]} > 0$ yields

$$E[U(E[X], \omega)] - E[U(X, \omega)]$$
$$= E\left[U'(E[X], \omega)\right]\left(\left(E[X] - E^{\mathbb{Q}_x}[X]\right) + E^{\mathbb{Q}_x}\left[-\frac{U''(\xi, \omega)}{U'(E[X], \omega)}\frac{(X - E[X])^2}{2}\right]\right).$$

This simplifies to

$$\mathscr{R}(U, X) = E^{\mathbb{Q}_x}\left[-\frac{U''(\xi, \omega)}{U'(E[X], \omega)}\frac{(X - E[X])^2}{2}\right] + \left(E[X] - E^{\mathbb{Q}_x}[X]\right).$$

For small $X - E[X]$, one can replace ξ with $E[X]$ in the above expression to obtain

$$\mathscr{R}(U, X) \approx E^{\mathbb{Q}_x}\left[-\frac{U''(E[X], \omega)}{U'(E[X], \omega)}\frac{(X - E[X])^2}{2}\right] + \left(E[X] - E^{\mathbb{Q}_x}[X]\right).$$

For a given U and X, if $-\frac{U''(E[X],\omega)}{U'(E[X],\omega)}$ increases for a set of $\omega \in A \in \mathscr{F}$ of strictly positive probability, holding all else constant, then $\mathscr{R}(U, X)$ increases.

This completes the proof.

Another common measure of risk aversion is called *relative risk aversion*, and it is defined as

$$RRA(U, X, \omega) = X \cdot ARA(U, X, \omega).$$

The inverse of the absolute risk aversion measure is called *risk tolerance* and denoted

$$RT(U, X, \omega) = \frac{1}{ARA(U, X, \omega)}.$$

Examples of utility functions satisfying the above assumptions include the following state independent utility functions.

Example 9.3 (State Independent Utility Functions)
(logarithmic) $U(x) = \ln(x)$ *for* $x > 0$.

This implies $U'(x) = \frac{1}{x} > 0$ for $x > 0$ and $(U')^{-1}(y) = \frac{1}{y}$ for $y > 0$.
Note that

$$U''(x) = -\frac{1}{x^2} < 0.$$

$$ARA(U, x) = -\frac{U''(x)}{U'(x)} = \frac{1}{x} > 0.$$

$$RT(U, x) = x > 0.$$

$$RRA(U, x) = -\frac{xU''(x)}{U'(x)} = 1.$$

(power) $U(x) = \frac{x^\rho}{\rho} > 0$ *for* $x > 0$ *and* $\rho < 1, \rho \neq 0$.

This implies $U'(x) = x^{\rho-1}$ for $x > 0$ and $(U')^{-1}(y) = y^{\frac{1}{\rho-1}}$ for $y > 0$.
Note that

$$U''(x) = (\rho - 1)x^{\rho-2} < 0.$$

$$ARA(U, x) = -\frac{U''(x)}{U'(x)} = -\frac{(\rho - 1)x^{\rho-2}}{x^{\rho-1}} = \frac{1 - \rho}{x} > 0.$$

$$RT(U, x) = \frac{x}{1 - \rho} > 0.$$

$$RRA(U, x) = -\frac{xU''(x)}{U'(x)} = 1 - \rho > 0.$$

(exponential) $U(x) = -e^{-\alpha x} < 0$ *for* $x > 0$ *and* $\alpha > 0$.

This implies $U'(x) = \alpha e^{-\alpha x}$ for $x > 0$ and $(U')^{-1}(y) = -\frac{1}{\alpha}\log\left(\frac{y}{\alpha}\right)$ for $y > 0$. Note that

$$U''(x) = -\alpha^2 e^{-\alpha x} < 0.$$

$$ARA(U,x) = -\frac{U''(x)}{U'(x)} = -\frac{-\alpha^2 e^{-\alpha x}}{\alpha e^{-\alpha x}} = \alpha > 0.$$

$$RT(U,x) = \frac{1}{\alpha} > 0.$$

$$RRA(U,x) = -\frac{xU''(x)}{U'(x)} = \alpha x > 0.$$

This completes the example.

Note that all of these utility functions' risk tolerances are linear in x. This leads to a wider class of state independent utility functions with linear risk tolerance.

Example 9.4 (State Independent Linear Risk Tolerance Utility Functions) Any state independent utility function U with $RT(U,x) = d_0 + d_1 x$ for $d_0, d_1 \in \mathbb{R}$ is said to have linear risk tolerance. It can be shown that this class of utility functions satisfy

$$U(x) = c_1 + c_2 \int e^{\int \frac{1}{d_0 + d_1 x} dx}$$

for $c_1, c_2 \in \mathbb{R}$.

Proof Define $\frac{-U''(x)}{U'(x)} = A(x)$.

Let $\mathcal{U}(x) = U'(x)$. Then, $\frac{-\mathcal{U}'(x)}{\mathcal{U}(x)} = A(x)$.

$$\frac{d\ln(\mathcal{U}(x))}{dx} = A(x).$$

Computing the indefinite integral yields

$$\ln(\mathcal{U}(x)) = c_0 + \int A(x)dx \text{ for } c_0 \in \mathbb{R}.$$

Or,

$$U'(x) = e^{c_0 + \int A(x)dx} = c_2 e^{\int A(x)dx} \text{ for } c_2 \in \mathbb{R}.$$

Computing the indefinite integral again gives

$$U(x) = c_1 + c_2 \int e^{\int A(x)dx} \text{ for } c_1 \in \mathbb{R}.$$

Since $A(x) = \frac{1}{RT(U)} = \frac{1}{d_0+d_1x}$ for linear risk tolerance, substitution completes the proof.

This completes the example.

9.4 State Dependent Utility Functions

We assume that preferences $\overset{\rho}{\succeq}$ have a state dependent EU representation with the following assumptions on the state dependent utility function U.

Assumption (Utility Function) *A utility function is a* $U : (0, \infty) \times \Omega \to \mathbb{R}$

such that for all $\omega \in \Omega$ *a.s.* $\mathbb{P}$,

(i) $U(x, \omega)$ *is* $\mathscr{B}(0, \infty) \otimes \mathscr{F}_T$*-measurable,*
(ii) $U(x, \omega)$ *is continuous and differentiable in* $x \in (0, \infty)$,
(iii) $U(x, \omega)$ *is strictly increasing and strictly concave in* $x \in (0, \infty)$, *and*
(iv) (Inada Conditions) $\lim_{x\downarrow 0} U'(x, \omega) = \infty$ and $\lim_{x\to\infty} U'(x, \omega) = 0$.

Condition (i) is a necessary measurability condition. Condition (ii) allows the use of calculus to find an optimal solution. Condition (iii) implies that preferences $\overset{\rho}{\succeq}$ are strictly monotone (see Lemma 9.3). Condition (iii) also states that the utility function is strictly concave, which was shown to be weaker than assuming the investor's preferences $\overset{\rho}{\succeq}$ are risk averse (see Theorem 9.1). The Inada conditions (iv) guarantee an interior optimal solution to the subsequently defined utility maximization problem (in the portfolio optimization Chaps. 10–12).

The state dependent utility function $U : (0, \infty) \times \Omega \to \mathbb{R}$ is defined over terminal wealth (or consumption in the portfolio optimization Chap. 12) denominated in units of the mma. The fact that this utility function is state dependent implies that this representation of a trader's preferences is very general (see Kreps [128, Chapter 3]). It includes, as special cases, recursive utility functions and habit formation (see Back [5, Chapter 21], and Skiadas [171, Chapter 6]).

Utility functions that both satisfy and violate the above assumption are provided in the following example.

Example 9.5 (State Independent Utility Functions)
(logarithmic) $U(x) = \ln(x)$ *for* $x > 0$.

This utility function satisfies the assumption. Indeed, from Example 9.3 above, we have that properties (i)–(iii) are satisfied. To show property (iv), note that $U'(x) = \frac{1}{x} > 0$ for $x > 0$ implies $\lim_{x\downarrow 0} U'(x, \omega) = \infty$ and $\lim_{x\to\infty} U'(x, \omega) = 0$.

(power) $U(x) = \frac{x^\rho}{\rho} > 0$ *for* $x > 0$ *and* $\rho < 1, \rho \neq 0$.

This utility function satisfies the assumption. Indeed, from Example 9.3 above, we have that properties (i)–(iii) are satisfied. To show property (iv), note that $U'(x) = x^{\rho-1}$ for $x > 0$ implies $\lim_{x \downarrow 0} U'(x, \omega) = \infty$ and $\lim_{x \to \infty} U'(x, \omega) = 0$.

(exponential) $U(x) = -e^{-\alpha x} < 0$ *for* $x > 0$ *and* $\alpha > 0$.

This utility function violates the assumption. From Example 9.3 above, we have that properties (i)–(iii) are satisfied. But, property (iv) is violated. Note that $U'(x) = \alpha e^{-\alpha x}$ for $x > 0$. This gives $\lim_{x \to \infty} U'(x, \omega) = 0$ but $\lim_{x \downarrow 0} U'(x, \omega) = \alpha < \infty$.

This completes the example.

We will need the following properties of the utility function's derivative in $x \in (0, \infty)$ for a given $\omega \in \Omega$.

Lemma 9.6 (Properties of U') *For every* $\omega \in \Omega$ a.s. $\mathbb{P}$,

(i) $U'(x, \omega) : (0, \infty) \times \Omega \to \mathbb{R}$ *is strictly decreasing in* $x \in (0, \infty)$*, and*

(ii) the inverse function, $I(y, \omega) \equiv \left(U'\right)^{-1}(y, \omega) : (0, \infty) \times \Omega \to \mathbb{R}$*, exists, it is* $\mathscr{B}(0, \infty) \otimes \mathscr{F}_T$*-measurable, and is strictly decreasing in* $y \in (0, \infty)$ *with* $I(0, \omega) = \infty$ *and* $I(\infty, \omega) = 0$.

Proof

(i) Strictly decreasing is implied by strict concavity.

(ii) These properties follow directly from the properties of U'.

This completes the proof.

We note that this lemma implies (suppressing the $\omega \in \Omega$ notation), by the definition of an inverse function, that

$$\begin{aligned} &U'(I(y)) = y \text{ for } 0 < y < \infty, \qquad \text{and} \\ &I(U'(x)) = x \text{ for } 0 < x < \infty. \end{aligned} \tag{9.3}$$

Remark 9.5 (Money Market Account Numeraire) In this utility function formulation, wealth (or consumption in the portfolio optimization Chap. 12) is normalized by the mma's value $\mathbb{B}_T$. This is without loss of generality. Indeed, suppose instead that one postulates the utility function is defined over wealth denominated in units of a single perishable consumption good, $U^c(W_T, \omega)$, where W_T denotes time T wealth in units of the consumption good. Let $\psi_T > 0$ be dollars per consumption good at time T. Then, nominal wealth is $\mathbb{X}_T = W_T \psi_T$. As noted above, we consider time T dollar wealth in units of the mma, defined by $X_T = \frac{\mathbb{X}_T}{\mathbb{B}_T} = \frac{W_T \psi_T}{\mathbb{B}_T}$. Then, because the utility function is state dependent, we can define an equivalent utility function over units of the mma by $U(X_T, \omega) \equiv U^c\left(X_T \frac{\mathbb{B}_T}{\psi_T}, \omega\right) = U^c(W_T, \omega)$. The utility U function defined over wealth in units of the mma will yield the same optimal trading strategy and terminal wealth, in units of the consumption good, as the utility function U^c defined over wealth using the consumption good as the numeraire. This completes the remark.

9.5 Conjugate Duality

To solve the trader's optimization problem defined in the portfolio optimization Chaps. 10–12 using convex optimization, we need to define the convex conjugate of the utility function U.

Definition 9.2 (Convex Conjugate) The *convex conjugate* of U is defined by

$$\tilde{U}(y,\omega) = \sup_{x>0}\,[U(x,\omega) - xy], \quad y > 0 \tag{9.4}$$

We state without proof the following properties of this convex conjugate. This lemma can be found in Mostovyi [144, p. 139].

Lemma 9.7 (Properties of the Convex Conjugate) *For all* $\omega \in \Omega$ a.s. $\mathbb{P}$,

(i) $\tilde{U}(y,\omega)$ *is* $\mathscr{B}(0,\infty)\otimes\mathscr{F}_T$*-measurable, differentiable, decreasing, and strictly convex in* $y \in (0,\infty)$ *with* $\tilde{U}(0,\omega) = U(\infty,\omega)$,

(ii) the derivative is

$$\tilde{U}'(y,\omega) = -(U')^{-1}(y,\omega) = -I(y,\omega), \quad y > 0,$$

(iii) the supremum in expression (9.4) is attained at $x = I(y,\omega)$*, i.e.*

$$\tilde{U}(y,\omega) = U(I(y,\omega),\omega) - yI(y,\omega), \quad y > 0,$$

(iv) the biconjugate relation holds

$$U(x,\omega) = \inf_{y>0}\left[\tilde{U}(y,\omega) + xy\right], \quad x > 0 \tag{9.5}$$

with the infimum attained at $y = U'(x,\omega)$*, i.e.*

$$U(x,\omega) = \tilde{U}(U'(x,\omega),\omega) + xU'(x,\omega), \quad x > 0.$$

Example 9.6 (Convex Conjugate Functions) This example uses the state independent utility functions in Example 9.5 that satisfy the assumption.
(logarithmic) $U(x) = \ln(x)$ for $x > 0$, gives

$$\tilde{U}(y) = -\ln y - 1.$$

Proof $\tilde{U}(y) = \sup_{x>0}[\ln(x) - xy]$. Setting the first derivative equal to zero yields $\frac{1}{x} - y = 0$. This implies $x = \frac{1}{y}$. This is the unique maximum since $\ln(x) - xy$ is strictly concave for all $y > 0$ (the second derivative is strictly negative). Substitution yields $\tilde{U}(y) = \ln\left(\frac{1}{y}\right) - 1$. Algebra completes the proof.

(*power*) $U(x) = \frac{x^\rho}{\rho}$ for $x > 0$ and $\rho < 1$, $\rho \neq 0$, gives

$$\tilde{U}(y) = \frac{y^{-\left(\frac{\rho}{1-\rho}\right)}}{\left(\frac{\rho}{1-\rho}\right)}.$$

Proof $\tilde{U}(y) = \sup_{x>0}[\frac{x^\rho}{\rho} - xy]$. Setting the first derivative equal to zero yields $x^{\rho-1} - y = 0$. This implies $x = y^{\left(\frac{1}{\rho-1}\right)}$. This is the unique maximum since $\frac{x^\rho}{\rho} - xy$ is strictly concave for all $y > 0$ (the second derivative is strictly negative). Substitution yields $\tilde{U}(y) = \frac{y^{\left(\frac{\rho}{\rho-1}\right)}}{\rho} - y^{\left(\frac{1}{\rho-1}\right)}y$. Algebra gives $\tilde{U}(y) = \frac{y^{\left(\frac{\rho}{\rho-1}\right)}}{\rho} - \frac{\rho y^{\left(\frac{\rho}{\rho-1}\right)}}{\rho}$. Simplification completes the proof.

9.6 Reasonable Asymptotic Elasticity

The next assumption is needed as a sufficient condition to guarantee the existence of a solution to the investor's portfolio problem as defined in Chaps. 10–12 below. It is "almost" necessary in that, if it is not satisfied, examples can be found where the investor's portfolio optimization problem has no solution (see Kramkov and Schachermayer [126]).

Assumption (Reasonable Asymptotic Elasticity (AE(U) < 1)) *For all $\omega \in \Omega$* a.s. $\mathbb{P}$,

$$AE(U, \omega) \equiv \limsup_{x\to\infty} \frac{xU'(x, \omega)}{U(x, \omega)} < 1. \tag{9.6}$$

For a given $\omega \in \Omega$, this assumption relates the marginal utility $U'(x)$ to the average utility $\frac{U(x)}{x}$ for large x via the ratio $U'(x) / (U(x)/x)$. *AE(U) < 1* states that for large enough wealth x, the marginal utility must be less than the average utility. If *AE(U) < 1* is violated, then as wealth gets large, a trader's marginal utility is greater than or equal to their average utility, implying more wealth is desired. Intuitively, this suggests that an optimal wealth may not exist for such a utility function.

An alternative characterization of the reasonable asymptotic elasticity assumption further clarifies its meaning. This characterization is based on the utility function's relative risk aversion: $RRA(U, x, \omega) = -\frac{xU''(x,\omega)}{U'(x,\omega)}$, which was discussed in Sect. 9.3 above.

Theorem 9.2 (Characterization of AE(U) < 1) *Suppose $U'(x, \omega)$ is differentiable on $x \in (0, \infty)$ and $ARRA(U, \omega) \equiv \lim_{x\to\infty}\left(-\frac{xU''(x,\omega)}{U'(x,\omega)}\right)$ exists a.s. $\mathbb{P}$. Then,*

$\lim_{x\to\infty} \frac{xU'(x,\omega)}{U(x,\omega)}$ *exists a.s.* $\mathbb{P}$ *and*

$$ARRA(U) > 0 \textit{ if and only if } AE(U) < 1.$$

Proof This proof is from Schachermayer [164].

(Step 1) Suppose $U(\infty) < \infty$. Then, for $x_0 > 0$,

$$0 \leq xU'(x) = (x - x_0)U'(x) + x_0U'(x).$$

Hence,

$$0 \leq \limsup_{x\to\infty} xU'(x) = \limsup_{x\to\infty} (x - x_0)U'(x) + \limsup_{x\to\infty} x_0U'(x).$$

But, $\lim_{x\to\infty} U'(x) = 0$. And, by the mean value theorem,

$$U(x) - U(x_0) = U'(z)(x - x_0) \text{ for some } z \in [x_0, x].$$

Since U' is decreasing, this implies $U'(x)(x - x_0) \leq U(x) - U(x_0)$. Hence,

$$\limsup_{x\to\infty} xU'(x) \leq \limsup_{x\to\infty} (U(x) - U(x_0)) = \left(\limsup_{x\to\infty} U(x)\right) - U(x_0).$$

This holds for all $x_0 \to \infty$, which implies that $\limsup_{x\to\infty} xU'(x) = 0$.

(Case a) If $U(\infty) = 0$, then applying L'Hospital's rule gives

$$\lim_{x\to\infty} \frac{xU'(x)}{U(x)} = \lim_{x\to\infty} \frac{U'(x) + xU''(x)}{U'(x)} = 1 - \lim_{x\to\infty} \left(-\frac{xU''(x)}{U'(x)}\right).$$

This completes the proof for Case (a).

(Case b) If $U(\infty) \neq 0$, then $\lim_{x\to\infty} \frac{xU'(x)}{U(x)} = 0$. In this case, define

$$\tilde{U}(x) \equiv U(x) - U(\infty).$$

Then,

$$\tilde{U}(\infty) = 0.$$

This is Case (a), thus

$$\lim_{x\to\infty}\frac{x\tilde{U}'(x)}{\tilde{U}(x)} = 1 - \lim_{x\to\infty}\left(-\frac{x\tilde{U}''(x)}{\tilde{U}'(x)}\right).$$

Substitution gives

$$\lim_{x\to\infty}\frac{xU'(x)}{U(x)-U(\infty)} = 1 - \lim_{x\to\infty}\left(-\frac{xU''(x)}{U'(x)}\right).$$

But,

$$\lim_{x\to\infty}\frac{xU'(x)}{U(\infty)} = 0,$$

hence

$$\lim_{x\to\infty}\frac{xU'(x)}{U(x)} = 1 - \lim_{x\to\infty}\left(-\frac{xU''(x)}{U'(x)}\right).$$

This completes the proof for Case (b).

(Step 2) Suppose $U(\infty) = \infty$.
For $x \geq 1$,

$$\begin{aligned} xU'(x) &= U'(1) + \int_1^x (zU'(z))'dz = U'(1) + \int_1^x (U'(x) + zU''(z))dz \\ &= U'(1) + U(x) + \int_1^x zU''(z)dz. \end{aligned}$$

So,

$$\frac{xU'(x)}{U(x)} = \frac{U'(1)}{U(x)} + 1 + \frac{\int_1^x zU''(z)dz}{U(x)}.$$

Taking limits

$$\begin{aligned} \lim_{x\to\infty}\frac{xU'(x)}{U(x)} &= \lim_{x\to\infty}\frac{U'(1)}{U(x)} + 1 + \lim_{x\to\infty}\frac{\int_1^x zU''(z)dz}{U(x)} \\ &= 1 + \lim_{x\to\infty}\frac{\int_1^x zU''(z)dz}{U(x)}. \end{aligned}$$

(Case a) If $\int_1^x zU''(z)dz = \infty$, then applying L'Hospital's rule gives

$$\lim_{x\to\infty} \frac{\int_1^x zU''(z)dz}{U(x)} = \lim_{x\to\infty} \frac{xU''(x)}{U'(x)}.$$

Hence,

$$\lim_{x\to\infty} \frac{xU'(x)}{U(x)} = 1 - \lim_{x\to\infty} \left(-\frac{xU''(x)}{U'(x)}\right).$$

This completes the proof of Case (a).

(Case b) If $\int_1^\infty zU''(z)dz < \infty$, then

$$\lim_{x\to\infty} \frac{\int_1^x zU''(z)dz}{U(x) - U(1)} = \lim_{x\to\infty} \frac{\int_1^x zU''(z)dz}{\int_1^x U'(z)dz} = 0.$$

By the mean value theorem, $\frac{\int_1^x zU''(z)dz}{\int_1^x U'(z)dz} = \frac{\xi U''(\xi)(x-1)}{U'(\delta)(x-1)}$ for some $\delta, \xi \in [1, x]$.

Let $\eta = \frac{\delta}{\xi}$. Then,

$$\frac{\int_1^x zU''(z)dz}{\int_1^x U'(z)dz} = \frac{\xi U''(\xi)}{U'(\eta\xi)}.$$

Taking the limit as $\xi \to \infty$ for fixed η,
we get

$$\lim_{\xi\to\infty} \frac{\int_1^x zU''(z)dz}{\int_1^x U'(z)dz} = \lim_{x\to\infty} \frac{\int_1^x zU''(z)dz}{\int_1^x U'(z)dz}$$

and

$$\lim_{\xi\to\infty} \frac{\xi U''(\xi)}{U'(\eta\xi)} = \lim_{\xi\to\infty} \frac{\xi U''(\xi)}{U'(\xi)}.$$

Combined these give $\lim\limits_{\xi\to\infty} \frac{\xi U''(\xi)}{U'(\xi)} = 0$, which completes the proof of Case (b).

This completes the proof.

In terms of relative risk aversion, the reasonable asymptotic elasticity assumption states that as wealth becomes infinitely large, the investor remains risk averse.

Example 9.7 ($AE(U) < 1$ and $AE(U) = 1$) Applying this definition to the state independent utility functions in Example 9.5 shows that they satisfy the reasonable asymptotic elasticity assumption.

(logarithmic) $U(x) = \ln(x)$ *for* $x > 0$.

$$AE(U) = \limsup_{x\to\infty}\left(\frac{x\cdot\frac{1}{x}}{\ln(x)}\right) = \limsup_{x\to\infty}\left(\frac{1}{\ln(x)}\right) = 0.$$

This utility function satisfies $AE(U) < 1$.
(power) $U(x) = \frac{x^\rho}{\rho}$ *for* $x > 0$ *and* $\rho < 1$, $\rho \neq 0$.

$$AE(U) = \limsup_{x\to\infty}\left(\frac{x\cdot x^{\rho-1}}{\frac{x^\rho}{\rho}}\right) = \limsup_{x\to\infty}\rho = \rho < 1.$$

This utility function satisfies $AE(U) < 1$.

Here is an example that violates the reasonable asymptotic elasticity condition.
(violates AE(U)<1) $U(x) = \frac{x}{\ln x}$ *for* $x > 0$.

$$AE(U) = \limsup_{x\to\infty}\frac{x\left(\frac{1}{\ln(x)} - \frac{x}{(\ln x)^2}\frac{1}{x}\right)}{\frac{x}{\ln(x)}} = \limsup_{x\to\infty}\left(1 - \frac{1}{\ln x}\right) = 1.$$

This completes the example.

9.7 Differential Beliefs

This section discusses the changes to the utility function framework needed when we introduce differential beliefs in subsequent chapters, where the trader's beliefs differ from the statistical probability measure $\mathbb{P}$. Suppose that $\mathbb{P}$ represents the statistical probability measure and $\mathbb{P}_i$ represents the trader's beliefs where $\mathbb{P}_i$ and $\mathbb{P}$ are equivalent probability measures. Thus $\frac{d\mathbb{P}_i}{d\mathbb{P}} > 0$ is an $\mathscr{F}_T$-measurable random variable and $E\left[\frac{d\mathbb{P}_i}{d\mathbb{P}}\right] = 1$.

It will be convenient to consider the modified utility function

$$\mathscr{U}(x,\omega) = \frac{d\mathbb{P}_i}{d\mathbb{P}}U(x,\omega). \tag{9.7}$$

Because $\frac{d\mathbb{P}^*}{d\mathbb{P}} > 0$, $\mathscr{U}$ inherits all of the assumed properties of U and those given in Lemma 9.6. For example, the first derivative is

$$\mathscr{U}'(x,\omega) = \frac{d\mathbb{P}_i}{d\mathbb{P}}U'(x,\omega) > 0. \tag{9.8}$$

The convex conjugate of the modified utility function is

$$\tilde{\mathscr{U}}(y,\omega) = \sup_{x>0}\,[\mathscr{U}(x,\omega) - xy] = \frac{d\mathbb{P}_i}{d\mathbb{P}}\tilde{U}\left(y\frac{d\mathbb{P}}{d\mathbb{P}_i}\right). \tag{9.9}$$

Proof

$$\begin{aligned}
\tilde{\mathscr{U}}(y,\omega) &= \sup_{x>0}\,[\mathscr{U}(x,\omega) - xy] \\
&= \sup_{x>0}\left[\frac{d\mathbb{P}^*}{d\mathbb{P}}U(x,\omega) - xy\right] \\
&= \sup_{x>0}\left[\frac{d\mathbb{P}_i}{d\mathbb{P}}U(x,\omega) - \frac{d\mathbb{P}_i}{d\mathbb{P}}xy\frac{d\mathbb{P}}{d\mathbb{P}_i}\right] \\
&= \frac{d\mathbb{P}_i}{d\mathbb{P}}\sup_{x>0}\left[U(x,\omega) - xy\frac{d\mathbb{P}}{d\mathbb{P}_i}\right] = \frac{d\mathbb{P}_i}{d\mathbb{P}}\tilde{U}\left(y\frac{d\mathbb{P}}{d\mathbb{P}_i}\right).
\end{aligned}$$

This completes the proof.

9.8 Notes

The characterization of an agent's preferences is a well studied topic in both economics and statistical decision theory. A classical book is Fishburn [61]. For excellent presentations in microeconomic texts, see Kreps [128] and Mas-Colell et al. [135]; in finance texts, see Back [5], Follmer and Schied [63], and Skiadas [171]; and in statistical decision theory texts, see Berger [11] and DeGroot [43].

Chapter 10
Complete Markets (Utility over Terminal Wealth)

This chapter studies an individual's portfolio optimization problem. In this optimization, the solution differs depending on whether the market is complete or incomplete. This chapter investigates the optimization problem in a complete markets setting, and the next chapter analyzes incomplete markets.

10.1 The Set-Up

Given is a normalized market $(S, (\mathscr{F}_t), \mathbb{P})$ where the value of the money market account $B_t \equiv 1$. To make the optimization problem meaningful, we assume that there are no arbitrage opportunities in the market, i.e.

Assumption (NFLVR) $\mathfrak{M}_l \neq \emptyset$
where $\mathfrak{M}_l = \{\mathbb{Q} \sim \mathbb{P} : S \text{ is a } \mathbb{Q}\text{-local martingale}\}$.

As mentioned in the introduction to this chapter, we assume that the market is complete.

Assumption (Complete Markets) *Choose a* $\mathbb{Q} \in \mathfrak{M}_l$.
The market $(S, (\mathscr{F}_t), \mathbb{P})$ *is complete with respect to* $\mathbb{Q}$.

By the Second Fundamental Theorem 2.4 in Chap. 2, this implies that the set of equivalent local martingale measures is a singleton, i.e.

$$\mathfrak{M}_l = \{\mathbb{Q}\}.$$

Remark 10.1 (Stochastic Processes in a Complete Market) The assumption of a complete market with respect to $\mathbb{Q}$ restricts the class of semimartingale processes S possible for the risky assets in the market. By the definition of a complete market (see the fundamental theorems Chap. 2), the stochastic process for S must be such that synthetic construction of an arbitrary $\mathbb{Q}$-integrable random time T payoff is

R. A. Jarrow, *Continuous-Time Asset Pricing Theory*, Springer Finance,
https://doi.org/10.1007/978-3-319-77821-1_10

possible where the resulting value process of the s.f.t.s. that generates the payoff must be a $\mathbb{Q}$-martingale. Not all stochastic processes for S satisfy this restriction. For example, in the finite-dimension Brownian motion market of Sect. 2.7 in Chap. 2, the volatility matrix of the risky asset price processes must have a rank equal to the number of Brownian motions in the market. When S has a discontinuous sample path process, this restriction implies that the probability density of the jump amplitude must be discrete with the number of possible jump magnitudes less than or equal to the number of traded risky assets (see Cont and Tankov [34, Chapter 9.2]). This completes the remark.

Denote the probability density function of the equivalent local martingale measure by $Y_T = \frac{d\mathbb{Q}}{d\mathbb{P}} \in M_l \subset D_l$. Y_T is the unique local martingale deflator M_l. Recall that

$$\mathscr{D}_l = \{Y \in \mathscr{L}_+^0 : Y_0 = 1,\ XY \text{ is a } \mathbb{P}\text{-local martingale}, \\ X = 1 + \int \alpha \cdot dS,\ (\alpha_0, \alpha) \in \mathscr{A}(1)\}$$

$$D_l = \{Y_T \in L_+^0 : \exists Z \in \mathscr{D}_l,\ Y_T = Z_T\}$$

$$\mathscr{M}_l = \left\{Y \in \mathscr{D}_l : \exists \mathbb{Q} \sim \mathbb{P},\ Y_T = \frac{d\mathbb{Q}}{d\mathbb{P}}\right\}$$

$$M_l = \{Y_T \in L_+^0 :\ \exists Z \in \mathscr{M}_l,\ Y_T = Z_T\}$$

(see the fundamental theorems Chap. 2 for a discussion of local martingale deflators). Note that $E[Y_T] = 1$ and $Y_t = E[Y_T | \mathscr{F}_t] \in \mathscr{M}_l$ is a $\mathbb{P}$-martingale.

10.2 Problem Statement

This section defines the investor's expected utility over terminal wealth optimization problem. We assume that the investor's preferences $\overset{\rho}{\succeq}$ have a state dependent EU representation with a utility function defined over terminal wealth $U(X_T, \omega)$ and satisfying the assumption given in the utility function Chap. 9, repeated here for convenience.

Assumption (Utility Function) *A utility function is a* $U : (0, \infty) \times \Omega \to \mathbb{R}$ *such that for all* $\omega \in \Omega$ *a.s.* $\mathbb{P}$,

(i) $U(x, \omega)$ *is* $\mathscr{B}(0, \infty) \otimes \mathscr{F}_T$*-measurable,*
(ii) $U(x, \omega)$ *is continuous and differentiable in* $x \in (0, \infty)$,
(iii) $U(x, \omega)$ *is strictly increasing and strictly concave in* $x \in (0, \infty)$, *and*
(iv) *(Inada Conditions)* $\lim_{x \downarrow 0} U'(x, \omega) = \infty$ *and* $\lim_{x \to \infty} U'(x, \omega) = 0$.

We do not consider utility over intermediate consumption in this chapter. This simplification is imposed for pedagogical reasons. Indeed, if one understands the solution to the portfolio optimization problem for utility over terminal wealth, then the solution to the portfolio optimization problem with intermediate consumption is easier to understand and to implement.

The solution to the portfolio optimization problem with utility defined over both intermediate consumption and terminal wealth is discussed in Chap. 12. Because the complete market setting is a special case of the solution given in Chap. 12, there will not be a separate chapter studying complete markets with utility defined over intermediate consumption and terminal wealth.

Recall that the set of admissible s.f.t.s.'s is denoted by

$$\begin{aligned}\mathscr{A}(x) = \{(\alpha_0, \alpha) \in (\mathscr{O}, \mathscr{L}(S)) : \ & X_t = \alpha_0(t) + \alpha_t \cdot S_t, \ \exists c \leq 0, \\ & X_t = x + \textstyle\int_0^t \alpha_u \cdot dS_u \geq c, \ \forall t \in [0, T]\Big\}.\end{aligned}$$

For the investor's optimization problem, we restrict consideration to admissible s.f.t.s.'s where the lower bound $c = 0$, i.e. admissible s.f.t.s.'s where the wealth process is always nonnegative. This is the set of *nonnegative wealth s.f.t.s.'s* with initial wealth $x \geq 0$, denoted by

$$\begin{aligned}\mathscr{N}(x) = \{(\alpha_0, \alpha) \in (\mathscr{O}, \mathscr{L}(S)) : \ & X_t = \alpha_0(t) + \alpha_t \cdot S_t, \\ & X_t = x + \textstyle\int_0^t \alpha_u \cdot dS_u \geq 0, \ \forall t \in [0, T]\Big\}.\end{aligned}$$

Note that $\mathscr{N}(x) \subset \mathscr{A}(x)$ where $x \geq 0$.

To facilitate solving the investor's optimization problem, we need to define two sets. The first is the set of wealth processes generated by the nonnegative wealth s.f.t.s.'s with $x \geq 0$, denoted by

$$\mathscr{X}(x) = \left\{X \in \mathscr{L}_+^0 : \exists(\alpha_0, \alpha) \in \mathscr{N}(x), \ X_t = x + \int_0^t \alpha_u \cdot dS_u, \ \forall t \in [0, T]\right\}.$$

The second is the set of time T terminal wealths (random variables) generated by these wealth processes with $x \geq 0$, i.e.

$$\begin{aligned}\mathscr{C}(x) &= \{X_T \in L_+^0 : \exists(\alpha_0, \alpha) \in \mathscr{N}(x), \ x + \textstyle\int_0^T \alpha_t \cdot dS_t = X_T\} \\ &= \{X_T \in L_+^0 : \exists Z \in \mathscr{X}(x), \ X_T = Z_T\}.\end{aligned}$$

We use these two sets below.

Formally, the trader's portfolio optimization problem can now be stated. For simplicity of notation, we suppress the dependence of the state dependent utility function on the state $\omega \in \Omega$, writing $U(x) = U(x, \omega)$ everywhere below.

Problem 10.1 Utility Optimization (Choose the Optimal Trading Strategy (α_0, α) for $x \geq 0$)

$$v(x) = \sup_{(\alpha_0,\alpha)\in\mathcal{N}(x)} E\left[U\left(X_T\right)\right], \qquad \text{where}$$

$$\mathcal{N}(x) = \{(\alpha_0, \alpha) \in (\mathcal{O}, \mathcal{L}(S)) : X_t = \alpha_0(t) + \alpha_t \cdot S_t,\\ X_t = x + \textstyle\int_0^t \alpha_u \cdot dS_u \geq 0,\ \forall t \in [0, T]\Big\}.$$

In the classical literature this problem was solved using dynamic programming, which requires the additional assumption that the stock price process follows a Markov process. Although often a reasonable assumption, this problem can be solved using martingale methods and convex duality without adding this assumption. This alternative approach has the advantage that it builds on the insights obtained from the earlier chapters on derivative pricing and hedging. Consequently, we follow the second approach.

The second approach solves the problem in two steps. The two steps are:

1. to solve a static optimization problem which chooses the optimal time T wealth (the optimal "derivative"), and then
2. to determine the trading strategy that obtains this "derivative."

Step 2 is equivalent to the construction of a synthetic derivative. This problem was solved in the derivative pricing and hedging Sect. 2.6 in the fundamental theorems Chap. 2. Hence, the only new challenge in solving the investor's optimization problem is step 1. We now focus on solving step 1.

Problem 10.2 Utility Optimization (Choose the Optimal Derivative X_T for $x \geq 0$)

$$v(x) = \sup_{X_T \in C(x)} E\left[U\left(X_T\right)\right], \qquad \text{where}$$

$$\mathcal{C}(x) = \{X_T \in L^0_+ : \exists(\alpha_0, \alpha) \in \mathcal{N}(x),\ x + \int_0^T \alpha_t \cdot dS_t = X_T\}.$$

To solve this problem, we give an alternative characterization of the constraint set $\mathcal{C}(x)$, which represents those payoffs $X_T \in L^0_+$ that can be generated by a nonnegative wealth s.f.t.s. $(\alpha_0, \alpha) \in \mathcal{N}(x)$ with initial wealth $x \geq 0$. The intuition behind this alternative characterization uses risk neutral valuation. Consider the unique local martingale measure $\mathbb{Q} \in \mathfrak{M}_l$, which exists by assumption. Using risk neutral valuation in a complete market to compute present values (see Theorem 2.8 in Chap. 2), the constraint set $\mathcal{C}(x)$ should be the same as the set of all payoffs whose present values $E^{\mathbb{Q}}\left[X_T\right]$ are affordable at time 0, i.e. less than or equal to x. This is indeed the case, as the next theorem shows.

Theorem 10.1 (Budget Constraint Equivalence)

$$\mathscr{C}(x) = \left\{X_T \in L^0_+ : E^{\mathbb{Q}}[X_T] \le x\right\}.$$

Proof Define $\mathscr{C}_1(x) = \{X_T \in L^0_+ : E^{\mathbb{Q}}[X_T] \le x\}$.

Step 1: ($\mathscr{C}(x) \subset \mathscr{C}_1(x)$)

Take $X_T \in \mathscr{C}(x)$, then $\exists(\alpha_0, \alpha) \in \mathscr{N}(x)$, $x + \int_0^T \alpha_t \cdot dS_t = X_T$.

The stochastic process $x + \int_0^t \alpha_u \cdot dS_u \ge 0$ is a local martingale under $\mathbb{Q}$ that is bounded below by 0. Hence, by Lemma 1.3 in Chap. 1, it is a supermartingale. Thus, $E^{\mathbb{Q}}\left[x + \int_0^T \alpha_t \cdot dS_t\right] \le x$.

This implies $E^{\mathbb{Q}}[X_T] \le x$. Hence, $X_T \in \mathscr{C}_1(x)$.

Step 2: ($\mathscr{C}_1(x) \subset C(x)$)

Take $X_T \in \mathscr{C}_1(x)$. Then, $E^{\mathbb{Q}}[X_T] < \infty$. Since the market is complete with respect to $\mathbb{Q}$,

$\exists(\alpha_0, \alpha) \in \mathscr{N}(x)$ such that $x + \int_0^t \alpha_u \cdot dS_u = E^{\mathbb{Q}}[X_T | \mathscr{F}_t]$ for all $t \in [0, T]$.

Hence, $x + \int_0^T \alpha_t \cdot dS_t = X_T$.

This implies $X_T \in \mathscr{C}(x)$, which completes the proof.

Using this new budget constraint, we can rewrite the optimization problem using the local martingale deflator $Y_T = \frac{d\mathbb{Q}}{d\mathbb{P}} \in M_l$, which satisfies $E^{\mathbb{Q}}[X_T] = E[Y_T X_T]$.

Problem 10.3 Utility Optimization (Choose the Optimal Derivative X_T for $x \ge 0$)

$$v(x) = \sup_{\mathscr{C}(x)} E[U(X_T)], \qquad \text{where}$$

$$\mathscr{C}(x) = \left\{X_T \in L^0_+ : E[Y_T X_T] \le x\right\}$$

for $Y_T = \frac{d\mathbb{Q}}{d\mathbb{P}} \in M_l$.

To solve this problem, we define the *Lagrangian*

$$\mathscr{L}(X_T, y) = E[U(X_T)] + y(x - E[X_T Y_T]). \tag{10.1}$$

Using the Lagrangian transforms a constrained optimization problem to an unconstrained optimization, where the unconstrained optimization's first-order conditions can be used to characterize the solution. With this observation in mind, we first show that the original constrained optimization problem is equivalent to solving the *primal problem* using the Lagrangian, i.e.

$$\text{primal}: \quad \sup_{X_T \in L^0_+} \left(\inf_{y>0} \mathscr{L}(X_T, y)\right) = \sup_{X_T \in \mathscr{C}(x)} E[U(X_T)] = v(x).$$

Proof Note that

$$\inf_{y>0} \{y(x - E[X_T Y_T])\} = \begin{cases} 0 & \text{if} \quad x - E[Y_T X_T] \geq 0 \\ -\infty & \text{otherwise.} \end{cases}$$

Hence,

$$\inf_{y>0} \{E[U(X_T)] + y(x - E[X_T Y_T])\} = \begin{cases} E[U(X_T)] & \text{if} \quad E[Y_T X_T] \leq x \\ -\infty & \text{otherwise.} \end{cases}$$

$$\inf_{y>0} \mathscr{L}(X_T, y) = \begin{cases} E[U(X_T)] & \text{if} \quad E[Y_T X_T] \leq x \\ -\infty & \text{otherwise.} \end{cases}$$

So,

$$\begin{aligned} \sup_{X_T \in L_+^0} \left(\inf_{y>0} \mathscr{L}(X_T, y) \right) &= \sup_{X_T \in L_+^0} \begin{cases} E[U(X_T)] & \text{if} \quad E[Y_T X_T] \leq x \\ -\infty & \text{otherwise.} \end{cases} \\ &= \sup_{X_T \in C(x)} E[U(X_T)] = v(x). \end{aligned}$$

This completes the proof.

The *dual problem* using the Lagrangian is

$$\text{dual :} \quad \inf_{y>0} \left(\sup_{X_T \in L_+^0} \mathscr{L}(X_T, y) \right).$$

For optimization problems we know that the dual problem provides an upper bound on the value function of the primal problem, i.e.

$$\inf_{y>0} \left(\sup_{X_T \in L_+^0} \mathscr{L}(X_T, y) \right) \geq \sup_{X_T \in L_+^0} \left(\inf_{y>0} \mathscr{L}(X_T, y) \right) = v(x).$$

Proof We have $\mathscr{L}(X_T, y) = E[U(X_T)] + y(x - E[X_T Y_T])$.

Note that $\sup_{X_T \in L_+^0} \mathscr{L}(X_T, y) \geq \mathscr{L}(X_T, y) \geq \inf_{y>0} \mathscr{L}(X_T, y)$ for all (X_T, y).

Taking the infimum over y on the left side and the supremum over X_T on the right side gives

$$\inf_{y>0} \left(\sup_{X_T \in L_+^0} \mathscr{L}(X_T, y) \right) \geq \sup_{X_T \in L_+^0} \left(\inf_{y>0} \mathscr{L}(X_T, y) \right).$$

This completes the proof.

The primal and dual problems' value functions are equal if there is no *duality gap*. Often, solving the dual problem is easier than solving the primal problem directly. For our problem, this turns out to be the case. Hence, to solve the investor's optimization problem we will show that there is no duality gap and we solve the dual problem. This is the task to which we now turn.

10.3 Existence of a Solution

We need to show that there is no duality gap and that the optimum is attained, equivalently, the Lagrangian has a saddle point. To prove the existence of a saddle point, we need two additional assumptions.

Assumption (Non-trivial Optimization Problem) *There exists an* $x > 0$ *such that* $v(x) < \infty$.

Assumption (Reasonable Asymptotic Elasticity (AE(U)<1))

$$AE(U, \omega) \equiv \limsup_{x \to \infty} \frac{xU'(x, \omega)}{U(x, \omega)} < 1 \quad a.s.\ \mathbb{P}. \tag{10.2}$$

The motivation for the first assumption is obvious. The motivation for the second assumption was discussed in the utility function Chap. 9 when the definition of asymptotic elasticity was introduced. Given these assumptions, we can prove the following theorem.

Theorem 10.2 (Existence of a Unique Saddle Point and Characterization of $v(x)$ for $x \geq 0$) *Given the above assumptions, there exists a unique saddle point, i.e. there exists a unique* $(\hat{X}_T, \hat{y})$ *such that*

$$\inf_{y>0} \left(\sup_{X_T \in L^0_+} \mathscr{L}(X_T, y) \right) = \mathscr{L}(\hat{X}_T, \hat{y}) = \sup_{X_T \in L^0_+} \left(\inf_{y>0} \mathscr{L}(X_T, y) \right) = v(x). \tag{10.3}$$

Define $\tilde{v}(y) \equiv E[\tilde{U}(yY_T)]$ *where* $\tilde{U}(y, \omega) = \sup_{x>0} [U(x, \omega) - xy], \quad y > 0.$

Then, v *and* $\tilde{v}$ *are in conjugate duality, i.e.*

$$v(x) = \inf_{y>0} (\tilde{v}(y) + xy), \ \forall x > 0, \qquad and \tag{10.4}$$

$$\tilde{v}(y) = \sup_{x>0} (v(x) - xy), \ \forall y > 0. \tag{10.5}$$

In addition,

(i) $\tilde{v}$ *is strictly convex, decreasing, differentiable on* $\text{int}(\text{dom}(\tilde{v})) \neq \emptyset$,
(ii) v *is strictly concave, increasing, and differentiable on* $(0, \infty)$, *and*
(iii) defining $\hat{y}$ *to be where the infimum is attained in expression (10.4)*

we have

$$v(x) = \tilde{v}(\hat{y}) + x\hat{y}. \tag{10.6}$$

Proof This theorem is a special case of Theorem 11.2 in the portfolio optimization Chap. 11. This completes the proof.

10.4 Characterization of the Solution

This section characterizes the solution to the portfolio optimization problem.

10.4.1 The Characterization

To characterize the solution, we focus on the dual problem, i.e.

$$v(x) = \inf_{y>0} \left(\sup_{X_T \in L_+^0} \mathscr{L}(X_T, y) \right)$$

$$= \inf_{y>0} \left(\sup_{X_T \in L_+^0} E\left[U(X_T) - yX_T Y_T\right] + xy \right).$$

To solve, we first exchange the sup and expectation operator

$$v(x) = \inf_{y>0} E\left[\sup_{X_T \in L_+^0} (U(X_T) - yX_T Y_T) + xy \right]. \tag{10.7}$$

The justification for this step is proven in the appendix to this chapter. We now solve this problem, working from the inside out.

Step 1. (Solve for $X_T \in L^0_+$)

Fix y and solve for the optimal $\hat{X}_T \in L^0_+$, i.e.

$$\sup_{X_T \in L^0_+} \left(U(X_T) - yX_T Y_T\right).$$

To solve this problem, fix a $\omega \in \Omega$, and consider the related problem

$$\sup_{X_T(\omega) \in \mathbb{R}_+} \left(U(X_T(\omega), \omega) - yX_T(\omega)Y_T(\omega)\right).$$

This is a simple optimization problem on the real line. The first-order condition for an optimum gives

$$U'(X_T(\omega), \omega) - yY_T(\omega) = 0, \qquad \text{or}$$

$$\hat{X}_T(\omega) = I(yY_T(\omega), \omega) \quad \text{with} \quad I(\cdot, \omega) = (U'(\cdot, \omega))^{-1}.$$

Given the properties of $I(\cdot, \omega)$, $\hat{X}_T(\omega)$ is $\mathscr{F}_T$-measurable, hence $\hat{X}_T(\omega) \in L^0_+$. Note that $\hat{X}_T$ depends on y.

Step 2. (Solve for $y > 0$)

Given the optimal $\hat{X}_T$, next we solve for the optimal Lagrangian multiplier $\hat{y}$. Using conjugate duality (see Lemma 9.7 in the utility function Chap. 9), we have

$$\tilde{U}(yY_T) = \sup_{X_T \in L^0_+} [U(X_T) - yX_T Y_T] = [U(\hat{X}_T) - y\hat{X}_T Y_T].$$

This transforms the problem to

$$\inf_{y>0} E\left[\sup_{X_T \in L^0_+} \left(U(X_T) - yX_T Y_T\right) + xy\right] = \inf_{y>0} \left(E[\tilde{U}(yY_T)] + xy\right).$$

Taking the derivative of the right side with respect to y yields

$$E[\tilde{U}'(yY_T)Y_T] + x = 0. \tag{10.8}$$

This step requires taking the derivative underneath the expectation operator. The justification for this step is proven in the appendix to this chapter. Noting from Lemma 9.7 in the utility function Chap. 9 that $\tilde{U}'(y) = -I(y)$, we get that the optimal $\hat{y}$ satisfies

$$E\left[I(\hat{y}Y_T)Y_T\right] = E[\hat{X}_T Y_T] = E^{\mathbb{Q}}[\hat{X}_T] = x, \tag{10.9}$$

which is that the optimal terminal wealth $\hat{X}_T$ satisfies the budget constraint with an equality. Solving this equation gives the optimal $\hat{y}$.

Step 3. (Characterization of $\hat{X} \in \mathscr{X}(x)$ for $x \geq 0$)

The optimal time t wealth is denoted by $\hat{X}_t$. Given the market is complete with respect to $\mathbb{Q}$, we know that there exists a nonnegative wealth s.f.t.s. process generating $\hat{X}_T$ at time T, where $x + \int_0^t \alpha_u \cdot dS_u = \hat{X}_t$ is a $\mathbb{Q}$-martingale, i.e.

$$\hat{X}_t = E^{\mathbb{Q}}\left[\hat{X}_T \,|\mathscr{F}_t\right]. \tag{10.10}$$

This implies that the optimal wealth process $\hat{X} \in \mathscr{X}(x)$ when multiplied by the local martingale deflator process, $\hat{X}_t Y_t$, is a $\mathbb{P}$-martingale. Indeed,

$$E\left[\hat{X}_T Y_T \,|\mathscr{F}_t\right] = E\left[\hat{X}_T \frac{Y_T}{Y_t}\middle|\mathscr{F}_t\right] Y_t = E^{\mathbb{Q}}\left[\hat{X}_T \,|\mathscr{F}_t\right] Y_t = \hat{X}_t Y_t.$$

Step 4. (Nonnegativity of $\hat{X} \in \mathscr{X}(x)$)

In the optimization problem, time T wealth must be nonnegative. This is captured by the restriction that $\hat{X}_T \in L^0_+$, i.e. $\hat{X}_T \geq 0$. From expression (10.10), this implies that the optimal wealth process $\hat{X}_t$ must be nonnegative for all t as well, i.e.

$$E^{\mathbb{Q}}\left[\hat{X}_T \,|\mathscr{F}_t\right] = \hat{X}_t \geq 0.$$

Remark 10.2 (Random Variables Versus Stochastic Processes) The optimization problem determines the random variable $\hat{X}_T \in L^0_+$, which is the optimal time T wealth. But, since the market is complete with respect to $\mathbb{Q}$, there exists $(\alpha_0, \alpha) \in \mathscr{N}(x)$ with $x \geq 0$ such that $X_t = x + \int_0^t \alpha_u \cdot dS_u = E^{\mathbb{Q}}\left[\hat{X}_T \,|\mathscr{F}_t\right]$. This characterizes the optimal wealth (stochastic) process $X \in \mathscr{X}(x)$. From this optimal wealth process, the nonnegative wealth s.f.t.s. generating it can be determined using the methods in the fundamental theorems Chap. 2.

Similarly, the NFLVR and complete market assumptions, by the Second Fundamental Theorem 2.4 of asset pricing in Chap. 2, identify a unique equivalent local martingale probability measure $\mathbb{Q} \in \mathfrak{M}_l$, which in turn identifies the unique local martingale deflator $Y_T = \frac{d\mathbb{Q}}{d\mathbb{P}} \in M_l \subset D_l$. This local martingale deflator uniquely determines the local martingale deflator process $Y \in \mathscr{M}_l \subset \mathscr{D}_l$ via the $\mathbb{P}$-martingale condition

$$Y_t = E\left[Y_T \,|\mathscr{F}_t\right].$$

These observations will prove useful, for comparisons, when we study optimization in an incomplete market in Chap. 11 below. This completes the remark.

10.4.2 Summary

For easy reference we summarize the previous characterization results. The optimal value function

$$v(x) = \sup_{X_T \in \mathscr{C}(x)} E\left[U(X_T)\right]$$

for $x \geq 0$ has the solution for terminal wealth given by

$$\hat{X}_T = I(\hat{y}Y_T) \quad \text{with} \quad I = (U')^{-1} = -\tilde{U}',$$

where $Y_T = \frac{d\mathbb{Q}}{d\mathbb{P}} \in M_l \subset D_l$ is the local martingale deflator associated with the unique local martingale measure $\mathbb{Q} \in \mathfrak{M}_l$. This generates the local martingale deflator process $Y \in \mathscr{M}_l \subset \mathscr{D}_l$ via the expression

$$Y_t = E\left[Y_T \,|\mathscr{F}_t\right].$$

The Lagrangian multiplier $\hat{y}$ is the solution to the budget constraint

$$E\left[Y_T I(\hat{y}Y_T)\right] = x,$$

where the optimal wealth $\hat{X}_T$ satisfies the budget constraint with an equality

$$E\left[Y_T \hat{X}_T\right] = x.$$

The optimal wealth process $\hat{X} \in \mathscr{X}(x)$ for $x \geq 0$ exists because by market completeness with respect to $\mathbb{Q}$, there exists a nonnegative wealth s.f.t.s. $(\alpha_0, \alpha) \in \mathscr{N}(x)$ with time T payoff $\hat{X}_T$, where $\hat{X}_t = x + \int_0^t \alpha_u \cdot dS_u = E^{\mathbb{Q}}\left[\hat{X}_T \,|\mathscr{F}_t\right] \geq 0$ for all t. The optimal wealth process $\hat{X} \in \mathscr{X}(x)$, when multiplied by the local martingale deflator process, $\hat{X}_t Y_t \geq 0$, is a $\mathbb{P}$-martingale.

10.5 The Shadow Price

Using expression (10.6), we can obtain the shadow price of the budget constraint. The shadow price is the benefit, in terms of expected utility, of increasing the initial wealth x by 1 unit (of the mma). The shadow price equals the Lagrangian multiplier.

Theorem 10.3 (Shadow Price of the Budget Constraint)

$$\hat{y} = v'(x) = E[U'(\hat{X}_T)]. \tag{10.11}$$

Proof The first equality is obtained by taking the derivative of $v(x) = \tilde{v}(\hat{y}) + x\hat{y}$. The second equality follows from the first-order condition for an optimum $U'(\hat{X}_T) = \hat{y}Y_T$. Taking expectations gives $E[U'(\hat{X}_T)] = E[\hat{y}Y_T] = \hat{y} < \infty$, where $E[Y_T] = 1$. This completes the proof.

10.6 The State Price Density

A key insight of the individual optimization problem is the characterization of the local martingale deflator, in this case the probability density $Y_T = \frac{d\mathbb{Q}}{d\mathbb{P}} \in M_l$ of an equivalent local martingale measure $\mathbb{Q} \in \mathfrak{M}_l$, in terms of the individual's utility function and initial wealth. To obtain this characterization, using Theorem 10.3 above, we rewrite the first-order condition for the optimal wealth $\hat{X}_T$ as

$$Y_T = \frac{U'(\hat{X}_T)}{\hat{y}} = \frac{U'(\hat{X}_T)}{E[U'(\hat{X}_T)]}. \tag{10.12}$$

As seen, the local martingale deflator equals the investor's marginal utility of wealth, normalized by the expected marginal utility of terminal wealth.

Given investors trade in a competitive market, prices are taken as exogenous. And since the market is complete with respect to $\mathbb{Q}$, prices uniquely determine the equivalent local martingale measure $\mathbb{Q}$. At an optimum, the individual equates her marginal utility (normalized) to the market determined probability density $Y_T = \frac{d\mathbb{Q}}{d\mathbb{P}} \in M_l$ of the equivalent local martingale measure $\mathbb{Q} \in \mathfrak{M}_l$. Of course, in equilibrium (Part III of this book), the local martingale deflator is endogenously determined by all of the investors' collective trades and the market clearing conditions.

Since the local martingale deflator $Y_T \in M_l$ is a probability density with respect to $\mathbb{P}$, the state price density process is given by the $\mathbb{P}$-martingale relation

$$Y_t = E\left[Y_T \,|\mathscr{F}_t\right] = \frac{E[U'(\hat{X}_T)\,|\mathscr{F}_t]}{E[U'(\hat{X}_T)]}. \tag{10.13}$$

Remark 10.3 (Asset Price Bubbles) In this individual optimization problem, prices and hence the local martingale measure $\mathbb{Q}$ are taken as exogenous. The above analysis does not require that $\mathbb{Q}$ is a martingale measure. Hence, from the asset price bubbles Chap. 3, since $\mathbb{Q}$ can be a strict local martingale measure, this optimization problem applies to a market with asset price bubbles. This completes the remark.

Remark 10.4 (Systematic Risk) In this remark, we use the non-normalized market $((\mathbb{B}, \mathbb{S}), (\mathscr{F}_t), \mathbb{P})$ representation to characterize systematic risk. Recall that the state price density is the key input to the systematic risk return relation in the multiple-factor model Chap. 4, Theorem 4.4. To apply this theorem, we *assume that the local martingale deflator $Y_T \in M_l$ is a martingale deflator*, i.e. $Y_T \in M$.

Then, the risky asset returns over the time interval $[t, t+\triangle]$ satisfy

$$E\left[R_i(t)\,|\mathscr{F}_t\right] = r_0(t) - \text{cov}\left[R_i(t), \frac{E[U'(\hat{X}_T)\,|\mathscr{F}_{t+\triangle}]}{E[U'(\hat{X}_T)\,|\mathscr{F}_t]} \frac{\mathbb{B}_t}{\mathbb{B}_{t+\triangle}}(1+r_0(t)) \middle| \mathscr{F}_t \right] \tag{10.14}$$

where the ith risky asset return is $R_i(t) = \frac{\mathbb{S}_i(t+\Delta)-\mathbb{S}_i(t)}{\mathbb{S}_i(t)}$, $r_0(t) \equiv \frac{1}{p(t,t+\Delta)} - 1$ is the default-free spot rate of interest where $p(t, t+\Delta)$ is the time t price of a default-free zero-coupon bond maturing at time $t+\Delta$, and the state price density is

$$H_t = \frac{Y_t}{\mathbb{B}_t} = \frac{E\left[Y_T\,|\mathscr{F}_t\right]}{\mathbb{B}_t} = \frac{E[U'(\hat{X}_T)\,|\mathscr{F}_t]}{E[U'(\hat{X}_T)]\mathbb{B}_t}.$$

When the time step is very small, i.e. $\Delta \approx dt$, the expression $\frac{\mathbb{B}_t}{\mathbb{B}_{t+\triangle}}(1+r_0(t)) \approx 1$, and this relation simplifies to

$$E\left[R_i(t)\,|\mathscr{F}_t\right] \approx r_0(t) - \text{cov}\left[R_i(t), \frac{E[U'(\hat{X}_T)\,|\mathscr{F}_{t+\triangle}]}{E[U'(\hat{X}_T)\,|\mathscr{F}_t]} \middle| \mathscr{F}_t \right]. \tag{10.15}$$

This last expression clarifies the meaning of systematic risk. Recalling the discussion in the multiple-factor model Chap. 4, systematic risk is characterized by the right side of expression (10.15) where we have replaced the term $\frac{H_{t+\triangle}}{H_t}(1+r_0(t))$ with the ratio $\left(\frac{E[U'(\hat{X}_T)|\mathscr{F}_{t+\triangle}]}{E[U'(\hat{X}_T)|\mathscr{F}_t]}\right)$. The intuition for this ratio is as follows.

- When $\left(\frac{E[U'(\hat{X}_T)|\mathscr{F}_{t+\triangle}]}{E[U'(\hat{X}_T)|\mathscr{F}_t]}\right)$ is large, systematic risk is low (note the negative sign). This occurs when the marginal utility of wealth is large, hence wealth is scarce. An asset whose return is large when wealth is scarce is valuable. Such an asset is "anti-risky."
- When $\left(\frac{E[U'(\hat{X}_T)|\mathscr{F}_{t+\triangle}]}{E[U'(\hat{X}_T)|\mathscr{F}_t]}\right)$ is small, systematic risk is high. This occurs when the marginal utility of wealth is small, hence wealth is plentiful. An asset whose return is large when wealth is plentiful is less valuable then when wealth is scarce. Such an asset is "risky."

This completes the remark.

10.7 The Optimal Trading Strategy

This section solves step 2 of the optimization problem, which is to determine the optimal trading strategy $(\alpha_0, \alpha) \in \mathscr{N}(x)$ for $x \geq 0$. This is done using the standard techniques developed for the synthetic construction of derivatives in the fundamental theorems Chap. 2.

Since the market is complete with respect to $\mathbb{Q}$, we know there exists a nonnegative wealth s.f.t.s. $(\hat{\alpha}_0, \alpha) \in \mathscr{N}(x)$ such that

$$\hat{X}_t = x + \int_0^t \alpha_u \cdot dS_u = E^{\mathbb{Q}}[\hat{X}_T \,|\mathscr{F}_t\,] \tag{10.16}$$

for all t, where

$$x = \hat{X}_0 \qquad \text{and} \qquad \hat{X}_t = \alpha_0(t) + \alpha_t \cdot S_t. \tag{10.17}$$

We need to characterize this nonnegative wealth s.f.t.s. $(\alpha_0(t), \alpha_t)$ given the initial wealth $x \geq 0$. To illustrate this determination, suppose that S is a Markov diffusion process. Due to the diffusion assumption, S is a continuous process. Given this assumption, we can write the optimal wealth process as a deterministic function of time t and the risky asset prices S_t, i.e.

$$E^{\mathbb{Q}}[\hat{X}_T \,|\mathscr{F}_t\,] = E^{\mathbb{Q}}[\hat{X}_T \,|S_t\,] = \hat{X}_t(S_t).$$

Using Ito's formula (assuming the appropriate differentiability conditions hold), we have that

$$\hat{X}_t(S_u) = x + \int_0^t \frac{\partial \hat{X}_u}{\partial u} du + \frac{1}{2}\int_0^t \sum_{i=1}^n \sum_{j=1}^n \frac{\partial^2 \hat{X}_u}{\partial S_{ij}^2(u)} d\left\langle S_i(u), S_j(u)\right\rangle + \int_0^t \frac{\partial \hat{X}_u}{\partial S_u} \cdot dS_u, \tag{10.18}$$

where $\frac{\partial \hat{X}}{\partial S} = (\frac{\partial \hat{X}}{\partial S_1}, \ldots, \frac{\partial \hat{X}}{\partial S_n})' \in \mathbb{R}^n$. Equating the integrands of dS_t in the two stochastic integral equations immediately above, the holdings in the risky assets are easily determined to be

$$\frac{\partial \hat{X}_u(S_u)}{\partial S_u} = \alpha_u.$$

This is the standard "delta" used to construct a "synthetic derivative," in this case the optimal terminal wealth, $\hat{X}_T$. Finally, given the holdings in the risky assets α_t, the holdings in the mma are obtained from expression (10.17). If the underlying asset price process is not a Markov diffusion process, then for a large class of processes one can use Malliavin calculus to determine the trading strategy (see Detemple et al. [49] and Nunno et al. [51]).

10.8 An Example

This section presents an example to illustrate the abstract expressions of the previous sections. This example is from Pham [149, p. 196].

10.8.1 The Market

Suppose the normalized market ($r = 0$) consists of a single risky asset and an mma where the risky asset price process follows a geometric Brownian motion, i.e.

$$\begin{aligned} dS_t &= S_t\,(b dt + \sigma dW_t) \text{ or} \\ S_t &= S_0 e^{bt - \frac{1}{2}\sigma^2 t + \sigma W_t}, \end{aligned} \tag{10.19}$$

where b, σ are strictly positive constants and W_t is a standard Brownian motion with $W_0 = 0$ that generates the filtration $(\mathscr{F}_t)$.

In the BSM model Chap. 5 it was shown that for geometric Brownian motion, the market is complete with respect to $\mathbb{Q} \in \mathfrak{M}_l$. Hence, by the Second Fundamental Theorem 2.4 of asset pricing in Chap. 2, the equivalent local martingale measure is unique. In this case, it was also shown that $\mathbb{Q}$ is an equivalent martingale measure, and that its probability density is given by

$$Y_T = \frac{d\mathbb{Q}}{d\mathbb{P}} = e^{-\theta \cdot W_T - \frac{1}{2}\theta^2 T} > 0, \tag{10.20}$$

where $\theta = \frac{b}{\sigma}$. Thus, the evolution of the risky asset's price is a martingale under $\mathbb{Q}$, and it is given by

$$\begin{aligned} dS_t &= S_t \sigma dW_t^\theta \quad \text{ or} \\ S_t &= S_0 e^{-\frac{1}{2}\sigma^2 t + \sigma W_t^\theta}, \end{aligned} \tag{10.21}$$

where $dW^\theta(t) = dW(t) + \theta(t)dt$ is a Brownian motion under $\mathbb{Q}$.

This implies that under $\mathbb{Q}$, using $dW(t) = dW^\theta(t) - \theta(t)dt$, we have that the martingale deflator is

$$Y_T = e^{-\theta \cdot W_T^\theta + \theta^2 T - \frac{1}{2}\theta^2 T} = e^{-\theta \cdot W_T^\theta + \frac{1}{2}\theta^2 T} > 0. \tag{10.22}$$

10.8.2 The Utility Function

Let preferences be represented by a state independent power utility function

$$U(x) = \frac{x^\rho}{\rho}$$

for $x > 0$ and $\rho < 1$, $\rho \neq 0$. Recall that

$$I(y) = y^{\frac{1}{\rho - 1}}$$

for $y > 0$.

10.8.3 The Optimal Wealth Process

The optimal wealth process is

$$\hat{X}_T = I(\hat{y}Y_T) = (\hat{y}Y_T)^{-\kappa},$$

where $\kappa = \frac{1}{1-\rho} > 0$. And, for an arbitrary time t,

$$\hat{X}_t = E^{\mathbb{Q}}\left[I(\hat{y}Y_T)\,|\mathscr{F}_t\right] = E^{\mathbb{Q}}\left[(\hat{y}Y_T)^{-\kappa}\,|\mathscr{F}_t\right].$$

By algebra and substitution this is equal to

$$\begin{aligned}\hat{y}^{-\kappa}E^{\mathbb{Q}}\left[{Y_T}^{-\kappa}\,|\mathscr{F}_t\right] &= \hat{y}^{-\kappa}E^{\mathbb{Q}}\left[e^{\theta\kappa\cdot W_T^{\theta}-\frac{1}{2}\theta^2\kappa T}\,|\mathscr{F}_t\right]\\ &= \hat{y}^{-\kappa}e^{-\frac{1}{2}\theta^2\kappa T}e^{\theta\kappa\cdot W_t^{\theta}+\frac{1}{2}(\theta\kappa)^2(T-t)}.\end{aligned}$$

Recall that $\hat{y}$ is the solution to the budget constraint

$$\hat{X}_0 = E^{\mathbb{Q}}\left[I(\hat{y}Y_T)\right] = x.$$

Substitution for time 0 gives

$$x = \hat{y}^{-\kappa}e^{-\frac{1}{2}\theta^2\kappa T+\frac{1}{2}(\theta\kappa)^2 T}.$$

Combined we get that the optimal wealth process is

$$\hat{X}_t = xe^{\theta\kappa\cdot W_t^{\theta}-\frac{1}{2}(\theta\kappa)^2 t}. \tag{10.23}$$

10.8.4 The Optimal Trading Strategy

To get the optimal trading strategy we apply Ito's formula to expression (10.23), which yields

$$d\hat{X}_t = \hat{X}_t\theta\kappa dW_t^{\theta}.$$

Proof Dropping all the subscripts and superscripts,

$$\begin{aligned}dX &= \frac{\partial X}{\partial t}dt + \frac{1}{2}\frac{\partial^2 X}{\partial W^2}dt + \frac{\partial X}{\partial W}dW^{\theta}.\\ &= -X\frac{1}{2}(\theta\kappa)^2 dt + \frac{1}{2}X(\theta\kappa)^2 dt + X\theta\kappa dW^{\theta}.\end{aligned}$$

This completes the proof.

Since the market is complete with respect to $\mathbb{Q}$, we know that there exists a nonnegative wealth s.f.t.s. $(\hat{\alpha}_0, \hat{\alpha}) \in \mathcal{N}(x)$ with $x \geq 0$ such that

$$d\hat{X}_t = \hat{\alpha}_t dS_t = \hat{\alpha}_t S_t \sigma dW_t^{\theta}.$$

Equating the coefficients of the Brownian motion terms in the above two expressions for $d\hat{X}_t$ yields

$$\begin{aligned} \hat{\alpha}_t S_t \sigma &= \hat{X}_t \theta \kappa \text{ or} \\ \hat{\alpha}_t &= \frac{\hat{X}_t \theta \kappa}{S_t \sigma}. \end{aligned} \tag{10.24}$$

Note that the holdings in the mma are determined by the expression $\hat{\alpha}_0(t) = \hat{X}_t - \hat{\alpha}_t S_t$ for all t.

Expressing these holdings as a proportion of wealth, we get (using $\theta = \frac{b}{\sigma}$)

$$\hat{\pi}_t = \frac{\hat{\alpha}_t S_t}{\hat{X}_t} = \frac{\theta \kappa}{\sigma} = \frac{b}{\sigma^2(1-\rho)}. \tag{10.25}$$

This shows that for power utility functions in a geometric Brownian motion market, the optimal portfolio weight in the risky asset is a constant across time and independent of the level of wealth.

10.8.5 The Value Function

Last, to obtain the value function, note that

$$\begin{aligned} v(x) &= E\left[U(\hat{X}_T)\right] = E\left[U(xe^{\theta\kappa \cdot W_T - \frac{1}{2}(\theta\kappa)^2 T})\right] \\ &= E\left[\frac{(xe^{\theta\kappa \cdot W_T - \frac{1}{2}(\theta\kappa)^2 T})^{\rho}}{\rho}\right] = \frac{x^{\rho}}{\rho} e^{-\frac{1}{2}\rho(\theta\kappa)^2 T} E\left[e^{\theta\kappa\rho \cdot W_T}\right] \\ &= \frac{x^{\rho}}{\rho} e^{-\frac{1}{2}\rho(\theta\kappa)^2 T}\left[e^{\frac{1}{2}(\theta\kappa)^2\rho^2 T}\right] = \frac{x^{\rho}}{\rho} e^{-\frac{1}{2}\rho(1-\rho)(\theta\kappa)^2 T}. \end{aligned}$$

Substituting in $\kappa = \frac{1}{1-\rho} > 0$ and $\theta = \frac{b}{\sigma}$ gives

$$v(x) = \frac{x^{\rho}}{\rho} e^{-\frac{1}{2}\rho(1-\rho)(\frac{1}{1-\rho})^2 \frac{b^2}{\sigma^2} T},$$

or

$$v(x) = \frac{x^\rho}{\rho} e^{-\frac{1}{2}\frac{\rho}{1-\rho}\frac{b^2}{\sigma^2}T}. \tag{10.26}$$

This completes the example.

10.9 Notes

Optimization in a complete market is often studied as a special case of an incomplete market. Excellent references for solving the investor's optimization problem include Dana and Jeanblanc [42], Duffie [52], Karatzas and Shreve [118], Merton [140], and Pham [149].

Appendix

Portfolio Weights

Define the portfolio weights as

$$\pi_0(t) = \frac{\alpha_0(t)\mathbb{B}_t}{\mathbb{X}_t} = \frac{\alpha_0(t)}{X_t} \qquad \text{and}$$

$$\pi_i(t) = \frac{\alpha_i(t)\mathbb{S}_i(t)}{\mathbb{X}_t} = \frac{\alpha_i(t)\left(\frac{\mathbb{S}_i(t)}{\mathbb{B}_t}\right)}{\frac{\mathbb{X}_t}{\mathbb{B}_t}} = \frac{\alpha_i(t)S_i(t)}{X_t}$$

for all $i = 1, \ldots, n$.

Note that in percentage holdings, the normalization versus non-normalization definitions are identical. Of course, all quantities in the denominator as assumed to be nonzero.

Define $\pi_t = \pi(t) = (\pi_1(t), \ldots, \pi_n(t))' \in \mathbb{R}^n$.

Note that $\pi_0(t) + \pi_t \cdot \mathbf{1} = 1$ where $\mathbf{1} = (1, \ldots, 1)' \in \mathbb{R}^n$.

The following gives the correspondences between the two formulations

$$\begin{array}{rcl} \pi_i(t) = \frac{\alpha_i(t)S_i(t)}{X_t} & \Leftrightarrow & \alpha_i(t) = \frac{\pi_i(t)X_t}{S_i(t)} \\ X_t = x + \int_0^t \frac{\pi_u}{S_u} X_u \cdot dS_u & \Leftrightarrow & X_t = x + \int_0^t \alpha_u \cdot dS_u \\ 1 = \pi_0(t) + \pi_t \cdot \mathbf{1} & \Leftrightarrow & X_t = \alpha_0(t) + \alpha_t \cdot S_t \end{array}$$

where $\frac{\pi_t}{S_t} = \left(\frac{\pi_1(t)}{S_1(t)}, \ldots, \frac{\pi_n(t)}{S_n(t)}\right)' \in \mathbb{R}^n$ and (for later use) $\frac{dS_t}{S_t} = \left(\frac{dS_1(t)}{S_1(t)}, \ldots, \frac{dS_n(t)}{S_n(t)}\right)'$ $\in \mathbb{R}^n$.

Proof The proof of the last identification is

$$\begin{aligned} X_t &= \alpha_0(t) + \alpha_t \cdot S(t) \quad \Leftrightarrow \\ \frac{X_t}{X_t} &= \frac{\alpha_0(t)}{X_t} + \frac{\alpha_t \cdot S(t)}{X_t} \quad \Leftrightarrow \\ 1 &= \pi_0(t) + \pi_t \cdot 1. \end{aligned}$$

This completes the proof.

Proof of Expression (10.7)

For the use of this result in an incomplete market, the portfolio optimization Chap. 11, we note that the following proof holds for any $Y_T \in D_s$ as well.

Proof (Exchange of sup and E[·] Operator) It is trivial that

$$\sup_{X_T \in L^0_+} E\left[U(X_T) - yX_TY_T\right] \leq E\Big[\sup_{X_T \in L^0_+} (U(X_T) - yX_TY_T)\Big].$$

We want to prove the opposite inequality.

Since U is strictly concave, there exists a unique solution X^*_T to

$$\sup_{X_T \in L^0_+} \left[U(X_T) - yX_TY_T\right] = \left[U(X^*_T) - yX^*_TY_T\right].$$

But,

$$\begin{aligned} \sup_{X_T \in L^0_+} E\left[U(X_T) - yX_TY_T\right] &\geq E\left[U(X^*_T) - yX^*_TY_T\right] \\ &= E\Big[\sup_{X_T \in L^0_+} (U(X_T) - yX_TY_T)\Big], \end{aligned}$$

which completes the proof.

Proof of Expression (10.8)

For the use of this result in an incomplete market, the portfolio optimization Chap. 11, we note that the following proof holds for any $Y_T \in D_s$ as well.

Proof (Exchange of $E[\cdot]$ *and Derivative)* Since the derivative exists, we use the left derivative.

$$\lim_{\Delta \to 0} \frac{\tilde{U}((y+\Delta)Y_T) - \tilde{U}(yY_T)}{\Delta} = \lim_{\Delta Y_T \to 0} \frac{\tilde{U}((y+\Delta)Y_T) - \tilde{U}(yY_T)}{(y+\Delta)Y_T - yY_T} \cdot \lim_{\Delta \to 0} \frac{(y+\Delta)Y_T - yY_T}{\Delta}$$

$$= \lim_{\Delta Y_T \to 0} \frac{\tilde{U}((y+\Delta)Y_T) - \tilde{U}(yY_T)}{(y+\Delta)Y_T - yY_T} \cdot Y_T$$

$$= \lim_{\Delta \to 0} \tilde{U}'(\xi Y_T) Y_T \text{ a.s. } \mathbb{P}, \text{ where } \Delta < 0.$$

The last equality follows from the mean value theorem (Guler [66, p. 3]), i.e. there exists an $\xi \in [y + \Delta, y]$ such that

$$\tilde{U}((y+\Delta)Y_T) - \tilde{U}(yY_T) = \tilde{U}'(\xi Y_T)\left[(y+\Delta)Y_T - yY_T\right].$$

Thus, $\frac{\partial E[\tilde{U}(yY_T)]}{\partial y} = \lim_{\Delta \to 0} \frac{E\left[\tilde{U}((y+\Delta)Y_T) - \tilde{U}(yY_T)\right]}{\Delta} = \lim_{\Delta \to 0} E[\tilde{U}'(\xi Y_T)Y_T]$.

Now $E[\tilde{U}(yY_T)] < \infty$ because $\tilde{v}(y) < \infty$ and Y_T is the supermartingale deflator such that $\tilde{v}(y) = E[\tilde{U}(yY_T)]$.

By Kramkov and Schachermayer [126, Lemma 6.3 (iv) and (iii), p. 944], $AE(U) < 1$ implies there exists a constant C and $z_0 > 0$ such that

$$-\tilde{U}'(z)z < C\tilde{U}(z) \text{ for } 0 < z \le z_0,$$

and

$$\tilde{U}(\mu z) < K(\mu)\tilde{U}(z) \text{ for } 0 < \mu < 1 \text{ and } 0 < z \le z_0,$$

where $K(\mu)$ is a constant depending upon μ.

Combined, $-\tilde{U}'(\mu z)\mu z < C\tilde{U}(\mu z) < CK(\mu)\tilde{U}(z)$ implies that there exists a $z_0 > 0$ such that

$$-\tilde{U}'(\mu z)\mu z < \bar{K}(\mu)\tilde{U}(z) \text{ for } 0 < z \le z_0$$

where $\bar{K}(\mu)$ is a constant depending upon μ for $0 < \mu < 1$.

Letting $z = yY_T$ and $\mu = \frac{\xi}{y} < 1$, because $\Delta < 0$ so that $\xi < y$. Then,

$$-\tilde{U}'(\xi Y_T)Y_T < \frac{1}{\xi}\bar{K}\left(\frac{\xi}{y}\right)\tilde{U}(yY_T).$$

Since the right side is $\mathbb{P}$-integrable, using the dominated convergence theorem,

$$\lim_{\Delta\to 0} E[\tilde{U}'(\xi Y_T)Y_T] = E[\lim_{\Delta\to 0} \tilde{U}'(\xi Y_T)Y_T] = E[\tilde{U}'(yY_T)Y_T].$$

The last equality follows from the continuity of $\tilde{U}'(\cdot)$. The continuity of $\tilde{U}'(\cdot)$ follows because $\tilde{U}(\cdot)$ is strictly convex, hence $\tilde{U}'(\cdot)$ is a strictly increasing function, which is therefore differentiable a.s. $\mathbb{P}$ (see Royden [160, Theorem 2, p. 96]), and hence continuous. This completes the proof.

Chapter 11
Incomplete Markets (Utility over Terminal Wealth)

This chapter studies the investor's portfolio optimization problem in an incomplete market. The solution in this chapter parallels the solution for the complete market setting in the portfolio optimization Chap. 10.

11.1 The Set-Up

Given is a normalized market $(S, (\mathscr{F}_t), \mathbb{P})$ where the value of a money market account $B_t \equiv 1$. As before, we assume that there are no arbitrage opportunities in the market, i.e.

Assumption (NFLVR) $\mathfrak{M}_l \neq \emptyset$ *where* $\mathfrak{M}_l = \{\mathbb{Q} \sim \mathbb{P} : S \text{ is a } \mathbb{Q}\text{-local martingale}\}$.

In this chapter the market may be incomplete. In an incomplete market, given an equivalent local martingale measure $\mathbb{Q} \in \mathfrak{M}_l$, there exist integrable derivatives that cannot be synthetically constructed using a value process that is a $\mathbb{Q}$-martingale. If the set of equivalent local martingale measures is not a singleton, then the cardinality of $|\mathfrak{M}_l| = \infty$, i.e. the set contains an infinite number of elements. In this case, derivative prices are not uniquely determined because the equivalent local martingale measure in not unique. Here, the investor's portfolio optimization problem provides a method to identify the relevant local martingale measure for pricing derivatives.

11.2 Problem Statement

This section presents the investor's portfolio optimization problem. We assume that the investor's preferences $\overset{\rho}{\succeq}$ have a state dependent EU representation with the utility function defined over terminal wealth $U(X_T, \omega)$ and satisfying the assumption given in the utility function Chap. 9, repeated here for convenience.

R. A. Jarrow, *Continuous-Time Asset Pricing Theory*, Springer Finance,
https://doi.org/10.1007/978-3-319-77821-1_11

Assumption (Utility Function)
A utility function is a $U : (0, \infty) \times \Omega \to \mathbb{R}$ *such that for all* $\omega \in \Omega$ *a.s.* $\mathbb{P}$,

(i) $U(x, \omega)$ *is* $\mathscr{B}(0, \infty) \otimes \mathscr{F}_T$*-measurable,*
(ii) $U(x, \omega)$ *is continuous and differentiable in* $x \in (0, \infty)$,
(iii) $U(x, \omega)$ *is strictly increasing and strictly concave in* $x \in (0, \infty)$, *and*
(iv) (Inada Conditions) $\lim_{x \downarrow 0} U'(x, \omega) = \infty$ *and* $\lim_{x \to \infty} U'(x, \omega) = 0$.

We do not consider utility over intermediate consumption in this chapter. The solution to the portfolio optimization problem with intermediate consumption is discussed in the portfolio optimization Chap. 12.

Recall that the set of *nonnegative wealth s.f.t.s.'s* for $x \geq 0$ is

$$\mathscr{N}(x) = \Big\{(\alpha_0, \alpha) \in (\mathscr{O}, \mathscr{L}(S)) : X_t = \alpha_0(t) + \alpha_t \cdot S_t, \\ X_t = x + \int_0^t \alpha_u \cdot dS_u \geq 0, \ \forall t \in [0, T]\Big\}.$$

The set of value processes generated by the nonnegative wealth s.f.t.s.'s for $x \geq 0$ is

$$\mathscr{X}^e(x) = \left\{X \in \mathscr{L}^0_+ : \exists (\alpha_0, \alpha) \in \mathscr{N}(x), \ X_t = x + \int_0^t \alpha_u \cdot dS_u, \ \forall t \in [0, T]\right\},$$

and the set of time T terminal wealths generated by these value processes for $x \geq 0$ is

$$\mathscr{C}^e(x) = \{X_T \in L^0_+ : \exists (\alpha_0, \alpha) \in \mathscr{N}(x), \ x + \int_0^T \alpha_t \cdot dS_t = X_T\} \\ = \{X_T \in L^0_+ : \exists Z \in \mathscr{X}(x), \ X_T = Z_T\}$$

where the superscript "e" stands for equality in the constraint set.

The trader's problem is to choose a nonnegative wealth s.f.t.s. to maximize their expected utility of terminal wealth. For simplicity of notation, we suppress the dependence of the utility function on the state $\omega \in \Omega$, writing $U(x) = U(x, \omega)$ everywhere below.

Problem 11.1 (Utility Optimization (Choose the Optimal Trading Strategy (α_0, α) **for** $x \geq 0$**))**

$$v(x) = \sup_{(\alpha_0, \alpha) \in \mathscr{N}(x)} E\left[U\left(X_T\right)\right], \qquad \text{where}$$

$$\mathscr{N}(x) = \Big\{(\alpha_0, \alpha) \in (\mathscr{O}, \mathscr{L}(S)) : X_t = \alpha_0(t) + \alpha_t \cdot S_t, \\ X_t = x + \int_0^t \alpha_u \cdot dS_u \geq 0, \ \forall t \in [0, T]\Big\}.$$

Similar to a complete market, we first solve the static problem to choose the optimal time T wealth (a "derivative"). Then, after solving the static problem, we use the methods for the synthetic construction of a derivative (see Sect. 2.6.3 in the fundamental theorems Chap. 2) to determine the trading strategy that generates the optimal wealth.

Problem 11.2 (Utility Optimization (Choose the Optimal Derivative X_T for $x \geq 0$))

$$v(x) = \sup_{X_T \in \mathscr{C}(x)} E\left[U\left(X_T\right)\right], \qquad \text{where} \tag{11.1}$$

$$\mathscr{C}^e(x) = \{X_T \in L_+^0 : \exists (\alpha_0, \alpha) \in \mathscr{N}(x),\ x + \int_0^T \alpha_t \cdot dS_t = X_T\}.$$

To facilitate the determination of the solution, we solve an equivalent problem, where we allow some of the time T wealth to be discarded, called *free disposal.* The equivalent optimization problem is

$$v(x) = \sup_{X_T \in \mathscr{C}(x)} E\left[U\left(X_T\right)\right], \qquad \text{where} \tag{11.2}$$

$$\mathscr{C}(x) = \{X_T \in L_+^0 : \exists (\alpha_0, \alpha) \in \mathscr{N}(x),\ x + \int_0^T \alpha_t \cdot dS_t \geq X_T\}.$$

The equivalence of these two problems is guaranteed by the next lemma.

Lemma 11.1 (Free Disposal) *X_T is an optimal solution to problem (11.1) if and only if X_T is an optimal solution to problem (11.2).*

Proof

(Step 1) Since $\mathscr{C}^e(x) \subset \mathscr{C}(x)$, we have

$$\sup_{X_T \in \mathscr{C}^e(x)} E\left[U\left(X_T\right)\right] \leq \sup_{X_T \in \mathscr{C}(x)} E\left[U\left(X_T\right)\right].$$

(Step 2) Suppose $\exists (\alpha_0, \alpha) \in \mathscr{N}(x)$, $x + \int_0^T \alpha_t \cdot dS_t > \hat{X}_T$, where $\hat{X}_T$ is an optimal solution of $\sup_{X_T \in \mathscr{C}(x)} E\left[U\left(X_T\right)\right]$.

Let $Z_T = x + \int_0^T \alpha_t \cdot dS_t$. Then, $E\left[U\left(Z_T\right)\right] > E\left[U\left(\hat{X}_T\right)\right] = \sup_{X_T \in \mathscr{C}(x)} E\left[U\left(X_T\right)\right]$ since $U(\cdot)$ is strictly increasing.

But, $\sup_{X_T \in \mathscr{C}^e(x)} E\left[U\left(X_T\right)\right] \geq E\left[U\left(Z_T\right)\right]$.

This implies that $\sup_{X_T \in \mathscr{C}^e(x)} E\left[U\left(X_T\right)\right] > \sup_{X_T \in \mathscr{C}(x)} E\left[U\left(X_T\right)\right]$. This contradicts (Step 1).

Thus, the solution $\hat{X}_T$ to $\sup_{X_T \in \mathscr{C}(x)} E\left[U\left(X_T\right)\right]$ must satisfy $x + \int_0^T \alpha_t \cdot dS_t = \hat{X}_T$ for some $(\alpha_0, \alpha) \in \mathscr{N}(x)$.

This completes the proof.

Remark 11.1 (Self-Financing Trading Strategy with Cash Flows) For comparison with the portfolio optimization Chap. 12 when intermediate consumption is included, consider a cash flow process $A_t(\omega) : \Omega \times [0, T] \to [0, \infty)$, which is an adapted, right continuous, and nondecreasing process with $A_0 = 0$. Such a cash flow process was considered in Chap. 8 when discussing super- and sub-replication in an incomplete market. Let $\mathscr{V}_+$ be the set of adapted, right continuous, and nondecreasing processes with initial values equal to 0.

An *s.f.t.s. with positive cash flows* A is a nonnegative wealth s.f.t.s. $(\alpha_0, \alpha) \in \mathscr{N}(x)$ and a cash flow process $A_t \in \mathscr{V}_+$ such that the value process $X_t = \alpha_0(t) + \alpha_t \cdot S_t$ satisfies $X_t + A_t = x + \int_0^t \alpha_u \cdot dS_u$ for all t. As seen, the s.f.t.s. "finances" both the value process X_t and the cash flow process A_t.

Note that this implies that the nonnegative wealth s.f.t.s. $(\alpha_0, \alpha) \in \mathscr{N}(x)$ satisfies $x + \int_0^t \alpha_u \cdot dS_u \geq X_t$ for all t, i.e. $X_T \in \mathscr{C}(x)$. Using this new notation, it can easily be seen that

$$\mathscr{C}(x) = \{X_T \in L_+^0 : \exists(\alpha_0, \alpha) \in \mathscr{N}(x), \exists A \in \mathscr{V}_+, \ x + \int_0^T \alpha_t \cdot dS_t = X_T + A_T\}.$$

This completes the remark.

To facilitate the determination of the optimal solution, we seek an alternative characterization of the constraint set that is easier to use. Recall that we have the following sets of local martingale deflators and their relationships (see the fundamental theorems Chap. 2, Sect. 2.4.3)

$$\mathscr{D}_l = \big\{Y \in \mathscr{L}_+^0 : Y_0 = 1, \ XY \text{ is a } \mathbb{P}\text{-local martingale}, \\ X = 1 + \textstyle\int \alpha \cdot dS, \ (\alpha_0, \alpha) \in \mathscr{N}(1)\big\}$$

$$D_l = \{Y_T \in L_+^0 : \exists Z \in \mathscr{D}_l, \ Y_T = Z_T\}$$

$$\mathscr{M}_l = \left\{Y \in \mathscr{D}_l : \exists \mathbb{Q} \sim \mathbb{P}, \ Y_T = \frac{d\mathbb{Q}}{d\mathbb{P}}\right\}$$

$$M_l = \{Y_T \in L_+^0 : \exists Z \in \mathscr{M}_l, \ Y_T = Z_T\}$$

$$\mathscr{M} = \{Y \in \mathscr{L}_+^0 : \quad Y_T = \frac{d\mathbb{Q}}{d\mathbb{P}}, \ \mathbb{Q} \in \mathfrak{M}\}$$

$$M = \{Y \in L_+^0 : \exists Z \in \mathscr{M}, \ Y_T = Z_T\}$$

$$\text{stochastic processes} \qquad \mathscr{M} \subset \mathscr{M}_l \subset \mathscr{D}_l \subset \mathscr{L}_+^0$$

$$\text{random variables} \qquad M \subset M_l \subset D_l \subset L_+^0.$$

We use these sets of local martingale deflators below.

The constraint set $\mathscr{C}(x)$ represents those payoffs $X_T \in L^0_+$ that can be generated by a nonnegative wealth s.f.t.s. $(\alpha_0, \alpha) \in \mathscr{N}(x)$ with initial wealth x. Consider the set of local martingale deflators $Y_T \in M_l$. Each of these deflators generates a present value for the payoff X_T, $E[X_T Y_T]$. Intuitively, since it unknown which local martingale deflator should be used to compute the present value, $\mathscr{C}(x)$ should be the same as the set of all payoffs whose "worst case" present values are affordable at time 0, i.e. less than or equal to x. This is indeed the case, as the next theorem shows.

Theorem 11.1 (Budget Constraint Equivalence 1)

$$\mathscr{C}(x) = \{X_T \in L^0_+ : \sup_{Y_T \in M_l} E[X_T Y_T] \le x\}.$$

Proof Define $\mathscr{C}_1(x) = \{X_T \in L^0_+ : \sup_{Y_T \in M_l} E[X_T Y_T] \le x\}$.

(Step 1) Show $(\mathscr{C}(x) \subset \mathscr{C}_1(x))$.

Take $X_T \in \mathscr{C}(x)$, then $\exists(\alpha_0, \alpha) \in \mathscr{N}(x)$ such that $x + \int_0^T \alpha_t \cdot dS_t \ge X_T$.

Let $W_t \equiv x + \int_0^t \alpha_u \cdot dS_u$. Choose $Y_T \in M_l$. Then, there exists a $Z \in \mathscr{M}_l$ such that $Y_T = Z_T$.

By definition of $\mathscr{M}_l$, $W_t Z_t$ is a $\mathbb{P}$-local martingale bounded below by zero. Hence, by Lemma 1.3 in Chap. 1, $W_t Z_t$ is a $\mathbb{P}$-supermartingale. Thus, $E[Z_T W_T] \le E[Z_T x]$, or equivalently $E[Y_T W_T] \le E[Y_T x]$.

But, $Y_T \in M_l$ implies $E[Y_T] = E\left[\frac{d\mathbb{Q}}{d\mathbb{P}}\right] = 1$ for some $\mathbb{Q} \sim \mathbb{P}$.

This implies $E[Y_T X_T] \le E[Y_T W_T] \le x$.

Since this is true for all $Y_T \in M_l$, $X_T \in \mathscr{C}_1(x)$.

(Step 2) Show $(\mathscr{C}_1(x) \subset \mathscr{C}(x))$.

Take $Z_T \in \mathscr{C}_1(x)$.

Define $Z_t = \operatorname{ess\,sup}_{Y_T \in M_l} E[\frac{Y_T}{Y_t} Z_T | \mathscr{F}_t] = \operatorname{ess\,sup}_{\mathbb{Q} \in \mathfrak{M}_l} E^{\mathbb{Q}}[Z_T | \mathscr{F}_t] \ge 0$.

Part 1 of step 2 is to note that Z_t is a supermartingale for all $\mathbb{Q} \in \mathfrak{M}_l$, this follows by Theorem 1.4 in Chap. 1.

Part 2 of step 2 is to use the optional decomposition Theorem 1.5 in Chap. 1, which implies that

$\exists \alpha \in \mathscr{L}(S)$ and an adapted right continuous nondecreasing process A_t with $A_0 = 0$ such that $X_t \equiv Z_0 + \int_0^t \alpha_u \cdot dS_u - A_t = Z_t$ for all t.

Let α_t be the position in the risky assets. Define the position in the mma $\alpha_0(t)$ by the expression $X_t = \alpha_0(t) + \alpha_t \cdot S_t$ for all t. By the definition of X_t we get that $Z_0 + \int_0^t \alpha_u \cdot dS_u = \alpha_0(t) + \alpha_t \cdot S_t + A_t$. Hence, the trading strategy is self-financing with cash flow $A_t \ge 0$ (free disposal) and initial value $x = Z_0$.

We have $Z_0 + \int_0^t \alpha_u \cdot dS_u = X_t + A_t \ge Z_t \ge 0$ since $A_t \ge 0$ and $Z_t \ge 0$. Hence, $Z_0 + \int_0^t \alpha_u \cdot dS_u$ is bounded below by zero, i.e. $(\alpha_0, \alpha) \in \mathscr{N}(x)$ with $x = Z_0$. Thus, $X_T \in \mathscr{C}(x)$. This completes the proof.

Remark 11.2 (Alternative Version of the Budget Constraint) Notice that in this budget constraint characterization there are an infinite number of constraints, one for each $Y_T \in M_l$. Indeed, we can rewrite the constraint as

$$E[X_T Y_T] \leq x \quad \text{for all } Y_T \in M_l.$$

To prove this, note that this constraint is implied by $\sup_{Y_T \in M_l} E[X_T Y_T] \leq x$. Conversely, taking the supremum of the left side of the previous expression across all $Y_T \in M_l$ gives back the supremum constraint. This completes the remark.

Given this remark, we can write the optimization problem as

Problem 11.3 (Utility Optimization (Choose the Optimal Derivative X_T for $x \geq 0$))

$$v(x) = \sup_{X_T \in L_+^0} E[U(X_T)], \qquad \text{where}$$

$$E[X_T Y_T] \leq x \text{ for all } Y_T \in M_l.$$

As in the portfolio optimization Chap. 10 when solving the investor's optimization problem in a complete market, we are going to solve this optimization problem (the primal problem) by solving its dual problem and showing that there is no duality gap. Unfortunately, it can be shown that, in general, there does not exist an element $Y_T \in M_l$ such that the solution to the dual problem is attained (see Pham [149, p. 186] or Kramkov and Schachermayer [126]). To guarantee the existence of an optimal solution to the dual problem, we need to "fill-in the interior" of the set of local martingale probability measures M_l. This is the task to which we now turn.

Define the set

$$D_s \equiv \{Y_T \in L_+^0 : Y_0 = 1, \exists (Z(T)_n)_{n \geq 1} \in M_l, \ Y_T \leq \lim_{n \to \infty} Z_n(T) \ a.s.\}. \tag{11.3}$$

D_s is the smallest convex, solid, closed subset of L_+^0 (in the $\mathbb{P}$-convergence topology) that contains M_l (see Pham [149, p. 186]). Solid means that if $0 \leq \hat{Y}_T \leq Y_T$ and $Y_T \in D_s$, then $\hat{Y}_T \in D_s$. We note that the random variable $Y_T \in D_s$ need not be the density of a probability measure with respect to $\mathbb{P}$ since $E(Y_T) < 1$ is possible. Indeed, $E[Y_T] \leq E[\lim_{n \to \infty} Z_n(T)] \leq \lim_{n \to \infty} E[Z_n(T)] = 1$, where the second inequality is due to Fatou's lemma.

It can be shown that D_s is equal to the set of supermartingale deflators, i.e.

$$D_s = \{Y_T \in L_+^0 : \exists Z \in \mathscr{D}_s, \ Y_T = Z_T\},$$

where

$$\mathscr{D}_s \equiv \big\{Y \in \mathscr{L}_+^0 : Y_0 = 1,\ XY \text{ is a } \mathbb{P}\text{-supermartingale}, \\ X = 1 + \int \alpha \cdot dS,\ (\alpha_0, \alpha) \in \mathscr{N}(1)\big\}.$$

Proof (Also see Schachermayer [164]) This follows because by the bipolar theorem in Brannath and Schachermayer [21], $\mathscr{D}_s$ is the smallest convex, solid, closed subset of L_+^0 (in the $\mathbb{P}$-convergence topology) that contains M_l, which is expression (11.3). This completes the proof.

In an incomplete market, the supermartingale deflators take the role that the set of martingale deflators played in a complete market. Consequently, we need to understand their properties, especially in relation to the set of martingale deflators. First, we note that, in general, the set of supermartingale deflator processes depend on the probability measure $\mathbb{P}$. Indeed, by Girsanov's Theorem (see Theorem 1.3 in Chap. 1), an equivalent change of measure can change a supermartingale to a martingale or a submartingale.

Second, the supermartingale deflator $Y \in \mathscr{D}_s$ is itself a $\mathbb{P}$-supermartingale. This follows by considering the nonnegative wealth s.f.t.s. $(\alpha_0, \alpha) = (1, 0) \in \mathscr{N}(1)$, which represents a buy and hold trading strategy in only the mma. In this case $X_t = 1$ for all t, and the statement follows from the definition of the set $\mathscr{D}_s$. Since $Y_0 = 1$, this implies $E[Y_T] \le 1$. Similar to a martingale deflator, $Y \in \mathscr{D}_s$ need not be a probability density with respect to $\mathbb{P}$ (see the fundamental theorems Chap. 2, Sect. 2.4.2). If Y is a probability density with respect to $\mathbb{P}$, however, then $Y_t = E[Y_T | \mathscr{F}_t]$ is a $\mathbb{P}$-martingale. We will use this insight below when characterizing the state price density.

The set of local martingale deflators $\mathscr{D}_l$ is a strict subset of the set of supermartingale deflator processes $\mathscr{D}_s$. Indeed, this follows because a nonnegative local martingale is a supermartingale by Lemma 1.3 in Chap. 1. Hence, we have the follow relationships:

$$\text{stochastic processes} \qquad \mathscr{M}_l \subset \mathscr{D}_l \subset \mathscr{D}_s \subset \mathscr{L}_+^0$$

$$\text{random variables} \qquad M_l \subset D_l \subset D_s \subset L_+^0.$$

We can now prove the following equivalence.

Lemma 11.2 (Budget Constraint Equivalence 2)

$$\mathscr{C}(x) = \{X_T \in L_+^0 : E[X_T Y_T] \le x \textit{ for all } Y_T \in D_s\}.$$

Proof Let $\mathscr{C}_1(x) = \{X_T \in L_+^0 : E[X_T Y_T] \le x \text{ for all } Y_T \in D_s\}$.

(Step 1) Show $(\mathscr{C}_1(x) \subset \mathscr{C}(x))$.

If $X_T \in \mathscr{C}_1(x)$, then $E[X_T Y_T] \le x$ for all $Y_T \in D_s$.

Since $M_l \subset D_s$, $E[X_T Y_T] \le x$ for all $Y_T \in M_l$.

Hence $X_T \in \mathscr{C}(x)$.

(Step 2) Show $(\mathscr{C}(x) \subset \mathscr{C}_1(x))$.
If $X_T \in \mathscr{C}(x)$, then $E[X_T Z_n(T)] \leq x$ for all $Z_n(T) \in M_l$.
Choose a $Y_T \in D_s$.
Note that there exists a sequence $Z_n(T) \in M_l$ such that $Y_T \leq \lim_{n\to\infty} Z_n(T) \quad a.s.$
Thus, $X_T Y_T \leq \lim_{n\to\infty} X_T Z_n$ a.s. since $X_T \geq 0$.
Taking expectations, $E[X_T Y_T] \leq E[\lim_{n\to\infty} X_T Z_n(T)] \leq \lim_{n\to\infty} E[X_T Z_n(T)]$, where the second inequality is due to Fatou's lemma.
But, $E[X_T Z_n(T)] \leq \sup_{Y \in M_l} E[X_T Y_T]$ for all n.
Hence, $\lim_{n\to\infty} E[X_T Z_n(T)] \leq \sup_{Y_T \in M_l} E[X_T Y_T]$.
This implies that $E[X_T Y_T] \leq \sup_{Y_T \in M_l} E[X_T Y_T] \leq x$.
This is true for an arbitrary $Y_T \in D_s$, hence
$E[X_T Y_T] \leq x$ for all $Y_T \in D_s$.
This shows that $X_T \in \mathscr{C}_1(x)$. This completes the proof.

This implies that we can rewrite the optimization problem in a form where a solution to its dual problem can be proven to exist.

Problem 11.4 (Utility Optimization (Choose the Optimal Derivative X_T for $x \geq 0$))

$$v(x) = \sup_{X_T \in \mathscr{C}(x)} E[U(X_T)],$$

where

$$\mathscr{C}(x) = \{X_T \in L^0_+ : E[X_T Y_T] \leq x \text{ for all } Y_T \in D_s\}.$$

To solve this problem, define the Lagrangian function

$$\mathscr{L}(X_T, y, Y_T) = E[U(X_T)] + y(x - E[X_T Y_T]). \tag{11.4}$$

The *primal problem* can be written as

$$\text{primal}: \quad \sup_{X_T \in L^0_+} \left(\inf_{y>0, Y_T \in D_s} \mathscr{L}(X_T, y, Y_T) \right) = \sup_{X_T \in \mathscr{C}(x)} E[U(X_T)] = v(x).$$

Proof Note that

$$E[X_T Y_T] \leq x \text{ for all } Y_T \in D_s$$

if and only if $\sup_{Y_T \in D_s} E[X_T Y_T] \leq x$. Hence,

$$\sup_{Y_T \in D_s} (E[X_T Y_T] - x) \leq 0.$$

But,

$$\sup_{Y_T \in D_s} (E[X_T Y_T] - x) = -\inf_{Y_T \in D_s} (x - E[X_T Y_T]).$$

Substitution yields

$$\inf_{Y_T \in D_s} (x - E[X_T Y_T]) \geq 0.$$

Hence,

$$\inf_{y>0} y \left\{ \inf_{Y_T \in D_s} (x - E[X_T Y_T]) \right\} = \begin{cases} 0 & \text{if } \sup_{Y_T \in D_s} E[X_T Y_T] \leq x \\ -\infty & \text{otherwise.} \end{cases}$$

Or,

$$\inf_{y>0, Y_T \in D_s} \{ y (x - E[X_T Y_T]) \} = \begin{cases} 0 & \text{if } \sup_{Y_T \in D_s} E[X_T Y_T] \leq x \\ -\infty & \text{otherwise.} \end{cases}$$

This implies

$$\inf_{y>0, Y_T \in D_s} \{ E[U(X_T)] + y (x - E[X_T Y_T]) \}$$
$$= \begin{cases} E[U(X_T)] & \text{if } \sup_{Y_T \in D_s} E[X_T Y_T] \leq x \\ -\infty & \text{otherwise.} \end{cases}$$

Substitution yields

$$\inf_{y>0, Y_T \in D_s} \mathscr{L}(X_T, y, Y_T) = \begin{cases} E[U(X_T)] & \text{if } \sup_{Y_T \in D_s} E[X_T Y_T] \leq x \\ -\infty & \text{otherwise.} \end{cases}$$

So,

$$\sup_{X_T \in L^0_+} \left(\inf_{y>0, Y_T \in D_s} \mathscr{L}(X_T, y, Y_T) \right)$$

$$= \sup_{X_T \in L^0_+} \begin{cases} E\left[U(X_T)\right] \text{ if } & \sup_{Y_T \in D_s} E\left[X_T Y_T\right] \leq x \\ -\infty & \text{otherwise} \end{cases}$$

$$= \sup_{X_T \in \mathscr{C}(x)} E\left[U\left(X_T\right)\right] = v(x).$$

This completes the proof.

The *dual problem* is

$$\text{dual}: \quad \inf_{y>0, Y_T \in D_s} \left(\sup_{X_T \in L^0_+} \mathscr{L}(X_T, y, Y_T) \right).$$

For optimization problems we know that

$$\inf_{y>0, Y_T \in D_s} \left(\sup_{X_T \in L^0_+} \mathscr{L}(X_T, y, Y_T) \right) \geq \sup_{X_T \in L^0_+} \left(\inf_{y>0, Y_T \in D_s} \mathscr{L}(X_T, y, Y_T) \right) = v(x).$$

Proof We have $\mathscr{L}(X_T, y) = E\left[U(X_T)\right] + y\left(x - E\left[X_T Y_T\right]\right)$.

Note that $\sup_{X_T \in L^0_+} \mathscr{L}(X_T, y, Y_T) \geq \mathscr{L}(X_T, y, Y_T) \geq \inf_{y>0} \mathscr{L}(X_T, y, Y_T)$ for all (X_T, y, Y_T). Taking the infimum over (y, Y_T) on the left side and the supremum over X_T on the right side gives

$$\inf_{y>0, Y_T \in D_s} \left(\sup_{X_T \in L^0_+} \mathscr{L}(X_T, y, Y_T) \right) \geq \sup_{X_T \in L^0_+} \left(\inf_{y>0, Y_T \in D_s} \mathscr{L}(X_T, y, Y_T) \right).$$

This completes the proof.

The solutions to the primal and dual problems are equal if there is no duality gap. We solve the primal problem by solving the dual problem and showing that there is no duality gap.

11.3 Existence of a Solution

We need to show that there is no duality gap and that the optimum is attained, equivalently, a saddle point exists for the Lagrangian. To prove the existence of a saddle point, we need two additional assumptions. These are the same two assumptions used in the portfolio optimization Chap. 10 in a complete market to guarantee the existence of a solution to the trader's optimization problem.

Assumption (Non-trivial Optimization Problem) *There exists an* $x > 0$ *such that* $v(x) < \infty$.

Assumption (Reasonable Asymptotic Elasticity)

$$AE(U, \omega) = \limsup_{x \to \infty} \frac{xU'(x, \omega)}{U(x, \omega)} < 1 \quad a.s.\ \mathbb{P}.$$

We can now prove that the investor's optimization problem has a unique solution.

Theorem 11.2 (Existence of a Unique Saddle Point and Characterization of $v(x)$ **for** $x \geq 0$**)** *Given the above assumptions, there exists a unique saddle point, i.e. there exists a unique* $(\hat{X}_T, \hat{Y}_T, \hat{y})$ *such that*

$$\inf_{Y_T \in D_s,\, y>0} \left(\sup_{X_T \in L^0_+} \mathscr{L}(X_T, Y_T, y) \right) = \mathscr{L}(\hat{X}_T, \hat{Y}_T, \hat{y})$$

$$= \sup_{X_T \in L^0_+} \left(\inf_{Y_T \in D_s,\, y>0} \mathscr{L}(X_T, Y_T, y) \right) = v(x). \tag{11.5}$$

Define $\tilde{v}(y) = \inf_{Y_T \in D_s} E\left[\tilde{U}(yY_T)\right]$, *where* $\tilde{U}(y, \omega) = \sup_{x>0}[U(x, \omega) - xy], \quad y > 0.$
Then, v *and* $\tilde{v}$ *are in conjugate duality, i.e.*

$$v(x) = \inf_{y>0} \left(\tilde{v}(y) + xy\right), \ \forall x > 0, \tag{11.6}$$

$$\tilde{v}(y) = \sup_{x>0} \left(v(x) - xy\right), \ \forall y > 0. \tag{11.7}$$

In addition,

(i) $\tilde{v}$ *is strictly convex, decreasing, and differentiable on* $(0, \infty)$,
(ii) v *is strictly concave, increasing, and differentiable on* $(0, \infty)$,
(iii) defining $\hat{y}$ *to be where the infimum is attained in expression (11.6),*

$$v(x) = \tilde{v}(\hat{y}) + x\hat{y}, \tag{11.8}$$

(iv)

$$\tilde{v}(y) = \inf_{Y_T \in D_s} E[\tilde{U}(yY_T)] = \inf_{Y_T \in M_l} E[\tilde{U}(yY_T)].$$

Proof To prove this theorem, we show conditions 1 and 2 of the Lemma 11.3 in the appendix to this chapter hold. Under the above assumptions, these conditions follow from Zitkovic [178, Theorem 4.2]. To apply this theorem note the following facts. First, choosing the stochastic clock in Zitkovic [178] as in Example 2.6 (2), p. 757 gives the objective as the utility of terminal wealth. Second, choose the random endowment to be identically zero. Third, our hypotheses on the utility function defined above satisfy Zitkovic [178, Definition 2.3, p. 755] except for condition (b). But, for utility of terminal wealth satisfying the reasonable asymptotic utility assumption ($AE(U) < 1$), these conditions are automatically satisfied, see Karatzas and Zitkovic [119, Example 3.2, p. 1838]. Although Zitkovic [178] requires that S is locally bounded, this condition is unnecessary for utility of terminal wealth, as shown in Karatzas and Zitkovic [119]. This completes the proof.

Remark 11.3 (Alternative Sufficient Conditions for Existence) The sufficient conditions (i) there exists an $x > 0$ such that $v(x) < \infty$ and (ii) $AE(U) < 1$ can be replaced by the alternative sufficient conditions $\tilde{v}(y) < \infty$ for all $y > 0$ and $v(x) > -\infty$ for all $x > 0$. The same proof as given for Theorem 11.2 works but with Mostovyi [144, Theorem 2.3] replacing Zitkovic [178, Theorem 4.2]. In this circumstance, the utility function does not need to satisfy condition (b) of Zitkovic [178, Definition 2.3]. Here, Mostovyi [144] extends the domain of U to include $x = 0$ and the range of U to include $-\infty$ at $x = 0$. In addition, the assumption that the set of equivalent local martingale measures $\mathfrak{M}_l \neq \emptyset$ can be replaced by the set of local martingale deflator processes $\mathscr{D}_l \neq \emptyset$.

A sketch of the proof is as follows. The existence of a buy and hold nonnegative wealth s.f.t.s. strategy implies that the constraint set $\mathscr{C}^e(x)$ for v is nonempty, and $\mathscr{D}_l \neq \emptyset$ implies that the constraint set for $\tilde{v}$ is nonempty because $\mathscr{D}_l \subset \mathscr{D}_s \neq \emptyset$. Given the appropriate topologies for $\mathscr{C}^e(x)$ and $\mathscr{D}_l$, assuming that $\tilde{v}(y) < \infty$ for all $y > 0$ and $v(x) > -\infty$ for all $x > 0$ together imply, because of the strict convexity of $\tilde{v}(y)$ for all $y > 0$ and the strict concavity of $v(x)$ for all $x > 0$, that a unique solution exists to each of these problems. Then, the conditions of Lemma 11.3 in the appendix of the portfolio optimization Chap. 11 apply to obtain the remaining results. This completes the remark.

11.4 Characterization of the Solution

This section characterizes the solution to the investor's optimization problem.

11.4.1 The Characterization

To characterize the solution, we focus on the solution to the dual problem, i.e.

$$v(x) = \inf_{y>0, Y_T \in D_s} \left(\sup_{X_T \in L^0_+} \mathscr{L}(X_T, y, Y_T) \right)$$

$$= \inf_{y>0, Y_T \in D_s} \left(\sup_{X_T \in L^0_+} E\left[U(X_T) - yX_T Y_T\right] + xy \right).$$

To solve, we first exchange the sup and expectation operator

$$v(x) = \inf_{y>0, Y_T \in D_s} E\left[\sup_{X_T \in L^0_+} \left(U(X_T) - yX_T Y_T\right) + xy \right]. \tag{11.9}$$

The proof of this step is identical to that given in the appendix to the portfolio optimization Chap. 10. We now solve this problem, working from the inside out.

Step 1. (Solve for $X_T \in L^0_+$)
Fix y, Y_T and solve for the optimal wealth $\hat{X}_T \in L^0_+$:

$$\sup_{X_T \in L^0_+} \left[U(X_T) - yX_T Y_T\right].$$

To solve this problem, fix a $\omega \in \Omega$, and consider the related problem

$$\sup_{X_T(\omega) \in \mathbb{R}_+} \left(U(X_T(\omega), \omega) - yX_T(\omega) Y_T(\omega)\right).$$

This is a simple optimization problem on the real line. The first-order condition for an optimum gives

$$U'(X_T(\omega), \omega) - yY_T(\omega) = 0, \qquad \text{or}$$

$$\hat{X}_T(\omega) = I(yY_T(\omega), \omega) \quad \text{with} \quad I(\cdot, \omega) = (U'(\cdot, \omega))^{-1}.$$

Given the properties of $I(\cdot, \omega)$, $\hat{X}_T(\omega)$ is $\mathscr{F}_T$-measurable, hence $\hat{X}_T(\omega) \in L^0_+$. Note that $\hat{X}_T$ depends on y, Y_T.

Step 2. (Solve for $Y_T \in D_s$)

Given the optimal wealth $\hat{X}_T$ and y, solve for the optimal supermartingale deflator $\hat{Y}_T \in D_s$. Using conjugate duality (see Lemma 9.7 in the utility function Chap. 9), we have

$$\tilde{U}(yY_T) = \sup_{X_T \in L_+^0} [U(X_T) - X_T y Y_T] = [U(\hat{X}_T) - \hat{X}_T y Y_T].$$

This transforms the problem to

$$\inf_{y>0, Y_T \in D_s} E\left[\sup_{X_T \in L_+^0} (U(X_T) - yX_T Y_T) + xy\right] =$$

$$\inf_{y>0, Y_T \in D_s} (E[\tilde{U}(yY_T)] + xy) = \inf_{y>0}(\{\inf_{Y_T \in D_s} E[\tilde{U}(yY_T)]\} + xy).$$

We solve the inner most infimum problem first. Defining $\tilde{v}(y)$ as the value function for this infimum, let $\hat{Y}_T$ be the solution such that

$$\tilde{v}(y) = E[\tilde{U}(y\hat{Y}_T)] = \inf_{Y_T \in D_s} E[\tilde{U}(yY_T)].$$

Note that $\hat{Y}_T$ depends on y.

Step 3. (Solve for $y > 0$)

Given the optimal wealth $\hat{X}_T$ and supermartingale deflator $\hat{Y}_T$, solve for the optimal Lagrangian multiplier $y > 0$

$$\inf_{y>0}(\tilde{v}(y) + xy).$$

Taking the derivative of the right side with respect to y, one gets

$$\tilde{v}'(y) + x = 0.$$

Substituting the definition for $\tilde{v}$ and taking the derivative yields

$$E[\tilde{U}'(y\hat{Y}_T)\hat{Y}_T] + x = 0.$$

This step requires taking the derivative underneath the expectation operator. The same proof as in the appendix to the portfolio optimization Chap. 10 applies. Noting that $\tilde{U}'(y) = -I(y)$ from Lemma 9.7 in the utility function Chap. 9, we get

$$E\left[I(\hat{y}\hat{Y}_T)\hat{Y}_T\right] = E[\hat{X}_T \hat{Y}_T] = x, \tag{11.10}$$

which is that the optimal terminal wealth $\hat{X}_T$ satisfies the budget constraint with an equality. Solving this equation gives the optimal $\hat{y}$.

Step 4. (Characterization of $\hat{X} \in \mathscr{X}(x)$ **for** $x \geq 0$ **)**

Step 1 solves for the optimal wealth $\hat{X}_T \in L^0_+$, a random variable. Since $\hat{X}_T \in \mathscr{C}(x)$, there exists a stochastic process $Z \in \mathscr{X}(x)$ such that $Z_t = x + \int_0^t \alpha_u \cdot dS_u \in \mathscr{L}^0_+$ with $\hat{X}_T = Z_T$ for some $(\alpha_0, \alpha) \in \mathscr{N}(x)$. For convenience, let us label this process Z as $\hat{X} \in \mathscr{L}^0_+$. This gives the first characterization of the optimal wealth process $\hat{X}$ using a nonnegative wealth s.f.t.s. $(\alpha_0, \alpha) \in \mathscr{N}(x)$.

We can also obtain a second characterization. By the definition of the supermartingale deflator $\hat{Y}_T \in D_s$, there exists a stochastic process $Y \in \mathscr{D}_s$ with $\hat{Y}_T = Y_T$. Label this process $Y \in \mathscr{D}_s$ as $\hat{Y}$. Then, by the definition of $\mathscr{D}_s$, $\hat{Y}_t \hat{X}_t$ is a supermartingale under $\mathbb{P}$. Hence,

$$E\left[\hat{X}_T \hat{Y}_T \,|\mathscr{F}_t\right] \leq \hat{X}_t \hat{Y}_t.$$

But by expression (11.10) we have that

$$E[\hat{X}_T \hat{Y}_T] = x = \hat{X}_0 \hat{Y}_0.$$

The only way this can happen is if $E\left[\hat{X}_T \hat{Y}_T \,|\mathscr{F}_t\right] = \hat{X}_t \hat{Y}_t$ for all t, i.e. $\hat{X}_t \hat{Y}_t$ is a $\mathbb{P}$-martingale. Hence,

$$\hat{X}_t = E\left[\hat{X}_T \frac{\hat{Y}_T}{\hat{Y}_t} \middle| \mathscr{F}_t\right]. \tag{11.11}$$

Step 5. (Nonnegativity of $\hat{X} \in \mathscr{X}(x)$ **)**

As noted in Step 4 above, the wealth process $\hat{X} \geq 0$ because $\hat{X}_T \in \mathscr{C}(x)$, which is the set of random variables generated by nonnegative wealth s.f.t.s.'s in the set $\mathscr{N}(x)$.

Remark 11.4 (Random Variables Versus Stochastic Processes) The optimization problem determines the random variable $\hat{X}_T \in L^0_+$, which is the optimal time T wealth. But, since $\hat{X}_T \in \mathscr{C}(x)$, there exists a nonnegative stochastic process $Z \in \mathscr{X}(x)$ such that $Z_T = X_T$ and $Z_t = x + \int_0^t \alpha_u \cdot dS_u$ for some nonnegative wealth s.f.t.s. $(\alpha_0, \alpha) \in \mathscr{N}(x)$. Note that the equality in this statement is justified by the free disposal Lemma 11.1.

This characterization of the optimal wealth (stochastic) process $X \in \mathscr{X}(x)$ can be used to determine the nonnegative wealth s.f.t.s. generating it using the methods for constructing a synthetic derivative in the fundamental theorems Chap. 2.

The optimization problem also determines the random variable $\hat{Y}_T \in D_s$, which is the optimal supermartingale deflator. By the definition of D_s there exists a stochastic process $Y \in \mathscr{D}_s$ with $\hat{Y}_T = Y_T$. Label this process $Y \in \mathscr{D}_s$ as $\hat{Y}$. $\hat{Y}$

is the optimal supermartingale deflator process. Unlike the situation in a complete market, in an incomplete market the supermartingale deflator process can exhibit many different properties. In order of increasing restrictiveness, these properties are listed below.

1. $\hat{Y} \in \mathscr{D}_s$ is a supermartingale deflator process.
2. $\hat{Y} \in \mathscr{D}_s$, where $\hat{Y}$ is a supermartingale deflator process and $\hat{Y}_T = \frac{d\mathbb{Q}}{d\mathbb{P}}$ is a probability density for $\mathbb{P}$. Here, $X \in \mathscr{X}(x)$ is a $\mathbb{Q}$-supermartingale.
3. $\hat{Y} \in \mathscr{D}_l \subset \mathscr{D}_s$, where $\hat{Y}$ is a local martingale deflator process, i.e. given an $X \in \mathscr{X}(x)$, $X\hat{Y}$ is a $\mathbb{P}$-local martingale.
4. $\hat{Y} \in \mathscr{M}_l \subset \mathscr{D}_l$, where $\hat{Y}$ is a local martingale deflator process, $\hat{Y}_T = \frac{d\mathbb{Q}}{d\mathbb{P}}$ is a probability density with respect to $\mathbb{P}$, and $X \in \mathscr{X}(x)$ is a $\mathbb{Q}$-local martingale.
5. $\hat{Y} \in \mathscr{M} \subset \mathscr{M}_l$, where $\hat{Y}$ is a martingale deflator process, $\hat{Y}_T = \frac{d\mathbb{Q}}{d\mathbb{P}}$ is a probability density with respect to $\mathbb{P}$, and $X \in \mathscr{X}(x)$ is a $\mathbb{Q}$-martingale.

We note that for case (2), where $\hat{Y}_T \in D_s$ is a probability density for $\mathbb{P}$, $\hat{Y}$ is a $\mathbb{P}$-martingale. Necessary and sufficient conditions for case (5), where the supermartingale deflator is both a martingale deflator and a probability density with respect to $\mathbb{P}$, i.e. $\hat{Y}_T \in M$, are contained in Kramkov and Weston [127]. This completes the remark.

11.4.2 Summary

For easy reference we summarize the previous characterization results. The optimal value function

$$v(x) = \sup_{X_T \in \mathscr{C}(x)} E\left[U(X_T)\right]$$

for $x \geq 0$ has the solution for terminal wealth given by

$$\hat{X}_T = I(\hat{y}\hat{Y}_T) \quad \text{with} \quad I = (U')^{-1} = -\tilde{U}',$$

where $\hat{Y}_T \in D_s$ is the optimal supermartingale deflator. The optimal supermartingale deflator is the solution to

$$\tilde{v}(y) = \inf_{Y_T \in D_s} E[\tilde{U}(yY_T)].$$

By the definition of the set D_s, there is a supermartingale deflator process $\hat{Y} \in \mathscr{D}_s \subset \mathscr{L}_0^+$ such that its time t value is $\hat{Y}_T$.

The Lagrangian multiplier $\hat{y}$ is the solution to the budget constraint

$$E\left[\hat{Y}_T I(\hat{y}\hat{Y}_T)\right] = x.$$

The optimal wealth $\hat{X}_T$ satisfies the budget constraint with an equality

$$E\left[\hat{Y}_T \hat{X}_T\right] = x.$$

The optimal wealth process $\hat{X} \in \mathscr{X}(x)$ exists since $\hat{X}_T \in \mathscr{C}(x)$, i.e. there exists a nonnegative wealth s.f.t.s. $(\alpha_0, \alpha) \in \mathscr{N}(x)$ with time T payoff equal to that of $\hat{X}_T$, where $\hat{X}_t = x + \int_0^t \alpha_u \cdot dS_u$ for all t. The optimal portfolio wealth process $\hat{X} \in \mathscr{X}(x)$ when multiplied by the supermartingale deflator, $\hat{X}_t \hat{Y}_t \geq 0$, is a $\mathbb{P}$-martingale. This implies that for an arbitrary time $t \in [0, T]$, we have

$$\hat{X}_t = E\left[\hat{X}_T \frac{\hat{Y}_T}{\hat{Y}_t} \middle| \mathscr{F}_t \right] \geq 0.$$

It is important to note, for subsequent use, that the solution $\hat{Y}_T \in D_s$ depends on the utility function U.

11.5 The Shadow Price

Using the characterization of the value function, we can obtain the shadow price of the budget constraint. The shadow price is the benefit, in terms of expected utility, of increasing the initial wealth by 1 unit (of the mma). The shadow price equals the Lagrangian multiplier.

Theorem 11.3 (The Shadow Price of the Constraint)

$$\hat{y} = v'(x) \geq E[U'(\hat{X}_T)] \tag{11.12}$$

with equality if and only if the supermartingale deflator $\hat{Y}_T \in D_s$ is a probability density with respect to $\mathbb{P}$, i.e. $E[\hat{Y}_T] = 1$.

Proof The first equality is obtained by taking the derivative of $v(x) = \tilde{v}(\hat{y}) + x\hat{y}$. The second inequality follows from the first-order condition for an optimum $U'(\hat{X}_T) = \hat{y}\hat{Y}_T$. Taking expectations gives $E[U'(\hat{X}_T)] = E[\hat{y}\hat{Y}_T] = \hat{y}E[\hat{Y}_T] \leq \hat{y}$ because $E[\hat{Y}_T] \leq 1$. This completes the proof.

11.6 The Supermartingale Deflator

The key insight of the individual optimization problem is the characterization of the supermartingale deflator process $Y \in \mathscr{D}_s$ in terms of the individual's utility function and initial wealth. To obtain this characterization, using Theorem 11.3 above, we rewrite the first-order condition for $\hat{X}_T$:

$$\hat{Y}_T = \frac{U'(\hat{X}_T)}{\hat{y}} = \frac{U'(\hat{X}_T)}{v'(x)} \leq \frac{U'(\hat{X}_T)}{E[U'(\hat{X}_T)]} \tag{11.13}$$

with equality if and only if the supermartingale deflator $\hat{Y}_T \in D_s$ is a probability density with respect to $\mathbb{P}$, i.e. $E[\hat{Y}_T] = 1$.

This is similar to the result relating the local martingale measure to preferences in the solution to the investor's optimization problem in Chap. 10. Recall that in a complete market, the supermartingale deflator is a local martingale deflator, i.e. $\hat{Y}_T \in M_l \subset D_l \subset D_s$. The difference in an incomplete market is that the supermartingale deflator $\hat{Y}_T \in D_s$ need not be a local martingale deflator nor a probability density with respect to $\mathbb{P}$, i.e. $\hat{Y}_T \notin M_l$ is possible.

Assume now that $\hat{Y}_T \in M_l$, i.e. $\hat{Y}_T$ is a local martingale deflator and a probability density with respect to $\mathbb{P}$. However, $\hat{Y}_T$ need not be the probability density of a martingale measure. In this case, the individual's state price density is given by the $\mathbb{P}$-martingale,

$$\hat{Y}_t = E\left[\hat{Y}_T \,|\mathscr{F}_t\right] = \frac{E[U'(\hat{X}_T)\,|\mathscr{F}_t]}{v'(x)}. \tag{11.14}$$

Here, the supermartingale deflator equals the investor's marginal utility of wealth, normalized by the first derivative of the optimal value function. Given investors trade in a competitive market, prices are taken as exogenous. Because the market is incomplete, prices need not uniquely determine the equivalent local martingale measure $\mathbb{Q} \in \mathfrak{M}_l$. The individual's optimization problem determines a unique local martingale measure from among all $Y_T \in M_l$ that the investor can utilize to price derivatives. Different investors may have different local martingale deflators $Y_T \in M_l$.

Remark 11.5 (Asset Price Bubbles) In this individual's optimization problem, prices and hence the local martingale measure $\mathbb{Q}$ are taken as exogenous. The above analysis does not require the condition that $\mathbb{Q}$ is a martingale measure. In addition, even if the individual's supermartingale deflator is a state price density, i.e. $\hat{Y}_T \in M_l$, it need not be a martingale measure (it could be a strict local martingale measure). Hence, this analysis applies to a market with asset price bubbles (see Chap. 3). This completes the remark.

Remark 11.6 (Systematic Risk) In this remark, we use the non-normalized market $((\mathbb{B}, \mathbb{S}), (\mathscr{F}_t), \mathbb{P})$ representation to characterize systematic risk. Recall that the state price density is the key input to the systematic risk return relation in the multiple-factor model Chap. 4, Theorem 4.4. To apply this theorem, *we assume that the supermartingale deflator $\hat{Y}_T \in D_s$ is a martingale deflator, i.e. $\hat{Y}_T \in M$.* Then, the risky asset returns over the time interval $[t, t + \Delta]$ satisfy

$$E\left[R_i(t)\,|\mathscr{F}_t\right] = r_0(t) - \operatorname{cov}\left[R_i(t), \frac{E[U'(\hat{X}_T)\,|\mathscr{F}_{t+\Delta}]}{E[U'(\hat{X}_T)\,|\mathscr{F}_t]} \frac{\mathbb{B}_t}{\mathbb{B}_{t+\Delta}}(1 + r_0(t)) \middle| \mathscr{F}_t \right], \tag{11.15}$$

where the ith risky asset return is $R_i(t) = \frac{\mathbb{S}_i(t+\Delta)-\mathbb{S}_i(t)}{\mathbb{S}_i(t)}$, $r_0(t) \equiv \frac{1}{p(t,t+\Delta)} - 1$ is the default-free spot rate of interest where $p(t, t+\Delta)$ is the time t price of a default-free zero-coupon bond maturing at time $t + \Delta$, and the state price density is

$$H_t = \frac{Y_t}{\mathbb{B}_t} = \frac{E\left[Y_T\,|\mathscr{F}_t\right]}{\mathbb{B}_t} = \frac{E[U'(\hat{X}_T)\,|\mathscr{F}_t]}{v'(x)\mathbb{B}_t}.$$

When the time step is very small, i.e. $\Delta \approx dt$, the expression $\frac{\mathbb{B}_t}{\mathbb{B}_{t+\Delta}}(1+r_0(t)) \approx 1$, and this relation simplifies to

$$E\left[R_i(t)\,|\mathscr{F}_t\right] \approx r_0(t) - \operatorname{cov}\left[R_i(t), \frac{E[U'(\hat{X}_T)\,|\mathscr{F}_{t+\Delta}]}{E[U'(\hat{X}_T)\,|\mathscr{F}_t]} \middle| \mathscr{F}_t \right]. \tag{11.16}$$

This is the same expression that we obtained in the portfolio optimization Chap. 10 in a complete market. The same discussion regarding the interpretation of systematic risk therefore applies. This completes the remark.

11.7 The Optimal Trading Strategy

This section solves step 2 of the optimization problem, which is to determine the optimal trading strategy $(\alpha_0, \alpha) \in \mathscr{N}(x)$ for $x \geq 0$. This is done using the standard techniques developed for the synthetic construction of derivatives in the fundamental theorems Chap. 2.

Since the optimal wealth $\hat{X}_T \in \mathscr{C}(x)$, we know there exists a nonnegative wealth s.f.t.s. $(\alpha_0, \alpha) \in \mathscr{N}(x)$ such that

$$\hat{X}_T = x + \int_0^T \alpha_t \cdot dS_t,$$

and an optimal wealth process $\hat{X} \in \mathscr{X}(x)$ such that

$$\hat{X}_t = x + \int_0^t \alpha_u \cdot dS_u$$

for all t. This uses the free disposal Lemma 11.1.

We need to characterize this nonnegative wealth s.f.t.s. $(\alpha_0, \alpha) \in \mathcal{N}(x)$ given the initial wealth $x \geq 0$. To illustrate this determination, suppose that S is a Markov diffusion process. Due to the diffusion assumption, S is a continuous process. Given this assumption, we can write the optimal wealth process as a deterministic function of time t and the risky asset prices S_t, i.e.

$$E\left[\hat{X}_T \frac{\hat{Y}_T}{\hat{Y}_t} \middle| \mathscr{F}_t\right] = E\left[\hat{X}_T \frac{\hat{Y}_T}{\hat{Y}_t} \middle| S_t\right] = \hat{X}_t(S_t).$$

Using Ito's formula (assuming the appropriate differentiability conditions), we have

$$\hat{X}_T(S_T) = x + \int_0^T \frac{\partial \hat{X}_t}{\partial t} dt + \frac{1}{2} \int_0^T \sum_{i=1}^n \sum_{j=1}^n \frac{\partial^2 \hat{X}_t}{\partial S_{ij}^2} d\langle S_i(t), S_j(t)\rangle + \int_0^T \frac{\partial \hat{X}_t}{\partial S} \cdot dS_t, \tag{11.17}$$

where $\frac{\partial \hat{X}}{\partial S} = (\frac{\partial \hat{X}}{\partial S_1}, \ldots, \frac{\partial \hat{X}}{\partial S_n})' \in \mathbb{R}^n$. Equating the coefficients of dS_t in the two preceding stochastic integral equations implies that the trading strategy is

$$\frac{\partial \hat{X}_t(S_t)}{\partial S_t} = \alpha_t.$$

This is the standard "delta" used to construct the synthetic derivative $\hat{X}_T$ seen in the fundamental theorems Chap. 2. Finally, α_0 is determined, given α, by the value process expression

$$X_t = \alpha_0 + \alpha \cdot S_t.$$

If the underlying asset price process is not Markov, then for a large class of processes one can use Malliavin calculus to determine the trading strategy (see Detemple et al. [49] and Nunno et al. [51]).

11.8 An Example

This section presents an example to illustrate the abstract expressions of the previous sections. This example is obtained from Pham [149, p. 197].

11.8.1 The Market

We consider the normalized Brownian motion market given in Sect. 2.7 of the fundamental theorems Chap. 2. Suppose the normalized market ($r_t = 0$) consists of n risky assets and an mma where the risky asset price process evolves as

$$dS_i(t) = S_i(t)\left(b_i(t)dt + \sum_{d=1}^{D} \sigma_{id}(t)dW_d(t)\right) \tag{11.18}$$

for $i = 1, \ldots, n$, where $W_t = W(t) = (W_1(t), \ldots, W_D(t))' \in \mathbb{R}^D$ are independent Brownian motions with $W_d(0) = 0$ for all $d = 1, \ldots, D$, and

$$\sigma_t = \sigma(t) = \begin{bmatrix} \sigma_{11}(t) & \cdots & \sigma_{1D}(t) \\ \vdots & & \vdots \\ \sigma_{n1}(t) & \cdots & \sigma_{nD}(t) \end{bmatrix}_{n\times D}. \tag{11.19}$$

In vector notation, we can write the evolution of the stock price process as

$$dS_t = \text{diag}(S_t)b_t dt + \text{diag}(S_t)\sigma_t dW_t, \tag{11.20}$$

where $dS_t = (dS_1(t), \ldots, dS_n(t))' \in \mathbb{R}^n$ and $\text{diag}(S_t)$ equals the $n \times n$ diagonal matrix with elements $(S_1(t), \ldots, S_n(t))$ in the diagonal.

We assume that

1. $\text{rank}(\sigma_t) = n$ for all t a.s. $\mathbb{P}$, and
2. $\int_0^T \left\| \sigma_t' \left[\sigma_t \sigma_t'\right]^{-1} b_t \right\|^2 dy < \infty$.

Since $\text{rank}(\sigma_t) = \min\{D, n\}$, this implies that $D \geq n$.

By Theorems 2.9 and 2.10 in the fundamental theorems Chap. 2, the market satisfies NFLVR and by the First Fundamental Theorem 2.3 of asset pricing in Chap. 2, $\mathfrak{M}_l \neq \emptyset$.

It was shown in Theorem 2.11 in the fundamental theorems Chap. 2 that the set of local martingale measures is

$$\mathfrak{M}_l = \{\mathbb{Q}^\nu : \tfrac{d\mathbb{Q}^\nu}{d\mathbb{P}} = e^{-\int_0^T (\theta_t + \nu_t)\cdot dW_t - \frac{1}{2}\int_0^T \|\theta_t + \nu_t\|^2 dt} > 0, \quad E\left[\tfrac{d\mathbb{Q}^\nu}{d\mathbb{P}}\right] = 1, \ \nu \in K(\sigma)\},$$

where

$$\theta_t = \sigma_t' \left[\sigma_t \sigma_t'\right]^{-1} b_t$$

and

$$K(\sigma) = \{\nu \in \mathscr{L}(W) : \int_0^T \|\nu_t\|^2 dt < \infty, \ \sigma_t \nu_t = 0 \quad \text{for all } t\}.$$

We want the market to be incomplete. Using Theorem 2.12 in the fundamental theorems Chap. 2, we assume that $D > n$.

For later use, we need the following sets.

$$D_l = \{Y_T \in L^0_+ : Y_T = e^{-\int_0^T (\theta_t + \nu_t) \cdot dW_t - \frac{1}{2}\int_0^T \|\theta_t + \nu_t\|^2 dt} > 0, \ \nu \in K(\sigma)\}$$

is the set of local martingale deflators, and

$$M_l = \{Y_T \in L^0_+ : Y_T = e^{-\int_0^T (\theta_t + \nu_t) \cdot dW_t - \frac{1}{2}\int_0^T \|\theta_t + \nu_t\|^2 dt} > 0, \ E[Y_T] = 1, \ \nu \in K(\sigma)\}$$

is the set of local martingale deflators that are probability densities with respect to $\mathbb{P}$.

11.8.2 The Utility Function

Let preferences be represented by a state independent logarithmic utility function

$$U(x) = \ln(x) \qquad \text{for } x > 0.$$

Recall that

$$I(y) = \frac{1}{y} \qquad \text{and}$$

$$\tilde{U}(y) = -\ln y - 1$$

for $y > 0$.

11.8.3 The Optimal Supermartingale Deflator

The optimal supermartingale deflator $\hat{Y}_T$ is the solution to

$$\tilde{v}(y) = \inf_{Y_T \in D_s} E\left[\tilde{U}(yY_T)\right].$$

From Theorem 11.2 above we have that

$$\tilde{v}(y) = \inf_{Y_T \in D_l} E\left[\tilde{U}(yY_T)\right]$$

because $M_l \subset D_l$ and $\tilde{v}(y) = \inf_{Y_T \in M_l} E\left[\tilde{U}(yY_T)\right]$. Karatzas et al. [120] (see also Pham [149, p. 197]) show that for this Brownian motion market the solution is in the set D_l.

We solve for the optimal element in this set. In this regard, note that

$$\begin{aligned} E\left[\tilde{U}(yY_T)\right] &= -E\left[\ln(yY_T)\right] - 1 \\ &= -E\left[\ln\left(ye^{-\int_0^T(\theta_t+\nu_t)\cdot dW_t - \frac{1}{2}\int_0^T\|\theta_t+\nu_t\|^2 dt}\right)\right] - 1 \\ &= -E\left[\ln(y) - \int_0^T(\theta_t+\nu_t)\cdot dW_t - \frac{1}{2}\int_0^T\|\theta_t+\nu_t\|^2 dt\right] - 1. \end{aligned}$$

If $E\left[\int_0^T\|\theta_t+\nu_t\|^2 dt\right] = \infty$, then $E\left[\tilde{U}(yY_T)\right] = \infty$. Hence, we can restrict consideration to those ν such that $E\left[\int_0^T\|\theta_t+\nu_t\|^2 dt\right] < \infty$. In this case, $E\left[\int_0^T(\theta_t+\nu_t)\cdot dW_t\right] = 0$ since $\int_0^T(\theta_t+\nu_t)\cdot dW_t$ is a martingale (see Protter [151, p. 171]).

Hence, $E\left[\tilde{U}(yY_T)\right] = -\ln(y) + \frac{1}{2}E\left[\int_0^T\|\theta_t+\nu_t\|^2 dt\right] - 1$. Since $\|\theta_t+\nu_t\|^2 = \|\theta_t\|^2 + \|\nu_t\|^2$, inspection shows that the infimum is obtained at $\nu_t = 0$. Thus, the optimal supermartingale deflator is

$$\hat{Y}_T = e^{-\int_0^T\theta_t\cdot dW_t - \frac{1}{2}\int_0^T\|\theta_t\|^2 dt}.$$

We note that this solution is independent of y.

11.8.4 The Optimal Wealth Process

The optimal wealth process is

$$\hat{X}_T = I(\hat{y}\hat{Y}_T) = \frac{1}{\hat{y}\hat{Y}_T} = \frac{1}{\hat{y}}e^{\int_0^T\theta_t\cdot dW_t + \frac{1}{2}\int_0^T\|\theta_t\|^2 dt}.$$

And, for an arbitrary time t,

$$\hat{X}_t = E\left[\frac{1}{\hat{y}\hat{Y}_T}\frac{\hat{Y}_T}{\hat{Y}_t}\,\middle|\,\mathscr{F}_t\right].$$

Algebra gives

$$\hat{X}_t = \frac{1}{\hat{y}\hat{Y}_t} = \frac{1}{\hat{y}} e^{\int_0^t \theta_s \cdot dW_s + \frac{1}{2}\int_0^t \|\theta_s\|^2 ds}.$$

Recall that $\hat{y}$ is the solution to the budget constraint

$$\hat{X}_0 = E\left[I(\hat{y}\hat{Y}_T)\frac{\hat{Y}_T}{\hat{Y}_t} \right] = x.$$

Substitution for time 0 yields

$$x = \frac{1}{\hat{y}}.$$

Combined we get that the optimal wealth process satisfies

$$\hat{X}_t = x e^{\int_0^t \theta_s \cdot dW_s + \frac{1}{2}\int_0^t \|\theta_s\|^2 ds}. \tag{11.21}$$

11.8.5 The Optimal Trading Strategy

To get the optimal trading strategy we apply Ito's formula to expression (11.21), which yields

$$d\hat{X}_t = \hat{X}_t \left\|\theta_t\right\|^2 dt + \hat{X}_t \theta_t \cdot dW_t.$$

Proof Dropping all the subscripts and superscripts,

$$\begin{aligned} dX &= \frac{\partial X}{\partial t} dt + \frac{1}{2}\frac{\partial^2 X}{\partial W^2} dt + \frac{\partial X}{\partial W} dW. \\ &= X\frac{1}{2}\left\|\theta\right\|^2 dt + \frac{1}{2} X \left\|\theta\right\|^2 dt + X\theta \cdot dW. \end{aligned}$$

This completes the proof.

Since the optimal wealth $\hat{X}_T \in \mathscr{C}(x)$, there exists a nonnegative wealth s.f.t.s. $(\alpha_0, \hat{\alpha}) \in \mathscr{N}(x)$ with $x \geq 0$ such that

$$d\hat{X}_t = \alpha_t \cdot dS_t = \text{diag}\,(S_t)\,\alpha_t \cdot b_t dt + \text{diag}\,(S_t)\,\alpha_t \cdot \sigma_t dW_t.$$

Equating the coefficients of the Brownian motion terms yields

$$(\text{diag}\,(S_t)\,\alpha_t)'\,\sigma_t = \theta_t' \hat{X}_t,$$

where the holdings in the mma are determined by the expression

$$\hat{\alpha}_0(t) = \hat{X}_t - \hat{\alpha}_t \cdot S_t.$$

We solve this expression for α_t noting that $\theta_t = \sigma_t' \left[\sigma_t \sigma_t'\right]^{-1} b_t$.

The algebra is

$$\begin{aligned}
\sigma_t' \left(\alpha_t \text{diag}\left(S_t\right)\right) &= \theta_t \hat{X}_t, \\
\sigma_t' \left(\alpha_t \text{diag}\left(S_t\right)\right) &= \sigma_t' \left[\sigma_t \sigma_t'\right]^{-1} b_t X_t, \\
\sigma_t \sigma_t' \left(\alpha_t \text{diag}\left(S_t\right)\right) &= \sigma_t \sigma_t' \left[\sigma_t \sigma_t'\right]^{-1} b_t X_t, \\
\left[\sigma_t \sigma_t'\right] \left(\alpha_t \text{diag}\left(S_t\right)\right) &= b_t X_t.
\end{aligned}$$

Since σ_t is of full rank, which equals n, by Theil [175, p. 11], $\sigma_t \sigma_t'$ also has rank equal to n and is invertible. Hence, expressing the holdings in the risky assets as a proportion of wealth, we get

$$\hat{\pi}_t = \frac{\left(\alpha_t \text{diag}\left(S_t\right)\right)}{X_t} = \left[\sigma_t \sigma_t'\right]^{-1} b_t. \tag{11.22}$$

11.8.6 The Value Function

Last, to obtain the value function, note that

$$v(x) = E\left[U(\hat{X}_T)\right] = E\left[U(xe^{\int_0^t \theta_s \cdot dW_s + \frac{1}{2}\int_0^t \|\theta_s\|^2 ds})\right]$$

$$= \ln(x) + E\left[\int_0^T \theta_s \cdot dW_s + \frac{1}{2}\int_0^T \|\theta_s\|^2 ds\right] = \ln(x) + \frac{1}{2}E\left[\int_0^T \|\theta_s\|^2 ds\right].$$

Substituting $\theta_t = \sigma_t' \left[\sigma_t \sigma_t'\right]^{-1} b_t$ gives the optimal value function

$$v(x) = \ln(x) + \frac{1}{2}E\left[\int_0^T \left\|\sigma_t' \left[\sigma_t \sigma_t'\right]^{-1} b_t\right\|^2 ds\right]. \tag{11.23}$$

This completes the example.

11.9 Differential Beliefs

This section discusses the changes to the investor's optimization problem that are needed when we introduce differential beliefs in subsequent chapters, where the trader's beliefs differ from the statistical probability measure $\mathbb{P}$. Suppose, therefore, that $\mathbb{P}$ represents the statistical probability measure and $\mathbb{P}_i$ represents the trader i's beliefs, where $\mathbb{P}_i$ and $\mathbb{P}$ are equivalent probability measures. Thus $\frac{d\mathbb{P}_i}{d\mathbb{P}} > 0$ is an $\mathscr{F}_T$-measurable random variable and $E\left[\frac{d\mathbb{P}_i}{d\mathbb{P}}\right] = 1$.

The investor's objective function is

$$E_i\left[U(x,\omega)\right] = E\left[\frac{d\mathbb{P}_i}{d\mathbb{P}}U(x,\omega)\right]. \tag{11.24}$$

Recall from Sect. 9.7 in the utility function Chap. 9 that given the modified utility function $\frac{d\mathbb{P}_i}{d\mathbb{P}}U(x,\omega)$, its convex conjugate is

$$\sup_{x>0}\left(\frac{d\mathbb{P}_i}{d\mathbb{P}}U(x) - yx\right) = \frac{d\mathbb{P}_i}{d\mathbb{P}}\tilde{U}\left(y\frac{d\mathbb{P}}{d\mathbb{P}_i}\right). \tag{11.25}$$

The reason for this change is that

$$\begin{aligned}\mathscr{D}_s \equiv \big\{&Y \in \mathscr{L}_+^0 : Y_0 = 1,\ XY \text{ is a } \mathbb{P}\text{-supermartingale},\\ &X = 1 + \textstyle\int \alpha \cdot dS,\ (\alpha_0,\alpha) \in \mathscr{N}(1)\big\}\end{aligned}$$

depends on the probability measure $\mathbb{P}$, and we want to keep $\mathscr{D}_s$ the same across all investors.

The investor's optimal value function is given by

$$v_i(x) = \sup_{X_T \in \mathscr{C}(x)} E\left[\frac{d\mathbb{P}_i}{d\mathbb{P}}U(X_T)\right] \tag{11.26}$$

$$= \inf_{y>0, Y_T \in D_s} E\left[\sup_{X_T \in L_+^0}\left(\frac{d\mathbb{P}_i}{d\mathbb{P}}U(X_T) - yX_TY_T\right) + xy\right]$$

$$= \inf_{y>0}\left(\left\{\inf_{Y_T \in D_s} E\left[\frac{d\mathbb{P}_i}{d\mathbb{P}}\tilde{U}\left(yY_T\frac{d\mathbb{P}}{d\mathbb{P}_i}\right)\right]\right\} + xy\right).$$

The optimal solution $\hat{X}_T$ solves

$$\sup_{X_T \in L_+^0}\left(\frac{d\mathbb{P}_i}{d\mathbb{P}}U(X_T) - \hat{y}X_T\hat{Y}_T\right),$$

i.e.

$$\frac{d\mathbb{P}_i}{d\mathbb{P}} U'(\hat{X}_T) - \hat{y}\hat{Y}_T = 0, \tag{11.27}$$

or

$$\hat{X}_T = I\left(\hat{y}\hat{Y}_T \frac{d\mathbb{P}}{d\mathbb{P}_i}\right) \quad \text{with} \quad I = (U')^{-1} = -\tilde{U}'.$$

Furthermore, $\hat{Y}_T \in D_s$ solves

$$\tilde{v}_i(\hat{y}) = \inf_{Y_T \in D_s} E\left[\frac{d\mathbb{P}_i}{d\mathbb{P}} \tilde{U}\left(\hat{y} Y_T \frac{d\mathbb{P}}{d\mathbb{P}_i}\right)\right]. \tag{11.28}$$

As before, the solution $\hat{y}$ solves

$$\inf_{y>0} (\tilde{v}_i(y) + xy)$$

or

$$\tilde{v}_i'(\hat{y}) + x = 0.$$

Computing the derivative yields

$$E\left[\tilde{U}'\left(\hat{y}\hat{Y}_T \frac{d\mathbb{P}}{d\mathbb{P}_i}\right) \hat{Y}_T\right] + x = 0.$$

Using $I = (U')^{-1} = -\tilde{U}'$ and the optimal $\hat{X}_T$ shows that $\hat{y}$ satisfies the budget constraint, i.e.

$$E\left[I\left(\hat{y}\hat{Y}_T \frac{d\mathbb{P}}{d\mathbb{P}_i}\right) \hat{Y}_T\right] = E\left[\hat{X}_T \hat{Y}_T\right] = x.$$

Finally, using the fact that

$$v(x) = \tilde{v}(\hat{y}) + x\hat{y}$$

yields the shadow price of the constraint

$$v'(x) = \hat{y}.$$

The optimal supermartingale deflator

$$\hat{Y}_T = \frac{\frac{d\mathbb{P}_i}{d\mathbb{P}} U'(\hat{X}_T)}{\hat{y}} = \frac{\frac{d\mathbb{P}_i}{d\mathbb{P}} U'(\hat{X}_T)}{v'(x)}. \tag{11.29}$$

This completes the modifications to the investor's optimization problem for differential beliefs.

11.10 Notes

Excellent references for solving the investor's optimization problem in an incomplete market are Dana and Jeanblanc [42], Duffie [52], Karatzas and Shreve [118], Merton [140], and Pham [149].

Appendix

Lemma 11.3 (Existence of a Saddle Point) *Assume that*

(1) a solution $\hat{X}_T$ *exists to* $v(x) = \sup_{X_T \in \mathscr{C}^e(x)} E[U(X_T)]$ *and*

(2) a solution $\hat{Y}_T$ *exists to* $\tilde{v}(y) = \inf_{Y_T \in D_s} E[\tilde{U}(yY_T)]$.

Then, v *and* $\tilde{v}$ *are in conjugate duality, i.e.*

$$v(x) = \inf_{y>0} (\tilde{v}(y) + xy), \ \forall x > 0 \tag{11.30}$$

$$\tilde{v}(y) = \sup_{x>0} (v(x) - xy), \ \forall y > 0. \tag{11.31}$$

In addition,

(i) $\tilde{v}$ *is strictly convex, decreasing, and differentiable on* $(0, \infty)$,
(ii) v *is strictly concave, increasing, and differentiable on* $(0, \infty)$,
(iii) defining $\hat{y}$ *to be where the infimum is attained in expression (11.30),*

$$v(x) = \tilde{v}(\hat{y}) + x\hat{y}.$$

And, $(\hat{X}_T, \hat{Y}_T, \hat{y})$ *is a saddle point of* $\mathscr{L}(\hat{X}_T, \hat{Y}_T, \hat{y})$.

Proof (Part 1)

If a solution $\hat{X}_T$ exists to $v(x) = \sup_{X_T \in \mathscr{C}^e(x)} E[U(X_T)]$, then by the free disposal Lemma 11.1, it has the identical solution and value function as $v(x) = \sup_{X_T \in \mathscr{C}(x)} E[U(X_T)]$.

Given $v(x)$ define $v^*(y) = \sup_{x>0} (v(x) - xy)$, $\forall y > 0$.

First, because $v(x) < \infty$ for some $x > 0$, v is proper. v is increasing and strictly concave because $v(x) = \sup_{X_T \in \mathscr{C}(x)} E[U(X_T)]$ and U is increasing and strictly concave. This implies $v(x) < \infty$ for all $x > 0$ (see Pham [149, p. 181]). $v(x)$ is strictly concave on $(0, \infty)$, hence continuous on $(0, \infty)$, and therefore upper semi-continuous.

By Pham [149, Theorem B.2.3, p. 219], v and v^* are in conjugate duality, where $v(x) = \inf_{y>0} (v^*(y) + xy)$.

By Pham [149, Proposition B.2.4, p. 219], we get that $v^*(y)$ is differentiable on $\text{int}(\text{dom}(\tilde{v}))$.

By Pham [149, Proposition B.3.5, p. 219], since v^* is strictly convex, we get v is differential on $(0, \infty)$.

To complete the proof, we need to show that $v^*(y) = \tilde{v}(y)$.

(Step 1) By the definition of $\tilde{U}$ we have $U(x) \le \tilde{U}(y) + xy$ for all $x > 0$ and $y > 0$.

Thus, $U(X_T) \le \tilde{U}(yY_T) + X_T yY_T$ for $X_T \in \mathscr{C}(x)$. Taking expectations yields

$$E[U(X_T)] \le E[\tilde{U}(yY_T)] + yE[X_T Y_T] \le E[\tilde{U}(yY_T)] + yx.$$

The last inequality uses $X_T \in \mathscr{C}(x)$.

Taking the supremum on the left side, the infimum on the right side, and using the definition of $\tilde{v}(y) = \inf_{Y_T \in D_s} E[\tilde{U}(yY_T)]$, we get:

$$v(x) \le \tilde{v}(y) + xy \text{ for all } y > 0, \text{ or } v(x) \le \inf_{y>0} (\tilde{v}(y) + xy).$$

(Step 2) We have $\tilde{U}(y) = U(I(y)) - yI(y)$ for all $y > 0$.

Hence, $\tilde{U}(y\hat{Y}_T) = U(I(y\hat{Y}_T)) - yY_T I(y\hat{Y}_T)$ where $\hat{Y}_T$ attains $\tilde{v}(y)$.

Take expectations to get

$$\tilde{v}(y) = E\left[U(I(y\hat{Y}_T))\right] - yE\left[\hat{Y}_T I(y\hat{Y}_T)\right].$$

Choose y such that $X_T = I(y\hat{Y}_T) \in \mathscr{C}(x)$. Then, this equals $\tilde{v}(y) + xy = E[U(X_T)]$.

Taking the supremum on the right side yields
$\tilde{v}(y) + xy \le v(x)$. The infimum of the left side gives $\inf_{y>0} (\tilde{v}(y) + xy) \le v(x)$.

Combined steps 1 and 2 show $\inf_{y>0}(\tilde{v}(y)+xy)=v(x)$.

Because $\tilde{v}(x)<\infty$ for some $x>0$, $\tilde{v}$ is proper.

Because $\tilde{v}(y) = \inf_{Y_T\in D_s} E[\tilde{U}(yY_T)]$, we see that $\tilde{v}$ is strictly convex, hence continuous on int(dom($\tilde{v}$)).

By Pham [126, Theorem B.2.3, p. 219], $\tilde{v}(y)=\sup_{x>0}(v(x)-xy)$, $\forall y>0$.

Define $\hat{y}$ to be where the infimum is attained in $\inf_{y>0}(\tilde{v}(y)+xy)=v(x)$. It exists because $v(x)<\infty$ and $\tilde{v}$ is strictly convex. Then, $v(x)=\tilde{v}(\hat{y})+x\hat{y}$.

(Part 2) By Guler [66, p. 278], $\mathscr{L}(\hat{X}_T,\hat{Y}_T,\hat{y})$ is a saddle point if and only if the solution to the primal problem equals the solution to the dual problem, i.e.

$$v(x)=\sup_{X_T\in L^0_+}\left(\inf_{Y_T\in D_s,\,y>0}\mathscr{L}(X_T,Y_T,y)\right)=\inf_{Y_T\in D_s,\,y>0}\left(\sup_{X_T\in L^0_+}\mathscr{L}(X_T,Y_T,y)\right).$$

We show this later condition.

First, $v(x)=\inf_{y>0}(\tilde{v}(y)+xy)$. Using the definitions, this is equivalent to

$$\begin{aligned}v(x)&=\inf_{y>0}\left(\inf_{Y_T\in D_s}E[\tilde{U}(yY_T)]+xy\right)\\&=\inf_{y>0}\left(\inf_{Y_T\in D_s}E\left[\sup_{X_T\in L^0_+}U(X_T)-yX_TY_T\right]+xy\right).\end{aligned}$$

Exchange the sup and $E[\cdot]$ operator, which is justified by the same proof as in the appendix to the portfolio optimization Chap. 10.

$$\begin{aligned}v(x)&=\inf_{y>0}\left(\inf_{Y_T\in D_s}\left(\sup_{X_T\in L^0_+}E\left[U(X_T)-yX_TY_T\right]\right)+xy\right)\\&=\inf_{y>0}\left(\inf_{Y_T\in D_s}\left(\sup_{X_T\in L^0_+}E\left[U(X_T)\right]-y\left(E\left[X_TY_T\right]-x\right)\right)\right)\\&=\inf_{y>0}\left(\inf_{Y_T\in D_s}\left(\sup_{X_T\in L^0_+}\mathscr{L}(X_T,Y_T,y)\right)\right).\end{aligned}$$

This completes the proof.

Remark 11.7 (Extension to Chap. 12) This same proof works with $X_T \in \mathscr{C}(x) \subset L^0_+$ replaced by $(c, X_T) \in \mathscr{C}(x) \subset \mathscr{L}^0_+ \times L^0_+$ and $Y \in \mathscr{D}_s \subset \mathscr{L}^0_+$, where $v(x) = \sup_{(c,X_T)\in\mathscr{C}(x)} E\left[\int_0^T U_1(c_t)dt + U_2(X_T)\right]$ and $\tilde{v}(y) = \inf_{Y\in\mathscr{D}_s} E\left[\int_0^T \tilde{U}_1(yY_t)dt + \tilde{U}_2(yY_T)\right]$. This completes the remark.

Chapter 12
Incomplete Markets (Utility over Intermediate Consumption and Terminal Wealth)

This chapter studies the investor's optimization problem in an incomplete market where the investor has a utility function defined over both terminal wealth and intermediate consumption. The presentation parallels the portfolio optimization problem studied in Chap. 11. This chapter is based on Jarrow [92].

12.1 The Set-Up

We need to alter the previous set-up to include intermediate consumption. Given is a normalized market $(S, (\mathscr{F}_t), \mathbb{P})$ where the value of a money market account (mma) $B_t \equiv 1$. As in the previous two chapters, the utility function is defined over terminal wealth. Now, the utility function will also be defined over intermediate consumption. Both terminal wealth and consumption are denominated in units of the mma. As discussed in the utility function Chap. 9, this is without loss of generality.

In addition to choosing a trading strategy, now an investor also chooses a *consumption plan* $c : \Omega \times [0, T] \to [0, \infty)$, which is a nonnegative optional process. Thus, we consider a triplet, consisting of a trading strategy and consumption plan

$$(\alpha_0, \alpha, c) \in (\mathscr{O}, \mathscr{L}(S), \mathscr{O}).$$

The value of the trading strategy at time t is equal to

$$X_t = \alpha_0(t) + \alpha_t \cdot S_t. \tag{12.1}$$

R. A. Jarrow, *Continuous-Time Asset Pricing Theory*, Springer Finance,
https://doi.org/10.1007/978-3-319-77821-1_12

This is the same wealth process as in the previous portfolio optimization problems studied in Chaps. 10 and 11 without consumption. Next, define the *cumulative* time t consumption as

$$C_t = \int_0^t c_u du, \tag{12.2}$$

where $c_t \geq 0$ all t a.s. $\mathbb{P}$. Note that $\int_0^t c_s ds$ is well defined for all $t \in [0, T]$, although it could be $+\infty$. If cumulative consumption is finite a.s. $\mathbb{P}$, then $C \in \mathscr{L}_+^0$ (adapted and cadlag) is continuous and of finite variation (see Lemma 1.1 in Chap. 1). We next introduce the extension of a self-financing trading strategy when there is consumption.

Definition 12.1 (Self-Financing with Consumption) A trading strategy $(\alpha_0, \alpha, c) \in (\mathscr{O}, \mathscr{L}(S), \mathscr{O})$ is said to be a *self-financing trading strategy (s.f.t.s) with consumption* c if

$$X_t + C_t = x + \int_0^t \alpha_u \cdot dS_u \tag{12.3}$$

for all t a.s. $\mathbb{P}$.

Here, the change in value of the trading strategy "finances" both the value process X_t and the cumulative consumption C_t. We note that this trading strategy is not "truly" self-financing because there is a cash outflow for consumption. Note the analogy to the s.f.t.s. with cash flows as defined in the super- and sub-replication Chap. 8 and Remark 11.1 in the portfolio optimization Chap. 11.

We define the set of *nonnegative wealth s.f.t.s.'s with consumption* c for $x \geq 0$ as

$$\begin{aligned} \mathscr{N}(x) = \Big\{ & (\alpha_0, \alpha, c) \in (\mathscr{O}, \mathscr{L}(S), \mathscr{O}) : X_t = \alpha_0(t) + \alpha_t \cdot S_t, \\ & x + \int_0^t \alpha_u \cdot dS_u = X_t + C_t \geq 0, \ \forall t \in [0, T] \Big\}. \end{aligned} \tag{12.4}$$

We note that for the mathematics, $X_t + C_t = x + \int_0^t \alpha_u \cdot dS_u$ becomes the fundamental quantity for defining the relevant sets of deflators and deflator processes.

The set of wealth processes plus cumulative consumption generated by the nonnegative wealth s.f.t.s.'s with consumption is denoted

$$\begin{aligned} \mathscr{X}^e(x) = \Bigg\{ & (X + C) \in \mathscr{L}_+^0 : \exists (\alpha_0, \alpha, c) \in \mathscr{N}(x), X_t + C_t \\ & = x + \int_0^t \alpha_u \cdot dS_u, \forall t \in [0, T] \Bigg\} \end{aligned}$$

and the set of time T terminal wealths and consumption generated by these wealth processes is

$$\mathscr{C}^e(x) = \{(X_T + C_T) \in L^0_+ : \exists(\alpha_0, \alpha, c) \in \mathscr{N}(x),\ x + \int_0^T \alpha_t \cdot dS_t = X_T + C_T\}$$
$$= \{(X_T + C_T) \in L^0_+ : \exists Z \in \mathscr{X}(x),\ X_T + C_T = Z_T\}.$$

The sets of relevant local martingale and supermartingale deflators, analogous to those in the portfolio optimization Chap. 11, are given by

$$\mathscr{D}_s = \big\{Y \in \mathscr{L}^0_+ : Y_0 = 1,\ (X + C)Y \text{ is a } \mathbb{P}\text{-supermartingale},$$
$$X + C = 1 + \int \alpha \cdot dS,\ (\alpha_0, \alpha, c) \in \mathscr{N}(1)\big\}$$

$$D_s = \{Y_T \in L^0_+ : Y_0 = 1, \exists (Z(T)_n)_{n \geq 1} \in M_l,\ Y_T \leq \lim_{n\to\infty} Z_n(T)\ a.s.\}$$

$$= \{Y_T \in L^0_+ : \exists Z \in \mathscr{D}_s,\ Y_T = Z_T\}$$

$$\mathscr{D}_l = \big\{Y \in \mathscr{L}^0_+ : Y_0 = 1,\ (X + C)Y \text{ is a } \mathbb{P}\text{-local martingale},$$
$$X + C = 1 + \int \alpha \cdot dS,\ (\alpha_0, \alpha, c) \in \mathscr{N}(1)\big\}$$

$$D_l = \{Y_T \in L^0_+ : \exists Z \in \mathscr{D}_l,\ Y_T = Z_T\}$$

$$\mathscr{M}_l = \left\{Y \in \mathscr{D}_l : \exists \mathbb{Q} \sim \mathbb{P},\ Y_T = \frac{d\mathbb{Q}}{d\mathbb{P}}\right\}$$

$$M_l = \{Y_T \in L^0_+ : \exists Z \in \mathscr{M}_l,\ Y_T = Z_T\}$$

$$\mathscr{M} = \{Y \in \mathscr{L}^0_+ :\ Y_T = \frac{d\mathbb{Q}}{d\mathbb{P}},\ \mathbb{Q} \in \mathfrak{M}\}$$

$$M = \{Y \in L^0_+ : \exists Z \in \mathscr{M},\ Y_T = Z_T\}.$$

We have the following relationships among the various sets.

$$\text{stochastic processes} \qquad \mathscr{M} \subset \mathscr{M}_l \subset \mathscr{D}_l \subset \mathscr{D}_s \subset \mathscr{L}^0_+$$

$$\text{random variables} \qquad M \subset M_l \subset D_l \subset D_s \subset L^0_+$$

Note that since the s.f.t.s. with consumption plan $\alpha = 0, c = 0$ for all t with initial wealth $x = 1$ is a feasible nonnegative wealth s.f.t.s. with consumption, $Y \in \mathscr{D}_l$ implies that Y is a nonnegative $\mathbb{P}$-local martingale and a $\mathbb{P}$-supermartingale by Lemma 1.3 in Chap. 1. By the Doob–Meyer decomposition (Protter [151, Theorem 13, p. 115]) for supermartingales, Y is also a semi-martingale. We note that Y need not be a probability density with respect to $\mathbb{P}$ (and therefore not a $\mathbb{P}$-martingale).

As in the portfolio optimization Chap. 11, we assume there are no arbitrage opportunities in the market, i.e.

Assumption (NFLVR) *$\mathfrak{M}_l \neq \emptyset$ where $\mathfrak{M}_l = \{\mathbb{Q} \sim \mathbb{P} : S$ is a $\mathbb{Q}$-local martingale$\}$.*

In this chapter the market may be incomplete. In an incomplete market, given an equivalent local martingale measure $\mathbb{Q} \in \mathfrak{M}_l$, there exist integrable derivatives that cannot be synthetically constructed using a value process that is a $\mathbb{Q}$-martingale. If the set of equivalent local martingale measures is not a singleton, then the cardinality of $|\mathfrak{M}_l| = \infty$, i.e. the set contains an infinite number of elements. In this case, derivative prices are not uniquely determined. Here, the investor's portfolio optimization problem provides a method to uniquely identify the relevant local martingale measure for the investor to price derivatives.

12.2 Problem Statement

This section presents the investor's portfolio optimization problem. We assume that the investor's preferences $\overset{\rho}{\succeq}$ have a state dependent EU representation given by

$$E\left[\int_0^T U_1(c_t, t, \omega)dt + U_2(X_T, \omega)\right], \tag{12.5}$$

where U_1 and U_2 are utility functions with the properties assumed in the utility function Chap. 9, repeated here for convenience.

Assumption (Utility Function) *A utility function is a $U_i : (0, \infty) \times \Omega \to \mathbb{R}$ for $i = 1, 2$ such that for all $\omega \in \Omega$ a.s. $\mathbb{P}$,*

(i) $U_i(x, \omega)$ is $\mathscr{B}(0, \infty) \otimes \mathscr{F}_T$-measurable,
(ii) $U_i(x, \omega)$ is continuous and differentiable in $x \in (0, \infty)$,
(iii) $U_i(x, \omega)$ is strictly increasing and strictly concave in $x \in (0, \infty)$, and
(iv) (Inada Conditions) $\lim_{x \downarrow 0} U_i'(x, \omega) = \infty$ and $\lim_{x \to \infty} U_i'(x, \omega) = 0$.

And, with the following modifications:

1. $U_1(c, t, \omega)$ is an optional process in $(t, \omega) \in [0, T] \times \Omega$ for every $c > 0$, and
2. $\int_0^T |U_1(c, t, \omega)|\, dt < \infty$ a.s. $\mathbb{P}$ for every $c > 0$.

The investor's optimization problem is to choose a nonnegative wealth s.f.t.s. with consumption to maximize their expected utility of consumption and terminal wealth. The entire wealth and consumption process is constrained to be nonnegative. For simplicity of notation, we suppress the dependence of the utility functions on the state $\omega \in \Omega$, writing $U_1(c,t) = U_1(c,t,\omega)$ and $U_2(x) = U_2(x,\omega)$ everywhere below. Formally, the portfolio optimization problem is given by

Problem 12.1 (Utility Optimization (Choose the Optimal (α_0, α, c) **for** $x \geq 0$**))**

$$v(x) = \sup_{(\alpha_0,\alpha,c)\in\mathscr{N}(x)} E\left[\int_0^T U_1(c_t,t)dt + U_2(X_T)\right], \qquad \text{where}$$

$$\mathscr{N}(x) = \Big\{(\alpha_0,\alpha,c) \in (\mathscr{O}, \mathscr{L}(S), \mathscr{O}) : X_t = \alpha_0(t) + \alpha_t \cdot S_t,\ x + \int_0^t \alpha_u \cdot dS_u = X_t + C_t \geq 0,\ \forall t \in [0,T]\Big\}.$$

To solve this problem, as before, we can break the problem up into two steps. First, we solve for the optimal "derivative" $(c, X_T) \in \mathscr{O} \times L_+^0$ where a derivative is now a wealth process with intermediate cash flows. Of course, this is more consistent with actual markets because many derivatives have intermediate cash flows. Second, we find the trading strategy that generates the optimal "derivative." As before, the solution to the second step uses the insights from the fundamental theorems Chap. 2 on the synthetic construction of a derivative's payoffs.

Problem 12.2 (Utility Optimization (Choose the Optimal Derivative (c, X_T) **for** $x \geq 0$**))**

$$v(x) = \sup_{(c,X_T)\in\mathscr{C}^e(x)} E\left[\int_0^T U_1(c_t,t)dt + U_2(X_T)\right], \qquad \text{where}$$

$$\mathscr{C}^e(x) = \{(c, X_T) \in \mathscr{O} \times L_+^0 : \exists(\alpha_0,\alpha,c) \in \mathscr{N}(x),\ x + \textstyle\int_0^T \alpha_t \cdot dS_t = X_T + C_T\}.$$

As in the portfolio optimization Chap. 11, we solve an equivalent problem where we add free disposal. Lemma 11.1 in Chap. 11 justifies the equivalence (the same proof applies with consumption).

Problem 12.3 (Utility Optimization (Choose the Optimal Derivative (c, X_T) **for** $x \geq 0$**))**

$$v(x) = \sup_{(c,X_T)\in\mathscr{C}(x)} E\left[\int_0^T U_1(c_t,t)dt + U_2(X_T)\right], \qquad \text{where}$$

$$\mathscr{C}(x) = \{(c, X_T) \in \mathscr{O} \times L_+^0 : \exists(\alpha_0,\alpha,c) \in \mathscr{N}(x),\ x + \textstyle\int_0^T \alpha_t \cdot dS_t \geq X_T + C_T\}.$$

To facilitate the solution, we give an alternative characterization of the budget constraint. Note that $\mathscr{C}(x)$ represents those payoffs $X_T + C_T \in L_+^0$ that can be generated by a nonnegative wealth with consumption process $(\alpha_0,\alpha,c) \in \mathscr{N}(x)$ with initial wealth x. Consider the set of local martingale deflator processes $Y \in \mathscr{M}_l$. Each of these local martingale deflator processes generates a present value for the payoff $X_T + C_T$, $E[(X_T + C_T)Y_T]$. Since it is unknown which local martingale deflator process should be used, the constraint set $\mathscr{C}(x)$ should be the same as the set of all payoffs whose "worst case" present values are affordable at time 0, i.e. less than or equal to x. This is indeed the case, as the next theorem shows.

Theorem 12.1 (Budget Constraint Equivalence)

$$\mathscr{C}(x) = \left\{ (c, X_T) \in \mathscr{O} \times L_+^0 : \sup_{Y \in \mathscr{M}_l} E\left[(X_T + C_T) Y_T \right] \leq x \right\}.$$

Note that in the budget constraint the supremum is taken across all local martingale deflator processes $Y \in \mathscr{M}_l$.

Proof Define $\mathscr{C}_1(x) = \left\{ (c, X_T) \in \mathscr{O} \times L_+^0 : \sup_{Y \in \mathscr{M}_l} E\left[(X_T + C_T) Y_T \right] \leq x \right\}.$

(Step 1) Show $\mathscr{C}(x) \subset \mathscr{C}_1(x)$.

Let $(c, X_T) \in \mathscr{C}(x)$. Then, $\exists (\alpha_0, \alpha, c) \in \mathscr{N}(x)$ such that $X_T + \int_0^T c_u du \leq x + \int_0^T \alpha_u \cdot dS_u$.

Let $W_t \equiv x + \int_0^T \alpha_t \cdot dS_t$. Choose $Y \in \mathscr{M}_l$, which implies that $Y_T \in M_l$.

By definition of $\mathscr{M}_l$, $W_t Y_t$ is a $\mathbb{P}$-local martingale bounded below by zero. Hence, by Lemma 1.3 in Chap. 1, $W_t Y_t$ is a $\mathbb{P}$-supermartingale.

Thus, $E\left[Y_T W_T \right] \leq Y_0 x = x$.

Since this is true for all $Y \in \mathscr{M}_l$, $(c, X_T) \in \mathscr{C}_1(x)$.

(Step 2) Show $\mathscr{C}_1(x) \subset \mathscr{C}(x)$.

Take $(c, X_T) \in \mathscr{C}_1(x)$.

Define $X_t + C_t \equiv \operatorname{ess\,sup}_{Y_T \in M_l} E[\frac{Y_T}{Y_t}(X_T + C_T) | \mathscr{F}_t] = \operatorname{ess\,sup}_{\mathbb{Q} \in \mathfrak{M}_l} E^{\mathbb{Q}}[X_T + C_T | \mathscr{F}_t] \geq 0$.

Part 1 of step 2 is to note that $X_t + C_t$ is a supermartingale for all $\mathbb{Q} \in \mathfrak{M}_l$, see Theorem 1.4 in Chap. 1.

Part 2 of step 2 is to use the optional decomposition Theorem 1.5 in Chap. 1, which implies that

$\exists \alpha \in \mathscr{L}(S)$ and an adapted right continuous nondecreasing process A_t with $A_0 = 0$ such that $X_0 + \int_0^t \alpha_u \cdot dS_u - A_t = X_t + C_t$ for all t.

Since $A_t \geq 0$ and $X_t + C_t \geq 0$, this shows that $X_0 + \int_0^t \alpha_u \cdot dS_u \geq 0$ for all t. Hence, the trading strategy is bounded below by zero, i.e. $(\alpha_0, \alpha, c) \in \mathscr{N}(x)$ with $x = X_0$. Thus, $(c, X_T) \in \mathscr{C}(x)$, which completes the proof.

As in the portfolio optimization Chap. 11, in general, there does not exist an element $Y_T \in M_l$ such that the corresponding dual problem has a solution. To guarantee the existence of such an element, we need to "fill-in the interior" of the set of local martingale deflators M_l. To do this, we expand M_l to the set of supermartingale deflators D_s. In this regard, we have the following equivalence.

Theorem 12.2 (Budget Constraint Equivalence)

$$\mathscr{C}(x)=\left\{(c,X_T)\in\mathscr{O}\times L^0_+:\ \sup_{Y\in\mathscr{D}_s}E\left[(X_T+C_T)\,Y_T\right]\le x\right\}.$$

Proof Define $\mathscr{C}_1(x)=\left\{(c,X_T)\in\mathscr{O}\times L^0_+:\ \sup_{Y\in\mathscr{D}_s}E\left[(X_T+C_T)\,Y_T\right]\le x\right\}$.

(Step 1) Show $\mathscr{C}_1(x)\subset\mathscr{C}(x)$.
Take $(c,X_T)\in\mathscr{C}_1(x)$. Then, $E\left[(X_T+C_T)\,Y_T\right]\le x$ for all $Y\in\mathscr{D}_s$.
Since $\mathscr{M}_l\subset\mathscr{D}_s$, $E\left[(X_T+C_T)\,Y_T\right]\le x$ for all $Y\in\mathscr{M}_l$.
Hence, $(c,X_T)\in\mathscr{C}(x)$.

(Step 2) Show $\mathscr{C}(x)\subset\mathscr{C}_1(x)$.
Let $(c,X_T)\in\mathscr{C}(x)$. Then,
$E[(X_T+C_T)\,Z_n]\le x$ for all $Z_n\in\mathscr{M}_l$.
Choose a $Y\in\mathscr{D}_s$. Then, $Y_T\in D_s$. By the definition of D_s,
there exists a $Z_n(T)\in M_l$ such that $Y_T\le\lim_{n\to\infty}Z_n(T)$ a.s.
Thus, $(X_T+C_T)\,Y_T\le\lim_{n\to\infty}(X_T+C_T)\,Z_n(T)$ a.s. since $(X_T+C_T)\ge 0$.
Taking expectations yields

$$\begin{aligned}E[(X_T+C_T)\,Y_T]&\le E[\lim_{n\to\infty}(X_T+C_T)\,Z_n(T)]\\&\le\lim_{n\to\infty}E[(X_T+C_T)\,Z_n(T)],\end{aligned}$$

where the second inequality is due to Fatou's lemma.
But, $E[(X_T+C_T)\,Z_n(T)]\le\sup_{Y_T\in M_l}E\left[(X_T+C_T)\,Y_T\right]$ for all n.
Hence, $\lim_{n\to\infty}E[(X_T+C_T)\,Z_n]\le\sup_{Y_T\in M_l}E\left[(X_T+C_T)\,Y_T\right]$.
This implies that $E[(X_T+C_T)\,Y_T]\le\sup_{Y_T\in M_l}E\left[(X_T+C_T)\,Y_T\right]\le x$.
Since $Y\in\mathscr{D}_s$ is arbitrary, this shows that

$$E[(X_T+C_T)\,Y_T]\le x\quad\text{for}\quad\text{all }Y\in\mathscr{D}_s.$$

Hence, $(c,X_T)\in\mathscr{C}_1(x)$. This completes the proof.

One more transformation generates the final representation of the budget constraint used to solve the investor's optimization problem.

Theorem 12.3 (Budget Constraint Equivalence)

$$\mathscr{C}(x)=\left\{(c,X_T)\in\mathscr{O}\times L^0_+:\ \sup_{Y\in\mathscr{D}_s}E\left[X_TY_T+\int_0^T Y_uc_u du\right]\le x\right\}.$$

Proof

(Step 1) Let $V_t = V_0 + \int_0^t \alpha_u dZ_u$, where Z is a semimartingale under $\mathbb{P}$.
By the numeraire invariance Theorem 12.7 given in the appendix to this chapter, $YV = \alpha \bullet (YZ)$, where $\bullet$ denotes stochastic integration.

(Step 2) Let $\alpha = 1$ and $Z = C$. This implies $V = C$ by the definition of V.
Substitution gives $YC = 1 \bullet (YC) = 1 \bullet Y \bullet C = Y \bullet C$.
The second equality is the associate law for stochastic integrals, see Protter [151, Theorem 19, p. 62].
Or, $Y_t C_t = \int_0^t Y_u dC_u = \int_0^t Y_u c_u du$ since $Y_0 C_0 = 1 \cdot 0$.
This completes the proof.

Using this characterization of the budget constraint, we get the final problem statement.

Problem 12.4 (Utility Optimization (Choose the Optimal Derivative (c, X_T) for $x \geq 0$))

$$v(x) = \sup_{(c,X_T) \in \mathscr{C}(x)} E\left[\int_0^T U_1(c_t, t)dt + U_2(X_T)\right], \qquad \text{where}$$

$$\mathscr{C}(x) = \left\{(c, X_T) \in \mathscr{O} \times L_+^0 : E\left[X_T Y_T + \int_0^T Y_u c_u du\right] \leq x \text{ for all } Y \in \mathscr{D}_s\right\}.$$

To solve this problem, we define the Lagrangian function

$$\mathscr{L}(c, X_T, y, Y) = E\left[\int_0^T U_1(c_t, t)dt + U_2(X_T)\right] + y\left(x - E\left[\int_0^T c_t Y_t dt + X_T Y_T\right]\right).$$

The *primal problem* can be written as

$$\text{primal}: \quad \sup_{(c,X_T) \in \mathscr{O} \times L_+^0} \left(\inf_{y>0, Y \in \mathscr{D}_s} \mathscr{L}(c, X_T, y, Y)\right)$$

$$= \sup_{(c,X_T) \in \mathscr{C}(x)} E\left[\int_0^T U_1(c_t, t)dt + U_2(X_T)\right] = v(x).$$

Proof Note that

$$E\left[X_T Y_T + \int_0^T Y_u c_u du\right] \leq x \text{ for all } Y \in \mathscr{D}_s$$

if and only if

$$\sup_{Y\in\mathscr{D}_s} E\left[X_T Y_T + \int_0^T Y_u c_u du\right] \le x.$$

Hence,

$$\sup_{Y\in\mathscr{D}_s}\left(E\left[X_T Y_T + \int_0^T Y_u c_u du\right] - x\right) \le 0.$$

But,

$$\sup_{Y\in\mathscr{D}_s}\left(E\left[X_T Y_T + \int_0^T Y_u c_u du\right] - x\right) = -\inf_{Y\in\mathscr{D}_s}\left(x - E\left[X_T Y_T + \int_0^T Y_u c_u du\right]\right).$$

Substitution yields

$$\inf_{Y\in\mathscr{D}_s}\left(x - E\left[X_T Y_T + \int_0^T Y_u c_u du\right]\right) \ge 0.$$

Hence,

$$\inf_{y>0}\ y\left\{\inf_{Y\in\mathscr{D}_s}\left(x - E\left[X_T Y_T + \int_0^T Y_u c_u du\right]\right)\right\}$$
$$= \begin{cases} 0 & \text{if} \quad \sup_{Y\in\mathscr{D}_s} E\left[X_T Y_T + \int_0^T Y_u c_u du\right] \le x \\ -\infty & \text{otherwise.} \end{cases}$$

Or,

$$\inf_{y>0, Y\in\mathscr{D}_s}\left\{y\left(x - E\left[\int_0^T c_t Y_t dt + X_T Y_T\right]\right)\right\}$$
$$= \begin{cases} 0 & \text{if} \quad \sup_{Y\in\mathscr{D}_s} E\left[X_T Y_T + \int_0^T Y_u c_u du\right] \le x \\ -\infty & \text{otherwise.} \end{cases}$$

This implies

$$\inf_{y>0, Y\in\mathscr{D}_s}\left\{E\left[\int_0^T U_1(c_t, t)dt + U_2(X_T)\right] + y\left(x - E\left[\int_0^T c_t Y_t dt + X_T Y_T\right]\right)\right\}$$
$$= \begin{cases} E\left[\int_0^T U_1(c_t, t)dt + U_2(X_T)\right] & \text{if} \quad \sup_{Y\in\mathscr{D}_s} E\left[X_T Y_T + \int_0^T Y_u c_u du\right] \le x \\ -\infty & \text{otherwise.} \end{cases}$$

Substitution yields

$$\inf_{y>0, Y\in\mathscr{D}_s} \mathscr{L}(X_T, y, Y_T)$$

$$= \begin{cases} E\left[\int_0^T U_1(c_t, t)dt + U_2(X_T)\right] & \text{if} \quad \sup_{Y\in\mathscr{D}_s} E\left[X_T Y_T + \int_0^T Y_u c_u du\right] \leq x \\ -\infty & \text{otherwise.} \end{cases}$$

So,

$$\sup_{(c,X_T)\in\mathscr{O}\times L^0_+} \left(\inf_{y>0, Y\in\mathscr{D}_s} \mathscr{L}(X_T, y, Y_T) \right) = \sup_{(c,X_T)\in\mathscr{O}\times L^0_+} \Theta(c, X_T)$$

where

$$\Theta(c, X_T) = \begin{cases} E\left[\int_0^T U_1(c_t, t)dt + U_2(X_T)\right] & \text{if } \sup_{Y\in\mathscr{D}_s} E\left[X_T Y_T + \int_0^T Y_u c_u du\right] \leq x \\ -\infty & \text{otherwise} \end{cases}$$

$$= \sup_{(c,X_T)\in\mathscr{C}(x)} E\left[\int_0^T U_1(c_t, t)dt + U_2(X_T)\right] = v(x).$$

This completes the proof.

The *dual problem* is

$$\text{dual}: \quad \inf_{y>0, Y\in\mathscr{D}_s} \left(\sup_{(c,X_T)\in\mathscr{O}\times L^0_+} \mathscr{L}(c, X_T, y, Y) \right).$$

For optimization problems we know that

$$\inf_{y>0, Y\in\mathscr{D}_s} \left(\sup_{(c,X_T)\in\mathscr{O}\times L^0_+} \mathscr{L}(c, X_T, y, Y) \right) \geq$$

$$\sup_{(c,X_T)\in\mathscr{O}\times L^0_+} \left(\inf_{y>0, Y\in\mathscr{D}_s} \mathscr{L}(c, X_T, y, Y) \right) = v(x).$$

Proof We have

$$\mathscr{L}(c, X_T, y, Y) = E\left[\int_0^T U(c_t, t)dt + U(X_T)\right] + y\left(x - E\left[\int_0^T c_t Y_t dt + X_T Y_T\right]\right).$$

Note that

$$\sup_{(c,X_T)\in\mathscr{O}\times L^0_+} \mathscr{L}(c, X_T, y, Y) \geq \mathscr{L}(c, X_T, y, Y) \geq \inf_{y>0, Y\in\mathscr{D}_s} \mathscr{L}(c, X_T, y, Y)$$

for all (c, X_T, y, Y).

Taking the infimum over (y, Y) on the left side and the supremum over (c, X_T) on the right side gives

$$\inf_{y>0, Y\in\mathscr{D}_s} \left(\sup_{(c,X_T)\in\mathscr{O}\times L^0_+} \mathscr{L}(c, X_T, y, Y) \right) \geq \sup_{(c,X_T)\in\mathscr{O}\times L^0_+} \left(\inf_{y>0, Y\in\mathscr{D}_s} \mathscr{L}(c, X_T, y, Y) \right).$$

This completes the proof.

The solutions to the primal and dual problems are equal if there is no duality gap. We will solve the primal problem by solving the dual problem and showing that there is no duality gap.

12.3 Existence of a Solution

We need to show that there is no duality gap and that the optimum is attained, or equivalently, a saddle point exists. To prove this existence, we need three additional assumptions. The first two assumptions are identical to those used in the portfolio optimization Chap. 11 to guarantee the existence of a solution when the utility function is only defined over terminal wealth. The third assumption is needed because the utility function now includes intermediate consumption.

Assumption (Non-trivial Optimization Problem) *There exists an* $x > 0$ *such that* $v(x) < \infty$.

Assumption (Reasonable Asymptotic Elasticity)

$$AE(U_1, \omega) \equiv \limsup_{c\to\infty} \left(\operatorname*{ess\,sup}_{t,\omega} \frac{cU_1'(c, t, \omega)}{U_1(c, t, \omega)} \right) < 1 \qquad \textit{and}$$

$$AE(U_2, \omega) \equiv \limsup_{x\to\infty} \frac{xU_2'(x, \omega)}{U_2(x, \omega)} < 1 \quad a.s.\ \mathbb{P}.$$

Assumption (Boundedness of Utility) *There are nonrandom, continuous, strictly decreasing, functions* $K_i : (0, \infty) \to (0, \infty)$, *i=1,2 such that for all* $t \in [0, T]$ *and* $c > 0$,

$$K_1(c) \leq U_1'(c, t, \omega) \leq K_2(c) \quad a.s.\ \mathbb{P},$$

there exist constants $k_1 < k_2$ such that for all $t \in [0, T]$,

$$k_1 \leq U_1(1, t, \omega) \leq k_2 \quad a.s.\ \mathbb{P}, \qquad and$$

$$\limsup_{c \to \infty} \left(\operatorname*{ess\,inf}_{t,\omega} U_1(c, t, \omega) \right) > 0.$$

We can now prove that the investor's optimization problem has a unique solution.

Theorem 12.4 (Existence of a Unique Saddle Point and Characterization of $v(x)$) *Given the above assumptions, there exists a unique saddle point, i.e. there exists a unique $(\hat{c}, \hat{X}_T, \hat{Y}_T, \hat{y})$ such that*

$$\inf_{Y \in \mathscr{D}_s,\, y>0} \Big(\sup_{(c,X_T) \in \mathscr{O} \times L^0_+} \mathscr{L}(c, X_T, Y, y)\Big) = \mathscr{L}(\hat{c}, \hat{X}_T, \hat{Y}, \hat{y})$$

$$= \sup_{(c,X_T) \in \mathscr{O} \times L^0_+} \Big(\inf_{Y \in \mathscr{D}_s,\, y>0} \mathscr{L}(c, X_T, Y, y)\Big) = v(x). \tag{12.6}$$

Define $\tilde{v}(y) = \inf_{Y \in \mathscr{D}_s} E\left[\int_0^T \tilde{U}_1(yY_t, t)dt + \tilde{U}_2(yY_T)\right]$, *where*

$$\tilde{U}_i(y, \omega) = \sup_{x>0} \left[U_i(x, \omega) - xy\right], \quad y > 0 \text{ for } i = 1, 2.$$

Then, v and $\tilde{v}$ are in conjugate duality, i.e.

$$v(x) = \inf_{y>0} \left(\tilde{v}(y) + xy\right), \ \forall x > 0, \tag{12.7}$$

$$\tilde{v}(y) = \sup_{x>0} \left(v(x) - xy\right), \ \forall y > 0. \tag{12.8}$$

In addition,

(i) $\tilde{v}$ is strictly convex, decreasing, and differentiable on $(0, \infty)$,
(ii) v is strictly concave, increasing, and differentiable on $(0, \infty)$,
(iii) defining $\hat{y}$ to be where the infimum is attained in expression (12.7),

$$v(x) = \tilde{v}(\hat{y}) + x\hat{y}, \tag{12.9}$$

(iv)

$$\tilde{v}(y) = \inf_{Y \in \mathscr{D}_s} E\left[\int_0^T \tilde{U}_1(yY_t, t)dt + \tilde{U}_2(yY_T)\right]$$

$$= \inf_{Y \in \mathscr{M}_l} E\left[\int_0^T \tilde{U}_1(yY_t, t)dt + \tilde{U}_2(yY_T)\right].$$

Proof To prove this theorem, we show conditions 1 and 2 of the Lemma 11.3 in the appendix in the portfolio optimization Chap. 11 hold, as modified by adding consumption. Under the above assumptions, these conditions follow from Zitkovic [178, Theorem 4.2]. To apply this theorem, first we need to choose the random endowment to be identically zero. For utility of consumption U_1, the above assumptions are the hypotheses of Zitkovic [178, Theorem 4.2]. For the utility of terminal wealth U_2, choose the stochastic clock in Zitkovic [178] as in Example 2.6 (2), p. 757. Second, for utility of terminal wealth satisfying the reasonable asymptotic utility assumption ($AE(U) < 1$), the boundedness of utility assumption (above) is automatically satisfied, see Karatzas and Zitkovic [119, Example 3.2, p. 1838]. Although Zitkovic [178] requires that S is locally bounded, as shown in Karatzas and Zitkovic [119] this condition is not necessary for the choice of the stochastic clock given above, which is deterministic. This completes the proof.

Remark 12.1 (Alternative Sufficient Condition) The sufficient conditions (i) there exists an $x > 0$ such that $v(x) < \infty$, (ii) $AE(U) < 1$, and (iii) the boundedness of utility assumptions can be replaced by the alternative sufficient conditions $\tilde{v}(y) < \infty$ for all $y > 0$ and $v(x) > -\infty$ for all $x > 0$. The same proof as given for Theorem 12.4 works but with Mostovyi [144, Theorem 2.3] replacing Zitkovic [178, Theorem 4.2]. Here, Mostovyi [144] extends the domain of U_1 to include $c = 0$ and the range of U_1 to include $-\infty$ at $c = 0$. In addition, the assumption that the set of local martingale measures $\mathfrak{M}_l \neq \emptyset$ can be replaced by the set of local martingale deflator processes $\mathscr{D}_l \neq \emptyset$.

A sketch of the proof is as follows. The existence of a buy and hold nonnegative wealth s.f.t.s. strategy implies that the constraint set $\mathscr{C}(x)$ for v is nonempty, and $\mathscr{D}_l \neq \emptyset$ implies that the constraint set for $\tilde{v}$ is nonempty because $\mathscr{D}_l \subset \mathscr{D}_s \neq \emptyset$. Given the appropriate topologies for $\mathscr{C}(x)$ and $\mathscr{D}_l$, assuming that $\tilde{v}(y) < \infty$ for all $y > 0$ and $v(x) > -\infty$ for all $x > 0$ together imply, because of the strict convexity of $\tilde{v}(y)$ for all $y > 0$ and the strict concavity of $v(x)$ for all $x > 0$, that a unique solution exists to each of these problems. Then, the conditions of Lemma 11.3 in the appendix of the portfolio optimization Chap. 11 apply to obtain the remaining results. This completes the remark.

Remark 12.2 (No Terminal Wealth) A special case of this theorem is the case where the utility of terminal wealth $U_2(X_T) \equiv 0$ and the trader's objective function is $E\left[\int_0^T U_1(c_t, t)dt\right]$. This completes the remark.

12.4 Characterization of the Solution

To solve this general problem, we will divide the solution procedure into three steps: (i) we characterize the optimal consumption problem, assuming that utility of terminal wealth is zero ($U_2 \equiv 0$), (ii) we characterize the optimal wealth problem, assuming utility of consumption is zero ($U_1 \equiv 0$), and then (iii) we combine the two characterizations.

12.4.1 Utility of Consumption ($U_2 \equiv 0$)

The simplified problem is given by

Problem 12.5 (Utility Optimization (Choose the Optimal Derivative c))

$$v_1(x_1) = \sup_{c\in\mathscr{C}_1(x_1)} E\left[\int_0^T U_1(c_t,t)dt\right], \qquad \text{where}$$

$$\mathscr{C}_1(x_1) = \left\{c \in \mathscr{O} : E\left[\int_0^T c_t Y_1(t)dt\right] \le x_1 \text{ for all } Y \in \mathscr{D}_s\right\}.$$

To solve this problem, we define the Lagrangian function

$$\mathscr{L}(c, y_1, Y_1) = E\left[\int_0^T U_1(c_t,t)dt+\right] + y_1\left(x - E\left[\int_0^T c_t Y_1(t)dt\right]\right).$$

We focus on solving the dual problem, i.e.

$$\text{dual}: \quad v_1(x_1) = \inf_{y_1>0, Y_1\in\mathscr{D}_s}\left(\sup_{c\in\mathscr{O}} \mathscr{L}(c, y_1, Y_1)\right).$$

12.4.1.1 The Solution

To solve this problem, we work from the inside of the right side to the outside.

Step 1. (Solve for $c \in \mathscr{O}$)
Fix y_1, Y_1 and solve for the optimal consumption $c \in \mathscr{O}$,

$$\sup_{c\in\mathscr{O}} \mathscr{L}(c, y_1, Y_1) = \sup_{c\in\mathscr{O}} E\left[\int_0^T (U_1(c_t,t) - y_1 c_t Y_1(t))\, dt\right] + y_1 x_1.$$

The sup and the $E\left[\int(\cdot)\right]$ operator can be interchanged.

Proof First, it is always true that

$$E\left[\int_0^T (U_1(c_t,t) - y_1 c_t Y_1(t))\, dt\right] \le E\left[\int_0^T \left(\sup_{c\in\mathscr{O}} (U_1(c_t,t) - y_1 c_t Y_1(t))\right) dt\right]$$

for all possible $c \in \mathscr{O}$. Taking the sup of the left side gives

$$\sup_{c\in\mathscr{O}} E\left[\int_0^T (U_1(c_t,t) - y_1 c_t Y_1(t))\,dt\right] \leq E\left[\int_0^T \left(\sup_{c\in\mathscr{O}} (U_1(c_t,t) - y_1 c_t Y_1(t))\right) dt\right].$$

Next, let c^* be such that

$$\left(U_1(c_t^*,t) - y_1 c_t^* Y_1(t)\right) = \sup_{c\in\mathscr{O}} (U_1(c_t,t) - y_1 c_t Y_1(t)).$$

Since c^* is a feasible,

$$E\left[\int_0^T \left(U_1(c_t^*,t) - y_1 c_t^* Y_1(t)\right) dt\right] \leq \sup_{c\in\mathscr{O}} E\left[\int_0^T (U_1(c_t,t) - y_1 c_t Y_1(t))\,dt\right],$$

i.e.

$$E\left[\int_0^T \left(\sup_{c\in\mathscr{O}} (U_1(c_t,t) - y_1 c_t Y_1(t))\right) dt\right] \leq \sup_{c\in\mathscr{O}} E\left[\int_0^T (U_1(c_t,t) - y_1 c_t Y_1(t))\,dt\right].$$

This completes the proof.

This implies that

$$\sup_{c\in\mathscr{O}} E\left[\int_0^T (U_1(c_t,t) - y_1 c_t Y_1(t))\,dt\right] = E\left[\int_0^T \left(\sup_{c\in\mathscr{O}} (U_1(c_t,t) - y_1 c_t Y_1(t))\right) dt\right].$$

Consider the optimization problem within the integral,

$$\sup_{c\in\mathscr{O}} (U_1(c_t,t) - y_1 c_t Y_1(t)).$$

To solve this problem, fix a $\omega \in \Omega$, and consider the related problem

$$\sup_{c_t(\omega)\in\mathbb{R}_+} (U_1(c_t(\omega),t,\omega) - y_1 c_t(\omega) Y_1(t,\omega)).$$

This is a simple optimization problem on the real line. The first-order condition for an optimum gives

$$U_1'(c_t(\omega),t,\omega) - y_1 Y_1(t,\omega) = 0, \qquad \text{or}$$

$$\hat{c}_t(\omega) = I_1(y_1 Y_1(t,\omega),t,\omega) \quad \text{with} \quad I_1(\cdot,t,\omega) = (U_1'(\cdot,t,\omega))^{-1}.$$

Given the properties of $I_1(\cdot,t,\omega)$, $\hat{c}_t(\omega) \in \mathscr{O}$.

Note that we have

$$\tilde{U}_1(y_1 Y_1(t), t) = \sup_{c \in \mathcal{O}} [U_1(c_t, t) - y_1 c_t Y_1(t)] = [U_1(\hat{c}_t, t) - y_1 \hat{c}_t Y_1(t)].$$

This yields

$$\sup_{c \in \mathcal{O}} \mathcal{L}(c, y_1, Y_1) = E\left[\int_0^T \left(\sup_{c \in \mathcal{O}} (U_1(c_t, t) - y_1 c_t Y_1(t))\right) dt\right] + x_1 y_1$$

$$= E\left[\int_0^T \tilde{U}_1(y_1 Y_1(t), t) dt\right] + x_1 y_1.$$

Hence,

$$\inf_{y_1 > 0, Y_1 \in \mathcal{D}_s} \left(\sup_{c \in \mathcal{O}} \mathcal{L}(c, y_1, Y_1)\right) = \inf_{y_1 > 0, Y_1 \in \mathcal{D}_s} \left(E\left[\int_0^T \tilde{U}_1(y_1 Y_1(t), t) dt\right] + x_1 y_1\right)$$

$$= \inf_{y_1 > 0} \left\{\inf_{Y_1 \in \mathcal{D}_s} \left(E\left[\int_0^T \tilde{U}_1(y_1 Y_1(t), t) dt\right] + x_1 y_1\right)\right\}.$$

Step 2. (Solve for $Y_1 \in \mathcal{D}_s$ **)**

Given the optimal consumption $\hat{c}$ and y_1, solve for the optimal supermartingale deflator $Y_1 \in \mathcal{D}_s$. This is the problem

$$\inf_{Y_1 \in \mathcal{D}_s} \left(E\left[\int_0^T \tilde{U}_1(y_1 Y_1(t), t) dt\right]\right).$$

Defining $\tilde{v}_1(y_1)$ as the value function for this infimum, let $\hat{Y}_1$ be the solution such that

$$\tilde{v}_1(y_1) = E\left[\int_0^T \tilde{U}_1(y_1 \hat{Y}_1(t), t) dt\right] = \inf_{Y_1 \in \mathcal{D}_s} E\left[\int_0^T \tilde{U}_1(y_1 Y_1(t), t) dt\right].$$

Note that $\hat{Y}_1$ depends on y_1.

Step 3. (Solve for $y_1 > 0$**)**

Given the optimal consumption process $\hat{c}$ and the optimal supermartingale deflator $\hat{Y}_1$, solve for the optimal Lagrangian multiplier $y_1 > 0$,

$$\inf_{y_1 > 0} (\tilde{v}_1(y_1) + x_1 y_1).$$

Taking the derivative of the right side with respect to y_1, one gets

$$\tilde{v_1}'(y_1) + x_1 = 0.$$

Substituting the definition for $\tilde{v}_1$ and taking the derivative yields

$$E\left[\int_0^T \tilde{U}_1'(y_1\hat{Y}_1(t), t)\hat{Y}_1(t)dt\right] + x_1 = 0.$$

This step requires taking the derivative underneath the expectation operator. The same proof as in the appendix to the portfolio optimization in Chap. 10 applies. From Lemma 9.7 in the utility function Chap. 9, we have $\tilde{U_1}'(y_1, t) = -I_1(y_1, t)$. Thus,

$$E\left[\int_0^T I_1(\hat{y}_1\hat{Y}_1(t), t)\hat{Y}_1(t)dt\right] = x_1.$$

Solving this equation gives the optimal $\hat{y}_1$.

Using the solutions from the first step, we write this as

$$E\left[\int_0^T \hat{c}_t\hat{Y}_1(t)dt\right] = x_1,$$

which is that the optimal consumption process $\hat{c}$ satisfies the budget constraint with an equality.

Step 4. (Characterization of $(\hat{X}^{(1)} + \hat{C}) \in \mathscr{X}(x)$ for $x \geq 0$)

Over the time horizon $[0, T]$, the trader invests in assets to obtain their optimal consumption. This implies that the wealth process $\hat{X}^{(1)}$ terminates with $\hat{X}_T^{(1)} = 0$. We now implicitly characterize this wealth process.

Step 1 solves for the optimal consumption process $\hat{c} \in \mathscr{O}$. Since $\hat{c} \in \mathscr{C}_1(x_1)$, by the definition of $\mathscr{C}_1(x_1)$, there exists a nonnegative wealth s.f.t.s. $(\alpha_0^{(1)}, \alpha^{(1)}, \hat{c}) \in \mathscr{N}(x)$ with consumption such that $x_1 + \int_0^T \alpha_u^{(1)} \cdot dS_u = \hat{C}_T$. The equality uses the free disposal Lemma 11.1 in the portfolio optimization Chap. 11. The time t value of the nonnegative wealth s.f.t.s. with consumption is $x_1 + \int_0^t \alpha_u^{(1)} \cdot dS_u = \hat{X}_t^{(1)} + \hat{C}_t$.

Since $\hat{Y}_1 \in \mathscr{D}_s$, we have that $\hat{Y}_1\left(x_1 + \int_0^t \alpha_u^{(1)} \cdot dS_u\right)$ is a supermartingale under $\mathbb{P}$. This implies that $\hat{Y}_1(t)(\hat{X}_t^{(1)} + \hat{C}_t) = \hat{Y}_1(t)\hat{X}_t^{(1)} + \int_0^t \hat{c}_u\hat{Y}_1(u)du$ is a $\mathbb{P}$-supermartingale.

Hence, $E\left[\hat{Y}_1(T)\hat{C}_T\right] = E\left[\hat{Y}_1(T)(\hat{X}_T^{(1)} + \hat{C}_T)\right] \leq Y_0X_0^{(1)} + Y_0C_0 = x_1$, since $Y_0 = 1$, $X_0^{(1)} = x_1$, and $C_0 = 0$. But, from the budget constraint

$$E\left[\hat{Y}_1(T)\hat{C}_T\right] = E\left[\int_0^T \hat{c}_u\hat{Y}_1(u)du\right] = x_1.$$

The first equality was given in the proof of Theorem 12.3.

The only way this last equality can happen is if $\hat{Y}_1(t)(\hat{X}_t^{(1)} + \hat{C}_t)$ is a martingale under $\mathbb{P}$, i.e.

$$E\left[\hat{Y}_1(T)(\hat{X}_T^{(1)} + \hat{C}_T)\,|\mathscr{F}_t\right] = \hat{Y}_1(t)(\hat{X}_t^{(1)} + \hat{C}_t) \tag{12.10}$$

for all t. This completes the characterization of $\hat{X}^{(1)} + \hat{C}$.

Step 5. (Nonnegativity of $\hat{C}_t$)

As noted in Step 4 above, the cumulative consumption process $\hat{C} \geq 0$ because $\hat{c} \in \mathscr{C}_1(x_1)$, which is the set of consumption processes generated by s.f.t.s.'s in the set $\mathscr{N}(x)$ of nonnegative wealth plus cumulative consumption processes with $x \geq 0$.

Remark 12.3 (Random Variables Versus Stochastic Processes) The optimization problem determines the stochastic process $\hat{c} \in \mathscr{O}$, which is the optimal consumption process. Since $\hat{c} \in \mathscr{C}_1(x)$, there exists a nonnegative stochastic process $Z \in \mathscr{X}(x)$ such that $Z_T = \hat{C}_T + \hat{X}_T^{(1)}$ with $\hat{X}_T^{(1)} = 0$ and $Z_t = x + \int_0^t \alpha_u \cdot dS_u$ for all t for some $(\alpha_0, \alpha, c) \in \mathscr{N}(x)$ with $x \geq 0$. Note that the equality in this statement is justified by the free disposal Lemma 11.1 in the portfolio optimization Chap. 11. We denote $Z = \hat{X}^{(1)} + \hat{C} \in \mathscr{X}(x)$. These expressions implicitly characterize the optimal wealth (stochastic) process $\hat{X}^{(1)}$ and it can be used to determine the nonnegative wealth s.f.t.s. $(\alpha_0, \alpha, c) \in \mathscr{N}(x)$ with consumption generating it.

As noted earlier, a second characterization of the optimal wealth process is given in expression (12.10). This characterization depends on the optimal supermartingale deflator (stochastic) process $\hat{Y}_1 \in \mathscr{D}_s$. Unlike the situation in a complete market, the supermartingale deflator can exhibit many different properties. In order of increasing restrictiveness these are properties 1–5 below.

1. $\hat{Y}_1 \in \mathscr{D}_s$ is a supermartingale deflator process.
2. $\hat{Y}_1 \in \mathscr{D}_s$, where $\hat{Y}_1$ is a supermartingale deflator process, $\hat{Y}_1(T) = \frac{d\mathbb{Q}}{d\mathbb{P}}$ is a probability density for $\mathbb{P}$, and $(X + C) \in \mathscr{X}(x)$ is a $\mathbb{Q}$-supermartingale.
3. $\hat{Y}_1 \in \mathscr{D}_l \subset \mathscr{D}_s$, where $\hat{Y}_1$ is a local martingale deflator process and given a $(X + C) \in \mathscr{X}(x)$, $(X + C)\hat{Y}_1$ is a $\mathbb{P}$-local martingale.
4. $\hat{Y}_1 \in \mathscr{M}_l \subset \mathscr{D}_l$, where $\hat{Y}_1$ is a local martingale deflator process, $\hat{Y}_1(T) = \frac{d\mathbb{Q}}{d\mathbb{P}}$ is a probability density for $\mathbb{P}$, and $(X + C) \in \mathscr{X}(x)$ is a $\mathbb{Q}$-local martingale.
5. $\hat{Y}_1 \in \mathscr{M} \subset \mathscr{M}_l$, where $\hat{Y}_1$ is a martingale deflator process, $\hat{Y}_1(T) = \frac{d\mathbb{Q}}{d\mathbb{P}}$ is a probability density for $\mathbb{P}$, and $(X + C) \in \mathscr{X}(x)$ is a $\mathbb{Q}$-martingale.

We note that for case (2), where $\hat{Y}_1 \in \mathscr{D}_s$ is a probability density for $\mathbb{P}$, $\hat{Y}_1$ is a $\mathbb{P}$-martingale. Necessary and sufficient conditions for case (5), where the supermartingale deflator is both a martingale deflator and a probability density with respect to $\mathbb{P}$, i.e. $\hat{Y}_1 \in \mathscr{M}$, are contained in Kramkov and Weston [127]. This completes the remark.

12.4.1.2 The Shadow Price

Using the characterization of the value function, we can obtain the shadow price of the budget constraint. The shadow price is the benefit, in terms of expected utility, of increasing the initial wealth by 1 unit (of the mma). The shadow price equals the Lagrangian multiplier.

Theorem 12.5 (Shadow Price of the Budget Constraint)

$$\hat{y}_1 = v_1'(x_1) \geq E[U_1'(\hat{c}_t, t)] \text{ for all } t$$

with equality if and only if the supermartingale deflator $\hat{Y}_1(T) \in D_s$ *is a probability density with respect to* $\mathbb{P}$, *i.e.* $E[\hat{Y}_1(T)] = 1$.

Proof The first equality is obtained by taking the derivative of $v_1(x_1) = \tilde{v_1}(\hat{y}_1) + x_1\hat{y}_1$.

From the first-order conditions for an optimal consumption we have $U_1'(\hat{c}_t, t) - \hat{y}_1\hat{Y}_1(t) = 0$. Taking expectations gives

$$E[U_1'(\hat{c}_t, t)] = E[\hat{y}_1\hat{Y}_1(t)] \leq \hat{y}_1 \text{ since } E[\hat{Y}_1(T)] \leq 1.$$

This completes the proof.

12.4.1.3 The Supermartingale Deflator Process

To obtain the characterization of the supermartingale deflator process $\hat{Y}_1 \in \mathscr{D}_s$, we rewrite the first-order condition for consumption as follows

$$\hat{Y}_1(t) = \frac{U_1'(\hat{c}_t, t)}{\hat{y}_1} = \frac{U_1'(\hat{c}_t, t)}{v_1'(x_1)} \leq \frac{U_1'(\hat{c}_t, t)}{E[U_1'(\hat{c}_t, t)]},$$

with equality if and only if the supermartingale deflator $\hat{Y}_1(T) \in D_s$ is a probability density with respect to $\mathbb{P}$, i.e. $E[\hat{Y}_1(T)] = 1$.

Assume now that $\hat{Y}_1(T) \in M_l$, *i.e.* $\hat{Y}_1(T)$ *is a local martingale deflator and a probability density* with respect to $\mathbb{P}$. In this case, $\hat{Y}_1 \in \mathscr{M}_l$ is a $\mathbb{P}$-martingale, i.e.

$$\hat{Y}_1(t) = E\left[\hat{Y}_1(T) \,|\mathscr{F}_t\right].$$

Here, at an optimum, the local martingale deflator process satisfies

$$\hat{Y}_1(t) = \frac{U_1'(\hat{c}_t, t)}{E[U_1'(\hat{c}_t, t)]} = E\left[\left.\frac{U_1'(\hat{c}_T, T)}{E[U_1'(\hat{c}_T, T)]}\right| \mathscr{F}_t\right]. \tag{12.11}$$

Remark 12.4 (Asset Price Bubbles) In this individual's optimization problem, prices and hence the local martingale measure $\mathbb{Q}$ are taken as exogenous. The above results never impose the condition that $\mathbb{Q}$ is a martingale measure. If the individual's supermartingale deflator is a state price density, i.e. $\hat{Y}_1(T) \in M_l$, which is a local martingale deflator and a probability density with respect to $\mathbb{P}$, it need not be a martingale measure (it could generate a strict local martingale measure). Hence, this analysis applies to a market with asset price bubbles (see Chap. 3). This completes the remark.

Remark 12.5 (Systematic Risk) In this remark, we use the non-normalized market $((\mathbb{B}, \mathbb{S}), (\mathscr{F}_t), \mathbb{P})$ representation to characterize systematic risk. Recall that the state price density is the key input to systematic risk return relation in the multiple-factor model Chap. 4, Theorem 4.4. To apply this theorem, we assume that the supermartingale deflator $\hat{Y}_1(T) \in D_s$ is a martingale deflator, i.e. $\hat{Y}_1(T) \in M$. Then, the risky asset returns over the time interval $[t, t + \triangle]$ satisfy

$$E\left[R_i(t)\,|\mathscr{F}_t\right] = r_0(t) - \operatorname{cov}\left[\left. R_i(t), \frac{U_1'(\hat{c}_t, t)}{E[U_1'(\hat{c}_t, t)]} \frac{\mathbb{B}_t}{\mathbb{B}_{t+\triangle}}(1 + r_0(t)) \right| \mathscr{F}_t\right], \tag{12.12}$$

where the ith risky asset return is $R_i(t) = \frac{\mathbb{S}_i(t+\Delta) - \mathbb{S}_i(t)}{\mathbb{S}_i(t)}$, $r_0(t) \equiv \frac{1}{p(t,t+\Delta)} - 1$ is the default-free spot rate of interest where $p(t, t+\Delta)$ is the time t price of a default-free zero-coupon bond maturing at time $t + \Delta$, and the state price density is

$$H_t = \frac{\hat{Y}_1(t)}{\mathbb{B}_t} = \frac{U_1'(\hat{c}_t, t)}{E[U_1'(\hat{c}_t, t)]\mathbb{B}_t} = E\left[\left. \frac{U_1'(\hat{c}_T, T)}{E[U_1'(\hat{c}_T, T)]} \right| \mathscr{F}_t\right] \frac{1}{\mathbb{B}_t}.$$

When the time step is very small, i.e. $\Delta \approx dt$, the expression $\frac{\mathbb{B}_t}{\mathbb{B}_{t+\triangle}}(1+r_0(t)) \approx 1$, and this relation simplifies to

$$E\left[R_i(t)\,|\mathscr{F}_t\right] \approx r_0(t) - \operatorname{cov}\left[\left. R_i(t), \frac{U_1'(\hat{c}_t, t)}{E[U_1'(\hat{c}_t, t)]} \right| \mathscr{F}_t\right]. \tag{12.13}$$

This is similar to the expression obtained in the portfolio optimization Chap. 11, except that the conditional expectation of the marginal utility terminal wealth is replaced by the marginal utility of current consumption. This completes the remark.

12.4.1.4 The Optimal Trading Strategy

We can now solve for the optimal trading strategy $(\alpha_0^{(1)}, \alpha^{(1)}, \hat{c}) \in \mathscr{N}(x)$ with consumption using the standard techniques developed for the synthetic construction of derivatives in the fundamental theorems Chap. 2. We are given an optimal wealth and consumption process $\hat{X}_t^{(1)} + \hat{C}_t = \hat{X}_t^{(1)} + \int_0^t \hat{c}_u du$, where $\hat{c} \in \mathscr{C}_1(x_1)$ and $\hat{X}_T^{(1)} = 0$. By the definition of the budget constraint set $\mathscr{C}_1(x_1)$, there exists a nonnegative wealth s.f.t.s. with consumption $(\alpha_0^{(1)}, \alpha^{(1)}, \hat{c}) \in \mathscr{N}(x)$ such that $x_1 + \int_0^T \alpha_u^{(1)} \cdot dS_u = \hat{C}_T$ and where $x_1 + \int_0^t \alpha_u^{(1)} \cdot dS_u = \hat{C}_t + \hat{X}_t$ for all t.

To illustrate the procedure, assume that S is a Markov diffusion process. The diffusion assumption implies that S is a continuous process. Then, we can write the conditional expectation as a deterministic function of the risky asset price vector S_t and time t, i.e.

$$\hat{X}_t^{(1)} + \hat{C}_t = E\left[\frac{\hat{Y}_1(T)}{\hat{Y}_1(t)} (\hat{X}_T^{(1)} + \hat{C}_T) \middle| \mathscr{F}_t \right] = E\left[\frac{\hat{Y}_1(T)}{\hat{Y}_1(t)} (\hat{X}_T^{(1)} + \hat{C}_T) \middle| S_t \right] = g(t, S_t),$$

where $g(t, S_t)$ denotes a deterministic function of S_t and t.

As shown in the portfolio optimization Chaps. 10 and 11, one can use Ito's formula on $g(t, S_t)$ to determine the trading strategy $\alpha^{(1)}$ by matching the integrands of the stochastic integral with respect to S_t. Given the holdings in the risky asset $\alpha^{(1)}$, the value process expression $X_t = \alpha_0(t) + \alpha_t \cdot S_t$ determines the holdings in the mma $\alpha_0^{(1)}$. This completes the identification.

12.4.1.5 Summary

For easy reference we summarize the previous characterization results. The optimal value function is

$$v_1(x_1) = \sup_{c \in \mathscr{C}_1(x_1)} E\left[\int_0^T U_1(c_t, t) dt \right], \qquad \text{where}$$

$$\mathscr{C}_1(x_1) = \left\{ c \in \mathscr{O} : \sup_{Y_1 \in \mathscr{D}_s} E\left[\int_0^T c_t Y_1(t) dt \right] \leq x_1 \right\}.$$

The optimal solution is

- consumption process $\hat{c}_t \in \mathscr{O}$: $\quad \hat{c}_t = I_1(\hat{y}_1 \hat{Y}_1(t), t) \quad$ with $\quad I_1 = (U_1')^{-1}$,
- supermartingale deflator process $\hat{Y}_1(t) \in \mathscr{D}_s$:

$$\tilde{v}_1(y_1) = E\left[\int_0^T \tilde{U}_1(y_1 \hat{Y}_1(t), t) dt \right] = \inf_{Y_1 \in \mathscr{D}_s} E\left[\int_0^T \tilde{U}_1(y_1 Y_1(t), t) dt \right],$$

- shadow price $\hat{y}_1 > 0$: $\quad E\left[\int_0^T I_1(\hat{y}_1 \hat{Y}_1(t), t) \hat{Y}_1(t) dt \right] = x_1,$

- wealth plus consumption process $(\hat{X}^{(1)} + \hat{C}) \in \mathscr{X}(x)$:

$$\hat{Y}_1(t)(\hat{X}_t^{(1)} + \hat{C}_t) \quad \text{is a } \mathbb{P}\text{-martingale,}$$

- $\hat{y}_1 = v_1'(x_1)$, and
- $\hat{Y}_1(t) = \frac{U_1'(\hat{c}_t, t)}{\hat{y}_1} = \frac{U_1'(\hat{c}_t, t)}{v_1'(x_1)}$ for all t. In particular, $\hat{Y}_1(T) = \frac{U_1'(\hat{c}_T, T)}{v_1'(x_1)}$.

12.4.2 Utility of Terminal Wealth ($U_1 \equiv 0$)

This optimization problem was solved in the portfolio optimization Chap. 11. We summarize the solution here with the modified notation for easy reference.

The optimal value function is

$$v_2(x_2) = \sup_{X_T \in \mathscr{C}_2(x_2)} E\left[U_2(X_T)\right], \qquad \text{where}$$

$$\mathscr{C}_2(x_2) = \{X_T \in L_+^0 : E\left[X_T Y_2(T)\right] \leq x_2 \text{ for all } Y_2(T) \in D_s\}.$$

The optimal solution is

- terminal wealth $\hat{X}_T^{(2)} \in L_+^0$: $\quad \hat{X}_T^{(2)} = I_2(y_2 Y_2(T)) \quad$ with $\quad I_2 = (U_2')^{-1}$,
- supermartingale deflator $\hat{Y}_2(T) \in D_s$:

$$\tilde{v}_2(y_2) = E[\tilde{U}_2(y_2 \hat{Y}_2(T))] = \inf_{Y_2(T) \in D_s} E[\tilde{U}_2(y_2 Y_2(T))],$$

- shadow price $\hat{y}_2 > 0$: $\quad E\left[I_2(\hat{y}_2 \hat{Y}_2(T))\hat{Y}_2(T)\right] = x_2$,
- wealth characterization $\hat{X}^{(2)} \in \mathscr{X}(x)$: $\quad \hat{X}_t^{(2)} \hat{Y}_2(t)$ is a $\mathbb{P}$-martingale,
- $\hat{y}_2 = v_2'(x_2)$, and
-

$$\hat{Y}_2(T) = \frac{U_2'(\hat{X}_T^{(2)})}{\hat{y}_2} = \frac{U'(\hat{X}_T^{(2)})}{v_2'(x_2)}.$$

Remark 12.6 ($Y(t)$ Based on $X(T)$) Note that the supermartingale deflator $Y_2(T) \in D_s$ in the utility of wealth problem determines the supermartingale deflator process $\hat{Y}_2 \in \mathscr{D}_s$ by the definition of D_s. This completes the remark.

12.4.3 Utility of Consumption and Terminal Wealth

This section combines the two previous solutions to characterize the combined problem including both intermediate consumption and terminal wealth. This approach to solving the general problem is based on Karatzas and Shreve [118].

Recall that the combined problem is to solve

$$v(x) = \sup_{(c,X_T)\in\mathscr{C}(x)} E\left[\int_0^T U_1(c_t,t)dt + U_2(X_T)\right], \qquad \text{where}$$

$$\mathscr{C}(x) = \{(c, X_T) \in \mathscr{O} \times L_+^0 : E\left[X_T Y_T + \int_0^T c_t Y(t)dt\right] \le x \text{ for all } Y \in \mathscr{D}_s\}.$$

Denote the optimal solutions as $\hat{c},\ \hat{X},\ \hat{\alpha},\ \hat{Y},\ \hat{y}$.

We claim that the solution to this combined problem can be obtained by solving a simpler problem, given the solutions to the previous two. The simpler problem is to find that proportion of the initial wealth $x \ge 0$ to allocate to the consumption problem and the remaining proportion of initial wealth to allocate to the terminal wealth problem such that these initial wealth allocations optimize the *sum* of both the utility of consumption and the utility of terminal wealth problems, respectively. This insight is formalized in the next theorem.

Theorem 12.6 (Equivalence of Solutions) *Let the initial wealth proportion allocated to consumption, $x_1^* \ge 0$, be the solution to*

$$V(x) = \sup_{x_1\in[0,x]} \left(v_1(x_1) + v_2(x - x_1)\right).$$

Then, this wealth allocation obtains the value function for the original optimization problem, i.e.

$$V(x) = \left(v_1(x_1^*) + v_2(x - x_1^*)\right) = v(x).$$

Proof First, the optimal solution x_1^* satisfies

$$v_1'(x_1^*) - v_2'(x - x_1^*) = 0. \text{ Or, } v_1'(x_1^*) = v_2'(x - x_1^*).$$

There are two remaining steps.

(Step 1) Show $V(x) \le v(x)$.

Note that $X^{(1)} + X^{(2)} = X$ and c for x_1^* and $x_2 = x - x_1^*$ are a feasible solution for the combined problem.

Since $v(x)$ is the supremum across all possible solutions, we get the stated result.

(Step 2) Show $V(x) \geq v(x)$.

We give a proof by contradiction. Suppose $\hat{X}, \hat{c}$ are the optimal solutions to the combined problem where

$$v(x) > v_1(x_1^*) + v_2(x - x_1^*).$$

Here $v(x) = E\left[\int_0^T U_1(\hat{c}_t, t)dt\right] + E\left[U_2\left(\hat{X}_T\right)\right]$ and

$$E\left[\int_0^T \hat{c}_t \hat{Y}(t)dt\right] + E\left[\hat{X}_T \hat{Y}(T)\right] = x,$$

where $\hat{Y}$ is the optimal supermartingale deflator for the combined problem.

Note that this is an equality at the optimum.

Define x_1 by $E\left[\int_0^T \hat{c}_t \hat{Y}(t)dt\right] = x_1$. Then, $E\left[\hat{X}_T \hat{Y}(T)\right] = x - E\left[\int_0^T \hat{c}_t \hat{Y}(t)dt\right] = x - x_1$.

We see that $\hat{c}_t$ is a feasible solution for $\mathscr{C}_1(x_1)$.

This implies $v_1(x_1) \geq E\left[\int_0^T U_1(\hat{c}_t, t)dt\right]$.

Similarly, $\hat{X}_T$ is a feasible solution for $\mathscr{C}_2(x - x_1)$.

This implies that $v_2(x_2) \geq E\left[U_2\left(\hat{X}_T\right)\right]$.

Combined, we get

$$\begin{aligned} v(x) > v_1(x_1^*) + v_2(x - x_1^*) &= \sup_{x_1 \in [0,x]} (v_1(x_1) + v_2(x - x_1)) \\ &\geq E\left[\int_0^T U_1(\hat{c}_t, t)dt\right] + E\left[U_2\left(\hat{X}_T\right)\right] \\ &= v(x). \end{aligned}$$

This contradiction completes the proof.

Given this theorem, we can now determine the combined solution.

Corollary 12.1 (General Solution) *For simplicity of notation, the optimal solutions for the two separate problems are denoted without "hats"*

$$\begin{aligned} \hat{\alpha}_t &= \alpha_t^{(1)} + \alpha_t^{(2)}, \\ \hat{X}_t &= X_t^{(1)} + X_t^{(2)}, \\ \hat{X}_T &= X_T^{(2)} \qquad \qquad \textit{since } X_T^{(1)} = 0, \\ \hat{c}_t &= c_t, \\ \hat{y} &= y_1 = y_2, \\ \hat{Y}(t) &= Y_1(t) = \tfrac{U_1(\hat{c}_t, t)}{v'(x)}, \\ \hat{Y}(T) = Y_1(T) = Y_2(T) &= \tfrac{U_1(\hat{c}_T, T)}{v'(x)} = \tfrac{U_2(\hat{X}_T)}{v'(x)}. \end{aligned}$$

Proof We have $v_1(x_1^*) + v_2(x - x_1^*) = v(x)$.

(Step 1) Taking derivatives $v_2'(x - x_1^*) = v'(x)$. But, $v_2'(x - x_1^*) = v_1'(x_1^*)$. Since $y_1 = v_1'(x_1^*)$, $y_2 = v_2'(x - x_1^*)$, $\hat{y} = v'(x)$, we get $y_1 = y_2 = \hat{y}$.

(Step 2) The value functions imply

$$E\left[\int_0^T U_1(c_t, t)dt\right] + E\left[U_2(X_T^{(2)})\right] = E\left[\int_0^T U_1(\hat{c}_t, t)dt\right] + E\left[U_2(\hat{X}_T)\right].$$

By uniqueness of the solutions to all the problems

$$\begin{aligned} \alpha_t^{(1)} + \alpha_t^{(2)} &= \hat{\alpha}_t, \\ X_t^{(1)} + X_t^{(2)} &= \hat{X}_t, \qquad \text{where} \\ X_T^{(2)} &= \hat{X}_T \qquad \text{since } X_T^{(1)} = 0, \text{ and} \\ c_t &= \hat{c}_t. \end{aligned}$$

(Step 3) Last,

$$Y_1(t) = \frac{U_1'(c_t, t)}{y_1} = \frac{U_1'(\hat{c}_t, t)}{v'(x)}.$$

$$Y_2(T) = \frac{U_2'(X_T^{(2)})}{y_2} = \frac{U_2'(\hat{X}_T)}{v'(x)}.$$

These use $y_1 = y_2 = \hat{y}$ and $\hat{y} = v'(x)$.

(Step 4) From the combined problem

$$E\left[\int_0^T \hat{c}_t \hat{Y}(t)dt\right] + E\left[\hat{X}_T \hat{Y}(T)\right] = x,$$

where $\hat{Y}$ is unique.

Summing the separate problems

$$E\left[\int_0^T c_t Y_1(t)dt\right] + E\left[X_T^{(2)} Y_2(T)\right] = x.$$

Or,

$$E\left[\int_0^T \hat{c}_t Y_1(t)dt\right] + E\left[\hat{X}_T Y_2(T)\right] = x.$$

This shows $\hat{Y}(t) = Y_1(t)$ and $\hat{Y}(T) = Y_1(T) = Y_2(T)$, which completes the proof.

12.5 Notes

Excellent references for solving the investor's optimization problem in an incomplete market are Dana and Jeanblanc [42], Duffie [52], Karatzas and Shreve [118], Merton [140], and Pham [149].

Appendix

Theorem 12.7 (Numeraire Invariance) *Let Z_t and Y_t be semimartingales under $\mathbb{P}$.*

Consider $V_t = V_0 + \int_0^t \alpha_u dZ_u$.

Then $VY = \alpha \bullet (Yz)$, where $\bullet$ denotes stochastic integration.

Proof For the notation, we use the convention in Protter [151, p. 60], related to time 0 values.

By the integration by parts formula Theorem 1.1 in Chap. 1 we have

$$VY = Y_- \bullet V + V_- \bullet Y + [V, Y].$$

For the first term we have

$$Y_- \bullet V = Y_- \bullet (\alpha \bullet Z) = (Y_-\alpha) \bullet Z$$

by the associate law for stochastic integrals, Protter [151, Theorem 19, p. 62].

$$= (\alpha Y_-) \bullet Z = \alpha \bullet (Y_- \bullet Z).$$

For the second term,

$$V_- \bullet Y = (\alpha \bullet Z_-) \bullet Y = \alpha \bullet (Z_- \bullet Y).$$

For the third term we have

$$[V, Y] = [\alpha \bullet z, Y] = \alpha \bullet [z, Y],$$

see Protter [151, Theorem 29, p. 75].

Combined,

$$VY = \alpha \bullet (Y_- \bullet z) + \alpha \bullet (z_- \bullet Y) + \alpha \bullet [Z, Y].$$

Or, $VY = \alpha \bullet ((Y_- \bullet Z) + (Z_- \bullet Y) + [z, Y])$.

But, $YZ = (Y_- \bullet Z) + (Z_- \bullet Y) + [Z, Y]$ by the integration by parts formula, Theorem 1.1 in Chap. 1.

Hence, $VY = \alpha \bullet (YZ)$. This completes the proof.

Part III
Equilibrium

Overview

This part of the book extends the previous analysis to study the notion of an economic equilibrium. The question studied is:

how do investors trading in markets determine prices?

We employ the competitive market paradigm, meaning that we assume that traders act as price-takers. Here investors do not trade strategically anticipating the impact of their trades on the price. This is not a new assumption because it has been utilized in all of the previous chapters. Equilibrium prices are determined such that the outstanding supply of each of the risky assets equals the sum of the traders' optimal demands.

As such, equilibrium endogenously determines the risky asset price processes and, hence, characterizes the equivalent local martingale measure in terms of the economy's fundamentals (beliefs, preferences, and endowments). This is the last step in our understanding of equivalent local martingale measures. In this characterization of an economic equilibrium, the traditional consumption capital asset pricing model (CCAPM) and the intertemporal capital asset pricing model (ICAPM) are deduced as special cases.

Interestingly, the individual optimization problem (Part II) is almost sufficient to characterize an equilibrium if a *representative trader economy* equilibrium exists that characterizes the equilibrium in the original economy. A representative trader is a hypothetical trader whose trades represent the collective actions of all traders in the economy. A representative trader economy equilibrium exists when the representative trader's optimal demands exactly equal supply, clearing the market.

If such a representative trader economy equilibrium exists, then characterizing the equilibrium is straightforward. First, we solve for the representative trader's optimal demands. Second, we impose one additional constraint—supply equals the representative trader's demand. This market clearing constraint endogenously determines prices, and hence endogenously determines the local martingale measure in terms of the representative trader's beliefs, preferences, and endowment. In this characterization, aggregate market wealth—the market portfolio—plays a crucial role.

If a representative trader economy equilibrium does not exist, then to characterize the equilibrium one must first aggregate the different traders' optimal demands. Second, equating these aggregate demands to aggregate supply endogenously determines the equilibrium price process. Given this equilibrium price process, each trader's optimal demands determines their local martingale measures. Finally, these individual local martingale measures need to be aggregated (under strong assumptions) to obtain a "market"-based local martingale measure and equilibrium risk return relation.

Fortunately, however, we will not need to pursue this alternative and more complex approach to characterizing an economic equilibrium. Instead, we invoke a mild set of hypotheses under which a representative trader economy equilibrium that characterizes the original economy's equilibrium always exists. This is a relatively unused result in asset pricing theory due to Cuoco and He [39], see also Jarrow and Larsson [98].

Chapter 13
Equilibrium

This chapter presents the description of an economy, the definition of an economic equilibrium, and some necessary conditions implied by the existence of an economic equilibrium.

13.1 The Set-Up

An economy consists of a normalized frictionless (unless otherwise noted) and competitive market $(S, (\mathscr{F}_t), \mathbb{P})$ where the value of a money market account $B_t \equiv 1$. Recall that a competitive market is one where all traders act as price-takers. This is the same structure used in all the previous chapters.

To study equilibrium prices, we need an additional assumption.

Assumption (Liquidating Cash Flows) *There exists an exogenous random payout vector*

$$\xi = (\xi_1, \ldots, \xi_n) : \Omega \to \mathbb{R}^n_{++} \text{ at time } T \text{ such that } S_T = \xi > 0,$$

where $\mathbb{R}^n_{++}$ *is the n-fold product of* $(0, \infty)$.

The vector ξ represents the liquidating cash flows (dividends) to the risky assets. By assumption, the risky assets have no intermediate cash flows (dividends) over $[0, T)$. The existence of liquidating dividends are needed in an equilibrium setting to determine market prices prior to time T. This assumption is only used below when proving the existence of an equilibrium.

In addition, to characterize an economy, we also need a structure that identifies the outstanding supply of the traded assets, the preferences of the traders in the market, and the traders' endowments. These quantities are necessary to characterize traders' optimal demands and the aggregate supply.

R. A. Jarrow, *Continuous-Time Asset Pricing Theory*, Springer Finance,
https://doi.org/10.1007/978-3-319-77821-1_13

13.1.1 Supply of Shares

The supply of shares outstanding for the money market account and the risky assets are

$$N_0(t) \equiv 0$$
$$N_j(t) \equiv N_j > 0, \ j = 1, \ldots, n$$

for all $t \in [0, T]$ a.s. $\mathbb{P}$.

The supply of the money market account is zero. The interpretation is that the "longs" and "shorts" in the market exactly offset. This is a financial asset. The supply of the risky assets is positive. The interpretation is that these represent share ownership in physical assets that exist in positive quantities. For simplicity, we assume that the supply of shares outstanding is constant across time. This assumption could be relaxed with additional complications.

13.1.2 Traders in the Economy

There are $\mathscr{I}$ traders in the economy with preferences $\overset{\rho}{\succeq}_i$ assumed to have a state dependent EU representation with utility functions $U_i : (0, \infty) \times \Omega \to \mathbb{R}$ defined over terminal wealth for $i = 1, \ldots, \mathscr{I}$. The last section in this chapter discusses the necessary changes needed to include intermediate consumption. We assume that the traders' utility functions satisfy the properties given in the utility function Chap. 9, repeated here for convenience.

Assumption (Utility Function) *A utility function is a* $U_i : (0, \infty) \times \Omega \to \mathbb{R}$ *such that for all* $\omega \in \Omega$ *a.s.* $\mathbb{P}$,

(i) $U_i(x, \omega)$ *is* $\mathscr{B}(0, \infty) \otimes \mathscr{F}_T$*-measurable,*
(ii) $U_i(x, \omega)$ *is continuous and differentiable in* $x \in (0, \infty)$,
(iii) $U_i(x, \omega)$ *is strictly increasing and strictly concave in* $x \in (0, \infty)$, *and*
(iv) (Inada Conditions) $\lim_{x \downarrow 0} U_i'(x, \omega) = \infty$ *and* $\lim_{x \to \infty} U_i'(x, \omega) = 0$.

In addition, we assume that preferences satisfy the reasonable asymptotic elasticity condition $AE(U_i) < 1$ for all i. In conjunction, these assumptions (nearly) guarantee the existence of each trader's optimal portfolio.

We let the trader's beliefs be represented by the probability measure $\mathbb{P}_i$ defined on $(\Omega, \mathscr{F})$, which is equivalent to $\mathbb{P}$ for all i, i.e. all the trader's agree on zero probability events, and these zero probability events are the same as those under the statistical probability measure $\mathbb{P}$. This is a necessary condition to impose on an economy, otherwise traders will disagree on the notions of arbitrage opportunities and dominated assets, in particular NA, NUPBR, NFLVR, and ND, which are all invariant with respect to a change in equivalent probability measures (see the fundamental theorems Chap. 2), and an economic equilibrium will not exist.

The filtration $(\mathscr{F}_t)$ corresponds to the information set possessed by each trader, hence, we assume that there is symmetric information in the economy. The generalization of this economy to include differential information is given when discussing market efficiency in Chap. 16. Differential information economies are more complex to analyze because the set of admissible s.f.t.s.'s in the definitions of a complete market, NA, NUPBR, NFLVR, and ND all change with a change in the filtration $(\mathscr{F}_t)$.

Each trader is endowed at time 0 with shares (number of units) of the money market account and the risky assets. These are denoted by e_0^i and e_j^i for $j = 1, \ldots, n$, respectively. We assume that these endowments are such that each trader has a strictly positive wealth at time 0, denoted by $x_i > 0$. A trader's time T wealth is denoted by $X_T^i \geq 0$.

Given a price process S, an accounting identity relates the time 0 wealth to the value of the endowed shares

$$x_i = e_0^i + e^i \cdot S_0 = e_0^i + \sum_{j=1}^{n} e_j^i S_j(0) > 0. \tag{13.1}$$

In addition, it must be the case that in aggregate, the total endowed shares equal the outstanding supply, i.e.

$$\begin{aligned} N_0 &= 0 = \textstyle\sum_{i=1}^{\mathscr{I}} e_0^i \\ N_j &= \textstyle\sum_{i=1}^{\mathscr{I}} e_j^i \quad \text{for } j = 1, \ldots, n. \end{aligned} \tag{13.2}$$

13.1.3 Aggregate Market Wealth

Given a price process S, the aggregate market wealth at time $t \in [0, T]$ is denoted

$$m_t = m(t) \equiv N_0 1 + \sum_{j=1}^{n} N_j S_j(t) = \sum_{j=1}^{n} N_j S_j(t). \tag{13.3}$$

The aggregate market wealth represents the value of the *market portfolio*. The market portfolio is defined to be that portfolio whose percentage holding in each asset is equal to the proportion of wealth that each asset represents in the economy, i.e.

$$\text{market portfolio weight}_j(t) \equiv \frac{N_j S_j(t)}{m_t} \quad \text{for } j = 1, \ldots, n,$$

where the market portfolio weights sum to one

$$\sum_{j=1}^{n} \frac{N_j S_j(t)}{m_t} = 1.$$

For subsequent use, we point out that the aggregate wealth of all traders in the economy is the aggregate market wealth.

Lemma 13.1 (Time 0 Aggregate Market Wealth)

$$\sum_{i=1}^{\mathscr{I}} x_i = m_0.$$

Proof

$$x_i = e_0^i + e^i \cdot S_0 = e_0^i + \sum_{j=1}^{n} e_j^i S_j(0).$$

$$\begin{aligned}
\sum_{i=1}^{\mathscr{I}} x_i &= \sum_{i=1}^{\mathscr{I}} e_0^i + \sum_{i=1}^{\mathscr{I}} \sum_{j=1}^{n} e_j^i S_j(0) \\
&= \sum_{i=1}^{\mathscr{I}} e_0^i + \sum_{j=1}^{n} \left[\sum_{i=1}^{\mathscr{I}} e_j^i \right] S_j(0) \\
&= \sum_{j=1}^{n} N_j S_j(0) = m_0 \text{ since } \sum_{i=1}^{\mathscr{I}} e_0^i = 0.
\end{aligned}$$

This completes the proof.

13.1.4 Trading Strategies

Superscripts identify the different trader's nonnegative wealth s.f.t.s.'s $(\alpha_0^i, \alpha^i) \in \mathscr{N}(x_i)$ for complete and incomplete markets (see the portfolio optimization Chaps. 10 and 11) where

$$\begin{aligned}
\mathscr{N}(x_i) = \{(\alpha_0, \alpha) \in (\mathscr{O}, \mathscr{L}(S)) : \ & X_t = \alpha_0(t) + \alpha_t \cdot S_t, \\
& X_t = x_i + \textstyle\int_0^t \alpha_u \cdot dS_u \geq 0, \ \forall t \in [0, T] \big\}
\end{aligned}$$

and

$$\alpha_0^i(t), \ \alpha_t^i = \alpha^i(t) = \left(\alpha_1^i(t), \ldots, \alpha_n^i(t)\right)' \in \mathbb{R}^n$$

for $i = 1, \ldots, \mathscr{I}$.

Hence, given a price process S, the trader's time 0 and time $t \in (0, T]$ budget constraints are

$$x_i = \alpha_0^i(0) + \alpha_0^i \cdot S_0 = \alpha_0^i(0) + \sum_{j=1}^{n} \alpha_j^i(0) S_j(0) > 0, \tag{13.4}$$

$$X_t^i = \alpha_0^i(t) + \alpha_t^i \cdot S_t = \alpha_0^i(t) + \sum_{j=1}^{n} \alpha_j^i(t) S_j(t) \geq 0,$$

where

$$\begin{array}{ll} X_t^i = x_i + \int_0^t \alpha_u^i \cdot dS_u & \text{or} \\ dX_t^i = \alpha_t^i \cdot dS_t & \text{for all } t. \end{array} \tag{13.5}$$

To simplify the subsequent analysis, we add the following assumption.

Assumption (Non-redundant Assets) *For a given price process S, the trading strategy $(\alpha_0^i, \alpha^i) \in \mathcal{N}(x_i)$ in expression (13.5) generating X_T^i is unique for all $i = 1, \ldots, \mathcal{I}$.*

This assumption removes redundant assets from the economy, and consequently is without-loss-of-generality. By redundant we mean that the stochastic process $S_j(t)$ with initial value $S_j(0)$ cannot be generated by an admissible s.f.t.s. using all the other risky assets ($i = 1, \cdots, n$ and $i \neq j$). In the context of the finite-dimension Brownian motion market in the fundamental theorems Chap. 2, Sect. 2.7, this just implies that the volatility matrix in expression (2.31) is an $n \times D$ matrix where $n \leq D$ with a column rank equal to n.

13.1.5 An Economy

Given this additional structure, we can now define an economy. An *economy* is defined to be a collection consisting of a market, excluding the specification of the price process, in conjunction with a listing of the traders' beliefs, utility functions, and endowments. We omit the price process S from the definition of an economy because the price process will be determined endogenously. In symbols, the following collection represents an economy:
(An Economy)

$$\left\{((\mathscr{F}_t), \mathbb{P}), (N_0 = 0, N), \left(\mathbb{P}_i, U_i, \left(e_0^i, e^i\right)\right)_{i=1}^{\mathcal{I}}\right\}.$$

13.2 Equilibrium

The subsequent definition of an equilibrium applies to the two types of market structures—complete and incomplete markets.

Definition 13.1 (An Equilibrium) Fix an economy $\{((\mathscr{F}_t), \mathbb{P}), (N_0 = 0, N), \left(\mathbb{P}_i, U_i, \left(e_0^i, e^i\right)\right)_{i=1}^{\mathscr{I}}\}$.

The economy is in *equilibrium* if there exists a price process S and demands $(\alpha_0^i(t), \alpha_t^i) : i = 1, \ldots, \mathscr{I}$ such that

(i) $(\alpha_0^i(t), \alpha_t^i)$ are optimal for $i = 1, \ldots, \mathscr{I}$,
(ii) $N_0 = 0 = \sum_{i=1}^{\mathscr{I}} \alpha_0^i(t)$ for all $t \in [0, T]$ a.s. $\mathbb{P}$, and
(iii) $N_j = \sum_{i=1}^{\mathscr{I}} \alpha_j^i(t)$ for $j = 1, \ldots, n$ for all $t \in [0, T]$ a.s. $\mathbb{P}$.

An economy is in equilibrium if there exists a price process S such that each trader's demands for the traded assets are optimal, and the supply of shares equals aggregate demand for all times $t \in [0, T]$ and all states a.s. $\mathbb{P}$. This is a very strong condition. In essence, for all future dates and realizations of uncertainty, the economy is in equilibrium. It provides the conceptual ideal against which actual economies can be contrasted. This is called a Radner [155] equilibrium.

13.3 Theorems

This section proves two theorems related to the existence of an economic equilibrium. The first relates to an economy's aggregate optimal wealth, and the second relates the existence of an equilibrium to the First and Third Fundamental Theorems 2.3 and 2.5 of asset pricing given in Chap. 2.

Theorem 13.1 (Equilibrium Aggregate Optimal Wealths) *Fix an economy* $\left\{((\mathscr{F}_t), \mathbb{P}), (N_0 = 0, N), \left(\mathbb{P}_i, U_i, \left(e_0^i, e^i\right)\right)_{i=1}^{\mathscr{I}}\right\}$.

Let S be an equilibrium price process with X_T^i the optimal time T wealth of trader $i = 1, \ldots, \mathscr{I}$. Then,

$$\sum_{i=1}^{\mathscr{I}} X_T^i = N_0 + \sum_{j=1}^{n} N_j S_j(T) = m_T.$$

Proof

$$\begin{aligned}\sum_{i=1}^{\mathscr{I}} X_T^i &= \sum_{i=1}^{\mathscr{I}} \alpha_0^i(T) + \sum_{i=1}^{\mathscr{I}} \sum_{j=1}^{n} \alpha_j^i(T) S_j(T) \\ &= \sum_{i=1}^{\mathscr{I}} \alpha_0^i(T) + \sum_{j=1}^{n} \left[\sum_{i=1}^{\mathscr{I}} \alpha_j^i(T)\right] S_j(T) \\ &= N_0 + \sum_{j=1}^{n} N_j S_j(T).\end{aligned}$$

This completes the proof.

Theorem 13.2 (Necessary Conditions for an Equilibrium) *Suppose there exists an equilibrium with price process S, then*

(i) NFLVR and
(ii) ND holds.

Proof This proof is based on Jarrow and Larsson [96]. First, we need a definition. An admissible s.f.t.s. $(\alpha_0(t), \alpha_t) \in \mathscr{A}(x)$ with initial wealth $x \geq 0$ is *maximal* if for every other admissible s.f.t.s. $(\beta_0(t), \beta_t) \in \mathscr{A}(x)$ such that $\int_0^T \beta_t \cdot dS_t \geq \int_0^T \alpha_t \cdot dS_t$, $\int_0^T \beta_t \cdot dS_t = \int_0^T \alpha_t \cdot dS_t$ holds.

The proof proceeds via a sequence of lemmas.

Lemma 1 (Delbaen and Schachermayer [46, Theorem 5.12]) $(\alpha_0(t), \alpha_t) \in \mathscr{A}(0)$ *is maximal if and only if there exists a* $\mathbb{Q} \in \mathfrak{M}_l$ *such that* $\int_0^t \alpha_s \cdot dS_s$ *is a* $\mathbb{Q}$*-martingale.*

This theorem in Delbaen and Schachermayer [46] applies to our market because local martingales and σ-martingales coincide when the risky asset prices are nonnegative.

Lemma 2 (Delbaen and Schachermayer [45, Theorem 2.14]) *Finite sums of maximal strategies are again maximal.*

This theorem in Delbaen and Schachermayer [45] applies to our market because the proof in Delbaen and Schachermayer [45] never uses the local boundedness assumption.

Lemma 3 *Let* $(\alpha_0(t), \alpha_t) \in \mathscr{N}(x)$ *be an optimal trading strategy for investor i, then* $(\alpha_0(t), \alpha_t)$ *is maximal in* $\mathscr{N}(x) \subset \mathscr{A}(x)$.

The proof is by contradiction. If $(\alpha_0(t), \alpha_t)$ is not maximal in $\mathscr{N}(x)$, then there is a $(\beta_0, \beta) \in \mathscr{N}(x)$ such that $V_T = x + \int_0^T \beta_t \cdot dS_t \geq \int_0^T \alpha_t \cdot dS_t + x = X_T$ and $\mathbb{P}\left(x + \int_0^T \beta_t \cdot dS_t > x + \int_0^T \alpha_t \cdot dS_t\right) > 0$. Hence, $E\left[U_i(V_T)\right] > E\left[U_i(X_T)\right]$, contradicting the optimality of $(\alpha_0(t), \alpha_t) \in \mathscr{N}(x)$. This completes the proof of Lemma 3.

Lemma 4 *Suppose investor i has an optimal trading strategy* $(\alpha_0, \alpha) \in \mathscr{N}(x)$ *with terminal wealth* X_T*. Then S satisfies NFLVR.*

The proof uses Theorem 2.2 in the fundamental theorems Chap. 2 that NFLVR holds if and only if NA and NUPBR hold.

(Case 1) Show NA.

Assume not NA in the nonnegative wealth s.f.t.s., i.e. there exists a simple arbitrage opportunity $(\beta_0, \beta) \in \mathscr{N}(0)$ with initial wealth 0 and time T value process $V_T = \int_0^T \beta_t \cdot dS_t \geq 0, \mathbb{P}(V_T > 0) > 0$.

Then, the s.f.t.s. $(\alpha_0 + \beta_0, \alpha + \beta) \in \mathscr{N}(x)$ is feasible and $E\left[U_i(X_T + V_T)\right] > E\left[U_i(X_T)\right]$, contradicting the optimality of the trading strategy.

Note that NA in $\mathcal{N}(x)$ implies NA in $\mathcal{A}(x)$. Indeed, if the lower bound in the admissible s.f.t.s. is $c < 0$, add c units of the mma to any simple arbitrage opportunity in $\mathcal{A}(x)$ to get one in $\mathcal{N}(x)$. This completes the proof of Case 1.

(Case 2) Show NUPBR.

By Karatzas and Kardaras [116, Proposition 4.19, p. 476], S satisfies NUPBR.

The same argument as in Case 1 for NA implies that NUPBR in $\mathcal{N}(x)$ implies NUPBR in $\mathcal{A}(x)$ as well.

In conjunction, Cases 1 and 2 prove NFLVR in $\mathcal{A}(x)$.

This completes the proof of Lemma 4.

Lemma 5 *In equilibrium, the market portfolio represents a maximal trading strategy.*

Proof In equilibrium, each trader has an optimal portfolio $\left(\alpha_0^i(t), \alpha_t^i\right) \in \mathcal{N}(x)$.

By Lemma 3, $\left(\alpha_0^i(t), \alpha_t^i\right)$ is maximal for all i.

Consider the market portfolio: $(0, N) = \left(\sum_{i=1}^{\mathcal{I}} \alpha_0^i(t), \sum_{i=1}^{\mathcal{I}} \alpha_t^i\right)$.

This represents a buy and hold nonnegative wealth s.f.t.s.

By Lemma 2 this s.f.t.s. is maximal.

This completes the proof of Lemma 5.

Lemma 6 *In equilibrium, the nonnegative wealth s.f.t.s.*

$\left(\gamma_0^j(t), \gamma_t^j\right) \equiv (0, (0, \ldots, 1, \ldots, 0))$ with a 1 in the jth place is maximal for all j.

Proof First, when $j = 0$, the definition of NA is equivalent to the statement that this trading strategy is maximal.

Next, consider $j \in \{1, \ldots, n\}$.

By Lemma 5, the trading strategy $(0, N) = \left(\sum_{i=1}^{\mathcal{I}} \alpha_0^i(t), \sum_{i=1}^{\mathcal{I}} \alpha_t^i\right)$ representing the market portfolio is maximal.

Consider the nonnegative wealth s.f.t.s.

$$\frac{1}{\sum_{i=1}^{\mathcal{I}} \alpha_j^i(t)} \left(\sum_{i=1}^{\mathcal{I}} \alpha_0^i(t), \sum_{i=1}^{\mathcal{I}} \alpha_t^i\right) = \left(\frac{\sum_{i=1}^{\mathcal{I}} \alpha_0^i(t)}{\sum_{i=1}^{\mathcal{I}} \alpha_j^i(t)}, \frac{\sum_{i=1}^{\mathcal{I}} \alpha_t^i}{\sum_{i=1}^{\mathcal{I}} \alpha_j^i(t)}\right) \text{ for } j = 1, \ldots, n.$$

This trading strategy is maximal because multiplying a maximal trading strategy by a positive constant maintains maximality.

The time t value process of this maximal trading strategy is

$$V_t = S_j(t) + \sum_{k=1, k \neq j}^{n} \left(\frac{\sum_{i=1}^{\mathcal{I}} \alpha_s^i}{\sum_{i=1}^{\mathcal{I}} \alpha_j^i(s)}\right) \cdot S_k(t) \text{ because } \sum_{i=1}^{\mathcal{I}} \alpha_0^i(t) = 0.$$

By Lemma 4, NFLVR holds and by the First Fundamental Theorem 2.3 of asset pricing in Chap. 2, $\mathfrak{M}_l$ is nonempty.

By Lemma 1, there exists a $\mathbb{Q} \in \mathfrak{M}_l$ such that the value process V_t is a $\mathbb{Q}$-martingale.

But, $V_t = S_j(t) + \sum_{k=1,k\neq j}^{n} \left(\frac{\sum_{i=1,}^{\mathscr{I}} \alpha_s^i}{\sum_{i=1}^{\mathscr{I}} \alpha_j^i(s)} \right) \cdot S_k(t) \geq S_j(t) \geq 0$.

Under this $\mathbb{Q}$, $S_j(t)$ is a $\mathbb{Q}$-local martingale.

By Theorem 1.4 in Chap. 1, $S_j(t)$ is a $\mathbb{Q}$-martingale.

By Lemma 1 again, $\left(\gamma_0^j(t), \gamma_t^j\right)$ is maximal for all j. This completes the proof of Lemma 6.

Lemma 7 *In equilibrium, NFLVR and ND holds.*

Proof Lemma 4 implies NFLVR.

Lemma 6 and the definition of ND implies ND. This completes the proof of Lemma 7.

This completes the proof.

This theorem shows that NFLVR and ND are necessary conditions for the existence of an equilibrium. Indeed, if NFLVR is violated in an economy, then every trader would see an FLVR, implying the nonexistence of an optimal portfolio, contradicting the existence of an equilibrium. Similarly, if ND is violated, aggregate supply will exceed aggregate demand for the dominated risky asset, again contradicting the existence of an equilibrium. Hence, the results of Parts I and II of this book are implied by equilibrium pricing models. The results of Parts I and II are more robust in that they do not require the additional structure necessary to define an economy, nor the existence of an economic equilibrium.

Applying the First and Third Fundamental Theorems of asset pricing (Theorems 2.3 and 2.5 in Chap. 2) we obtain the following corollary.

Corollary 13.1 (Necessary Conditions for an Equilibrium) *Suppose there exists an equilibrium with price process S, then*

there exists a $\mathbb{Q} \in \mathfrak{M}$ for the price process S
where $\mathfrak{M} = \{\mathbb{Q} \sim \mathbb{P} : S \text{ is a } \mathbb{Q}\text{-martingale}\}$.

Remark 13.1 (Trader's Supermartingale Deflator) It is important to note that in an incomplete market, the above theorem shows that an equilibrium implies that NFLVR and ND hold for the risky asset price process S, which implies that there exists an equivalent martingale probability measure $\mathbb{Q} \in \mathfrak{M}$. This does not imply, however, that a trader's supermartingale deflator Y is the probability density of a martingale measure. The reason for this difference is that in an incomplete market, there are potentially an infinite number of local martingale measures, local martingale deflators, and supermartingale deflators. The investor's supermartingale deflator may be one of these.

The implication of this observation is that in equilibrium, traders may view the risky asset price process as exhibiting a price bubble (see Chap. 3), even though a martingale measure exists. The existence of asset price bubbles for an equilibrium risky asset price process in an incomplete market will be discussed more thoroughly when the equilibrium price process is characterized in Chap. 15 below.

In a complete market, since there is only one local martingale measure (uniqueness by the Second Fundamental Theorem 2.4 of asset pricing in Chap. 2), if the trader's supermartingale deflator Y is a probability density with respect to $\mathbb{P}$ and a local martingale deflator, then it must be the martingale measure. Hence, in a complete market, no trader believes that there are asset price bubbles. This completes the remark.

13.4 Intermediate Consumption

This section discusses the changes in the model's structure necessary to include intermediate consumption. For the complete details, see Jarrow [92]. When we add consumption flows, we need to add the supply of consumption goods. This is done by adding an exogenous endowment of consumption goods for each trader, which can be a stochastic process. The utility function also needs to be augmented to include utility from intermediate consumption. Here, in addition to reasonable asymptotic elasticity, an additional assumption on the boundedness of the utility function is needed, see the portfolio optimization Chap. 12, which includes intermediate consumption.

13.4.1 Supply of the Consumption Good

Define $\epsilon_t^i \geq 0$ as a stochastic endowment flow of the consumption good for trader i. For simplicity, we assume that this is a constant. Aggregate endowment of the consumption good at time t is

$$\sum_{i=1}^{\mathcal{I}} \epsilon_t^i .$$

13.4.2 Demand for the Consumption Good

Of course, the time t trader i's demand for the consumption good is c_t^i. Aggregate demand for the consumption good at time t is

$$\sum_{i=1}^{\mathcal{I}} c_t^i .$$

13.4.3 An Economy

Given a price process S, trader i's time 0 wealth is

$$x_i = \alpha_0^i(0) + \alpha_0^i \cdot S_0.$$

All wealth is carried across time via holdings in the mma and the risky assets, i.e.

$$X_t^i = \alpha_0^i(t) + \alpha_t^i \cdot S_t.$$

Including the consumption and endowment processes, we get the following evolution for a trader's wealth

$$\begin{aligned} X_t^i &= x_i + \textstyle\int_0^t \alpha_u^i \cdot dS_u - \int_0^t c_u^i dt + \int_0^t \epsilon_u^i du \qquad \text{or} \\ dX_t^i &= \alpha_t^i \cdot dS_t - c_t^i dt + \epsilon_t^i dt \qquad \text{for all } t. \end{aligned}$$

An *economy* is defined to be a market in conjunction with a complete specification of the trader's utility functions and endowments, i.e.
(An Economy)

$$\left\{ ((\mathscr{F}_t), \mathbb{P}), (N_0 = 0, N), \left(\mathbb{P}_i, U_i, \left(e_0^i, e^i\right), \epsilon_t^i\right)_{i=1}^{\mathscr{I}} \right\}.$$

Definition 13.2 (An Equilibrium) Fix an economy $\{((\mathscr{F}_t), \mathbb{P}), (N_0 = 0, N), \left(\mathbb{P}_i, U_i, \left(e_0^i, e^i\right), \epsilon_t^i\right)_{i=1}^{\mathscr{I}}\}$.

The economy is in *equilibrium* if there exists a price process S and demands $\left((\alpha_0^i(t), \alpha_t^i), c_t^i\right)$ such that

(i) $\left((\alpha_0^i(t), \alpha_t^i), c_t^i\right)$ are optimal for $i = 1, \ldots, \mathscr{I}$,
(ii) $N_0 = 0 = \sum_{i=1}^{\mathscr{I}} \alpha_0^i(t)$ for all $t \in [0, T]$ a.s. $\mathbb{P}$,
(iii) $N_j = \sum_{i=1}^{\mathscr{I}} \alpha_j^i(t)$ for $j = 1, \ldots, n$ for all $t \in [0, T]$ a.s. $\mathbb{P}$, and
(iv) $\sum_{i=1}^{\mathscr{I}} \epsilon_t^i = \sum_{i=1}^{\mathscr{I}} c_t^i$ for all $t \in [0, T]$ a.s. $\mathbb{P}$.

Expression (iv) is the consumption goods market clearing condition. With these modifications, the previous theorems can all be extended.

13.5 Notes

Additional references for the determination of economic equilibrium in dynamic markets include Dana and Jeanblanc [42], Duffie [52], Karatzas and Shreve [118], and Merton [140].

Chapter 14
A Representative Trader Economy

To characterize the equilibrium and to facilitate existence proofs, this chapter introduces the notion of a representative trader. A representative trader is a hypothetical individual whose trades, in a sense to be made precise below, reflect the aggregate trades of all individuals in the economy. A representative trader is defined by her beliefs, utility function, and endowments.

For simplicity of presentation, this chapter focuses on traders having preferences only over terminal wealth. The last section in this chapter discusses the necessary changes needed to include intermediate consumption.

14.1 The Aggregate Utility Function

The representative trader's utility function over terminal wealth is the aggregate utility function defined next. It reflects the optimal behavior of the "average" trader in the economy.

Given is an economy

$$\left\{((\mathscr{F}_t), \mathbb{P}), (N_0 = 0, N), \left(\mathbb{P}_i, U_i, \left(e_0^i, e^i\right)\right)_{i=1}^{\mathscr{I}}\right\}.$$

Definition 14.1 (Aggregate Utility Function with Weightings λ) Define the *aggregate utility function* $U : (0, \infty) \times L_{++}^0 \times \Omega \to \mathbb{R}$ by

$$U(x, \lambda, \omega) \equiv \sup_{\{x_1, \ldots, x_{\mathscr{I}}\} \in \mathbb{R}^{\mathscr{I}}} \left\{ \sum_{i=1}^{\mathscr{I}} \lambda_i(\omega) \frac{d\mathbb{P}_i}{d\mathbb{P}} U_i(x_i, \omega) : x = \sum_{i=1}^{\mathscr{I}} x_i \right\},$$

R. A. Jarrow, *Continuous-Time Asset Pricing Theory*, Springer Finance,
https://doi.org/10.1007/978-3-319-77821-1_14

where $\lambda_i : \Omega \to \mathbb{R}$ is $\mathscr{F}_T$-measurable, $\lambda_i > 0$ for all $i = 1, \ldots, \mathscr{I}$, and $\lambda \equiv (\lambda_1, \ldots, \lambda_{\mathscr{I}})$.

The first lemma studies properties of the aggregate utility function.

Lemma 14.1 (The Aggregate Utility Function with Weightings λ) *For fixed $\lambda \in L^0_{++}$,*
$U(x, \lambda, \omega)$ satisfies the properties of a utility function from Chap. 9, i.e.

(i) $U(x, \lambda, \omega)$ is $\mathscr{B}(0, \infty) \otimes \mathscr{F}_T$-measurable,
(ii) U is continuous and differentiable in x on $(0, \infty)$ a.s. $\mathbb{P}$,
(iii) U is strictly increasing and strictly concave in x on $(0, \infty)$ a.s. $\mathbb{P}$,
(iv) (Inada Conditions) $U'(0, \lambda, \omega) \equiv \lim_{x \downarrow 0} U'(x, \lambda, \omega) = \infty$, and

$$U'(\infty, \lambda, \omega) \equiv \lim_{x \to \infty} U'(x, \lambda, \omega) = 0 \text{ a.s. } \mathbb{P}.$$

Proof Property (i) follows from the fact that U_i are jointly measurable for all i and $\frac{d\mathbb{P}_i}{d\mathbb{P}}$ is $\mathscr{F}_T$-measurable.

Note that $\sum_{i=1}^{\mathscr{I}} \lambda_i(\omega) \frac{d\mathbb{P}_i}{d\mathbb{P}} U_i(x_i, \omega)$, considered as a function of $x_i > 0$, is continuous, differentiable, strictly increasing, and strictly concave.

To prove the remaining properties we consider the Lagrangian

$$L(x, \mu) = \sum_{i=1}^{\mathscr{I}} \lambda_i \frac{d\mathbb{P}_i}{d\mathbb{P}} U_i(x_i) + \mu_x (x - \sum_{i=1}^{\mathscr{I}} x_i),$$

where $\mu_x \geq 0$ is the Lagrangian multiplier, which depends on x.

For fixed $(\lambda_i)_{i \in \mathscr{I}}$, a saddle point exists (see Ruszczynski [162, p. 127]) and the first-order conditions uniquely characterize the solution

$$\begin{aligned} \lambda_i \tfrac{d\mathbb{P}_i}{d\mathbb{P}} U_i'(x_i) &= \mu_x \text{ for } i = 1, \ldots, \mathscr{I} \text{ and} \\ x &= \textstyle\sum_{i=1}^{\mathscr{I}} x_i. \end{aligned} \tag{14.1}$$

Standard results (see Ruszczynski [162, p. 192]) imply properties (ii) and (iii).

In addition, we also have that $U'(x, \lambda, \omega) = \mu_x$ a.s. $\mathbb{P}$ (see Ruszczynski [162, p. 151]). Note that given the properties of U proven above, $\mu_x > 0$. This implies $\lambda_i \frac{d\mathbb{P}_i}{d\mathbb{P}} U_i'(x_i) = U'(x, \lambda)$ a.s. $\mathbb{P}$ for all i. Hence, condition (iv) holds as well. This completes the proof.

Remark 14.1 (Asymptotic Elasticity) The aggregate utility function need not satisfy the asymptotic elasticity condition

$$AE(U, \lambda, \omega) = \limsup_{x \to \infty} \frac{xU'(x, \lambda, \omega)}{U(x, \lambda, \omega)} < 1 \quad \text{a.s. } \mathbb{P}.$$

This is due to the summation and the random weightings λ. This completes the remark.

Remark 14.2 (Normalization of Aggregate Utility Function Weightings λ) Without loss of generality, the aggregate utility function can be modified to

$$U^*(x,\lambda,\omega) = \sup_{\{x_1,\ldots,x_{\mathscr{I}}\}\in\mathbb{R}^{\mathscr{I}}} \left\{ \sum_{i=1}^{\mathscr{I}} \xi(\omega)\lambda_i(\omega)\frac{d\mathbb{P}_i}{d\mathbb{P}}U_i(x_i,\omega) : x = \sum_{i=1}^{\mathscr{I}} x_i \right\},$$

where $\xi(\omega) > 0$ is a strictly positive $\mathscr{F}_T$-measurable random variable without changing the optimal solution $\{x_1^*,\ldots,x_{\mathscr{I}}^*\} \in \mathbb{R}^{\mathscr{I}}$ to

$$U(x,\lambda,\omega) = \sup_{\{x_1,\ldots,x_{\mathscr{I}}\}\in\mathbb{R}^{\mathscr{I}}} \left\{ \sum_{i=1}^{\mathscr{I}} \lambda_i(\omega)\frac{d\mathbb{P}_i}{d\mathbb{P}}U_i(x_i,\omega) : x = \sum_{i=1}^{\mathscr{I}} x_i \right\}. \tag{14.2}$$

Indeed, to prove this fact let $\{x_1^*,\ldots,x_{\mathscr{I}}^*\} \in \mathbb{R}^{\mathscr{I}}$ be the optimal solution to expression (14.2). Note that

$$\begin{aligned} U^*(x,\lambda,\omega) &= \sup_{\{x_1,\ldots,x_{\mathscr{I}}\}\in\mathbb{R}^{\mathscr{I}}} \left\{ \sum_{i=1}^{\mathscr{I}} \xi(\omega)\lambda_i(\omega)\frac{d\mathbb{P}_i}{d\mathbb{P}}U_i(x_i,\omega) : x = \sum_{i=1}^{\mathscr{I}} x_i \right\} \\ &= \xi(\omega)\left[\sup_{\{x_1,\ldots,x_{\mathscr{I}}\}\in\mathbb{R}^{\mathscr{I}}} \left\{ \sum_{i=1}^{\mathscr{I}} \lambda_i(\omega)\frac{d\mathbb{P}_i}{d\mathbb{P}}U_i(x_i,\omega) : x = \sum_{i=1}^{\mathscr{I}} x_i \right\}\right] \\ &= \xi(\omega)\left\{ \sum_{i=1}^{\mathscr{I}} \lambda_i(\omega)\frac{d\mathbb{P}_i}{d\mathbb{P}}U_i(x_i^*,\omega) \right\} = \left\{ \sum_{i=1}^{\mathscr{I}} \xi(\omega)\lambda_i(\omega)\frac{d\mathbb{P}_i}{d\mathbb{P}}U_i(x_i^*,\omega) \right\}. \end{aligned}$$

Hence, $\{x_1^*,\ldots,x_{\mathscr{I}}^*\} \in \mathbb{R}^{\mathscr{I}}$ is a solution to the modified problem as well, completing the proof of the assertion.

Possible useful normalizations include:

(i) setting $\xi(\omega) = \frac{1}{\lambda_1(\omega)\frac{d\mathbb{P}_1}{d\mathbb{P}}} > 0$ so that $\xi(\omega)\lambda_1(\omega)\frac{d\mathbb{P}_1}{d\mathbb{P}} = 1$, and

(ii) setting $\xi(\omega) = \frac{1}{\sum_{i=1}^{\mathscr{I}} \lambda_i(\omega)\frac{d\mathbb{P}_i}{d\mathbb{P}}} > 0$ so that $\sum_{i=1}^{\mathscr{I}} \xi(\omega)\lambda_i(\omega)\frac{d\mathbb{P}_i}{d\mathbb{P}} = 1$.

The first of these normalizations is used in the next example when considering state independent logarithmic and power utility functions. This completes the remark.

Example 14.1 (Logarithmic and Power Utility Functions) This example illustrates the construction of the aggregate utility function for two different economies each consisting of two traders. In the first economy, the traders have state independent logarithmic utility functions. In the second economy, they have state independent power utility functions.

(Logarithmic)
Consider two traders with state independent logarithmic utility functions

$$U_i(z) = \ln(z) \qquad \text{for} \quad i = 1, 2$$

both with beliefs $\mathbb{P}_i$ equal to $\mathbb{P}$. Recall that the conjugate utility function is

$$\tilde{U}_i(y) = -\ln(y) - 1,$$

see Example 9.6 in the utility function Chap. 9.

The aggregate utility function satisfies

$$U(x, \lambda) = \sup_{x_1 + x_2 = x} \left(\ln(x_1) + \lambda \ln(x_2)\right). \tag{14.3}$$

To solve, remove the constraint, which yields the identical aggregate utility function

$$U(x, \lambda) = \sup_{x_2} \left(\ln(x - x_2) + \lambda \ln(x_2)\right).$$

The first-order condition, which is necessary and sufficient for a solution, is

$$-\frac{1}{x - x_2} + \lambda \frac{1}{x_2} = 0.$$

Along with the constraint $x_1 + x_2 = x$, this gives the solution

$$x_2^* = \frac{\lambda}{(1+\lambda)} x \quad \text{and}$$
$$x_1^* = \frac{1}{1+\lambda} x.$$

Substitution and algebra yields the aggregate utility function

$$U(x, \lambda) = (1 + \lambda)\ln(x) + \lambda \ln(\lambda) - (1 + \lambda)\ln(1 + \lambda) \tag{14.4}$$

with

$$U'(x, \lambda) = (1 + \lambda)\frac{1}{x} > 0$$

and

$$U'', \lambda) = -(1 + \lambda) < 0.$$

Next, we compute the conjugate aggregate utility function. By definition,

$$\tilde{U}(y, \lambda) = \sup_{x > 0} [U(x, \lambda) - xy], \quad y > 0.$$

Substitution yields

$$\tilde{U}(y,\lambda) = \sup_{x>0} [(1+\lambda)\ln(x) + \lambda \ln(\lambda) - (1+\lambda)\ln(1+\lambda) - xy]. \tag{14.5}$$

The first-order condition, which is necessary and sufficient for a solution, is

$$(1+\lambda)\frac{1}{x} - y = 0 \qquad \text{or}$$

$$x^* = \frac{(1+\lambda)}{y}.$$

Substitution and algebra gives the conjugate aggregate utility function

$$\tilde{U}(y,\lambda) = -(1+\lambda)\ln(y) + \lambda\ln(\lambda) - (1+\lambda) \tag{14.6}$$

with

$$\tilde{U}'(y,\lambda) = -\frac{(1+\lambda)}{y} < 0$$

and

$$\tilde{U}''(y,\lambda) = \frac{(1+\lambda)}{y^2} > 0.$$

This completes the example.

(Power)
Consider two traders with state independent power utility functions

$$U_i(z) = \frac{1}{\rho} z^{\rho}, \quad \rho < 0, \rho \neq 0 \qquad \text{for} \quad i = 1, 2$$

both with beliefs $\mathbb{P}_i$ equal to $\mathbb{P}$. Recall that the conjugate utility function is

$$\tilde{U}_i(y) = \left[\frac{1}{\rho} - 1\right] y^{\frac{\rho}{\rho-1}},$$

see Example 9.6 in the utility function Chap. 9.

The aggregate utility function satisfies

$$U(x,\lambda) = \sup_{x_1+x_2=x} \left(\frac{1}{\rho}x_1^{\rho} + \lambda\frac{1}{\rho}x_2^{\rho}\right). \tag{14.7}$$

To solve, remove the constraint, which yields the identical aggregate utility function

$$U(x,\lambda) = \sup_{x_2}\left(\frac{1}{\rho}(x - x_2)^{\rho} + \lambda\frac{1}{\rho}x_2^{\rho}\right).$$

The first-order condition, which is necessary and sufficient for a solution, is

$$-(x - x_2)^{\rho-1} + \lambda x_2{}^{\rho-1} = 0.$$

Along with the constraint $x_1 + x_2 = x$, this gives the solution

$$x_2^* = \frac{x}{1+\lambda^{\frac{1}{\rho-1}}} \quad \text{and}$$
$$x_1^* = \frac{\lambda^{\frac{1}{\rho-1}}}{1+\lambda^{\frac{1}{\rho-1}}}x.$$

Substitution and algebra yields the aggregate utility function

$$U(x,\lambda) = \frac{x^{\rho}}{\rho}\frac{\lambda}{(1+\lambda^{\frac{1}{\rho-1}})^{\rho-1}} \tag{14.8}$$

with

$$U'(x,\lambda) = x^{\rho-1}\frac{\lambda}{(1+\lambda^{\frac{1}{\rho-1}})^{\rho-1}} > 0$$

and

$$U''(x,\lambda) = (\rho-1)\,x^{\rho-2}\frac{\lambda}{(1+\lambda^{\frac{1}{\rho-1}})^{\rho-1}} < 0.$$

Next, we compute the conjugate aggregate utility function. By definition,

$$\tilde{U}(y,\lambda) = \sup_{x>0}\left[U(x,\lambda) - xy\right], \quad y > 0.$$

Substitution yields

$$\tilde{U}(y,\lambda) = \sup_{x>0}\left[\frac{x^{\rho}}{\rho}\frac{\lambda}{(1+\lambda^{\frac{1}{\rho-1}})^{\rho-1}} - xy\right], \quad y > 0. \tag{14.9}$$

The first-order condition, which is necessary and sufficient for a solution, is

$$x^{\rho-1}\frac{\lambda}{(1+\lambda^{\frac{1}{\rho-1}})^{\rho-1}} - y = 0 \qquad \text{or}$$

$$x^* = y^{\frac{1}{\rho-1}} k^{\frac{-1}{\rho-1}},$$

where $k = \frac{\lambda}{(1+\lambda^{\frac{1}{\rho-1}})^{\rho-1}}$. Substitution and algebra gives the conjugate aggregate utility function

$$\tilde{U}(y,\lambda) = -y^{\frac{\rho}{\rho-1}}\left(\frac{1+\lambda^{\frac{1}{\rho-1}}}{\lambda^{\frac{1}{\rho-1}}}\right)\left(\frac{\rho-1}{\rho}\right) \tag{14.10}$$

with

$$\tilde{U}'(y,\lambda) = -y^{\frac{1}{\rho-1}}\left(\frac{1+\lambda^{\frac{1}{\rho-1}}}{\lambda^{\frac{1}{\rho-1}}}\right) < 0$$

and

$$\tilde{U}''(y,\lambda) = -\left(\frac{1}{\rho-1}\right)y^{\frac{2-\rho}{\rho-1}}\left(\frac{1+\lambda^{\frac{1}{\rho-1}}}{\lambda^{\frac{1}{\rho-1}}}\right) > 0.$$

This completes the example.

14.2 The Portfolio Optimization Problem

A *representative trader* is defined as a collection $(\mathbb{P}, U(\lambda), (N_0 = 0, N))$, where $N = (N_1, \ldots, N_n)$. A representative trader has beliefs $\mathbb{P}$ (the statistical probability measure), the aggregate utility function $U(x, \lambda)$ depending on λ, and an initial endowment of shares equal to the outstanding supply of shares in the economy for the mma and stocks $(N_0 = 0, N)$.

Given a price process S, the representative trader's initial wealth equals the aggregate initial wealth in the economy, i.e. $x = N \cdot S_0 = \sum_{i=1}^{\mathcal{I}} x_i$ (total aggregate wealth), see Lemma 13.1 in the equilibrium Chap. 13. Using the aggregate utility function, the representative trader's optimal portfolio problem is analogous to that of any trader.

Problem 14.1 (Representative Trader's Utility Optimization (Choose the Optimal Trading Strategy (α_0, α)**))** Given a price process S,

$$v(x, \lambda, S) = \sup_{(\alpha_0,\alpha)\in\mathcal{N}(x)} E\left[U\left(X_T, \lambda\right)\right], \qquad \text{where}$$

$$\mathcal{N}(x) = \{(\alpha_0, \alpha) \in (\mathcal{O}, \mathcal{L}(S)) : X_t = \alpha_0(t) + \alpha_t \cdot S_t,$$
$$X_t = x + \int_0^t \alpha_u \cdot dS_u \geq 0, \ \forall t \in [0, T]\Big\}.$$

As before we solve this problem by splitting the solution into two steps. The first step determines the optimal time T wealth, and the second step determines the trading strategy that achieves this wealth. For the first step we solve the following problem.

Problem 14.2 (Representative Trader's Utility Optimization (Choose the Optimal Derivative X_T**))** Given a price process S,

$$v(x, \lambda, S) = \sup_{X_T\in\mathcal{C}(x)} E\left[U\left(X_T, \lambda\right)\right], \qquad \text{where} \tag{14.11}$$

(Complete Market)

$$\mathcal{C}(x) = \left\{X_T \in L^0_+ : E^{\mathbb{Q}}\left[X_T\right] \leq x\right\},$$

(Incomplete Market)

$$\mathcal{C}(x) = \{X_T \in L^0_+ : E\left[X_T Y_T\right] \leq x \text{ for all } Y_T \in D_s\}.$$

Using the results from the portfolio optimization Chaps. 10–12, assuming that

(i) *(reasonable asymptotic elasticity)* $AE(U, \lambda, \omega) = \limsup_{x\to\infty} \frac{xU'(x,\lambda,\omega)}{U(x,\lambda,\omega)} < 1$ a.s. $\mathbb{P}$,
(ii) *(non-trivial optimization)* $v(x, \lambda, S) < \infty$ for some $x > 0$, and
(iii) *(NFLVR)* $\mathfrak{M}_l \neq \emptyset$ for the price process S where

$$\mathfrak{M}_l = \{\mathbb{Q} \sim \mathbb{P} : S \text{ is a } \mathbb{Q}\text{-local martingale}\}$$

a unique solution exists to the representative trader's optimization problem.

Theorem 14.1 (Representative Trader's Optimal Wealth) *The representative trader's optimal wealth is characterized by the following expressions.*

(Complete Market)

$$\hat{X}_T^{\lambda} = I(\hat{y}Y_T, \lambda) \quad with \quad I = (U')^{-1} = -\tilde{U}',$$

where

$$\tilde{U}(y,\omega) = \sup_{x>0}\,[U(x,\omega) - xy] \; for \; all \quad y > 0,$$

$Y_T \in M_l$ is the unique local martingale measure, and $\hat{y}$ is the solution to the budget constraint

$$E\left[Y_T I(\hat{y}Y_T, \lambda)\right] = x.$$

Finally, the optimal portfolio wealth process $\hat{X}^\lambda \in \mathscr{X}(x)$ when multiplied by the local martingale density, $\hat{X}_t^\lambda Y_t$, is a $\mathbb{P}$-martingale.

(Incomplete Market)

$$\hat{X}_T^\lambda = I(\hat{y}\hat{Y}_T^\lambda, \lambda) \quad with \quad I = (U')^{-1} = -\tilde{U}',$$

where $\hat{Y}_T^\lambda \in D_s$ is the solution to

$$\tilde{v}(y,\lambda,S) = \inf_{Y_T \in D_s} E[\tilde{U}(yY_T,\lambda)], \tag{14.12}$$

where

$$\tilde{U}(y,\omega) = \sup_{x>0}\,[U(x,\omega) - xy] \; for \; all \quad y > 0,$$

and $\hat{y}$ is the solution to the budget constraint

$$E\left[\hat{Y}_T^\lambda I(\hat{y}\hat{Y}_T^\lambda, \lambda)\right] = x.$$

It is important to note, for subsequent use, that the solution $\hat{Y}_T^\lambda \in D_s$ depends on the utility function $U(\lambda)$.

Finally, the optimal portfolio wealth process $\hat{X}^\lambda \in \mathscr{X}(x)$ when multiplied by the supermartingale deflator, $\hat{X}_t^\lambda \hat{Y}_t^\lambda$, is a $\mathbb{P}$-martingale.

From these chapters, the solution to the second step, the optimal trading strategy is given in the next theorem.

Theorem 14.2 (Representative Trader's Optimal Trading Strategy)

(Complete and Incomplete Markets)
Given a price process S, there exists a unique $(\alpha_0, \alpha) \in \mathscr{N}(x)$ such that

$$\hat{X}_T^\lambda = x + \int_0^T \alpha_t \cdot dS_t. \tag{14.13}$$

Proof This follows directly from the uniqueness of the optimal terminal wealth and the assumption that the traded assets are non-redundant (see Chap. 13 on equilibrium, Sect. 13.1). This completes the proof.

For the representative trader, buy and hold trading strategies will play a key role.

Theorem 14.3 (Buy and Hold Optimal Trading Strategies) *Given a price process S, suppose that the representative trader's optimal portfolio is a buy and hold strategy, i.e.*

$$\hat{X}_t^\lambda = \theta + \eta \cdot S_t \quad \textit{for all } t,$$

where $\theta \in \mathbb{R}$ *and* $\eta = (\eta_1, \ldots, \eta_n) \in \mathbb{R}^n_{++}$.

If (i) $\theta \neq 0$ *or (ii) if* $\theta = 0$ *and* $\hat{Y}_T^\lambda$ *is a probability density with respect to* $\mathbb{P}$*, then* $\hat{Y}_T^\lambda$ *is the probability density for an equivalent martingale probability measure.*

Proof Because the optimal portfolio is a buy and hold strategy, by Theorem 14.1, $\hat{X}_t^\lambda \hat{Y}_t^\lambda = (\theta + \eta \cdot S_t)\, \hat{Y}_t^\lambda$ is a martingale under $\mathbb{P}$.

Note that $\theta \hat{Y}_t^\lambda$, $(\eta \cdot S_t)\hat{Y}_t^\lambda$, and $S_j(t)\hat{Y}_t^\lambda$ for all j are supermartingales under $\mathbb{P}$ because each is the wealth process of a particular buy and hold trading strategy. The first represents buying just θ units of the mma, the second is just buying η units of the risky assets, and the third is buying one share of the jth risky asset.

If $\theta \neq 0$, then this implies $\theta \hat{Y}_t^\lambda$ and $(\eta \cdot S_t)\hat{Y}_t^\lambda$ are martingales under $\mathbb{P}$. $\theta \hat{Y}_t^\lambda$ being a martingale under $\mathbb{P}$ implies that $\hat{Y}_t^\lambda$ is a probability density with respect to $\mathbb{P}$.

If $\theta = 0$, let $\hat{Y}_T^\lambda$ be a probability density with respect to $\mathbb{P}$.

Note that in both cases, we have that $(\eta \cdot S_t)\hat{Y}_t^\lambda$ is a martingale under $\mathbb{P}$.

Since $\eta_i > 0$ for all i, the only way $(\eta \cdot S_t)\hat{Y}_t^\lambda$ can be a martingale under $\mathbb{P}$ is if each $S_j(t)\hat{Y}_t^\lambda$ is a martingale under $\mathbb{P}$.

Hence, $\hat{Y}_T^\lambda$ is the density for an equivalent martingale probability measure. This completes the proof.

This theorem has the following corollary.

Corollary 14.1 (Buy and Hold Trading Strategies) *Let the representative trader's optimal portfolio be a buy and hold trading strategy with strictly positive holdings in all of the risky assets and where* $\hat{Y}_T^\lambda$ *is a probability density with respect to* $\mathbb{P}$ *if the holdings in the mma are zero. Then,*

(Existence of an Equivalent Martingale Measure)
there exists a $\mathbb{Q} \in \mathfrak{M}$ *for the price process S*
where $\mathfrak{M} = \{\mathbb{Q} \sim \mathbb{P} : S \textit{ is a } \mathbb{Q}\textit{-martingale}\}$*, and*

(No Bubbles)
under the representative trader's risk adjusted beliefs $\left(\frac{d\mathbb{Q}}{d\mathbb{P}} = \hat{Y}_T^\lambda\right)$*, there is no asset price bubble for S.*

Proof Define $\frac{d\mathbb{Q}}{d\mathbb{P}} = \hat{Y}_T^\lambda$. By the previous theorem, this defines a probability density with respect to $\mathbb{P}$ under which S is a martingale. Hence, there is no bubble under this equivalent probability measure. This completes the proof.

Remark 14.3 (Bubbles Require Retrading) The above corollary is consistent with the result that an asset price bubble exists if and only if after purchasing the asset at time 0, there is a re-trading time before time T that strictly increases the value of the asset above liquidating the position at time T (see the asset price bubbles Chap. 3). This completes the remark.

14.3 Representative Trader Economy Equilibrium

This section studies an economy where there is only one trader—the representative trader. We have an analogous definition of an equilibrium for this economy. Here, the representative trader must be endowed with the market's aggregate wealth at time 0, and the representative trader's time T wealth is (by construction) the aggregate market wealth as well. Since the representative trader cannot trade with anyone else in the market, the notion of an equilibrium must reflect this fact. This implies that in equilibrium, no trades can take place. This is called a *no-trade* equilibrium. With this insight, the following definitions are straightforward.

In symbols, the following collection represents a representative trader economy indexed by the aggregate utility function's weights λ.

(A Representative Trader Economy)

$$\{((\mathscr{F}_t), \mathbb{P}), (N_0 = 0, N), (\mathbb{P}, U(\lambda), (N_0 = 0, N))\}$$

Note that in this collection, there is only one trader, represented by the triplet $(\mathbb{P}, U(\lambda), (N_0 = 0, N))$ and indexed by the aggregate utility function's weights λ.

Definition 14.2 (Representative Trader Economy Equilibrium) Fix a representative trader economy

$$\{((\mathscr{F}_t), \mathbb{P}), (N_0 = 0, N), (\mathbb{P}, U(\lambda), (N_0 = 0, N))\}.$$

The representative trader economy is in *equilibrium* if there exists a price process S and demands $(\alpha_0^\lambda(t), \alpha_t^\lambda)$ such that

(i) $(\alpha_0^\lambda(t), \alpha_t^\lambda)$ are optimal for $\left(\mathbb{P}, U(\lambda), \left(N_0 = 0, N_j\right)\right)$,
(ii) $N_0 = 0 = \alpha_0^\lambda(t)$ for all $t \in [0, T]$ a.s. $\mathbb{P}$, and
(iii) $N_j = \alpha_j^\lambda(t)$ for $j = 1, \ldots, n$ for all $t \in [0, T]$ a.s. $\mathbb{P}$.

In this definition, the representative trader's endowment is given by $(N_0 = 0, N)$, which is equal to all the shares outstanding in the market. Second, this is a *no-trade equilibrium*. This is because the representative trader's optimal trading strategy

$(\alpha_0^\lambda(t), \alpha_t^\lambda) = (0, N)$ represents the same holdings for all $t \in [0, T]$ a.s. $\mathbb{P}$. Note that this trading strategy generates a time T wealth equal to the aggregate market wealth, starting from an initial endowment that is also equal to aggregate wealth, i.e.

$$X_T^\lambda = \sum_{j=1}^{n} N_j S_j(0) + \int_0^T N \cdot dS_t = \sum_{j=1}^{n} N_j S_j(T), \tag{14.14}$$

where $N = (N_1, \ldots, N_n) \in \mathbb{R}^n$.

The key observation is that in a representative trader economy equilibrium, the endowment of the representative trader at times 0 and T are equal to the aggregate market wealths at times 0 and T, i.e.

$$m_0 = \sum_{j=1}^{n} N_j S_j(0) \quad \text{and} \quad m_T = \sum_{j=1}^{n} N_j S_j(T).$$

We now study the relation between equilibrium in the economy

$$\left\{((\mathscr{F}_t), \mathbb{P}), (N_0 = 0, N), \left(\mathbb{P}_i, U_i, \left(e_0^i, e^i\right)\right)_{i=1}^{\mathscr{I}}\right\}$$

and equilibrium in a hypothetical representative trader economy purposefully constructed to characterize the original economy's equilibrium. To facilitate understanding, we introduce two related definitions.

Definition 14.3 (Representative Trader Economy Reflects the Economy's Optimal Wealths and Asset Demands) Given an economy

$$\left\{((\mathscr{F}_t), \mathbb{P}), (N_0 = 0, N), \left(\mathbb{P}_i, U_i, \left(e_0^i, e^i\right)\right)_{i=1}^{\mathscr{I}}\right\}$$

with price process S, we say that the representative trader economy

$$\{((\mathscr{F}_t), \mathbb{P}), (N_0 = 0, N), (\mathbb{P}, U(\lambda), (N_0 = 0, N))\}$$

reflects the economy's aggregate optimal wealths with price process S if

$$\hat{X}_T^\lambda = \sum_{i=1}^{\mathscr{I}} \hat{X}_T{}^i \quad \text{for all } t \text{ a.s. } \mathbb{P},$$

where $\hat{X}_T^\lambda, \hat{X}_T{}^i$ correspond to the optimal wealth of the representative trader and the ith trader, respectively.

We say that the representative trader economy

$$\{((\mathscr{F}_t), \mathbb{P}), (N_0 = 0, N), (\mathbb{P}, U(\lambda), (N_0 = 0, N))\}$$

reflects the economy's aggregate optimal asset demands for the price process S if

$$\hat{\alpha}_0^{\lambda}(t) = \sum_{i=1}^{\mathscr{I}} \hat{\alpha}_0{}^i(t) \quad \text{and} \quad \hat{\alpha}_t^{\lambda} = \sum_{i=1}^{\mathscr{I}} \hat{\alpha}_t{}^i \quad \text{for all } t \text{ a.s. } \mathbb{P} \tag{14.15}$$

where $(\alpha_0^{\lambda}(t), \hat{\alpha}_t^{\lambda})$, $(\alpha_0^i(t), \hat{\alpha}_t{}^i)$ correspond to the optimal trading strategy of the representative trader and the ith trader, respectively.

Definition 14.4 (Representative Trader Economy's Equilibrium Reflecting an Economy's Equilibrium) Given are an economy

$$\left\{((\mathscr{F}_t), \mathbb{P}), (N_0 = 0, N), \left(\mathbb{P}_i, U_i, \left(e_0^i, e^i\right)\right)_{i=1}^{\mathscr{I}}\right\},$$

an equilibrium price process S with demands $(\alpha_0^i(t), \hat{\alpha}_t{}^i)_{i=1}^{\mathscr{I}}$ for all t, and a representative trader economy

$$\{((\mathscr{F}_t), \mathbb{P}), (N_0 = 0, N), (\mathbb{P}, U(\lambda), (N_0 = 0, N))\}$$

with equilibrium price process S^*.

We say that the economy's equilibrium is *reflected by this representative trader economy's equilibrium* if

1. $S^* = S$ and
2. the representative trader economy reflects the original economy's aggregate optimal asset demands.

These three definitions make precise when a representative trader economy reflects the economy's optimal wealths, asset demands, and equilibrium. We now apply these definitions to prove two theorems which, in conjunction, guarantee the existence of a representative trader economy equilibrium that reflects the original economy's equilibrium. These theorems are formulated for an incomplete market, with a complete market a special case.

Theorem 14.4 (Existence of a Representative Trader Economy that Reflects the Original Economy's Optimal Wealths) *Given an economy*

$$\left\{((\mathscr{F}_t), \mathbb{P}), (N_0 = 0, N), \left(\mathbb{P}_i, U_i, \left(e_0^i, e^i\right)\right)_{i=1}^{\mathscr{I}}\right\}$$

and a collection of representative trader economies

$$\{((\mathscr{F}_t), \mathbb{P}), (N_0 = 0, N), (\mathbb{P}, U(\lambda), (N_0 = 0, N))\}$$

indexed by λ.

Let S *be the same price process in all of these economies.*

Assume that

(i) $U(x, \lambda^*, \omega)$ *has an asymptotic elasticity strictly less than one a.s.* $\mathbb{P}$,

(ii) $v(x, \lambda^*, S) < \infty$ *for some* $x > 0$, *where*

$$\lambda_i^* = \frac{v_1'(x_1, S)Y_T^1(S)}{v_i'(x_i, S)Y_T^i(S)}$$

with $Y_T^i(S)$ *the solution to trader i's dual problem and* $v_i(x_i, S)$ *trader i's value function, given the initial wealth* x_i *and the price process* S, *for all* $i = 1, \ldots, \mathscr{I}$.

Then, the representative trader economy

$$\left\{((\mathscr{F}_t), \mathbb{P}), (N_0 = 0, N), \left(\mathbb{P}, U(\lambda^*), (N_0 = 0, N)\right)\right\}$$

reflects the economy's aggregate optimal wealths.

First, note that this theorem relates to the economy's aggregate optimal wealths, and not the economy's asset demands. Second, note the dependence of the individual and representative trader's value functions on the price process S, which is exogenously specified in the statement of this theorem. This makes sense because all traders act as price-takers when forming their optimal demands. We emphasize that the price process S given in the hypothesis to this theorem is not assumed to be an equilibrium price process.

This theorem states that under the hypotheses (i)–(ii) there exists a representative trader economy indexed by the aggregate utility function weightings λ^* as given in hypothesis (ii), where the aggregate utility function's optimal wealth reflects the aggregate optimal wealths of the individuals in the economy.

Proof For the given price process S, from Sect. 11.9 in the portfolio optimization Chap. 11, for each trader i the optimal wealth X_T^i, which maximizes $E_i\left[U_i(X_T^i)\right] = E\left[\frac{d\mathbb{P}_i}{d\mathbb{P}} U_i(X_T^i)\right]$, is characterized by the expression

$$\frac{d\mathbb{P}_i}{d\mathbb{P}} U_i'\left(X_T^i\right) = v_i'(x_i, S)Y_T^i(S), \tag{14.16}$$

where $v_i(x_i, S)$ is the value function for the investor's optimization problem and $Y_T^i(S)$ is the solution to

$$\tilde{v}_i(y, S) = \inf_{Y_T \in D_s(S)} E\left[\frac{d\mathbb{P}_i}{d\mathbb{P}} \tilde{U}_i\left(yY_T \frac{d\mathbb{P}}{d\mathbb{P}_i}\right)\right],$$

where $D_s(S)$ is the set of supermartingale deflators with respect to $\mathbb{P}$. We use this change of measure in the investor's objective function because $D_s(S)$ depends on the probability measure $\mathbb{P}$ (see the portfolio optimization Chap. 11 after the definition of $\mathcal{D}_s$). We make explicit the dependence on the price process S because this process will change in subsequent arguments. The optimal supermartingale deflator $Y_T^i(S)$ also depends on the price process S because $D_s(S)$ depends on S.

For a given λ, suppose that $U(x, \lambda, \omega)$ has an asymptotic elasticity strictly less than one a.s. $\mathbb{P}$ and $v(x, \lambda, S) < \infty$ for some $x > 0$ (we note for later use that this supposition is satisfied for $\lambda = \lambda^*$ by hypotheses (i) and (ii)). Optimizing the representative trader's wealth, we have

$$U'(X_T^\lambda, \lambda) = v'(x, \lambda, S)Y_T^\lambda(S), \tag{14.17}$$

where the individual components $\tilde{X}_T^i$ satisfy

$$\lambda_i \frac{d\mathbb{P}_i}{d\mathbb{P}} U_i'(\tilde{X}_T^i) = \mu_{X_T^\lambda} \text{ for } i = 1, \ldots, \mathcal{I}$$

$$X_T^\lambda = \sum_{i=1}^{\mathcal{I}} \tilde{X}_T^i$$

and (the shadow price of the constraint)

$$U'(X_T^\lambda, \lambda) = \mu_{X_T^\lambda}.$$

Combined these give

$$U'(X_T^\lambda, \lambda) = \lambda_i \frac{d\mathbb{P}_i}{d\mathbb{P}} U_i'(\tilde{X}_T^i) \quad \text{for all } i. \tag{14.18}$$

We want to show $X_T^i = \tilde{X}_T^i$ for all i. Using expression (14.16), this will be satisfied if and only if

$$\frac{d\mathbb{P}_i}{d\mathbb{P}} U_i'(\tilde{X}_T^i) = v_i'(x_i, S)Y_T^i(S) \quad \text{for all } i.$$

Using expression (14.18), this will be satisfied if and only if

$$\lambda_1 v_1'(x_1, S) Y_T^1(S) = \lambda_i v_i'(x_i, S) Y_T^i(S) \quad \text{for } i = 2, \ldots, \mathscr{I}.$$

It is easy to show that an $\mathscr{F}_T$-measurable solution $\lambda^* = (\lambda_1^*, \ldots, \lambda_{\mathscr{I}}^*)$ to this system of $(\mathscr{I} - 1)$ functional equations in $\mathscr{I}$ unknowns is

$$\lambda_i^* = \frac{v_1'(x_1, S) Y_T^1(S)}{v_i'(x_i, S) Y_T^i(S)} \quad \text{for } i = 1, 2, \ldots, \mathscr{I}.$$

This completes the proof.

Remark 14.4 (Complete Market with respect to $\mathbb{Q} \in \mathfrak{M}_l(S)$) Suppose the market is complete with respect to any $\mathbb{Q} \in \mathfrak{M}_l(S)$. Note that we make the dependence of the equivalent local martingale measure on the price process S explicit in the notation. Then, by the Second Fundamental Theorem 2.4 of asset pricing in Chap. 2, $\mathbb{Q} \in \mathfrak{M}_l(S)$ is unique. Define $Y_T = \frac{d\mathbb{Q}}{d\mathbb{P}}$. Recall that in the investor's optimization problem in Chap. 11, $\tilde{v}_i(y, S) = \inf_{Y_T \in D_s(S)} E\left[\frac{d\mathbb{P}_i}{d\mathbb{P}} \tilde{U}_i\left(y Y_T \frac{d\mathbb{P}}{d\mathbb{P}_i}\right)\right] = \inf_{Y_T \in M_l(S)} E\left[\frac{d\mathbb{P}_i}{d\mathbb{P}} \tilde{U}_i\left(y Y_T \frac{d\mathbb{P}}{d\mathbb{P}_i}\right)\right] = E\left[\frac{d\mathbb{P}_i}{d\mathbb{P}} \tilde{U}_i\left(y Y_T \frac{d\mathbb{P}}{d\mathbb{P}_i}\right)\right]$. The last equality holds in this set of equations because $M_l(S)$ is a singleton set. Combined, this implies that

$$Y_T^i(S) = arg\ min \left\{ E\left[\frac{d\mathbb{P}_i}{d\mathbb{P}} \tilde{U}_i\left(y Y_T \frac{d\mathbb{P}}{d\mathbb{P}_i}\right)\right] : Y_T \in D_s(S) \right\} = Y_T.$$

Consequently, in a complete market, the aggregate utility function's weights in Theorem 14.4 simplify to

$$\lambda_i^* = \frac{v_1'(x_1, S)}{v_i'(x_i, S)} \quad \text{for } i = 1, 2, \ldots, \mathscr{I}.$$

This completes the remark.

We can now derive the key theorem in this section which provides sufficient conditions for the existence of a representative trader economy equilibrium that reflects the original economy's equilibrium.

Theorem 14.5 (Existence of a Representative Trader Economy Equilibrium Reflecting the Original Economy's Equilibrium) *Given an economy*

$$\left\{((\mathscr{F}_t), \mathbb{P}), (N_0 = 0, N), \left(\mathbb{P}_i, U_i, \left(e_0^i, e^i\right)\right)_{i=1}^{\mathscr{I}}\right\},$$

let S be an equilibrium price process with optimal trading strategies $(\hat{\alpha}_0{}^i(t), \hat{\alpha}_t{}^i)_{i=1}^{\mathscr{I}}$.

Consider the collection of representative trader economies

$$\{((\mathscr{F}_t), \mathbb{P}), (N_0 = 0, N), (\mathbb{P}, U(\lambda), (N_0, N))\}$$

indexed by λ.

Assume that

(i) $U(x, \lambda^*, \omega)$ *has an asymptotic elasticity strictly less than one a.s.* $\mathbb{P}$,
(ii) $v(x, \lambda^*, S) < \infty$ *for some* $x > 0$, *where*

$$\lambda_i^* = \frac{v_1'(x_1, S)Y_T^1(S)}{v_i'(x_i, S)Y_T^i(S)}$$

with Y_T^i *the solution to trader i's dual problem and* $v_i(x_i, S)$ *trader i's value function, given the initial wealth* x_i *and the price process S, for all* $i = 1, \ldots, \mathscr{I}$.

Then, the representative trader economy

$$\{((\mathscr{F}_t), \mathbb{P}), (N_0 = 0, N), (\mathbb{P}, U(\lambda^*), (N_0 = 0, N))\}$$

reflects the original economy's equilibrium.

Proof By Theorem 14.4, given the equilibrium price process S for the original economy, the representative trader economy

$$\{((\mathscr{F}_t), \mathbb{P}), (N_0 = 0, N), (\mathbb{P}, U(\lambda^*), (N_0 = 0, N))\}$$

reflects the original economy's optimal wealths, i.e. $\hat{X}_T^{\lambda^*} = \sum_{i=1}^{\mathscr{I}} \hat{X}_T{}^i$ for all t a.e.$\mathbb{P}$. A trading strategy that achieves this wealth is

$$\hat{\alpha}_0^{\lambda^*}(t) = \sum_{i=1}^{\mathscr{I}} \hat{\alpha}_0^i(t) \quad \text{and} \quad \hat{\alpha}_t^{\lambda^*} = \sum_{i=1}^{\mathscr{I}} \hat{\alpha}_t{}^i \quad \text{for all } t \text{ a.s. } \mathbb{P}.$$

In the original economy's equilibrium

$$0 = \sum_{i=1}^{\mathscr{I}} \hat{\alpha}_0^i(t) \quad \text{and} \quad N = \sum_{i=1}^{\mathscr{I}} \hat{\alpha}_t{}^i \quad \text{for all } t \text{ a.s. } \mathbb{P}.$$

This implies that the buy and hold trading strategy $\hat{\alpha}_0^{\lambda^*}(t) = 0$ and $\hat{\alpha}_t^{\lambda^*} = N$ for all t a.s. $\mathbb{P}$ generates the representative trader's optimal wealth. By uniqueness of the trading strategy generating any terminal wealth (by the non-redundant assets assumption in the equilibrium Chap. 13), this is the representative trader's optimal trading strategy. Hence, the representative trader economy that reflects the original economy's optimal demands is in equilibrium with price process S. This completes the proof.

The difference in the hypothesis of this Theorem 14.5, as contrasted with Theorem 14.4, is that the price process S is assumed to be an equilibrium price process for the original economy. This is an important theorem for use in asset pricing theory because it enables one to characterize an economic equilibrium in a complex economy via an equivalent equilibrium in a simpler representative trader economy. This characterization is used extensively in Chap. 15 below.

14.4 Pareto Optimality

This section relates the notion of a representative trader and economic equilibrium to Pareto optimality. To facilitate this discussion, we need to introduce some definitions. First, fix an economy

$$\left\{((\mathscr{F}_t), \mathbb{P}), (N_0 = 0, N), \left(\mathbb{P}_i, U_i, \left(e_0^i, e^i\right)\right)_{i=1}^{\mathscr{I}}\right\}$$

with price process S.

A wealth allocation $(X_T^i : i = 1, \ldots, \mathscr{I})$ is called *feasible* if market wealth equals aggregate demand, i.e. $m_T = \sum_{i=1}^{\mathscr{I}} X_T^i$ where $m_T = \sum_{j=1}^{n} N_j S_j(T)$.

A feasible wealth allocation $(X_T^i : i = 1, \ldots, \mathscr{I})$ is *Pareto optimal* (efficient) if there exists no alternative feasible allocation $(\tilde{X}_T^i : i = 1, \ldots, \mathscr{I})$ such that $E_i[U_i(\tilde{X}_T^i)] \geq E_i[U_i(X_T^i)]$ for all $i = 1, \ldots, \mathscr{I}$ and $E_i[U_i(\tilde{X}_T^i)] > E_i[U_i(X_T^i)]$ for at least one $i = 1, \ldots, \mathscr{I}$.

It is easy to see that this definition is equivalent to the following restatement.

A given wealth allocation $(X_T^i : i = 1, \ldots, \mathscr{I})$ is Pareto optimal if for all $i = 1, \ldots, \mathscr{I}$, $(X_T^1, \ldots, X_T^{\mathscr{I}})$ maximizes

$$E_i[U_i(\tilde{X}_T^i)] \qquad \text{subject to}$$

$$\begin{aligned} &E_j[U_j(\tilde{X}_T^j)] \leq E_j[U_j(X_T^j)] \text{ for all } j,\ j \neq i \\ &\textstyle\sum_{j=1}^{\mathscr{I}} \tilde{X}_T^j = m_T. \end{aligned}$$

To solve this problem, for all $i = 1, \ldots, \mathscr{I}$, define the Lagrangian

$$\begin{aligned} L_i(\tilde{X}_T^1, \ldots, \tilde{X}_T^{\mathscr{I}}) = E_i[U_i(\tilde{X}_T^i)] + \textstyle\sum_{j=1, j\neq i}^{n} \gamma_j^i \left(E_j[U_j(X_T^j)] - E_j[U_j(\tilde{X}_T^j)]\right) \\ + \mu^i (m_T - \textstyle\sum_{j=1}^{\mathscr{I}} \tilde{X}_T^j) \end{aligned}$$

for $\gamma_j^i \geq 0$ all $j = 1, \ldots, \mathscr{I}$ and $\mu^i \geq 0$. We note that the Lagrangian multipliers γ_j^i and μ^i are constants. The first-order conditions for the optimal solution $(X_T^1, \ldots, X_T^{\mathscr{I}})$ are necessary and sufficient under our hypotheses (see Ruszczynski [162, p. 127]). These conditions are for all $i = 1, \ldots, \mathscr{I}$, the constraints are satisfied with equality,

$$\frac{\partial E_i[U_i(X_T^i)]}{\partial X_T^i} - \mu^i = 0, \qquad \text{and}$$

$$\gamma_j^i \frac{\partial E_j[U_j(X_T^j)]}{\partial X_T^j} - \mu^i = 0 \qquad \text{for } j = 1, \ldots, \mathscr{I},\ j \neq i.$$

Or, for all $i = 1, \ldots, \mathscr{I}$,

$$\frac{\frac{\partial E_i[U_i(X_T^i)]}{\partial X_T^i}}{\frac{\partial E_j[U_j(X_T^j)]}{\partial X_T^j}} = \gamma_j^i > 0 \qquad \text{for } j = 1, \ldots, \mathscr{I},\ j \neq i. \tag{14.19}$$

Next, consider the optimization problem

$$\max_{(X_T^1, \ldots, X_T^{\mathscr{I}})} \sum_{i=1}^{\mathscr{I}} \tilde{\lambda}_i E_i[U_i(X_T^i)] \qquad \text{subject to} \tag{14.20}$$

$$\sum_{i=1}^{\mathscr{I}} X_T^i = m_T,$$

where $\tilde{\lambda}_i > 0$ for $i = 1, \ldots, \mathscr{I}$ are constants.

To solve this problem, define the Lagrangian

$$L(X_T^1, \ldots, X_T^{\mathscr{I}}) = \sum_{i=1}^{\mathscr{I}} \tilde{\lambda}_i E_i[U_i(X_T^i)] + \delta(m_T - \sum_{i=1}^{\mathscr{I}} X_T^i)$$

for $\delta \geq 0$. The first-order conditions for the optimal solution $(X_T^1, \ldots, X_T^{\mathscr{I}})$, which are necessary and sufficient under our hypotheses, are that the constraint is satisfied with equality and

$$\tilde{\lambda}_i \frac{\partial E_i[U_i(X_T^i)]}{\partial X_T^i} - \delta = 0 \qquad \text{for } i = 1, \ldots, \mathscr{I}.$$

Or, for all $i = 1, \ldots, \mathscr{I}$,

$$\frac{\frac{\partial E_i[U_i(X_T^i)]}{\partial X_T^i}}{\frac{\partial E_j[U_j(X_T^j)]}{\partial X_T^j}} = \frac{\tilde{\lambda}_j}{\tilde{\lambda}_i} > 0 \qquad \text{for } j = 1, \ldots, \mathscr{I},\ j \neq i. \tag{14.21}$$

Note that expressions (14.19) and (14.21) are identical. This proves that an allocation is Pareto optimal if and only if it solves the optimization problem in expression (14.20) where $\tilde{\lambda}_i > 0$ for $i = 1, \ldots, \mathscr{I}$ are constants. It is convenient to rewrite the objective function in expression (14.20) as

$$\sum_{i=1}^{\mathscr{I}} \tilde{\lambda}_i E_i[U_i(X_T^i)] = \sum_{i=1}^{\mathscr{I}} E\left[\tilde{\lambda}_i U_i(X_T^i)\frac{d\mathbb{P}_i}{d\mathbb{P}}\right] = E\left[\sum_{i=1}^{\mathscr{I}} \tilde{\lambda}_i U_i(X_T^i)\frac{d\mathbb{P}_i}{d\mathbb{P}}\right]. \tag{14.22}$$

As written, the Pareto optimality objective function is almost identical to the aggregate utility function as given in Definition 14.1 for the representative trader. The key difference is that the aggregate utility function's weights $\lambda_i(\omega)$ are $\mathscr{F}_T$-measurable random variables, where in the Pareto optimality objective function the weights $\tilde{\lambda}_i$ are constants. Otherwise, the objective functions and the optimization problems are identical (see expression (14.11).

This implies that we can apply our insights from economic equilibrium to study when an equilibrium allocation is Pareto optimal. First, we assume that the economy is in equilibrium. By Theorem 13.2 in the equilibrium Chap. 13, the market satisfies NFLVR, hence, there exists an equivalent local martingale measure $\mathbb{Q} \in \mathfrak{M}_l$. Second, under the hypotheses of Theorem 14.5 we have that there exists a representative trader economy equilibrium with the aggregate utility function weights $\lambda_i^*(\omega)$ as given in the theorem, i.e. for all i,

$$\lambda_i^* = \frac{v_1'(x_1, S)Y_T^1(S)}{v_i'(x_i, S)Y_T^i(S)}.$$

From the above characterization of a Pareto optimal allocation, we have that the equilibrium allocation is Pareto optimal if and only if the aggregate utility function weights $\lambda_i^*(\omega)$ given in the previous expression are constants. We investigate the implications of this observation.

(i) Suppose the market is complete with respect to any $\mathbb{Q} \in \mathfrak{M}_l$. Then, by the Second Fundamental Theorem 2.4 of asset pricing in Chap. 2, $\mathbb{Q} \in \mathfrak{M}_l$ is unique. Define $Y_T = \frac{d\mathbb{Q}}{d\mathbb{P}}$. Then, for all investors $i = 1, \ldots, \mathscr{I}$, $Y_T^i = Y_T$. Substitution yields $\lambda_i^* = \frac{v_1'(x_1,S)}{v_i'(x_i,S)}$ for all $i = 1, \ldots, \mathscr{I}$. These are constants, hence, the equilibrium allocation is Pareto optimal.

(ii) In an incomplete market, an equilibrium allocation is in general not Pareto efficient. Indeed, in an incomplete market (see the portfolio optimization Chap. 11), $\lambda_i^* = \frac{v_1'(x_1,S)Y_T^1(S)}{v_i'(x_i,S)Y_T^i(S)}$ for all $i = 1, \ldots, \mathscr{I}$ are random due to $Y_T^i(S)$ being random. This implies that in an incomplete market, an equilibrium allocation is Pareto optimal if and only if $Y_T^i(S) = Y_T^1(S)$ for all $i = 1, \ldots, \mathscr{I}$.

This restrictive condition is difficult to satisfy in an incomplete market economy. This condition is satisfied, for example, trivially if all traders have identical beliefs, preferences, and initial endowments. If traders have identical beliefs and preferences, but different initial endowments, then this condition is satisfied if $Y_T^i(S)$ is independent of the initial wealth (because all other investor characteristics are identical). State independent utility functions that satisfy this condition include the logarithmic and power utility functions (see Pham [149, p. 198]). These utility functions exhibit linear risk tolerance (see Example 9.4 in the utility function Chap. 9) and they are related to Gorman aggregation (see Back [5, p. 55]), which is a well known sufficient condition that generates Pareto optimal equilibrium allocations.

14.5 Existence of an Equilibrium

This section uses the notion of a representative trader to prove the existence of an equilibrium under a strong set of sufficient conditions.

Theorem 14.6 (Existence of an Equilibrium) *Given an economy*

$$\left\{((\mathscr{F}_t), \mathbb{P}), (N_0 = 0, N), \left(\mathbb{P}_i, U_i, \left(e_0^i, e^i\right)\right)_{i=1}^{\mathscr{I}}\right\}.$$

Let S be an equilibrium price process for the representative trader economy

$$\{((\mathscr{F}_t), \mathbb{P}), (N_0 = 0, N), (\mathbb{P}, U(\lambda), (N_0 = 0, N))\}$$

with aggregate utility function weightings

$$\lambda_i = \frac{v_1'(x_1, S)Y_T^1(S)}{v_i'(x_i, S)Y_T^i(S)}$$

with $Y_T^i(S)$ the solution to trader i's dual problem and $v_i(x_i, S)$ trader i's value function, given the initial wealth x_i and the price process S, for all $i = 1, \ldots, \mathscr{I}$.

Then, there exists an equilibrium in the original economy with price process S.

Proof As in the proof of Theorem 14.4, these weightings imply that the optimal demands X_T^i generated by the aggregate utility function $U(\lambda) = \sum_{i=1}^{\mathscr{I}} \lambda_i \frac{d\mathbb{P}_i}{d\mathbb{P}} U_i$ are optimal for the individual traders as well.

We note that the individual traders' time 0 endowments sum to the representative trader's initial endowment. In addition the individual traders' optimal time T wealths sum to the representative trader's time T wealth, i.e. $X_T^\lambda = \sum_{i=1}^{\mathscr{I}} X_T^i$. In the representative trader economy equilibrium, $x = m_0$ and $X_T^\lambda = m_T$.

For the economy, the individual trader's asset holdings $\left((\alpha_0^i(t), \alpha_t^i) : i = 1, \ldots, \mathscr{I}\right)$ are uniquely implied by their initial endowments $\left((e_0^i, e^i) : i = 1, \ldots, \mathscr{I}\right)$ and optimal wealths $(X_T^i : i = 1, \ldots, \mathscr{I})$. This is due to the assumption of nonredundant assets in Chap. 13 on equilibrium. Since $X_T^\lambda = \sum_{i=1}^{\mathscr{I}} X_T^i$, the individual share holdings must sum to those of the representative trader for all t a.s. $\mathbb{P}$, i.e. $0 = N_0 = \sum_{i=1}^{\mathscr{I}} \alpha_0^i(t)$ and $N_j = \sum_{i=1}^{\mathscr{I}} \alpha_j^i(t)$ for $j = 1, \ldots, n$ for all $t \in [0, T]$. This implies that the economy is in equilibrium. This completes the proof.

This theorem provides a convenient method to prove that an equilibrium exists in an economy. As shown, to prove an equilibrium exists, it is sufficient to prove that an equilibrium exists in a representative trader economy with the aggregate utility function weightings as in Theorem 14.4 (e.g., see Karatzas and Shreve [118], Cuoco and He [39], Basak and Cuoco [8], and Hugonnier [73]). We now provide a set of sufficient conditions for the existence of an equilibrium. This set of sufficient conditions applies to both a complete and incomplete market.

Theorem 14.7 (Sufficient Conditions for an Equilibrium) *Given an economy*

$$\left\{((\mathscr{F}_t), \mathbb{P}), (N_0 = 0, N), \left(\mathbb{P}_i, U_i, \left(e_0^i, e^i\right)\right)_{i=1}^{\mathscr{I}}\right\},$$

assume that

(i) for all λ, $U(x, \lambda, \omega)$ has an asymptotic elasticity strictly less than one a.s. $\mathbb{P}$, $E\left[U'(N \cdot \xi, \lambda)\right] < \infty$, and $E\left[U(N \cdot \xi, \lambda)\right] < \infty$ where ξ is the exogenous liquidating dividends for the risky assets.

Define

$$S_t^\lambda = E_\lambda\left[\xi \,|\mathscr{F}_t\right], \qquad \textit{where} \qquad \frac{d\mathbb{Q}_\lambda}{d\mathbb{P}} = \frac{U'(N \cdot \xi, \lambda)}{E\left[U'(N \cdot \xi, \lambda)\right]} > 0 \tag{14.23}$$

and $E_\lambda[\cdot]$ is expectation with respect to $\mathbb{Q}_\lambda$,

(ii) for all λ and i, $v_i(x_i, S^\lambda) < \infty$ for some $x_i > 0$, and
(iii) there exists a $\tilde{\lambda}$ such that

$$\tilde{\lambda}_i = \frac{v_1'(x_1, S^{\tilde{\lambda}}) Y_T^1(S^{\tilde{\lambda}})}{v_i'(x_i, S^{\tilde{\lambda}}) Y_T^i(S^{\tilde{\lambda}})} \tag{14.24}$$

with $Y^i(S^{\tilde{\lambda}})$ the optimal solution to trader i's dual problem and $v_i(x_i, S^{\tilde{\lambda}})$ trader i's value function, given the initial wealth x_i and the price process $S^{\tilde{\lambda}}$, for all $i = 1, \ldots, \mathscr{I}$.

Then,

(1) $S^{\tilde{\lambda}}$ is an equilibrium price process for the economy,
(2) $S^{\tilde{\lambda}}$ is an equilibrium price process for the representative trader economy $\left\{((\mathscr{F}_t), \mathbb{P}), (N_0 = 0, N), \left(\mathbb{P}, U(\tilde{\lambda}), (N_0, N)\right)\right\}$, and
(3) this representative trader economy reflects the economy's equilibrium.

Proof (Step 1) Consider the collection of representative trader economies $\{((\mathscr{F}_t), \mathbb{P}), (N_0 = 0, N), (\mathbb{P}, U(\lambda), (N_0 = 0, N))\}$ indexed by λ.

We first prove using the price process S^λ that the representative trader's optimal trading strategy exists and it is a buy and hold with zero units in the mma and N shares in the stocks, i.e. $\hat{\alpha}_0^\lambda(t, S^\lambda) = 0$, and $\hat{\alpha}_t^\lambda(S^\lambda) = N$ for all t a.s. $\mathbb{P}$, where $\hat{X}_T^\lambda(S^\lambda)$ denotes the representative trader's optimal demands. Note the explicit dependence of all the previous quantities on λ and S^λ.

By the definition of S^λ, $\mathfrak{M} \neq \emptyset$. Hence, $\mathfrak{M}_l \neq \emptyset$ for the price process S^λ.

Therefore any nonnegative wealth self-financing trading strategy $X \in \mathscr{X}(x)$ is a supermartingale under $\mathbb{Q}_\lambda$.

Consider the buy and hold trading strategy $(0, N)$. Note that the wealth of this trading strategy $N \cdot S^\lambda \in \mathscr{X}(x)$. Since N is constant, $N \cdot S^\lambda$ is a $\mathbb{Q}_\lambda$-martingale. This implies that $x = E\left[\frac{U'(N \cdot \xi, \lambda)}{E[U'(N \cdot \xi, \lambda)]} N \cdot \xi\right] = E\left[\frac{U'(N \cdot S_T, \lambda)}{E[U'(N \cdot S_T, \lambda)]} N \cdot S_T\right]$. We now show that this buy and hold trading strategy is optimal.

Consider an arbitrary $X \in \mathscr{X}(x)$. Then,

$$E\left[U(X_T, \lambda) - U(N \cdot S_T, \lambda)\right] \leq E\left[U'(N \cdot S_T, \lambda)(X_T - N \cdot S_T)\right]$$

by the concavity of U. But,

$$E\left[U'(N \cdot S_T, \lambda)(X_T - N \cdot S_T)\right] = E\left[U'(N \cdot S_T, \lambda)\right] E\left[\frac{d\mathbb{Q}_\lambda}{d\mathbb{P}}(X_T - N \cdot S_T)\right] \leq 0$$

since

$$E\left[\frac{d\mathbb{Q}_\lambda}{d\mathbb{P}}(X_T - N \cdot S_T)\right] \leq x - x = 0.$$

This completes the proof of Step 1.

The proof shows that $v(x, \lambda, S^\lambda) < \infty$ for all $x > 0$.

(Step 2) Fix a λ. Consider an economy

$$\left\{((\mathscr{F}_t), \mathbb{P}), (N_0 = 0, N), \left(\mathbb{P}_i, U_i, \left(e_0^i, e^i\right)\right)_{i=1}^{\mathscr{I}}\right\}$$

with the price process S^λ.

Hypothesis (ii) in conjunction with $\mathfrak{M}_l \neq \emptyset$ for the price process S^λ implies that the individual traders have an optimal wealth. For each λ, let $X_T^i(S^\lambda)$ denote the optimal demands for trader i in the economy with the price process S^λ. Note the explicit dependence of the optimal demands on the price process S^λ.

(Step 3) Fix the $\tilde{\lambda}$ that satisfies hypothesis (iii).

Jointly consider the economy

$$\left\{((\mathscr{F}_t), \mathbb{P}), (N_0 = 0, N), \left(\mathbb{P}_i, U_i, \left(e_0^i, e^i\right)\right)_{i=1}^{\mathscr{I}}\right\}$$

and the collection of representative trader economies

$$\{((\mathscr{F}_t), \mathbb{P}), (N_0 = 0, N), (\mathbb{P}, U(\lambda), (N_0 = 0, N))\}$$

indexed by λ using the *same* price process $\tilde{S} \equiv S^{\tilde{\lambda}}$. Note that this is a fixed and identical price process for all the representative trader economies indexed by λ.

Since $\tilde{S} = S^{\tilde{\lambda}}$, $\mathfrak{M}_l \neq \emptyset$ for the price process $\tilde{S}$.

By Theorem 14.4, $\tilde{\lambda}$ is such that the representative trader economy

$$\left\{((\mathscr{F}_t), \mathbb{P}), (N_0 = 0, N), (\mathbb{P}, U(\tilde{\lambda}), (N_0 = 0, N))\right\}$$

reflects the aggregate optimal wealths in the original economy given the price process $\tilde{S}$, i.e.

$$\hat{X}_T^{\tilde{\lambda}}(\tilde{S}) = \sum_{i=1}^{\mathscr{I}} \hat{X}_T{}^i(\tilde{S}) \text{ a.s. } \mathbb{P}. \tag{14.25}$$

We note that the trading strategy $\left(\sum_{i=1}^{\mathscr{I}} \hat{\alpha}_0{}^i(t, \tilde{S}), \sum_{i=1}^{\mathscr{I}} \hat{\alpha}_t{}^i(\tilde{S})\right)$ generates $\sum_{i=1}^{\mathscr{I}} \hat{X}_T{}^i(\tilde{S})$ and, therefore, $\hat{X}_T^{\tilde{\lambda}}(\tilde{S})$.

(Step 4) We claim that $\tilde{S} = S^{\tilde{\lambda}}$ is an equilibrium price for the economy and the representative trader economy

$$\left\{((\mathscr{F}_t), \mathbb{P}), (N_0 = 0, N), \left(\mathbb{P}, U(\tilde{\lambda}), (N_0 = 0, N)\right)\right\}.$$

(Part a) First, by (Step 1) for the price process $S^{\tilde{\lambda}}$ the representative trader with aggregate utility $U(\tilde{\lambda})$'s trading strategy is a buy and hold with zero units in the mma and N shares in the stocks, i.e. $\hat{\alpha}_0^{\tilde{\lambda}}(t, S^{\tilde{\lambda}}) = 0$, and $\hat{\alpha}_t^{\tilde{\lambda}}(S^{\tilde{\lambda}}) = N$ for all t a.s. $\mathbb{P}$. This implies that $S^{\tilde{\lambda}}$ is an equilibrium price process for the representative trader economy $\left\{((\mathscr{F}_t), \mathbb{P}), (N_0 = 0, N), \left(\mathbb{P}, U(\tilde{\lambda}), (N_0 = 0, N)\right)\right\}$.

(Part b) By expression (14.25), we get

$$\hat{X}_T^{\tilde{\lambda}}(S^{\tilde{\lambda}}) = \sum_{i=1}^{\mathscr{I}} \hat{X}_T{}^i(S^{\tilde{\lambda}}) \text{ a.s. } \mathbb{P}.$$

And, by the assumption of nonredundant assets in Chap. 13 on equilibrium, uniqueness of the representative trader's optimal trading strategy generating $\hat{X}_T^{\tilde{\lambda}}(S^{\tilde{\lambda}})$ implies

$$0 = \sum_{i=1}^{\mathscr{I}} \hat{\alpha}_0{}^i(t, S^{\tilde{\lambda}}) \quad \text{and} \quad N = \sum_{i=1}^{\mathscr{I}} \hat{\alpha}_t{}^i(S^{\tilde{\lambda}}) \text{ for all } t \text{ a.s. } \mathbb{P},$$

i.e. aggregate demand equals aggregate supply in the original economy. This proves that $S^{\tilde{\lambda}}$ is an equilibrium price process for the original economy

$$\left\{((\mathscr{F}_t), \mathbb{P}), (N_0 = 0, N), \left(\mathbb{P}_i, U_i, \left(e_0^i, e^i\right)\right)_{i=1}^{\mathscr{I}}\right\}.$$

This completes the proof.

In this theorem, sufficient conditions (i)–(ii) are mild restrictions on an economy. These hypotheses are needed for the existence of an individual traders' and representative trader's optimal portfolio, plus the integrability of the representative trader's utility and marginal utility for aggregate wealth. The remaining sufficient condition (iii) is quite strong, effectively assuming the existence of a set of aggregate utility function weights $\tilde{\lambda}$ for which an equilibrium exists in the economy. Independently verifying this hypothesis involves solving a difficult fixed point problem in the infinite-dimensional space of nonnegative adapted and cadlag processes. Finding a set of sufficient conditions on the primitives of the economy such that solution λ to expression (14.24) exists is an important open research question. Section 14.6 below provides examples of economies and sufficient conditions in those economies such that aggregate utility function weightings $\tilde{\lambda}$ satisfying expression (14.24) can be shown to exist, proving that the set of solutions to expression (14.24) is not vacuous.

Recent papers providing sufficient conditions for the existence of an equilibrium in complete and incomplete markets, but using different methods than those discussed above, include Karatzas and Shreve [118], Kardaras, Xing, and Zitkovic

[122], Kramkov [125], Larsen [132], Choi and Larsen [31], Christensen and Larsen [32], Larsen and Sae-Sue [133], and Zitkovic [179, 180]. The theorem presented above is for a more general economy then the existence proofs contained in Karatzas and Shreve [118] and Kramkov [125], both of which study a Brownian motion-based economy.

Remark 14.5 (Uniqueness of the Equilibrium) The equilibrium price process in expression (14.23) is unique in the following sense. Given an economy $\left\{((\mathscr{F}_t), \mathbb{P}), (N_0 = 0, N), \left(\mathbb{P}_i, U_i, \left(e_0^i, e^i\right)\right)_{i=1}^{\mathscr{I}}\right\}$, let S be an equilibrium price process. Second, let the equilibrium in the representative trader economy $\{((\mathscr{F}_t), \mathbb{P}), (N_0 = 0, N), (\mathbb{P}, U(\lambda), (N_0 = 0, N))\}$ with price process S reflect the economy's equilibrium. Then, S must satisfy expression (14.23).

Proof By the hypotheses in this statement, the representative trader's optimal trading strategy is $\alpha_0^\lambda(t) = 0$, and $\alpha_t^\lambda = N$ for all t. Since $Y_T^\lambda(S) = \frac{U'(N \cdot \xi, \lambda)}{E[U'(N \cdot \xi, \lambda)]}$ is a probability density with respect to $\mathbb{P}$, by Theorem 14.3, S is a $\mathbb{Q}_\lambda$-martingale. Noting that $S_T = \xi$ completes the proof.

This completes the remark.

Remark 14.6 (Uniqueness of the Equilibrium Supermartingale Deflators) Given an equilibrium price process S, this remarks relates to the uniqueness of the individual trader's and the representative trader's supermartingale deflators.

(Complete Market)

In a complete market with respect to any $\mathbb{Q} \in \mathfrak{M}_l(S)$, from the portfolio optimization Chap. 10, these supermartingale deflators are all local martingale deflators which are probability densities with respect to $\mathbb{P}$. By Remark 14.4 above, they are all equal since there is a unique local martingale measure $\mathbb{Q} \in \mathfrak{M}_l(S)$ by the Second Fundamental Theorem 2.4 of asset pricing in Chap. 2. Using Corollary 13.1 in Chap. 13 on equilibrium, we have that the local martingale measure is also a martingale measure, i.e. $\mathbb{Q} \in \mathfrak{M}(S)$.

Hence, the expression for investor i's supermartingale deflator is given by expression (11.29) in Sect. 11.9 in the portfolio optimization Chap. 11, i.e.

$$\frac{d\mathbb{Q}}{d\mathbb{P}} = Y_T^i(S) = \frac{\frac{d\mathbb{P}_i}{d\mathbb{P}} U_i'(X_T^i(S))}{v_i'(x_i, S)} \in M(S) \text{ for all } i = 1, \ldots, \mathscr{I}.$$

The change of measure in the numerator is needed since the set of martingale deflators $M(S)$ depends on the probability measure $\mathbb{P}$.

The representative trader's optimal supermartingale deflator is given by

$$\frac{d\mathbb{Q}}{d\mathbb{P}} = Y_T^\lambda(S) = \frac{U'(m_T, \lambda)}{E[U'(m_T, \lambda)]} \in M(S),$$

where the aggregate market wealth is

$$X_T^{\lambda} = \sum_{i=1}^{\mathscr{I}} X_T^i = \sum_{j=1}^{n} N_j S_j(T) = m_T.$$

(Incomplete Market)
In an incomplete market, the supermartingale deflators are

$$Y_T^{\lambda}(S) = \frac{U'(m_T,\lambda)}{v'(m_0,\lambda,S)} \in D_s(S) \qquad \text{and}$$

$$Y_T^i(S) = \frac{\frac{d\mathbb{P}_i}{d\mathbb{P}} U_i'(X_T^i(S))}{v_i'(x_i,S)} \in D_s(S) \text{ for all } i = 1,\ldots,\mathscr{I},$$

where the aggregate market wealth is $X_T^{\lambda} = \sum_{i=1}^{\mathscr{I}} X_T^i = \sum_{j=1}^{n} N_j S_j(T) = m_T$ and $x = \sum_{i=1}^{\mathscr{I}} x_i = \sum_{j=1}^{n} N_j S_j(0) = m_0$. These supermartingale deflators can all be different.

We note that the set of supermartingale deflators $D_s(S)$ depends on the probability measure $\mathbb{P}$ (see the portfolio optimization Chap. 11 after the definition of $\mathscr{D}_s$). This is the reason investor i's supermartingale deflator has the change of measure in the numerator, see expression (11.29), Sect. 11.9 in the portfolio optimization Chap. 11. Second, since the representative trader's optimal trading strategy in the representative trader economy equilibrium is a buy and hold trading strategy with zero units in the mma, by Corollary 14.1 in the representative trader Chap. 14, if Y_T^{λ} is a probability density with respect to $\mathbb{P}$, then $Y_T^{\lambda} = \frac{U'(m_T,\lambda)}{v'(m_0,\lambda,S)} \in M(S)$ is a martingale deflator as well. This completes the remark.

14.6 Examples

This section provides, as examples, two economies where sufficient condition (iii) in Theorem 14.7 can be shown to have a solution, thereby proving that the set of economies satisfying expression (14.24) is not vacuous. Another example proving the existence of an equilibrium in an incomplete market, using the approach of Theorem 14.7, is contained in Chap. 17 on the static CAPM, Theorem 17.8.

14.6.1 Identical Traders

The first economy studied is one in which the market is incomplete, but all traders have identical beliefs, utility functions, and endowments, i.e.

$$\left(\mathbb{P}_i, U_i, \left(e_0^i, e^i\right)\right) = \left(\mathbb{P}, U_1, \left(e_0^1, e^1\right)\right) \quad \text{for all } i.$$

Assume hypotheses (i) and (ii) in Theorem 14.7 are satisfied. Then, since all traders are identical, expression (14.24) has the trivial solution

$$\tilde{\lambda}_i = 1 \quad \text{for all } i \qquad \text{or}$$

$$U(x, \mathbf{1}, \omega) = \sum_{i=1}^{\mathscr{I}} U_i\left(\frac{x}{\mathscr{I}}, \omega\right) = \mathscr{I} \cdot U_1\left(\frac{x}{\mathscr{I}}, \omega\right),$$

where $\mathbf{1} = (1, \ldots, 1)$ is the $\mathscr{I}$-tuple consisting of all ones.

Each trader is effectively a "representative trader in this economy." Trader i's optimal nonnegative wealth s.f.t.s. is the buy and hold trading strategy $(\alpha_0(t) = 0, \alpha_t^i = \frac{N}{\mathscr{I}})$. This implies that the equilibrium price process

$$S_t^\lambda = E^{\mathbb{Q}}\left[\xi \,|\mathscr{F}_t\right],$$

where

$$\frac{d\mathbb{Q}}{d\mathbb{P}} \equiv Y_T^\lambda(S^\lambda) = \frac{U'(N \cdot \xi, 1)}{E\left[U'(N \cdot \xi, 1)\right]},$$

is independent of λ as well.

Hence, we have just proven that an equilibrium exists for this incomplete market economy with identical traders, under hypotheses (i) and (ii) alone, with the price process $S_t = E^{\mathbb{Q}}\left[\xi \,|\mathscr{F}_t\right]$.

14.6.2 Logarithmic Preferences

The second economy studied is an incomplete market with

$$\left(\mathbb{P}_i, U_i, \left(e_0^i, e^i\right)\right)_{i=1}^{\mathscr{I}} = \left(\mathbb{P}_i, \ln(\cdot), \left(e_0^i, e^i\right)\right)_{i=1}^{\mathscr{I}},$$

where traders have state independent logarithmic utility functions $U_i(x) = \ln(x)$ with $x > 0$ for all i. The traders have different beliefs and endowments.

Assume hypotheses (i) and (ii) in Theorem 14.7 are satisfied. In this economy, given weights λ, the aggregate utility function satisfies

$$U(x, \lambda, \omega) = \sup_{x = x_1 + \cdots + x_{\mathscr{I}}} \left(\sum_{i=1}^{\mathscr{I}} \lambda_i \frac{d\mathbb{P}_i}{d\mathbb{P}} \ln(x_i)\right).$$

As discussed in Remark 14.2 above, without loss of generality we normalize the aggregate utility function weights such that

$$\sum_{i=1}^{\mathcal{I}} \lambda_i \frac{d\mathbb{P}_i}{d\mathbb{P}} = 1. \tag{14.26}$$

To determine the aggregate utility function, define the Lagrangian

$$\mathcal{L} \equiv \sum_{i=1}^{\mathcal{I}} \lambda_i \frac{d\mathbb{P}_i}{d\mathbb{P}} \ln(x_i) + \mu(x - x_1 - \cdots - x_{\mathcal{I}})$$

with $\mu \geq 0$. The first-order conditions, which are necessary and sufficient for an optimum, are

$$\lambda_i \frac{d\mathbb{P}_i}{d\mathbb{P}} \frac{1}{x_i} = \mu \qquad \text{for all } i.$$

The solution is

$$x_i = \frac{\lambda_i \frac{d\mathbb{P}_i}{d\mathbb{P}}}{\lambda_1 \frac{d\mathbb{P}_1}{d\mathbb{P}}} x_1 \qquad \text{for all } i. \tag{14.27}$$

Using the constraint $x - x_1 - \cdots - x_{\mathcal{I}} = 0$, substitution and algebra yields

$$x_1 = \frac{\lambda_1 \frac{d\mathbb{P}_1}{d\mathbb{P}} x}{\sum_{i=1}^{\mathcal{I}} \lambda_i \frac{d\mathbb{P}_i}{d\mathbb{P}}} = \lambda_1 \frac{d\mathbb{P}_1}{d\mathbb{P}} x,$$

where the last equality uses expression (14.26). Substituting back into expression (14.27) gives the solution

$$x_i = \lambda_i \frac{d\mathbb{P}_i}{d\mathbb{P}} x \qquad \text{for all } i.$$

Hence, the aggregate utility function is

$$U(x,\lambda) = \sum_{i=1}^{\mathcal{I}} \lambda_i \frac{d\mathbb{P}_i}{d\mathbb{P}} \ln\left(\lambda_i \frac{d\mathbb{P}_i}{d\mathbb{P}} x\right) = \sum_{i=1}^{\mathcal{I}} \lambda_i \frac{d\mathbb{P}_i}{d\mathbb{P}} \ln\left(\lambda_i \frac{d\mathbb{P}_i}{d\mathbb{P}}\right) + \ln(x), \tag{14.28}$$

where we have used expression (14.26) again. Note that

$$U'(x,\lambda) = \frac{1}{x}$$

is independent of λ.

The conjugate aggregate utility function is

$$\tilde{U}(y) = \sup_{x>0}\left[\sum_{i=1}^{\mathcal{I}} \lambda_i \frac{d\mathbb{P}_i}{d\mathbb{P}} \ln\left(\lambda_i \frac{d\mathbb{P}_i}{d\mathbb{P}}\right) + \ln(x) - xy\right].$$

The first-order condition, which is necessary and sufficient for an optimal solution, is

$$\tilde{U}'(y) = \frac{1}{x} - y = 0.$$

This gives $x = \frac{1}{y}$ and

$$\tilde{U}(y) = \sum_{i=1}^{\mathcal{I}} \lambda_i \frac{d\mathbb{P}_i}{d\mathbb{P}} \ln\left(\lambda_i \frac{d\mathbb{P}_i}{d\mathbb{P}}\right) - \ln(y) - 1.$$

The representative trader's supermartingale deflator is

$$\frac{d\mathbb{Q}_\lambda}{d\mathbb{P}} = Y_T^\lambda = \frac{U'(N \cdot \xi, \lambda)}{E\left[U'(N \cdot \xi, \lambda)\right]} = \frac{\frac{1}{N \cdot \xi}}{E\left[\frac{1}{N \cdot \xi}\right]},$$

which is independent of λ. The supermartingale deflator process is

$$Y_t^\lambda = E\left[Y_T^\lambda \,|\mathscr{F}_t\right] = \frac{E\left[\frac{1}{N \cdot \xi} \,|\mathscr{F}_t\right]}{E\left[\frac{1}{N \cdot \xi}\right]},$$

which is independent of λ. Consider the price process

$$S_t^\lambda = E_\lambda\left[\xi \,|\mathscr{F}_t\right] = E\left[\xi \frac{Y_T^\lambda}{Y_t^\lambda} \,|\mathscr{F}_t\right] = \frac{E\left[\frac{\xi}{N \cdot \xi} \middle| \mathscr{F}_t\right]}{E\left[\frac{1}{N \cdot \xi} \,|\mathscr{F}_t\right]}.$$

Since the price process is independent of λ, we write $S_t^\lambda = S_t$.

Now, from Example 14.1 and using the modification for differential beliefs in Sect. 9.7 in the utility function Chap. 9, we have that the ith trader's conjugate utility function is

$$\frac{d\mathbb{P}_i}{d\mathbb{P}} \tilde{U}_i\left(y_i Y_T^i \frac{d\mathbb{P}}{d\mathbb{P}_i}\right) = \frac{d\mathbb{P}_i}{d\mathbb{P}}\left(-\ln(y_i) - \ln(Y_T^i) + \ln\left(\frac{d\mathbb{P}}{d\mathbb{P}_i}\right) - 1\right).$$

Given y_i, the supermartingale deflator $Y_T^i(S)$ is the solution to

$$\inf_{Y_T^i \in D_s(S)} E\left[\frac{d\mathbb{P}_i}{d\mathbb{P}}\left(-\ln(y_i) - \ln(Y_T^i) + \ln\left(\frac{d\mathbb{P}}{d\mathbb{P}_i}\right) - 1\right)\right].$$

This observation proves that the optimal solution $Y_T^i(S)$ is independent of y_i.

Next, given $Y_T^i(S)$, y_i is determined as the solution to

$$x_i = e_0^i(0) + e_0^i \cdot S_0^\lambda = E\left[I_i\left(y_i Y_T^i(S)\frac{d\mathbb{P}}{d\mathbb{P}_i}\right) Y_T^\lambda\right],$$

where $I_i(y) = \frac{1}{y}$ for $y > 0$. Substitution for S_0^λ and Y_T^λ yields

$$e_0^i(0) + e_0^i \cdot \frac{E\left[\frac{\xi}{N\cdot\xi}\right]}{E\left[\frac{1}{N\cdot\xi}\right]} = E\left[\frac{1}{y_i Y_T^i(S)\frac{d\mathbb{P}}{d\mathbb{P}_i}}\left(\frac{\frac{\xi}{N\cdot\xi}}{E\left[\frac{1}{N\cdot\xi}\right]}\right)\right].$$

Algebra gives the solution

$$y_i = \left(\frac{E\left[\frac{1}{Y_T^i(S)\frac{d\mathbb{P}}{d\mathbb{P}_i}}\frac{\xi}{N\cdot\xi}\right]}{e_0^i(0)E\left[\frac{1}{N\cdot\xi}\right] + e_0^i \cdot E\left[\frac{\xi}{N\cdot\xi}\right]}\right),$$

which is independent of λ.

Next, we examine expression (14.24) to see if a solution exists. Recall that $y_i = v_i'(x_i, S^\lambda)$. In this case λ needs to satisfy

$$\lambda_i = \frac{y_1 Y_T^1(S)}{y_i Y_T^i(S)} \qquad \text{for all } i.$$

Since the right side is independent of λ, this is the explicit solution. Hence, we have just proven that an equilibrium exists for this incomplete market economy under hypotheses (i) and (ii) alone, with the price process

$$S_t = \frac{E\left[\frac{\xi}{N\cdot\xi}\middle|\mathscr{F}_t\right]}{E\left[\frac{1}{N\cdot\xi}|\mathscr{F}_t\right]}.$$

This completes the set of examples proving the existence of aggregate utility function weightings λ that satisfy expression (14.24).

14.7 Intermediate Consumption

To include intermediate consumption, the representative trader's utility function U as defined earlier needs to be extended to include intermediate consumption $U(c, t, \lambda)$ as given in the portfolio optimization Chap. 12. With this extension, analogous results to those obtained in the previous section can be derived by replacing the representative trader's supermartingale deflator based on optimizing the expected utility of terminal wealth with the representative trader's supermartingale deflator based on optimizing the expected utility of intermediate consumption, see Jarrow [92].

The characterization of the representative trader's supermartingale deflator is analogous to that given in the portfolio optimization Chap. 12. Indeed, the representative trader's optimal supermartingale deflator process is

$$Y_t^{\lambda}(S) = \frac{U'(c_t, t, \lambda)}{v'(x, \lambda, S)} \in \mathscr{D}_s(S), \tag{14.29}$$

where $c_t = \sum_{i=1}^{\mathscr{I}} c_t^i$ is the representative trader's optimal consumption. If S is an equilibrium price process, then in equilibrium, the representative trader's optimal consumption equals aggregate consumption.

14.8 Notes

For references discussing a representative trader in the context of Pareto optimality, see Back [5], Huang and Litzenberger [72], Dana and Jeanblanc [42], and Duffie [52]. For existence proofs in a complete market using the representative trader approach, see Dana and Jeanblanc [42] and Karatzas and Shreve [118].

Chapter 15
Characterizing the Equilibrium

Assuming that an equilibrium exists, this chapter characterizes the economic equilibrium. For simplicity of presentation, this chapter focuses on traders having preferences only over terminal wealth. The last section in this chapter discusses the necessary changes needed to include intermediate consumption. The key result in this chapter is a characterization of the equilibrium supermartingale deflator as a function of the economy's primitives: beliefs, preferences, and endowments. Indeed, using a representative trader economy equilibrium that reflects the equilibrium in the original economy, an equilibrium supermartingale deflator is characterized as a function of the representative trader's (aggregate) utility function and aggregate market wealth. Finally, this chapter derives the intertemporal capital asset pricing model (ICAPM) and the consumption capital asset pricing model (CCAPM) as special cases of this characterization.

15.1 The Set-Up

Given is an economy

$$\left\{((\mathscr{F}_t), \mathbb{P}), (N_0 = 0, N), \left(\mathbb{P}_i, U_i, \left(e_0^i, e^i\right)\right)_{i=1}^{\mathscr{I}}\right\}.$$

As in Chap. 13 on equilibrium, we assume that there are $\mathscr{I}$ traders in the economy with preferences $\overset{\rho}{\succeq}_i$ assumed to have a state dependent EU representation with utility functions U_i defined over terminal wealth that satisfy the properties in the utility function Chap. 9 including reasonable asymptotic elasticity $AE(U_i) < 1$ for all i.

R. A. Jarrow, *Continuous-Time Asset Pricing Theory*, Springer Finance,
https://doi.org/10.1007/978-3-319-77821-1_15

We assume that an equilibrium price process S exists. Hence, under the hypotheses of Theorem 14.4 in the representative trader Chap. 14, there exists a representative trader economy

$$\{((\mathscr{F}_t), \mathbb{P}), (N_0 = 0, N), (\mathbb{P}, U(\lambda), (N_0 = 0, N))\}$$

such that the representative trader economy equilibrium reflects the original economy's equilibrium. In particular, this implies that both the optimal asset demands and the equilibrium price process S is the same in the representative trader economy equilibrium as in the original economy's equilibrium.

15.2 The Supermartingale Deflator

This section characterizes the equilibrium supermartingale deflator.

Remark 15.1 (Choice of Characterization) Given an equilibrium price process S, there are many possible characterizations of the equilibrium supermartingale deflator. The different characterizations correspond to the representative trader's and the individual traders' supermartingale deflators, determined from their portfolio optimization problems. In this regard, we note that the representative trader's optimal trading strategy in the representative trader economy equilibrium is necessarily a buy and hold trading strategy with zero units in the mma. Consequently, by Corollary 14.1 in the representative trader Chap. 14, letting Y_T^λ denote the representative trader's optimal supermartingale deflator,

if Y_T^λ is a probability density with respect to $\mathbb{P}$, then the probability measure defined by $\frac{d\mathbb{Q}}{d\mathbb{P}} = Y_T^\lambda$ is an equivalent martingale measure, i.e. $Y_T^\lambda \in M$.

We will use this observation repeatedly below.

(Complete Market)

In a complete market with respect to any $\mathbb{Q} \in \mathfrak{M}_l$, where

$$\mathfrak{M}_l = \{\mathbb{Q} \sim \mathbb{P} : \ S \text{ is a } \mathbb{Q}\text{-local martingale}\},$$

by the Second Fundamental Theorem 2.4 of asset pricing in Chap. 2, the local martingale measure $\mathbb{Q} \in \mathfrak{M}_l$ is unique. By Remark 14.6 in the representative trader Chap. 14, all of the traders' and representative trader's supermartingale deflators are martingale deflators which are probability densities with respect to $\mathbb{P}$ (i.e. they are in the set M). Hence, all of these martingale densities are equal, i.e.

$$\frac{d\mathbb{Q}}{d\mathbb{P}} = Y_T = \frac{U'(m_T, \lambda)}{E[U'(m_T, \lambda)]} = \frac{\frac{d\mathbb{P}_i}{d\mathbb{P}} U_i'(X_T^i)}{E\left[\frac{d\mathbb{P}_i}{d\mathbb{P}} U_i'(X_T^i)\right]} \in M$$

for all $i = 1, \ldots, \mathscr{I}$, where the aggregate market wealth $m_T = \sum_{j=1}^n N_j S_j(T)$.

The expression for investor i's supermartingale deflator is given by expression (11.29), Sect. 11.9 in the portfolio optimization Chap. 11. The change of measure in the numerator is needed since the set of martingale deflators M depends on the probability measure $\mathbb{P}$.

(Incomplete Market)
In an incomplete market, the supermartingale deflators need not be unique, hence each may yield a different characterization of an equilibrium supermartingale deflator,

$$Y_T^{\lambda}(S) = \frac{U'(m_T,\lambda)}{v'(m_0,\lambda,S)} \in D_s(S) \qquad \text{and}$$

$$Y_T^i(S) = \frac{\frac{d\mathbb{P}_i}{d\mathbb{P}} U_i'(X_T^i(S))}{v_i'(x_i,S)} \in D_s(S) \;\; \text{for all } i = 1, \ldots, \mathscr{I},$$

where the aggregate market wealth $m_t = \sum_{j=1}^{n} N_j S_j(t)$ for $t = 0, T$.

If Y_T^{λ} is a probability density with respect to $\mathbb{P}$, then by the observation above

$$Y_T^{\lambda} = \frac{U'(m_T,\lambda)}{v'(m_0,\lambda,S)} \in M.$$

For purposes of empirical implementation, the characterization of the supermartingale deflator in terms of the representative trader's aggregate utility function is preferred. This is for two reasons. First, for the supermartingale deflators of the individual traders, their wealth at times 0 and T, (x_i, X_T^i), are unobservable. In contrast, for the supermartingale deflator of the representative trader, the market wealth process (m_0, m_T) is observable. In terms of preferences, both utility function representations U and U_i are equally unobservable. Second, the representative trader's supermartingale deflator, if a probability density with respect to $\mathbb{P}$, is a martingale deflator, in contract to those of the individual traders. This observation has important implications for the existence of asset price bubbles in equilibrium under the representative trader's martingale probability density. This is discussed shortly. This completes the remark.

Given these remarks and using Theorem 14.1 in the representative trader Chap. 14, the next theorem characterizes an equilibrium supermartingale deflator and state price density.

Theorem 15.1 (An Equilibrium Supermartingale Deflator (and State Price Density)) *Given an equilibrium price process S, an equilibrium supermartingale deflator is*

(Complete Market)

$$Y_T = \frac{U'(m_T,\lambda)}{E[U'(m_T,\lambda)]} \in M$$

(Incomplete Market)

$$Y_T^\lambda = \frac{U'(m_T, \lambda)}{v'(m_0, \lambda, S)} \in D_s \qquad \textit{and}$$

if Y_T^λ *is a probability density with respect to* $\mathbb{P}$, *then*

$$Y_T^\lambda = \frac{U'(m_T, \lambda)}{v'(m_0, \lambda, S)} \in M,$$

where the aggregate market wealth $m_t = \sum_{j=1}^n N_j S_j(t)$ *for* $t = 0, T$, $U(\lambda)$ *is the representative trader's utility function, and* $v(m_0, \lambda, S)$ *is the representative trader's value function.*

As shown in this theorem, an equilibrium supermartingale deflator is characterized as a function of the representative trader's marginal utility and aggregate market wealth at times 0 and T. The equilibrium supermartingale deflator is only unique in the case of a complete market where it is the probability density of a martingale measure with respect to $\mathbb{P}$. This statement yields the next corollary.

Corollary 15.1 (An Equilibrium State Price Density Process)
If Y_T^λ *is a probability density with respect to* $\mathbb{P}$, *then an equilibrium state price density process is given by*

$$Y_t^\lambda = E\left[Y_T^\lambda \,|\mathscr{F}_t\right] = \frac{E[U'(m_T, \lambda)\,|\mathscr{F}_t]}{v'(m_0, \lambda, S)} \in \mathscr{M}. \tag{15.1}$$

These expressions manifest the importance of the market portfolio in the pricing of all risky assets (and derivatives) in an economic equilibrium.

Remark 15.2 (Pricing Derivatives) When pricing derivatives in an incomplete market using risk-neutral valuation (see Sect. 2.6 in the fundamental theorems Chap. 2), the set of equivalent martingale measures contains an infinite number of elements. Derivatives that are non-attainable via an admissible s.f.t.s. whose value process is a $\mathbb{Q}$-martingale do not have a unique price.

Since the equilibrium supermartingale deflator is not unique in an incomplete market, choosing the representative trader's state price density (as in the above theorem), gives a rule for obtaining an element in this set. However, the derivative price determined in this fashion is different from that obtained by choosing an individual trader's supermartingale deflator. The individual traders would want to use their own supermartingale deflators to price derivatives, and not the representative trader's supermartingale deflator. We emphasize this insight with the following observation:

the representative trader's supermartingale deflator does not provide the relevant probability measure for a trader to use to price derivatives.

The representative trader's supermartingale deflator is useful, but only for characterizing the equilibrium asset price process S. This completes the remark.

15.3 Asset Price Bubbles

This section studies whether asset price bubbles can exist in a rational equilibrium. The answer is: (i) no in a complete market (for the representative trader and all individual traders), and (ii) no for the representative trader in an incomplete market, but yes for individual traders in an incomplete market.

To prove this claim, we start with an observation. By Theorem 13.2 in the equilibrium Chap. 13, for either a complete or incomplete market, we know that both NFLVR and ND hold. Next, by the Third Fundamental Theorem 2.5 of asset pricing in Chap. 2, we get that there exists an equivalent martingale measure $\mathbb{Q} \in \mathfrak{M}$.

15.3.1 Complete Markets

For a complete market with respect to any $\mathbb{Q} \in \mathfrak{M}_l$, since the equivalent local martingale measure is unique by the Second Fundamental Theorem 2.4 of asset pricing in Chap. 2, this implies that the unique local martingale measure is a martingale measure since

$$Y_T = \frac{U'(m_T, \lambda)}{E[U'(m_T, \lambda)]} \in M \subset M_l.$$

Thus, there can be no asset price bubbles. This provides an alternative proof of Theorem 3.6 in the asset price bubbles Chap. 3.

15.3.2 Incomplete Markets

For an incomplete market, the local martingale measure is not necessarily unique. By the observation at the start of this section, if Y_T^{λ} is a probability density for $\mathbb{P}$, then the representative trader's supermartingale deflator generates an equivalent martingale measure (see Corollary 14.1 in the representative trader Chap. 14). This implies that the representative trader sees no price bubble. In contrast, an individual trader's optimal supermartingale deflator need not be a martingale measure. Hence, individual traders may believe that the asset price exhibits a price bubble. In addition, it is possible that some investors see the price process as exhibiting an asset price bubble while others do not (see Jarrow [94] for an elaboration).

Remark 15.3 (Implications for Empirical Testing) The implications of these observations about asset price bubbles for the empirical testing of equilibrium asset price models are two. First, when estimating the equilibrium market price processes for the traded risky assets using time series data, one can use the representative trader's supermartingale deflator to evaluate risk. If Y_T^{λ} is a probability density for $\mathbb{P}$, then the representative trader's supermartingale deflator generates an equivalent martingale measure. Under this equivalent martingale measure, the equilibrium asset price process exhibits no price bubble. Hence, empirical estimation of the equilibrium systematic risk return relation (discussed in the next section) can exclude the consideration of asset price bubbles.

Second, when evaluating a trader's optimal trading strategy, the trader's supermartingale deflator is the relevant supermartingale deflator, even in an economic equilibrium. This implies that each trader must explicitly consider asset price bubbles when forming their optimal portfolio. Hence, the trader must also explicitly consider asset price bubbles when valuing derivatives in equilibrium in an incomplete market (in a complete market, as discussed above, bubbles do not exist in equilibrium).

It is also interesting to observe that in an incomplete market the equilibrium systematic risk return relation is not unique. Any trader's supermartingale deflator $Y_T^i \in D_s^i$ or the representative trader's supermartingale deflator Y_T^λ can be used to characterize systematic risk, using Theorem 4.2 in the multiple-factor model Chap. 4 (as discussed in the next section). This is because all of these supermartingale deflators are consistent with the equilibrium price process S. This completes the remark.

15.4 Systematic Risk

In this section we use the non-normalized market $((\mathbb{B}, \mathbb{S}), (\mathscr{F}_t), \mathbb{P})$. Given is an equilibrium price process S. Assuming the hypotheses of Theorem 14.4 in the representative trader Chap. 14 hold, there exists a representative trader economy

$$\{((\mathscr{F}_t), \mathbb{P}), (N_0 = 0, N), (\mathbb{P}, U(\lambda), (N_0 = 0, N))\}$$

such that the representative trader economy equilibrium reflects the original economy's equilibrium.

This section revisits the characterization of systematic risk using the equilibrium state price density. For this section, as in the multiple-factor model Chap. 4, Theorem 4.2, we *assume that the representative trader's equilibrium supermartingale deflator is a probability density with respect to* $\mathbb{P}$. Hence, by the observation given above, it is a *martingale deflator*, i.e. it is the probability density of an equivalent martingale measure with respect to $\mathbb{P}$. This theorem applies for complete and incomplete markets.

Next, we partition $[0, T]$ into a collection of sub-intervals of length $\Delta > 0$. Fix a time interval $[t, t + \Delta] \subset [0, T]$, where $t \geq 0$ aligns with one of these partitions. To simplify the presentation we assume that the risky assets contain default-free zero-coupon bonds paying \$1 at times $t = \Delta, \ldots, T$. From Theorem 4.4 in the multiple-factor model Chap. 4, we have that the time t conditional expected return of any traded asset satisfies

Theorem 15.2 (The Equilibrium Risk Return Relation)

$$E\left[R_i(t)\,|\mathscr{F}_t\right] = r_0(t) - \operatorname{cov}\left[R_i(t), \frac{E[U'(m_T,\lambda)\,|\mathscr{F}_{t+\Delta}]}{E[U'(m_T,\lambda)\,|\mathscr{F}_t]} \frac{\mathbb{B}_t}{\mathbb{B}_{t+\Delta}}(1+r_0(t)) \middle| \mathscr{F}_t \right], \tag{15.2}$$

where $r_0(t) = \frac{1}{p(t,t+\Delta)} - 1$ *is the return on a default-free zero-coupon bond that matures at time* $t + \Delta$ *and* $R_i(t) = \frac{\mathbb{S}_i(t+\Delta)-\mathbb{S}_i(t)}{\mathbb{S}_i(t)}$ *is the return on the* i^{th} *risky asset.*

We note that when the time step is small, i.e. $\Delta \approx dt$, then $\frac{\mathbb{B}_t}{\mathbb{B}_{t+\Delta}}(1 + r_0(t)) \approx 1$, and this relation simplifies to

$$E\left[R_i(t)\,|\mathscr{F}_t\right] \approx r_0(t) - \operatorname{cov}\left[R_i(t), \frac{E[U'(m_T,\lambda)\,|\mathscr{F}_{t+\Delta}]}{E[U'(m_T,\lambda)\,|\mathscr{F}_t]} \middle| \mathscr{F}_t \right]. \tag{15.3}$$

This expression is the equilibrium risk return relation for traded assets. Systematic risk is measured by the covariance of any asset's return with the equilibrium state price density $\left(\frac{E[U'(m_T,\lambda)|\mathscr{F}_{t+\Delta}]}{E[U'(m_T,\lambda)|\mathscr{F}_t]}\right)$, which depends on aggregate market wealth and the representative trader's marginal utility function.

- When $\left(\frac{E[U'(m_T,\lambda)|\mathscr{F}_{t+\Delta}]}{E[U'(m_T,\lambda)|\mathscr{F}_t]}\right)$ is large, systematic risk is low (remember the negative sign). This occurs when the marginal utility of wealth is large, hence aggregate market wealth is scarce. An asset whose return is large when market wealth is scarce is valuable. Such an asset is "anti-risky."
- When $\left(\frac{E[U'(m_T,\lambda)|\mathscr{F}_{t+\Delta}]}{E[U'(m_T,\lambda)|\mathscr{F}_t]}\right)$ is small, systematic risk is high. This occurs when the marginal utility of wealth is small, hence aggregate market wealth is plentiful. An asset whose return is large when market wealth is plentiful is less valuable. Such an asset is "risky."

15.5 Consumption CAPM

This section provides a restatement of Theorem 15.2 to obtain the consumption capital asset pricing model (CCAPM). As in the previous section, we partition $[0, T]$ into a collection of sub-intervals of length $\Delta > 0$. Fix a time interval $[t, t + \Delta] \subset [0, T]$ where $t \geq 0$ aligns with one of these partitions.

Theorem 15.3 (CCAPM) *Define* $\mu_m(t) \equiv E[m_T\,|\mathscr{F}_t]$. *Assume* U *is three times continuously differentiable. Then,*

$$E\left[R_i(t)\,|\mathscr{F}_t\right] \approx r_0(t) - \frac{U''(\mu_m(t),\lambda)}{U'(\mu_m(t),\lambda)} \operatorname{cov}\left[R_i(t), (\mu_m(t+\Delta) - \mu_m(t))\,|\mathscr{F}_t\right], \tag{15.4}$$

where $r_0(t) = \frac{1}{p(t,t+\Delta)} - 1$ *is the return on a default-free zero-coupon bond that matures at time* $t + \Delta$ *and* $R_i(t) = \frac{\mathbb{S}_i(t+\Delta) - \mathbb{S}_i(t)}{\mathbb{S}_i(t)}$ *is the return on the* i^{th} *risky asset.*

Proof Use a Taylor series expansion of U' around $\mu_m(t)$.

$$U'(m_T, \lambda) = U'(\mu_m(t), \lambda) + U''(\mu_m(t), \lambda)(m_T - \mu_m(t))$$
$$+ \frac{1}{2}U'''(\theta, \lambda)(m_T - \mu_m(t))^2$$

for

$$\theta \in [\mu_m(t), m_T].$$

Use the approximation

$$U'(m_T, \lambda) \approx U'(\mu_m(t), \lambda) + U''(\mu_m(t), \lambda)(m_T - \mu_m(t)).$$

Take conditional expectations at time $t + \Delta$.

$$\begin{aligned} E\left[U'(m_T, \lambda) \,|\mathscr{F}_{t+\Delta}\right] &\approx U'(\mu_m(t), \lambda) + U''(\mu_m(t), \lambda)(E[m_T \,|\mathscr{F}_{t+\Delta}] - \mu_m(t)) \\ &= U'(\mu_m(t), \lambda) + U''(\mu_m(t), \lambda)(\mu_m(t + \Delta) - \mu_m(t)). \end{aligned}$$

Take conditional expectations at time t.

$$\begin{aligned} E\left[U'(m_T, \lambda) \,|\mathscr{F}_t\right] &\approx U'(\mu_m(t), \lambda) + U''(\mu_m(t), \lambda)(E[m_T \,|\mathscr{F}_t] - \mu_m(t)) \\ &= U'(\mu_m(t), \lambda) + U''(\mu_m(t), \lambda)(\mu_m(t) - \mu_m(t)) = U'(\mu_m(t), \lambda). \end{aligned}$$

Then,

$$\begin{aligned} &\operatorname{cov}\left[R_i(t), \frac{E[U'(m_T, \lambda) \,|\mathscr{F}_{t+\Delta}]}{E[U'(m_T, \lambda) \,|\mathscr{F}_t]} \,|\mathscr{F}_t\right] \\ &\approx \operatorname{cov}\left[R_i(t), \frac{U'(\mu_m(t), \lambda) + U''(\mu_m(t), \lambda)(\mu_m(t + \Delta) - \mu_m(t))}{U'(\mu_m(t), \lambda)} \,|\mathscr{F}_t\right] \\ &= \frac{U''(\mu_m(t), \lambda)}{U'(\mu_m(t), \lambda)} \operatorname{cov}\left[R_i(t), (\mu_m(t + \Delta) - \mu_m(t)) \,|\mathscr{F}_t\right]. \end{aligned}$$

This completes the proof.

To understand this theorem we need to understand the various components on the right side of expression (15.4). First, the ratio $A_m(t) \equiv -\frac{U''(\mu_m(t),\lambda)}{U'(\mu_m(t),\lambda)}$ corresponds to the absolute risk aversion coefficient for the representative trader, i.e. it is a measure of the market's risk aversion (see Lemma 9.5 in the utility function Chap. 9). As the market becomes more risk averse, everything else constant, $A_m(t)$ increases. This implies that as the market's risk aversion increases, risk premiums increase.

Second, the difference $\Delta\mu_m(t) \equiv (\mu_m(t+\Delta) - \mu_m(t))$ measures the change in *expected* aggregate wealth, which is consumed at time T. Hence, this measures the change in anticipated "aggregate consumption" over $[t, t+\Delta]$. Any asset that is positively correlated with changes in aggregate consumption is risky because the additional return is less valuable in states where aggregate consumption is large relative to states where it is small.

Using this new notation, the above expression becomes

$$E\left[R_i(t)\,|\mathscr{F}_t\right] \approx r_0(t) + A_m(t)\mathrm{cov}\left[R_i(t), \Delta\mu_m(t)\,|\mathscr{F}_t\right]. \tag{15.5}$$

In this form, this expression is easily recognized as the CCAPM (see Back [5, p. 265]).

15.6 Intertemporal CAPM

This section derives the intertemporal capital asset pricing model (ICAPM). Merton's [137] ICAPM has implications with respect to the optimal nonnegative wealth s.f.t.s., mutual fund theorems, and the systematic risk return relation among the risky assets. This section focuses only on the systematic risk return relation among the risky assets. In this regard, the ICAPM characterizes the equilibrium relation between the expected return on any asset and the expected returns on the market portfolio and additional (hedging) risk factor returns. This is the relation derived below.

As before, we partition $[0, T]$ into a collection of sub-intervals of length $\Delta > 0$. Fix a time interval $[t, t+\Delta] \subset [0, T]$, where $t \geq 0$ aligns with one of these partitions. As in the multiple-factor model Chap. 4, we assume that the risky assets contain default-free zero-coupon bonds paying \$1 at times $t = \Delta, \ldots, T$.

Define $r_m(t) \equiv \frac{m_{t+\Delta} - m_t}{m_t}$ to be return on aggregate market wealth, which is the return on the *market portfolio*.

For the ICAPM, the only implication of an economic equilibrium used in the subsequent derivation is the fact that the existence of an equilibrium implies the existence of an equivalent martingale measure $\mathbb{Q} \in \mathfrak{M}$ (see Theorem 13.2 in the equilibrium Chap. 13). Then, using Theorem 4.2 in the multiple-factor model Chap. 4, we get

Theorem 15.4 (Multiple-Factor Beta Model)

$$\begin{aligned} R_i(t) - r_0(t) &= \beta_{im}(t)\left(r_m(t) - r_0(t)\right) \\ &+ \textstyle\sum_{j \in \Phi_i} \beta_{ij}(t)\left(r_j(t) - r_0(t)\right), \end{aligned} \tag{15.6}$$

where $\beta_{ij}(t) \neq 0$ *for all* $i,\ j \in \Phi_i$, $r_0(t) = \frac{1}{p(t,t+\Delta)} - 1$ *is the return on a default-free zero-coupon bond that matures at time* $t+\Delta$, $R_i(t) = \frac{\mathbb{S}_i(t+\Delta)-\mathbb{S}_i(t)}{\mathbb{S}_i(t)}$ *is the return on the* i^{th} *risky asset, and* $r_j(t)$ *is the return on the* j^{th} *risk factor.*

Proof This is Theorem 4.2, adding the market portfolio as a risk factor. This can be done without loss of generality, see Simmons [170, p. 197]. However, for the market portfolio, it may be the case that $\beta_{im}(t) = 0$. This completes the proof.

Taking time t conditional expectations gives the ICAPM.

Corollary 15.2 (ICAPM)

$$\begin{aligned} E[R_i(t)\,|\mathscr{F}_t] - r_0(t) &= \beta_{im}(t)\,(E[r_m(t)\,|\mathscr{F}_t] - r_0(t)) \\ &+ \textstyle\sum_{j\in\Phi_i}\beta_{ij}(t)\left(E[r_j(t)\,|\mathscr{F}_t] - r_0(t)\right). \end{aligned} \tag{15.7}$$

This is a key implication of Merton's [137] model (see Back [5, p. 267]). Merton's derivation was under a more restrictive Markov diffusion process assumption for the risky asset price processes, which implies continuous sample paths and no jumps. The previous derivation does not require the Markov diffusion process assumption.

15.7 Intermediate Consumption

This section discusses the changes in the previous results that occur with the inclusion of intermediate consumption into a trader's utility function as given in the portfolio optimization Chap. 12. Because all of the previous implications are based on a characterization of the representative trader's equilibrium supermartingale deflator, we just replace it with the representative trader's supermartingale deflator with intermediate consumption to obtain the analogous results (see Jarrow [92] for a complete presentation of the subsequent results). Recall that when including intermediate consumption into an investor's optimization problem, the optimal supermartingale deflator, when a martingale deflator, is given by

$$Y_t^i = \frac{E\left[\frac{d\mathbb{P}_i}{d\mathbb{P}}\middle|\mathscr{F}_t\right]U_i'(c_t,t)}{E\left[\frac{d\mathbb{P}_i}{d\mathbb{P}}U_i'(c_t,t)\right]}$$

(see the portfolio optimization Chap. 12, Sect. 12.4.1.3). The change of measure in the numerator is needed since the set of supermartingale deflators depends on the probability measure $\mathbb{P}$ (see Sect. 11.9 in the portfolio optimization Chap. 11).

15.7.1 Systematic Risk

In this section we use the non-normalized market $((\mathbb{B}, \mathbb{S}), (\mathscr{F}_t), \mathbb{P})$. Given is an equilibrium price process S. We assume that the representative trader's supermartingale deflator Y_T^λ generates a probability density with respect to $\mathbb{P}$. Then, the representative trader's supermartingale deflator process is a martingale density process with respect to $\mathbb{P}$, i.e.

$$Y_t^\lambda = \frac{U'(c_t, t, \lambda)}{E\left[U'(c_t, t, \lambda)\right]} \in \mathscr{M},$$

where c_t is the representative trader's optimal consumption, which in equilibrium equals aggregate consumption. Given the representative trader's martingale deflator, we obtain the next theorem using Theorem 4.4 in the multiple-factor model Chap. 4.

Partition $[0, T]$ into a collection of sub-intervals of length $\Delta > 0$. Fix a time interval $[t, t + \Delta] \subset [0, T]$, where $t \geq 0$ aligns with one of these partitions.

Theorem 15.5 (The Equilibrium Risk Return Relation)

$$E\left[R_i(t) \,|\mathscr{F}_t\right] = r_0(t) - \operatorname{cov}\left[R_i(t), \frac{U'(c_{t+\Delta}, t, \lambda)}{U'(c_t, t, \lambda)} \frac{\mathbb{B}_t}{\mathbb{B}_{t+\Delta}}(1 + r_0(t)) \middle| \mathscr{F}_t \right], \tag{15.8}$$

where $r_0(t) = \frac{1}{p(t,t+\Delta)} - 1$ is the return on a default-free zero-coupon bond that matures at time $t + \Delta$ and $R_i(t) = \frac{\mathbb{S}_i(t+\Delta) - \mathbb{S}_i(t)}{\mathbb{S}_i(t)}$ is the return on the ith risky asset.

We note that when the time step is small, i.e. $\Delta \approx dt$, then $\frac{\mathbb{B}_t}{\mathbb{B}_{t+\Delta}}(1 + r_0(t)) \approx 1$, and this relation simplifies to

$$E\left[R_i(t) \,|\mathscr{F}_t\right] \approx r_0(t) - \operatorname{cov}\left[R_i(t), \frac{U'(c_{t+\Delta}, t, \lambda)}{U'(c_t, t, \lambda)} \middle| \mathscr{F}_t \right].$$

15.7.2 Consumption CAPM

Using the previous theorem, we can derive the CCAPM. Partition $[0, T]$ into a collection of sub-intervals of length $\Delta > 0$. Fix a time interval $[t, t + \Delta] \subset [0, T]$ where $t \geq 0$ aligns with one of these partitions.

Theorem 15.6 (CCAPM) *Assume $U(c, t, \lambda)$ is three times continuously differentiable in c. Then,*

$$E\left[R_i(t) \,|\mathscr{F}_t\right] \approx r_0(t) - \frac{U''(c_t, t, \lambda)}{U'(c_t, t, \lambda)} \operatorname{cov}\left[R_i(t), (c_{t+\Delta} - c_t) \,|\mathscr{F}_t\right], \tag{15.9}$$

where $r_0(t) = \frac{1}{p(t,t+\Delta)} - 1$ *is the return on a default-free zero-coupon bond that matures at time* $t + \Delta$, $R_i(t) = \frac{\mathbb{S}_i(t+\Delta)-\mathbb{S}_i(t)}{\mathbb{S}_i(t)}$ *is the return on the* i^{th} *risky asset, and* c_t *is aggregate consumption.*

Proof Use a Taylor series expansion of U' around $c(t)$.

$$U'(c_{t+\Delta}, t, \lambda) = U'(c_t, t, \lambda) + U''(c_t, t, \lambda)(c_{t+\Delta} - c_t) + \frac{1}{2}U'''(\theta, t, \lambda)(c_{t+\Delta} - c_t)^2$$

for $\theta \in [c_t, c_{t+\Delta}]$.

Use the approximation

$$U'(c_{t+\Delta}, t, \lambda) \approx U'(c_t, t, \lambda) + U''(c_t, t, \lambda)(c_{t+\Delta} - c_t).$$

Then,

$$\begin{aligned}\mathrm{cov}\left[R_i(t), \frac{U'(c_{t+\Delta}, t, \lambda)}{U'(c_t, t, \lambda)} \,|\mathscr{F}_t\right] &\approx \mathrm{cov}\left[R_i(t), \frac{U'(c_t, t, \lambda) + U''(c_t, t, \lambda)(c_{t+\Delta} - c_t)}{U'(c_t, t, \lambda)} \,|\mathscr{F}_t\right] \\ &= \frac{U''(c_t, t, \lambda)}{U'(c_t, t, \lambda)} \mathrm{cov}\left[R_i(t), (c_{t+\Delta} - c_t) \,|\mathscr{F}_t\right].\end{aligned}$$

This completes the proof.

15.7.3 Intertemporal CAPM

As noted previously, the ICAPM as contained in Theorem 15.4 only depends on the existence of an equivalent martingale measure $\mathbb{Q} \in \mathfrak{M}$. And, the existence of an equivalent martingale measure follows directly from the existence of an equilibrium. Hence, the previous derivation of the ICAPM applies unchanged when intermediate consumption is included into the model's structure, as long as an equilibrium including intermediate consumption exists. This completes the discussion of the ICAPM.

15.8 Notes

Characterizing asset market equilibrium expected returns is the realm of traditional asset pricing theory. Hence, there are numerous excellent references on this topic, see Back [5], Bjork [14], Dana and Jeanblanc [42], Duffie [52], Follmer and Schied [63], Huang and Litzenberger [72], Ingersoll [74], Karatzas and Shreve [118], Merton [140], Pliska [150], and Skiadas [171].

Chapter 16
Market Informational Efficiency

Market informational efficiency is a key concept used in financial economics, introduced by Fama [57] in the early 1970s. To formalize this concept, we need the solution to a trader's portfolio optimization problem (as in Part II of this book) and the meaning of an economic equilibrium (as in the equilibrium Chap. 13 and the representative trader Chap. 14 above). Given these insights, a rigorous definition of an efficient market can be formulated. This rigorous definition is contrasted with the intuitive definition originally provided in Fama [57]. It will be shown that this rigorous definition of an efficient market requires only the existence, and not the characterization of an economic equilibrium. Such a rigorous definition allows new insights into the testing of an informationally efficient market, which will be discussed below. This chapter is based on Jarrow and Larsson [96].

16.1 The Set-Up

Given is a normalized market $(S, (\mathscr{F}_t), \mathbb{P})$ with $B_t \equiv 1$ and a collection of economies $\left\{((\mathscr{F}_t), \mathbb{P}), (N_0 = 0, N), \left(\mathbb{P}_i, U_i, \left(e_0^i, e^i\right)\right)_{i=1}^{\mathscr{I}}\right\}$, where the economies differ with respect to investors' beliefs, preferences, and endowments $\left(\mathbb{P}_i, U_i, \left(e_0^i, e^i\right)\right)_{i=1}^{\mathscr{I}}$. To simplify the presentation, this chapter only considers traders with utility functions over terminal wealth. For subsequent discussion, we emphasize that this collection of economies is characterized by the traders having symmetric information represented by the filtration $(\mathscr{F}_t)$. For some economies, this structure is also consistent with traders having differential information, see Detemple and Murthy [50] for a model that includes differential information via differential beliefs. We will discuss the relaxation of this assumption further after introducing the definition of an efficient market.

R. A. Jarrow, *Continuous-Time Asset Pricing Theory*, Springer Finance,
https://doi.org/10.1007/978-3-319-77821-1_16

16.2 The Definition

This section presents the definition of an informationally efficient market. The original intuitive definition of market efficiency is given by Fama [57, p. 383] in his seminal paper:

> A market in which prices always 'fully reflect' available information is called 'efficient'.

Three information sets have been considered when discussing efficient markets: (i) historical prices (weak form efficiency), (ii) publicly available information (semi-strong efficiency), and (iii) private information (strong form efficiency). A market may or may not be efficient with respect to each of these information sets.

In quantifying this definition, for its use in testing market efficiency, it is commonly believed that one must first specify an equilibrium model (see Fama [59]). This is called the *joint-hypothesis* or the *bad-model* problem. Indeed, Fama states [58, p. 1575],

> The joint-hypothesis problem is more serious. Thus, market efficiency per se is not testable. It must be tested jointly with some model of equilibrium, an asset pricing model. This point, the theme of the 1970 review (Fama (1970)), says that we can only test whether information is properly reflected in prices in the context of a pricing model that defines the meaning of 'properly'.

In contrast, we quantify the original definition in such a manner that one can test market efficiency without specifying *a particular* equilibrium model. As such, our formulation overcomes the bad-model problem in the existing tests.

Definition 16.1 (Market Efficiency) *The market* $(S, (\mathscr{F}_t), \mathbb{P})$ *is efficient* with respect to the information set $(\mathscr{F}_t)$ if there exists an economy

$$\left\{((\mathscr{F}_t), \mathbb{P}), (N_0 = 0, N), \left(\mathbb{P}_i, U_i, \left(e_0^i, e^i\right)\right)_{i=1}^{\mathscr{I}}\right\}$$

for which S is an equilibrium price process.

It is important to note that this definition does not require the identification of the equilibrium economy generating the price process, only that one such economy exists. That is, we do not have to specify beliefs, preferences, or endowments across all traders, i.e. $\left(\mathbb{P}_i, U_i, \left(e_0^i, e^i\right)\right)_{i=1}^{\mathscr{I}}$. This definition is easily seen to be consistent with Fama's [57] original definition, and the manner in which market efficiency has been tested over the subsequent years using the joint model hypothesis (see Fama [58, 59]). Indeed, as mentioned before, the standard approach to testing efficiency is to assume a particular equilibrium model, and then show the model is consistent or inconsistent with historical observations of the price process and different information sets. If the model is consistent with the data, then informational efficiency is accepted. If inconsistent, then informational efficiency is not necessarily rejected because of the joint hypothesis. To reject efficiency, one must show that there is no equilibrium model consistent with the data; hence, the definition.

Remark 16.1 (Differential Information) The collection of economies in the definition of an efficient market is characterized by the traders having symmetric information $(\mathscr{F}_t)$. This definition can be extended to a collection of economies where traders have different information sets $(\mathscr{F}_t^i)$, where $\mathscr{F}_0^i = \{\emptyset, \Omega\}$ and $\mathscr{F}_T^i = \mathscr{F}$, called their *private information*. Recall that the definitions of NA, NUPBR, NFLVR, and ND are all invariant with respect to a change in equivalent probability measures (beliefs $\mathbb{P}_i$). However, these definitions and the definition of a complete market are not invariant with respect to a change in the filtration $\left(\mathscr{F}_t^i\right)$. Consequently, care must be taken in this extension.

With differential information an economy is defined as a collection

$$\left\{(\mathscr{F}, \mathbb{P}), (N_0 = 0, N), \left(\left(\mathscr{F}_t^i\right), \mathbb{P}_i, U_i, \left(e_0^i, e^i\right)\right)_{i=1}^{\mathscr{I}}\right\}.$$

In this extension, we require that the information set is generated by the price process, denoted $\sigma(S) \subset \left(\mathscr{F}_t^i\right)$ for all i. There are three information sets that are relevant in the economy. The first is the information set generated by the price process $\sigma(S)$, which was defined above. The second is *publicly available information* represented by the filtration $\mathscr{F}_t = \bigwedge_i \mathscr{F}_t^i$, which is the largest filtration contained in all of the individual trader's private information sets. This common information is known by all the traders. The third is *all private information* represented by the filtration $\bigvee_i \mathscr{F}_t^i$, which is the smallest information set containing all of the individual trader's private information sets.

The definition of an efficient market is modified as follows.

Definition (Market Efficiency) The market $(S, (\mathscr{F}_t), \mathbb{P})$ is *efficient* with respect to the information set $(\mathscr{F}_t)$ if there exists a symmetric information economy

$$\left\{(\mathscr{F}, \mathbb{P}), (N_0 = 0, N), \left((\mathscr{F}_t), \mathbb{P}_i, U_i, \left(e_0^i, e^i\right)\right)_{i=1}^{\mathscr{I}}\right\}$$

for which S is an equilibrium price process.

When defining an efficient market, the choice of the information set $(\mathscr{F}_t)$ can vary. The market is said to be *weak form efficient* if $\mathscr{F}_t = \sigma(S)$, it is *semi-strong form efficient* if $\mathscr{F}_t = \bigwedge_i \mathscr{F}_t^i$, and it is *strong form efficient* if $\mathscr{F}_t = \bigvee_i \mathscr{F}_t^i$. Given these identifications and qualifications, the subsequent results apply unchanged. This completes the remark.

Remark 16.2 (Rational Expectations Equilibrium (REE)) In the differential information extension of the previous remark, an equilibrium is related to the notion of a rational expectations equilibrium. Recall that in our definition of an equilibrium, each trader conjectures a price process S when choosing their optimal trading strategies. An equilibrium price process is one where this conjectured price process is the same for all traders, and it is consistent with trader optimality and market clearing. The equilibrium price process generates an information set, denoted $\sigma(S)$. We can now define two types of *rational expectations equilibrium* (REE).

A *fully revealing REE* is defined to be an equilibrium where the equilibrium price process S is such that $\sigma(S) = \bigvee_i \mathscr{F}_t^i$, the smallest information set containing all of the individual trader's private information sets. The information set $\bigvee_i \mathscr{F}_t^i$ represents all private information. Since by construction of the economy $\sigma(S) \subset (\mathscr{F}_t^i)$ for all i, this implies that all of this private information is (fully) revealed by the equilibrium price process. A *partially revealing REE* is defined to be an equilibrium where the equilibrium price process S is not fully revealing. This completes the remark.

16.3 The Theorem

This section provides a characterization of market efficiency.

Theorem 16.1 (Market Efficiency) *Let $(S, (\mathscr{F}_t), \mathbb{P})$ be a market. Then, $(S, (\mathscr{F}_t), \mathbb{P})$ is efficient with respect to $(\mathscr{F}_t)$ if and only if there exists a $\mathbb{Q} \in \mathfrak{M}$, where $\mathfrak{M} = \{\mathbb{Q} \sim \mathbb{P} : S \text{ is a } \mathbb{Q}\text{-martingale}\}$.*

Proof

($\Rightarrow$) Suppose the market is efficient. Then, there exists an equilibrium supporting (S). By Theorem 13.2 in Chap. 13 on equilibrium, NFLVR and ND hold. The Third Fundamental Theorem 2.5 of asset pricing in Chap. 2 gives the result.

($\Leftarrow$) Suppose there exists a $\mathbb{Q} \in \mathfrak{M}$. To prove the existence of an equilibrium supporting S, we need to construct an economy for which S is an equilibrium price process. The following construction is based on that in Jarrow and Larsson [97].

Consider the economy

$$\left\{(\mathscr{F}, \mathbb{P}), (N_0 = 0, N), \left((\mathscr{F}_t), \mathbb{P}_i, U_i, \left(e_0^i, e^i\right)\right)_{i=1}^{\mathscr{I}}\right\}$$

where the trader's beliefs $\mathbb{P}_i = \mathbb{P}^*$ for all i and $\frac{d\mathbb{Q}}{d\mathbb{P}^*} = \frac{Z_T^{1-\rho}}{E^{\mathbb{Q}}\left[Z_T^{1-\rho}\right]}$ with $Z_T = \frac{N \cdot S_T}{N \cdot S_0} > 0$. Note that $Z_t = \frac{N \cdot S_t}{N \cdot S_0} > 0$ is a strictly positive $\mathbb{Q}$-martingale with $E^{\mathbb{Q}}[Z_T] = 1$, where $N \cdot S_t$ is the aggregate market wealth at time t. Let the traders' preferences be $U_i(x) = \frac{x^\rho}{\rho}$ for all i, where $0 < \rho < 1$ and $x > 0$ (see the utility function Chap. 9). And, let the traders' initial endowments be $(e_0^i(0) = 0, e_0^i = \frac{N}{\mathscr{I}})$ for all i, i.e. each trader is endowed with the same number of shares in each risky asset.

We claim that each trader's optimal nonnegative wealth s.f.t.s. is to not trade. If this is the case, then S_t is an equilibrium price process because at any time $t \in [0, T]$, letting $(\hat{\alpha}_0^i(t) = 0, \hat{\alpha}_t^i = \frac{N}{\mathscr{I}})$ denote trader i's optimal s.f.t.s., summing across i gives the market clearing conditions

$$\sum_{i=1}^{\mathscr{I}} \hat{\alpha}_0^i(t) = 0, \qquad \sum_{i=1}^{\mathscr{I}} \hat{\alpha}_t^i = N.$$

To prove the claim, first observe that the time T expected utility of this trading strategy is

$$E^{\mathbb{P}^*}\left[U_i\left(\frac{N}{\mathscr{I}} \cdot S_T\right)\right] = E^{\mathbb{P}^*}\left[U_i(zZ_T)\right],$$

where $z = \frac{N \cdot S_0}{\mathscr{I}}$.

Next, let X_T be the terminal wealth of an arbitrary s.f.t.s. $(\alpha_0, \alpha) \in \mathscr{N}(z)$ with initial wealth z, where

$$X_T = \alpha_0(T) + \alpha_T \cdot S_T.$$

Then,

$$E^{\mathbb{P}^*}\left[U_i(X_T)\right] - E^{\mathbb{P}^*}\left[U_i(zZ_T)\right] \leq E^{\mathbb{P}^*}\left[U_i'(zZ_T)\left(X_T - zZ_T\right)\right]$$

by the concavity of U_i.

But, $U_i'(zZ_T) = (zZ_T)^{\rho-1}$. Rewriting,

$$\begin{aligned} E^{\mathbb{P}^*}\left[U_i'(zZ_T)\left(X_T - zZ_T\right)\right] &= z^{\rho-1} E^{\mathbb{P}^*}\left[{Z_T}^{\rho-1}\left(X_T - zZ_T\right)\right] \\ &= z^{\rho-1} E^{\mathbb{Q}}\left[Z_T^{1-\rho}\right] E^{\mathbb{P}^*}\left[\frac{d\mathbb{Q}}{d\mathbb{P}^*}\left(X_T - zZ_T\right)\right] \\ &= z^{\rho-1} E^{\mathbb{Q}}\left[Z_T^{1-\rho}\right] E^{\mathbb{Q}}\left[X_T - zZ_T\right]. \end{aligned}$$

But, $E^{\mathbb{Q}}[X_T] \leq z$ since X_T is a $\mathbb{Q}$-supermartingale (a nonnegative $\mathbb{Q}$-local martingale) with initial value $X_0 = z$.

And, $E^{\mathbb{Q}}[zZ_T] = z$ since $E^{\mathbb{Q}}[Z_T] = 1$. Substitution gives $E^{\mathbb{P}^*}[U_i(X_T)] - E^{\mathbb{P}^*}[U_i(zZ_T)] \leq 0$, which proves the claim and completes the proof.

Corollary 16.1 (Market Efficiency) *Let $(S, (\mathscr{F}_t), \mathbb{P})$ be a market. Then, $(S, (\mathscr{F}_t), \mathbb{P})$ is efficient with respect to $(\mathscr{F}_t)$ if and only if NFLVR and ND hold.*

This corollary is a direct application of the Third Fundamental Theorem 2.5 of asset pricing in Chap. 2. It shows that a market is efficient if and only if there are no "mispriced" assets trading in the economy, where mispriced is interpreted as the existence of an FLVR or dominated assets.

16.4 Information Sets and Efficiency

A market $(S, (\mathscr{F}_t), \mathbb{P})$ is defined to be efficient or not with respect to the market information set $(\mathscr{F}_t)$ used by the traders in the economy to form their trading strategies. This corresponds to the "publicly available" information set given earlier (semi-strong form efficiency). Suppose that the market is efficient with respect to $(\mathscr{F}_t)$. This section explores whether the market is also efficient with respect to smaller information sets $(\mathscr{G}_t) \subset (\mathscr{F}_t)$ and larger information sets $(\mathscr{H}_t)$ with $(\mathscr{F}_t) \subset (\mathscr{H}_t)$. For this comparison, we fix the price process S and we consider the new markets $(S, (\mathscr{G}_t), \mathbb{P})$ and $(S, (\mathscr{H}_t), \mathbb{P})$. It is easy to prove that if the market is efficient with respect to $(\mathscr{F}_t)$, then it is also efficient with respect to any smaller information set $(\mathscr{G}_t) \subset (\mathscr{F}_t)$.

Corollary 16.2 (Smaller Information) Let $(S, (\mathscr{F}_t), \mathbb{P})$ *be a market that is efficient with respect to* $(\mathscr{F}_t)$.

Consider $(\mathscr{G}_t) \subset (\mathscr{F}_t)$. *Then,* $(S, (\mathscr{G}_t), \mathbb{P})$ *is efficient with respect to* $(\mathscr{G}_t)$.

Proof Since $(\mathscr{G}_t) \subset (\mathscr{F}_t)$, all nonnegative wealth s.f.t.s.'s using only $(\mathscr{G}_t)$ are also $\mathscr{F}_t$-measurable, hence NFLVR and ND hold for $(\mathscr{G}_t)$ as well. This implies the result by Corollary 16.1. This completes the proof.

The converse does not hold. The market need not be efficient with respect to larger information sets. To see this suppose the larger information set $(\mathscr{H}_t)$ where $(\mathscr{F}_t) \subset (\mathscr{H}_t)$ includes knowledge of a risky asset's price at time T. Suppose that the risky asset's price is known to increase with probability one from its time 0 value. Then, given the price process (S), buying and holding this risky asset and shorting the money market account to finance this purchase is an FLVR. NFLVR does not hold, hence the market cannot be in an equilibrium and it cannot be efficient with respect to this $(\mathscr{H}_t)$. With this observation, the next corollary is immediate.

Corollary 16.3 (Larger Information) *Let* $(S, (\mathscr{F}_t), \mathbb{P})$ *be a market that is efficient with respect to* $(\mathscr{F}_t)$.

Consider $(\mathscr{H}_t)$ *where* $(\mathscr{F}_t) \subset (\mathscr{H}_t)$. *Then,*

$(S, (\mathscr{H}_t), \mathbb{P})$ *is efficient with respect to* $(\mathscr{H}_t)$

if and only if

all self-financing and nonnegative wealth trading strategies based on $(\mathscr{H}_t)$ *satisfy NFLVR and ND.*

16.5 Testing for Market Efficiency

This section discusses three methods that can be used to test for an efficient market. The first two approaches are standard in the literature. The third approach is new and based on Theorem 16.1 given above.

16.5.1 Profitable Trading Strategies

As given in Corollary 16.1, a market is efficient with respect to $(\mathscr{F}_t)$ if and only if NFLVR and ND hold. Hence, one can reject efficiency by finding self-financing and admissible trading strategies based on $(\mathscr{F}_t)$ that violate NFLVR or ND. This approach has been used in the empirical asset pricing literature to reject market efficiency (see Jensen [114] and references therein).

Although this approach can be used to reject efficiency, it is less useful for proving that the market is efficient. To do so, one must prove that all self-financing and admissible trading strategies based on $(\mathscr{F}_t)$ have been exhausted. To do this, the Third Fundamental Theorem 2.5 of asset pricing in Chap. 2 can be invoked, see the third method in Sect. 16.5.3 below. We note that this proves that an efficient market with respect to $(\mathscr{F}_t)$ can be rejected, thereby side-stepping the "curse of a joint hypothesis" noted earlier.

16.5.2 Positive Alphas

An application of Theorem 16.1 provides another method for rejecting market efficiency with respect to $(\mathscr{F}_t)$. To do this, we want to consider the returns on the risky assets. Partition $[0, T]$ into a collection of sub-intervals of length $\Delta > 0$. Fix a time interval $[t, t+\Delta] \subset [0, T]$ where $t \geq 0$ aligns with one of these partitions.

To simplify the methodology we assume that the risky assets contain default-free zero-coupon bonds paying 1 dollar at times $t = \Delta, \ldots, T$. Applying Theorem 4.3 in the multiple-factor model Chap. 4, which is based on only assuming the existence of an equivalent martingale measure $\mathbb{Q} \in \mathfrak{M}$, we have the following corollary.

Corollary 16.4 (Test for Efficiency) *The market is inefficient with respect to* $(\mathscr{F}_t)$ *if* $\alpha_i(t) \neq 0$ *for some* i,
where

$$R_i(t) - r_0(t) = \alpha_i(t) + \sum_{j \in \Phi_i} \beta_{ij}(t) \left(r_j(t) - r_0(t)\right) + \varepsilon_i(t), \tag{16.1}$$

$\beta_{ij}(t) \neq 0$ *for all* (i, j), $r_0(t) = \frac{1}{p(t,t+\Delta)} - 1$ *is the return on a default-free zero-coupon bond that matures at time* $t + \Delta$, $R_i(t) = \frac{\mathbb{S}_i(t+\Delta)-\mathbb{S}_i(t)}{\mathbb{S}_i(t)}$ *is the return on the* i^{th} *risky asset,* $r_j(t)$ *is the return on the* j^{th} *risk factor, and* $E[\varepsilon_i(t) \,|\, \mathscr{F}_t] = 0$.

This is the standard approach used in the empirical asset pricing literature for rejecting market efficiency via estimating a multiple-factor model and rejecting market efficiency if an alpha is nonzero (see Fama [58, 59]). Unlike the models used in the existing empirical literature which assume a particular equilibrium economy, this corollary is derived only assuming the existence of a martingale measure, equivalently, an efficient market with respect to $(\mathscr{F}_t)$. Its derivation does not require the specification of a particular economy. Hence, this approach to testing efficiency also side-steps the "curse of a joint hypothesis" noted earlier.

16.5.3 Asset Price Evolutions

Given a market $(S, (\mathscr{F}_t), \mathbb{P})$, to accept market efficiency with respect to $(\mathscr{F}_t)$, by Theorem 16.1, one must show that there exists an equivalent martingale measure $\mathbb{Q} \in \mathfrak{M}$. To do this, a two step procedure can be used. Step one in this procedure is to empirically validate a risky asset price process S using historical time series data. Such procedures are commonly invoked when pricing derivatives as in Part I of this book. Step two in this procedure is to determine if there exists an equivalent martingale measure for the validated price process S with respect to the filtration $(\mathscr{F}_t)$. We note that this is identical to the issue studied in the asset price bubbles Chap. 3 when testing for asset price bubbles.

For example, if one shows that the validated price process S can be represented by a geometric Brownian motion as in the BSM model, then as shown in the BSM Chap. 5, there exists a $\mathbb{Q} \in \mathfrak{M}$. Hence, the market is efficient with respect to $(\mathscr{F}_t) = \sigma(S)$. Other examples explored in the asset price bubbles Chap. 3 include the simple diffusion process in Example 3.1, the CEV process in Example 3.2, and Levy processes as in Theorem 3.4. As of yet, this method for testing and accepting market efficiency is unexplored in the empirical asset pricing literature.

16.6 Random Walks and Efficiency

There is some confusion in the literature regarding the relation between market efficiency and risky asset prices following a random walk. This section clarifies this relation. The fact is that there is no relation, i.e. (i) an efficient market does not imply that stock price returns follow a random walk, and (ii) stock price returns following a random walk does not imply the market is efficient.

16.6.1 The Set-Up

Consider the returns on the risky assets. Partition $[0, T]$ into a collection of sub-intervals of length $\Delta > 0$. Fix a time interval $[t, t + \Delta] \subset [0, T]$ where $t \geq 0$ aligns with one of these partitions.

To simplify the methodology we assume that the risky assets contain default-free zero-coupon bonds paying \$1 at times $t = \Delta, \ldots, T$.

Let $r_0(t) = \frac{1}{p(t,t+\Delta)} - 1$ be the return on a default-free zero-coupon bond that matures at time $t + \Delta$.

Let $R_i(t) = \frac{\mathbb{S}_i(t+\Delta) - \mathbb{S}_i(t)}{\mathbb{S}_i(t)}$ be the return on the ith risky asset.

16.6.2 Random Walk

The risky asset prices, or more formally the risky asset price returns are said to follow a *random walk* if $R_i(t)$ are independent of $\mathscr{F}_t$ and identically distributed random variables for all $t \in [0, \Delta, 2\Delta, \ldots, T - \Delta]$ and all i. For empirical testing this implies that an asset's returns have zero autocorrelation, i.e.

$$\operatorname{cov}[R_i(t+\Delta), R_i(t)\,|\mathscr{F}_t] = 0 \tag{16.2}$$

for all $t \in [0, \Delta, 2\Delta, \ldots, T - \Delta]$.

Proof First, by the independence of $\mathscr{F}_t$,

$$\begin{aligned}\operatorname{cov}[R_i(t+\Delta), R_i(t)\,|\mathscr{F}_t] &= \operatorname{cov}[R_i(t+\Delta), R_i(t)] \\ &= E\,[R_i(t+\Delta)R_i(t)] - E\,[R_i(t+\Delta)]\,E\,[R_i(t)]\,.\end{aligned}$$

Since $R_i(t)$ is $\mathscr{F}_{t+\Delta}$-measurable, $R_i(t+\Delta)$ is independent of $R_i(t)$, i.e. $\mathbb{P}\,(R_i(t+\Delta)|\,R_i(t)) = \mathbb{P}\,(R_i(t+\Delta))$.
This implies $E\,[R_i(t+\Delta)R_i(t)] = E\,[R_i(t+\Delta)]\,E\,[R_i(t)]$. Substitution gives $\operatorname{cov}[R_i(t+\Delta), R_i(t)\,|\mathscr{F}_t] = 0$, which completes the proof.

Remark 16.3 (Probability Theory and Random Walks) In probability theory, a random walk is a sequence of random variables $Y_1, Y_2, \ldots, Y_n, \ldots$ with $Y_n = \sum_{t=0}^{n} X_t$ such that X_0 is a constant and $X_1, X_2, \ldots$ are independent and identically distributed random variables with $E\,|X_t| < \infty$ for all t, see Ross [159, p. 165].

To see the connection to risky asset prices, make the identification

$$X_t = \log\left(\frac{S_i(t+\Delta)}{S_i(t)}\right) = \log\left(1 + R_i(t)\right).$$

Then, given $R_i(t)$ is independent of $\mathscr{F}_t$, $R_i(t)$ being $\mathscr{F}_{t+\Delta}$-measurable implies that $R_i(t+\Delta)$ is independent of $R_i(t)$, i.e. $X_1, X_2, \ldots$ are independent. And, $R_i(t)$ being identically distributed implies $X_1, X_2, \ldots$ are identically distributed. Hence, $Y_n = \sum_{t=0}^{n} X_t = \sum_{t=0}^{n} \log\left(\frac{S_i(t+\Delta)}{S_i(t)}\right) = \log\left(\frac{S_i(n+\Delta)}{S_i(0)}\right)$, the exponent of $S_i(n + \Delta) = S_i(0)e^{Y_n}$, follows a random walk. This completes the remark.

16.6.3 Market Efficiency $\nRightarrow$ Random Walk

To see what an efficient market with respect to $(\mathscr{F}_t)$ implies, we need the following math identities.

Lemma 16.1 (Excess Expected Return Math Identities) *Define the unexpected excess returns for risky asset i by*

$$\varepsilon_i(t) \equiv R_i(t) - E[R_i(t) \,|\mathscr{F}_t\,].$$

Then,

(1) $E\,[\varepsilon_i(t)\,|\mathscr{F}_t\,] = 0$, $\mathrm{cov}\,[\varepsilon_i(t+\Delta), \varepsilon_i(t)\,|\mathscr{F}_t\,] = 0$ *for all t, and*
(2) $\mathrm{cov}\,[R_i(t+\Delta), R_i(t)\,|\mathscr{F}_t\,] = 0$ *if and only if*

$$\mathrm{cov}\,[E\,[R_i(t+\Delta)\,|\mathscr{F}_{t+\Delta}\,], \varepsilon_i(t)\,|\mathscr{F}_t\,] = 0.$$

Proof (Step 1)

$$\begin{aligned} E\,[\varepsilon_i(t)\,|\mathscr{F}_t\,] &= E\,[R_i(t) - E[R_i(t)\,|\mathscr{F}_t\,]\,|\mathscr{F}_t\,] \\ &= E\,[R_i(t)\,|\mathscr{F}_t\,] - E[R_i(t)\,|\mathscr{F}_t\,] = 0. \end{aligned}$$

And,

$$\begin{aligned} \mathrm{cov}\,[\varepsilon_i(t+\Delta), \varepsilon_i(t)\,|\mathscr{F}_t\,] &= E\,[\varepsilon_i(t+\Delta)\varepsilon(t)\,|\mathscr{F}_t\,] \\ &= E\left[E\left[\varepsilon_i(t+\Delta)\varepsilon_i(t)\,\middle|\mathscr{F}_{t+\Delta}\right]|\mathscr{F}_t\right] \\ &= E\left[\varepsilon_i(t)E\left[\varepsilon_i(t+\Delta)\,\middle|\mathscr{F}_{t+\Delta}\right]|\mathscr{F}_t\right] = 0 \end{aligned}$$

since

$$E\left[\varepsilon_i(t+\Delta)\,\middle|\mathscr{F}_{t+\Delta}\right] = 0.$$

(Step 2)

$$\begin{aligned} \mathrm{cov}\,[R_i(t+\Delta), R_i(t)\,|\mathscr{F}_t\,] &= \mathrm{cov}\,[E\,[R_i(t+\Delta)\,|\mathscr{F}_{t+\Delta}\,] + \varepsilon_i(t+\Delta), E\,[R_i(t)\,|\mathscr{F}_t\,] + \varepsilon_i(t)\,|\mathscr{F}_t\,] \\ &= \mathrm{cov}\,[E\,[R_i(t+\Delta)\,|\mathscr{F}_{t+\Delta}\,] + \varepsilon_i(t+\Delta), \varepsilon_i(t)\,|\mathscr{F}_t\,] \\ &= \mathrm{cov}\,[E\,[R_i(t+\Delta)\,|\mathscr{F}_{t+\Delta}\,], \varepsilon_i(t)\,|\mathscr{F}_t\,] + \mathrm{cov}\,[\varepsilon_i(t+\Delta), \varepsilon_i(t)\,|\mathscr{F}_t\,] \\ &= \mathrm{cov}\,[E\,[R_i(t+\Delta)\,|\mathscr{F}_{t+\Delta}\,], \varepsilon_i(t)\,|\mathscr{F}_t\,]. \end{aligned}$$

This completes the proof.

This lemma shows that the unexpected excess returns to any risky asset always have zero mean and are uncorrelated across time. These facts have nothing to do with market efficiency. Furthermore, the returns themselves have zero autocorrelation if and only if the correlation between $E\,[R_i(t+\Delta)\,|\mathscr{F}_{t+\Delta}\,]$ and $\varepsilon_i(t)$ is zero for all t.

Given this lemma, we can now investigate the implications of an efficient market. First, by Theorem 16.1, the market being efficient with respect to $(\mathscr{F}_t)$ implies the existence of an equivalent martingale measure $\mathbb{Q} \in \mathfrak{M}$. Then, using Theorem 4.2 in the multiple-factor model Chap. 4, this implies that any risky asset's return satisfies

$$R_i(t) - r_0(t) = \sum_{j \in \Phi_i} \beta_{ij}(t) \left(r_j(t) - r_0(t) \right) \tag{16.3}$$

with $\beta_{ij}(t) \neq 0$ all $i, \ j \in \Phi_i$, where $r_0(t) = \frac{1}{p(t,t+\Delta)} - 1$ is the return on a default-free zero-coupon bond that matures at time $t + \Delta$ and $r_j(t)$ is the return on the jth risk factor.

It is important to note that this expression does not imply that $R_i(t)$ are independent of $\mathscr{F}_t$ for all t. It does not imply that $R_i(t)$ are identically distributed random variables for all t. Furthermore, it does not imply that the correlation between $E\left[R_i(t+\Delta)\,|\mathscr{F}_{t+\Delta}\right]$ and $\varepsilon_i(t)$ is zero for all t. Most likely, quite the contrary is true. It would be quite natural for a large excess return at time t to imply that time $t + \Delta$ expected returns, $E\left[R_i(t+\Delta)\,|\mathscr{F}_{t+\Delta}\right]$, increase. Hence, market efficiency with respect to $(\mathscr{F}_t)$ does not imply risky asset prices follow a random walk.

16.6.4 Random Walk $\nRightarrow$ Market Efficiency

Conversely, if asset returns follow a random walk under the probability measure $\mathbb{P}$, this does not imply that the market is efficient with respect to $(\mathscr{F}_t)$. The reason is that to be efficient, by Theorem 16.1, there must exist a martingale measure $\mathbb{Q}$ for the assets and this may not be the case, as the next example shows.

Example 16.1 (Random Walk in an Inefficient Market) This example gives a market where the risky asset returns follow a random walk, but the market is inefficient. Consider a market $(\mathbb{S}, (\mathscr{F}_t), \mathbb{P})$ consisting of two risky assets whose price processes follow geometric Brownian motion, i.e.

$$\begin{aligned} d\mathbb{S}_i(t) &= \mathbb{S}_i(t)\left(\mu_i dt + \sigma dW_t\right) \qquad &\text{or} \\ \mathbb{S}_i(t) &= e^{\mu_i t - \frac{1}{2}\sigma^2 t + \sigma W_t} \qquad &\text{for } i = 1, 2, \end{aligned} \tag{16.4}$$

where μ_1, μ_2, σ are strictly positive constants, $\mu_1 > \mu_2$ and W_t is a standard Brownian motion under $\mathbb{P}$ with $W_0 = 0$.

First, we show that these risky assets follow a random walk. Indeed,

$$\begin{aligned} R_i(t) = \frac{\mathbb{S}_i(t+\Delta) - \mathbb{S}_i(t)}{\mathbb{S}_i(t)} &= \frac{e^{\mu_i(t+\Delta) - \frac{1}{2}\sigma^2(t+\Delta) + \sigma W_{t+\Delta}}}{e^{\mu_i t - \frac{1}{2}\sigma^2 t + \sigma W_t}} - 1 \\ &= e^{\mu_i \Delta - \frac{1}{2}\sigma^2 \Delta + \sigma(W_{t+\Delta} - W_t)} - 1. \end{aligned} \tag{16.5}$$

Note that $R_i(t)$ are independent and identically distributed across t since the Brownian motion increments $(W_{t+\Delta} - W_t)$ are independent and identically distributed.

Second, this market is inefficient with respect to $(\mathscr{F}_t)$ because there exists an NFLVR (apply Theorem 16.1 above). Indeed, to see this consider the zero investment, buy and hold (self-financing) trading strategy consisting of 1 unit long $\mathbb{S}_1(t)$ and 1 unit short $\mathbb{S}_2(t)$, i.e. $(\alpha_0(t) = 0, \alpha_1(t) = 1, \alpha_2(t) = -1)$ for all $t \in [0, T]$.

The value process for this s.f.t.s. is

$$\mathbb{X}_t = \alpha_1(t)\mathbb{S}_1(t) + \alpha_2(t)\mathbb{S}_2(t) = \mathbb{S}_1(t) - \mathbb{S}_2(t).$$

Note that

$$\mathbb{X}_0 = 0 \text{ and } \mathbb{X}_T = \left(e^{\mu_1 t} - e^{\mu_2 t}\right) e^{-\frac{1}{2}\sigma^2 t + \sigma W_t} > 0$$

a.s. $\mathbb{P}$, which is a simple arbitrage opportunity (see the fundamental theorems Chap. 2, Definition 2.3).

This completes the example.

16.7 Notes

The definition of market efficiency as contained in this chapter is new to the literature. The use of this formal definition to explore alternative tests of market efficiency is a fruitful area for future research. The extension of Corollary 16.1 to an economy with transaction costs or trading constraints is an interesting area for future research. Jarrow and Larsson [97] extend the previous insights to markets with short sale constraints.

Chapter 17
Epilogue (The Static CAPM)

This chapter studies the static CAPM for two reasons. First, because it is of historical interest. Second, because it highlights the advances and insights obtained from the dynamic models studied in this book. This chapter provides a new derivation of the static CAPM that uses the martingale approach.

17.1 The Fundamental Theorems

This is a discrete-time model with two times $t \in \{0, 1\}$. Trading takes place at time 0 and all trades are liquidated at time 1. We are given a complete probability space $(\Omega, \mathscr{F}, \mathbb{P})$. Given is a normalized market $(S, \mathscr{F}, \mathbb{P})$ with $B_t \equiv 1$.

To simplify the notation, we let the risky asset prices be denoted by $S_j(0) \equiv s_j > 0$ and $S_j(1) = S_j \geq 0$ for $j = 1, \ldots, n$. Returns over $[0, 1]$ are defined by $R_j = \frac{S_j}{s_j} - 1$ for $j = 1, \ldots, n$. Without loss of generality, we assume that there are no cash flows to the risky assets at time 0. As before, one can interpret the time 1 asset price as a liquidating dividend.

We will also consider the non-normalized market below, in which case the notation is $(s_j, \mathbb{S}_j)$ for the risky assets $j = 1, \ldots, n$ and $(1, \mathbb{B}) = (1, 1 + r)$ for the money market account, where r is the default-free spot rate of interest. Note that since the money market account's value is unity at time 0, the normalized and non-normalized risky asset prices are identical at time 0. In the non-normalized market, the returns on the risky assets are defined by $\tilde{R}_j = \frac{\mathbb{S}_j}{s_j} - 1$ for $j = 1, \ldots, n$. Algebra yields the following result, which will prove useful below.

$$R_j = \frac{\tilde{R}_j - r}{1 + r} \qquad \text{for } j = 1, \ldots, n. \tag{17.1}$$

R. A. Jarrow, *Continuous-Time Asset Pricing Theory*, Springer Finance,
https://doi.org/10.1007/978-3-319-77821-1_17

Proof $\frac{\tilde{R}_j - r}{1+r} = \frac{\frac{\mathbb{S}_j}{s_j} - 1 - r}{1+r} = \frac{\frac{S_j(1+r)}{s_j} - (1+r)}{1+r} = R_j$. The second equality follows since $\mathbb{S}_j = S_j(1+r)$. This completes the proof.

We add the following assumption on the risky assets.

Assumption (Non-redundant Assets) $E[S_j^2] < \infty$, $\text{var}\left[S_j\right] > 0$ *for all* $j = 1, \ldots, n$, *and*

$$\begin{pmatrix} \text{cov}[S_1, S_1] & \text{cov}[S_1, S_2] & \cdots & \text{cov}[S_1, S_n] \\ \text{cov}[S_2, S_1] & \text{cov}[S_2, S_2] & \cdots & \text{cov}[S_2, S_n] \\ \vdots & \vdots & \ddots & \vdots \\ \text{cov}[S_n, S_1] & \text{cov}[S_n, S_2] & \cdots & \text{cov}[S_n, S_n] \end{pmatrix}_{n \times n}$$

is nonsingular.

In this assumption, $E[S_j^2] < \infty$ implies that $E[S_j] < \infty$ for all j (see Ash [3, p. 226]) and by the Cauchy–Schwarz inequality (see Ash [3, p. 82]) that $E[S_i S_j] < \infty$ for all i, j. Hence, $\text{cov}[S_i, S_j] = E[S_i S_j] - E[S_i]E[S_j] < \infty$ as well. The assumption that the risky assets have nonzero variances ($\text{var}\left[S_j\right] > 0$ for all j) means that the assets are indeed "risky" and none are equivalent to the mma. The non-singularity of the covariance matrix implies the next lemma.

Lemma 17.1 *Under the non-redundant assets assumption,*

(1)

$$\dim\left[\text{span}(1, S_1, \ldots, S_n)\right] = n + 1,$$

i.e. the asset prices (considered as random variables $S_i(\omega)$) are linearly independent, and

(2) $|\Omega| \geq n + 1$, *i.e. the cardinality of the set of states is greater than or equal to the number of traded securities.*

Proof (Step 1) Show $(1, S_1, \ldots, S_n)$ are linearly independent. Let $(\gamma_0, \gamma_1, \ldots, \gamma_n) \in \mathbb{R}^{n+1}$ be such that

$$\gamma_0 + \sum_{j=1}^{n} \gamma_j S_j = 0. \tag{17.2}$$

We will show that $(\gamma_0, \gamma_1, \ldots, \gamma_n) = 0$, which gives the desired result. First, taking the covariance of expression (17.2) with respect to S_1, then $S_2, \ldots$, then S_n gives the system of equations

$$\begin{aligned} \textstyle\sum_{j=1}^{n} \gamma_j \text{cov}[S_1, S_j] &= 0 \\ \textstyle\sum_{j=1}^{n} \gamma_j \text{cov}[S_2, S_j] &= 0 \\ &\vdots \\ \textstyle\sum_{j=1}^{n} \gamma_j \text{cov}[S_n, S_j] &= 0. \end{aligned}$$

In matrix form

$$\begin{pmatrix} \text{cov}[S_1, S_1] & \text{cov}[S_1, S_2] & \cdots & \text{cov}[S_1, S_n] \\ \text{cov}[S_2, S_1] & \text{cov}[S_2, S_2] & \cdots & \text{cov}[S_2, S_n] \\ \vdots & \vdots & \ddots & \vdots \\ \text{cov}[S_n, S_1] & \text{cov}[S_n, S_2] & \cdots & \text{cov}[S_n, S_n] \end{pmatrix} \begin{pmatrix} \gamma_1 \\ \gamma_2 \\ \vdots \\ \gamma_n \end{pmatrix} = \begin{pmatrix} 0 \\ 0 \\ \vdots \\ 0 \end{pmatrix}.$$

The nonsingularity of the covariance matrix implies that $(\gamma_1, \ldots, \gamma_n) = 0$.

Second, substitution of $(\gamma_1, \ldots, \gamma_n) = 0$ into expression (17.2) gives

$$\gamma_0 = 0,$$

which completes the proof of (Step 1).

(Step 2) If $|\Omega| = \infty$, then the result follows. If $|\Omega| \equiv m < \infty$, then write $\Omega = \{1, 2, \ldots, m\}$. Consider the matrix

$$\underset{(n+1) \times m}{\begin{pmatrix} 1 & 1 & \cdots & 1 \\ S_1(1) & S_1(2) & \cdots & S_1(m) \\ \vdots & \vdots & \ddots & \vdots \\ S_n(1) & S_n(2) & \cdots & S_n(m) \end{pmatrix}}$$

By (Step 1) the row rank of this matrix is $(n + 1)$. Since the row rank and column rank of a matrix are equal (see Theil [175, p. 11]), $m \geq n + 1$. This completes the proof of (Step 2) and the Lemma.

A *self-financing trading strategy* (s.f.t.s.) in a static model is just a buy and hold trading strategy $(\alpha_0, \alpha) \equiv (\alpha_0, \alpha_1, \ldots, \alpha_n) \in \mathbb{R}^{n+1}$ with time 0 and 1 values

$$x = \alpha_0 + \sum_{j=1}^{n} \alpha_j s_j \quad \text{and}$$

$$X = \alpha_0 + \sum_{j=1}^{n} \alpha_j S_j.$$

To see that this trading strategy is self-financing, solving for α_0 in the first expression and substituting the result into the second yields

$$X = x + \sum_{j=1}^{n} \alpha_j [S_j - s_j].$$

In this form, it is easily seen that the time 1 value of the trading strategy is determined without any cash flow in or out of the portfolio.

Remark 17.1 (Admissibility) In this setting, we do not need an admissibility condition because no doubling trading strategies are possible. This is because one cannot trade more than once in this static market. This completes the remark.

The return on an s.f.t.s. is $R_X = \frac{X}{x} - 1$. For subsequent use, we have the following lemma.

Lemma 17.2 (Returns on Portfolios X)

$$R_X = \beta_0 \cdot 0 + \sum_{j=1}^{n} \beta_j R_j,$$

where $\beta_0 = \frac{\alpha_0}{x}$, $\beta_j = \frac{\alpha_j s_j}{x}$ *for* $j = 1, \ldots, n$, *and* $\beta_0 + \sum_{j=1}^{n} \beta_j = 1$.

Proof $\frac{X}{x} = \frac{\alpha_0}{x} + \sum_{j=1}^{n} \frac{\alpha_j}{x} S_j = \frac{\alpha_0}{x} + \sum_{j=1}^{n} \frac{\alpha_j s_j}{x} \frac{S_j}{s_j}$.

Then

$$\frac{X}{x} - 1 = \frac{\alpha_0}{x} + \sum_{j=1}^{n} \frac{\alpha_j s_j}{x} \frac{S_j}{s_j} - 1 = \frac{\alpha_0}{x} + \sum_{j=1}^{n} \frac{\alpha_j s_j}{x} \frac{S_j}{s_j} - \frac{\alpha_0 + \sum_{j=1}^{n} \alpha_j s_j}{x}.$$

Algebra yields

$$\frac{X}{x} - 1 = \frac{\alpha_0}{x} - \frac{\alpha_0}{x} + \sum_{j=1}^{n} \frac{\alpha_j s_j}{x} \frac{S_j}{s_j} - \sum_{j=1}^{n} \frac{\alpha_j s_j}{x}.$$

Algebra completes the proof.

Remark 17.2 (Non-normalized Market) Given $R_X = \beta_0 \cdot 0 + \sum_{j=1}^{n} \beta_j R_j$. In the non-normalized market, using expression (17.1) and noting that the same expression applies to a portfolio, substitution yields

$$\frac{\tilde{R}_X - r}{1+r} = \sum_{j=1}^{n} \beta_j \left(\frac{\tilde{R}_j - r}{1+r} \right).$$

Or,

$$\tilde{R}_X - r = \sum_{j=1}^{n} \beta_j \left(\tilde{R}_j - r \right),$$

or

$$\tilde{R}_X = r\left(1 - \sum_{j=1}^{n} \beta_j\right) + \sum_{j=1}^{n} \beta_j \tilde{R}_j.$$

Noting that $\beta_0 = 1 - \sum_{j=1}^{n} \beta_j$ yields the final result

$$\tilde{R}_X = \beta_0 r + \sum_{j=1}^{n} \beta_j \tilde{R}_j.$$

This completes the remark.

In this static model, the First Fundamental Theorem 2.3 of asset pricing in Chap. 2 simplifies. Before stating the simplified theorem, we need to repeat some definitions.

Definition 17.1 (No Arbitrage (NA)) A trading strategy $(\alpha_0, \alpha) \in \mathbb{R}^{n+1}$ with initial wealth x and time 1 value X is a (simple) *arbitrage opportunity* if

(i) $x = 0$, (zero investment)
(ii) $X \geq 0$ with $\mathbb{P}$ probability one, and
(iii) $\mathbb{P}(X > 0) > 0$.

In the context of a static model, we say that a risky asset price (s_j, S_j) is a $\mathbb{Q}$-martingale if $s_j = E^{\mathbb{Q}}[S_j]$. Given these definitions, we can now state the theorem.

Theorem 17.1 (The First Fundamental Theorem) *NA if and only if $\mathfrak{M} \neq \emptyset$, where $\mathfrak{M} = \{\mathbb{Q} \sim \mathbb{P} : S$ is a $\mathbb{Q}$-martingale$\}$, i.e. there exists an equivalent martingale probability $\mathbb{Q}$.*

Proof This proof is based on Follmer and Schied [63, Theorem 1.6, p. 7].

Define $Z_i = S_i - s_i \geq 0$ for all i. In this case, the mma's difference in value becomes identically zero.

NA is equivalent to: for $\alpha \in \mathbb{R}^n$, $\alpha \cdot Z \geq 0$ implies $\alpha \cdot Z = 0$, where $Z = (Z_1, \ldots, Z_n)'$.

Since $Z_i \geq -s_i$, under any probability measure $\mathbb{Q}$, $E^{\mathbb{Q}}[Z_i]$ is well defined for all i, although it may be $+\infty$.

Then, $\mathbb{Q} \in \mathfrak{M}$ if and only if $E^{\mathbb{Q}}[Z_i] = 0$ for all i.

(Step 1) To show there exists a $\mathbb{Q} \sim \mathbb{P}$ such that $E^{\mathbb{Q}}[Z_i] = 0$ for all i implies NA.

Suppose NA is violated, i.e. there exists an $\alpha \in \mathbb{R}^n$ such that $\alpha \cdot Z \geq 0$ and $\mathbb{P}(\alpha \cdot Z > 0) > 0$. Because $\mathbb{Q} \sim \mathbb{P}$, $\mathbb{Q}(\alpha \cdot Z > 0) > 0$.

Hence $E^{\mathbb{Q}}[\alpha \cdot Z] > 0$. But, $E^{\mathbb{Q}}[\alpha \cdot Z] = \alpha \cdot E^{\mathbb{Q}}[Z] = 0$, which is a contradiction.

(Step 2) To show NA implies that there exists a $\mathbb{Q} \sim \mathbb{P}$ such that $E^{\mathbb{Q}}[Z_i] = 0$ for all i.

Let $\mathscr{P}$ denote the set of all probability measures $\mathbb{Q}$ equivalent to $\mathbb{P}$ such that $E^{\mathbb{Q}}[Z_i] < \infty$ for all i.

Note that $\mathscr{P}$ is a convex set. The set is nonempty since $\mathbb{P} \in \mathscr{P}$.

Define $\mathscr{Z} = \{E^{\mathbb{Q}}[Z_i] : \mathbb{Q} \in \mathscr{P}\}$.

Note that $\mathscr{Z}$ is a convex set. To see this, choose $E^{\mathbb{Q}_1}[Z_i], E^{\mathbb{Q}_1}[Z_i] \in \mathscr{Z}$ and $a \in [0, 1]$. Then,

$aE^{\mathbb{Q}_1}[Z_i] + (1-a)E^{\mathbb{Q}_1}[Z_i] = E^{a\mathbb{Q}_1+(1-a)\mathbb{Q}_2}[Z_i] \in \mathscr{Z}$, which proves convexity.

We claim $0 \in \mathscr{Z}$, which proves that there exists a $\mathbb{Q} \sim \mathbb{P}$ such that $E^{\mathbb{Q}}[Z_i] = 0$ for all i.

To prove this, suppose not, i.e. $0 \notin \mathscr{Z}$.

Then, by a separating hyperplane theorem, we obtain a vector $\alpha \in \mathbb{R}^n$ such that $\alpha \cdot x \geq 0$ for all $x \in \mathscr{Z}$ and $\alpha \cdot x_0 > 0$ for some $x_0 \in \mathscr{Z}$ (see Follmer and Schied [63, Proposition A1, p. 399]). Written out, there exists a $\mathbb{Q}_0 \in \mathscr{P}$ such that $\alpha \cdot E^{\mathbb{Q}_0}[Z] = E^{\mathbb{Q}_0}[\alpha \cdot Z] > 0$, which implies $\mathbb{Q}_0(\alpha \cdot Z > 0) > 0$. We claim that $\alpha \cdot Z \geq 0$, which implies that α is an arbitrage opportunity, yielding the contradiction.

To prove $\alpha \cdot Z \geq 0$, let $A = \{\omega \in \Omega : \alpha \cdot Z < 0\}$.

Define $f_n(\omega) = \left(1 - \frac{1}{n}\right) 1_A(\omega) + \frac{1}{n} 1_{A^c}(\omega) > 0$, which is $\mathscr{F}$-measurable and $\frac{d\mathbb{Q}_n}{d\mathbb{P}}(\omega) = \frac{f_n(\omega)}{E[f_n]} > 0$ for $n = 2, 3, \ldots$, which are densities for probability measures $\mathbb{Q}_n \sim \mathbb{P}$.

Given that $Z_i \frac{f_n(\omega)}{E[f_n]} \leq Z_i$ because $0 < f_n \leq 1$, $0 \leq E^{\mathbb{Q}_n}[Z_i] = E\left[Z_i \frac{f_n(\omega)}{E[f_n]}\right] \leq E[Z_i] < \infty$, which implies $\mathbb{Q}_n \in \mathscr{P}$.

By the dominated convergence theorem,

$\lim_{n\to\infty} E[\alpha \cdot Z f_n(\omega)] = E\left[\lim_{n\to\infty} \alpha \cdot Z f_n(\omega)\right] = E[\alpha \cdot Z 1_A] \geq 0.$

By the definition of A, $\alpha \cdot Z 1_A \leq 0$, which implies $\mathbb{P}(\alpha \cdot Z < 0) = 0$, i.e. $\alpha \cdot Z \geq 0$. This proves the claim, which completes the proof.

Remark 17.3 (NFLVR and Bubbles) In this static model, the simpler NA is used and NFLVR is not needed. There are no local martingale measures in this single period setting, hence there is no analogue for the Third Fundamental Theorem 2.5 of asset pricing in Chap. 2 and no asset price bubbles can exist (see the asset price bubbles Chap. 3). This is a limitation of the static model, not shared by the dynamic model. This completes the remark.

The definition of market completeness simplifies.

Definition 17.2 (Complete Market with respect to $\mathbb{Q} \in \mathfrak{M}_l$) Given $\mathfrak{M} \neq \emptyset$. (NA)

Choose a $\mathbb{Q} \in \mathfrak{M}$.

The market is *complete with respect to* $\mathbb{Q}$ if given any $X \in L^1_+(\mathbb{Q})$,

there exists an $x \geq 0$ and $(\alpha_0, \alpha) \in \mathbb{R}^{n+1}$ such that $x = \alpha_0 + \sum_{j=1}^n \alpha_j s_j$ and

$$x + \sum_{j=1}^n \alpha_j [S_j - s_j] = X.$$

Note that in the static model, the fact that S_j are $\mathbb{Q}$-martingales for all j implies that X is a $\mathbb{Q}$-martingale. Hence, we do not have to include this restriction in the definition itself, as was necessary in the continuous-time setting of Chap. 2. Given this definition, the Second Fundamental Theorem of asset pricing can now be stated.

Theorem 17.2 (The Second Fundamental Theorem) *Assume NA, i.e.* $\mathfrak{M} \neq \emptyset$.
The market is complete with respect to $\mathbb{Q} \in \mathfrak{M}$ *if and only if* $\mathfrak{M}$ *is a singleton, i.e. the equivalent martingale measure is unique.*

This is the same as Theorem 2.6 in the fundamental theorems Chap. 2. The same proof applies. We now give some examples that illustrate both the First and Second Fundamental Theorems of asset pricing.

Example 17.1 (Finite State Space Market) Let $|\Omega| \equiv m < \infty$ and write $\Omega = \{1, 2, \ldots, m\}$. Consider the payoff matrix

$$\underset{(n+1)\times m}{\begin{pmatrix} 1 & 1 & \cdots & 1 \\ S_1(1) & S_1(2) & \cdots & S_1(m) \\ \vdots & \vdots & \ddots & \vdots \\ S_n(1) & S_n(2) & \cdots & S_n(m) \end{pmatrix}}$$

where the first row corresponds to the payoffs to the mma and the last n rows correspond to the payoffs to the risky assets. By Lemma 17.1, we have that $m \geq n + 1$.

Using the First Fundamental Theorem 17.1 of asset pricing, we have that this market satisfies NA if and only if there exists a $(q_0, q_1, \ldots, q_n) \in \mathbb{R}^{n+1}_{++}$ such that

$$\underset{(n+1)\times m}{\begin{pmatrix} 1 & 1 & \cdots & 1 \\ S_1(1) & S_1(2) & \cdots & S_1(m) \\ \vdots & \vdots & \ddots & \vdots \\ S_n(1) & S_n(2) & \cdots & S_n(m) \end{pmatrix}} \underset{m\times 1}{\begin{pmatrix} q_0 \\ q_1 \\ \vdots \\ q_m \end{pmatrix}} = \underset{(n+1)\times 1}{\begin{pmatrix} 1 \\ s_1 \\ \vdots \\ s_n \end{pmatrix}}$$

where $\mathbb{R}^{n+1}_{++}$ denotes the strictly positive subspace of $\mathbb{R}^{n+1}$. The first equation in this matrix equation makes $(q_0, q_1, \ldots, q_n)$ probabilities. The remaining $(n + 1)$ equations are the martingale conditions, i.e. $E^q(S_i) = s_i$ for all $i = 1, \ldots, n$, where $E^q(\cdot)$ is expectation with respect to $(q_0, q_1, \ldots, q_n) \in \mathbb{R}^{n+1}_{++}$. Whether or not a solution exists to this matrix equation depends on the elements within the payoff matrix. For example, if $n = 1$ and $m = 2$, called a binomial model, a solution exists to this matrix equation if and only if $S_1(1) > 1 > S_1(2)$, see Jarrow and Chatterjea [95].

Using the Second Fundamental Theorem 17.2 of asset pricing, assuming such a solution exists (i.e. NA holds), the market is complete if and only if $n + 1 = m$.

This completes the example.

Example 17.2 (Lognormally Distributed Return Market) Consider a market consisting of the mma and only one traded risky asset with time 0 and 1 prices denoted s and S, respectively. We assume that the risky asset's payoffs are

$$S = se^{z},$$

where z is normally distributed ($E[z] = \mu, \text{Var}[z] = \sigma^2 > 0$) under $\mathbb{P}$. Note that, using the moment generating function of a normal random variable (see Mood, Graybill, Boes [143, p. 541]), we have

$$E[S] = se^{\mu+\frac{1}{2}\sigma^2}.$$

Consider the random variable $Y \equiv e^{y} > 1$, where y is normally distributed ($E[y] = -\frac{1}{2}\nu^2, \text{Var}[y] = \nu^2 > 0$) under $\mathbb{P}$ with $\frac{\text{Cov}(z,y)}{\sqrt{\text{Var}[y]\text{Var}[z]}} = \rho$. Then, $E[Y] = E[e^{y}] = e^{-\frac{1}{2}\nu^2+\frac{1}{2}\nu^2} = 1$. Define the equivalent change of measures

$$\frac{d\mathbb{Q}^{(\nu,\rho)}}{d\mathbb{P}} = Y^{(\nu,\rho)} > 0,$$

where the superscript indicates the parameters of y's joint normal distribution with z. We note that $z + y$ is normally distributed ($E[z + y] = \mu - \frac{1}{2}\nu^2, \text{Var}[z + y] = \sigma^2 + 2\rho\sigma\nu + \nu^2$). Using these equivalent measures we have that

$$E^{\mathbb{Q}^{(\nu,\rho)}}[S] = E\left[SY^{(\nu,\rho)}\right] = E\left[se^{z+y}\right] = se^{\mu+\frac{1}{2}\sigma^2+\rho\sigma\nu}.$$

Using the First Fundamental Theorem 17.1 of asset pricing, this economy satisfies NA because $Y^{(\nu,\rho)}$ is a martingale deflator when $\rho\nu = -\frac{1}{\sigma}(\mu + \frac{1}{2}\sigma^2)$, i.e. $E\left[SY^{(\nu,\rho)}\right] = s$.

By the Second Fundamental Theorem 17.2 of asset pricing, the economy is incomplete since there are many (ρ, ν) pairs such that this is true, hence the equivalent martingale measure is not unique. This completes the example.

17.2 Systematic Risk

This section studies the systematic risk return relation derived only using the assumption of NA. In the static model, the set of traded assets themselves $(1, S_1, \ldots, S_n)'$ are the risk factors because they form an algebraic basis for the set of traded portfolios $\mathscr{X} = \text{span}(1, S_1, \ldots, S_n)$. Given any traded portfolio, by Lemma 17.2, we have

Theorem 17.3 (Multiple-Factor Beta Model) *Given an arbitrary* $X \in \text{span}(1, S_1, \ldots, S_n)$*, there exist unique* $(\beta_0, \beta_1, \ldots, \beta_n)$ *such that*

$$R_X = \beta_0 \cdot 0 + \sum_{j=1}^{n} \beta_j R_j, \tag{17.3}$$

where $\beta_0 + \sum_{j=1}^{n} \beta_j = 1$.

Remark 17.4 (Non-normalized Market) In the non-normalized market we have

$$\tilde{R}_X = \beta_0 r + \sum_{j=1}^{n} \beta_j \tilde{R}_j.$$

Using $\beta_0 + \sum_{j=1}^{n} \beta_j = 1$, we can rewrite this as

$$\tilde{R}_X = r + \sum_{j=1}^{n} \beta_j (\tilde{R}_j - r).$$

This completes the remark.

To understand which risk factors have nonzero risk premium, i.e. nonzero excess expected returns, we need to add the following assumption.

Assumption (NA) $\mathfrak{M} \neq \emptyset$.

Let $Y = \frac{d\mathbb{Q}}{d\mathbb{P}} \in M$ be a fixed martingale deflator, which is also called the state price density (see the fundamental theorems Chap. 2).

Theorem 17.4 (The Risk Return Relation) *A traded portfolio's expected return over* $[0, 1]$ *satisfies*

$$E[R_X] = -\text{cov}\,[R_X, Y]\,. \tag{17.4}$$

Proof $E\,[YX] = x$.
$E\left[Y\frac{X}{x}\right] = 1.$
$E\left[Y\left(\frac{X}{x} - 1\right)\right] = 0$ since $E\,[Y] = 1$.
$E\,[Y]\,E\,(R_X) + \text{cov}\,[R_X, Y] = 0.$
Algebra completes the proof.

Remark 17.5 (Non-normalized Market) For future reference, we give the beta model in the non-normalized market. Using expression (17.1), substituting into expression (17.4) gives
$E\left[\frac{\tilde{R}_X - r}{1+r}\right] = -\text{cov}\left[\frac{\tilde{R}_X - r}{1+r}, Y\right]$. Algebra gives the final result.

$$E[\tilde{R}_X] = r - \text{cov}\left[\tilde{R}_X, Y\right]. \tag{17.5}$$

This completes the remark.

We can now characterize which risk factors in expression (17.3) have nonzero risk premium. Since the risk factors are traded portfolios, applying Theorem 17.4 we have that the expected return on any risk factor satisfies

$$E[R_j] = -\text{cov}\left[R_j, Y\right]. \tag{17.6}$$

If $\text{cov}\left[R_j, Y\right] \neq 0$, then the jth risk factor has a nonzero risk premium. This completes the characterization.

If the martingale deflator $Y \in M$ trades, i.e. $Y \in \text{span}(1, S_1, \ldots, S_n)$ we can prove a beta model. Before this, however, we need the following lemma.

Lemma 17.3 (*$Y \in M$* **trades**) *If $Y \in \text{span}(1, S_1, \ldots, S_n)$, then $E\left[Y^2\right] < \infty$,*

$$Y = \gamma_0 + \sum_{j=1}^{n} \gamma_j S_j, \qquad where \tag{17.7}$$

$$\begin{pmatrix} \gamma_1 \\ \gamma_2 \\ \vdots \\ \gamma_n \end{pmatrix} = - \begin{pmatrix} \text{cov}[S_1, S_1] & \text{cov}[S_1, S_2] & \cdots & \text{cov}[S_1, S_n] \\ \text{cov}[S_2, S_1] & \text{cov}[S_2, S_2] & \cdots & \text{cov}[S_2, S_n] \\ \vdots & \vdots & \ddots & \vdots \\ \text{cov}[S_n, S_1] & \text{cov}[S_n, S_2] & \cdots & \text{cov}[S_n, S_n] \end{pmatrix}^{-1} \begin{pmatrix} E[S_1 - s_1] \\ E[S_2 - s_2] \\ \vdots \\ E[S_n - s_n] \end{pmatrix},$$

$$\gamma_0 = 1 - \sum_{j=1}^{n} \gamma_j E[S_j], \qquad and$$

$$E\left[Y^2\right] = \gamma_0 + \sum_{j=1}^{n} \gamma_j s_j.$$

Proof (Step 1) Show $E\left[Y^2\right] < \infty$.

If $Y \in \text{span}(1, S_1, \ldots, S_n)$, then there exists $(\gamma_0, \gamma_1, \ldots, \gamma_n) \in \mathbb{R}^{n+1}$ such that

$$Y = \gamma_0 + \sum_{j=1}^{n} \gamma_j S_j. \tag{17.8}$$

Thus,

$$E\left[Y^2\right] = E\left[\left(\gamma_0 + \sum_{j=1}^{n} \gamma_j S_j\right)^2\right] < \infty$$

since $E[S_i S_j] < \infty$ all i, j. Because Y trades, its time 0 price equals $E[Y \cdot Y] = E[Y^2] = \gamma_0 + \sum_{j=1}^n \gamma_j E[S_j Y] = \gamma_0 + \sum_{j=1}^n \gamma_j s_j$. This completes the proof of (Step 1).

(Step 2) Taking the covariance of expression (17.8) with respect to S_1, then $S_2, \ldots,$ then S_n gives the system of equations

$$\begin{aligned}
\textstyle\sum_{j=1}^n \gamma_j \mathrm{cov}[S_1, S_j] &= \mathrm{cov}[S_1, Y] \\
\textstyle\sum_{j=1}^n \gamma_j \mathrm{cov}[S_2, S_j] &= \mathrm{cov}[S_2, Y] \\
&\vdots \\
\textstyle\sum_{j=1}^n \gamma_j \mathrm{cov}[S_n, S_j] &= \mathrm{cov}[S_n, Y].
\end{aligned}$$

Using expression (17.6), we have

$$E[R_j] = -\frac{1}{s_j}\mathrm{cov}[S_j, Y]$$

since $\mathrm{cov}[R_j, Y] = \mathrm{cov}\left[\frac{S_j}{s_j} - 1, Y\right] = \frac{1}{s_j}\mathrm{cov}\left[S_j, Y\right]$. Algebra gives

$$\mathrm{cov}[S_j, Y] = -E[R_j]s_j = -E[S_j - s_j].$$

Substitution into the system of equations, in matrix form, we have

$$\begin{pmatrix} \mathrm{cov}[S_1, S_1] & \mathrm{cov}[S_1, S_2] & \cdots & \mathrm{cov}[S_1, S_n] \\ \mathrm{cov}[S_2, S_1] & \mathrm{cov}[S_2, S_2] & \cdots & \mathrm{cov}[S_2, S_n] \\ \vdots & \vdots & \ddots & \vdots \\ \mathrm{cov}[S_n, S_1] & \mathrm{cov}[S_n, S_2] & \cdots & \mathrm{cov}[S_n, S_n] \end{pmatrix} \begin{pmatrix} \gamma_1 \\ \gamma_2 \\ \vdots \\ \gamma_n \end{pmatrix} = \begin{pmatrix} -E[S_1 - s_1] \\ -E[S_2 - s_2] \\ \vdots \\ -E[S_n - s_n] \end{pmatrix}.$$

Multiplying both sides of this equation by the inverse of the covariance matrix yields the first part of expression (17.7).

Second, taking the expectation of expression (17.8) gives

$$\gamma_0 + \sum_{j=1}^n \gamma_j E[S_j] = E[Y] = 1.$$

This completes the proof of (Step 2). This completes the proof.

Note that $E\left[Y^2\right] = \gamma_0 + \sum_{j=1}^n \gamma_j s_j$ is the time 0 price of the traded asset $Y \in M$. Then, the return on Y can be written as

$$R_Y = \frac{Y}{E[Y^2]} - 1.$$

We can now derive the beta model.

Theorem 17.5 (Beta Model) *If $Y \in \text{span}(1, S_1, \ldots, S_n)$, then*

$$E[R_j] = \frac{\text{cov}[R_j, R_Y]}{\text{var}[R_Y]} E[R_Y]. \tag{17.9}$$

Proof From expression (17.4), applied to Y, we get $E(R_Y) = -\text{cov}[R_Y, Y] = -E[Y^2]\text{cov}\left[R_Y, \frac{Y}{E[Y^2]}\right] = -E[Y^2]\text{cov}[R_Y, R_Y] = -E[Y^2]\text{var}[R_Y]$.

Algebra gives $-E[Y^2] = \frac{E[R_Y)]}{\text{var}[R_Y]}$.

Using expression (17.6), we have $E[R_j] = -\text{cov}\left[R_j, Y\right] = -E[Y^2] \text{cov}\left[R_j, \frac{Y}{E[Y^2]}\right] = -E[Y^2]\text{cov}\left[R_j, R_Y\right]$. Substitution for $E[Y^2]$ completes the proof.

17.3 Utility Functions

This section introduces the traders' utility functions. Let all traders have the same beliefs, i.e. $\mathbb{P}_i = \mathbb{P}$ for $i = 1, \ldots, \mathscr{I}$. In addition, assume that all investors have mean-variance utility functions defined over terminal wealth, i.e. $U_i : (0, \infty) \to \mathbb{R}$ with

$$U_i(x) = x - \frac{b_i}{2}x^2 \tag{17.10}$$

for $i = 1, \ldots, \mathscr{I}$, where $b_i > 0$ is a risk aversion parameter. This utility function is unique up to an affine transformation (see Lemma 9.2 in the utility function Chap. 9). We note that

$$U_i'(x) = 1 - b_i x = \begin{cases} > 0 \text{ if } 0 < x < \frac{1}{b_i} \\ = 0 \text{ if } x = \frac{1}{b_i} \\ < 0 \text{ if } x > \frac{1}{b_i}. \end{cases} \tag{17.11}$$

This utility function exhibits decreasing marginal utility for $x > \frac{1}{b_i}$. This is a well-known deficiency of mean-variance preferences. For the subsequent analysis to apply, we need to restrict ourselves to $0 < x < \frac{1}{b_i}$.

The inverse of the derivative U_i' over $0 < x < \frac{1}{b_i}$ is

$$I_i(y) = \frac{1 - y}{b_i} \quad \text{for } 0 < y < 1. \tag{17.12}$$

The conjugate function is

$$\begin{aligned} \tilde{U}_i(y) &= \sup_{\frac{1}{b_i} > x > 0} [U_i(x) - xy] \\ &= \sup_{\frac{1}{b_i} > x > 0} [x - \tfrac{b_i}{2}x^2 - xy] = \tfrac{1}{2}\tfrac{(1-y)^2}{b_i}, \quad 1 > y > 0, \end{aligned}$$

which is strictly concave with first derivative equal to

$$\tilde{U}_i'(y) = -\frac{(1-y)}{b_i} < 0.$$

Proof Solve the optimization problem using standard calculus.
$U_i'(x) - y = 1 - b_i x - y = 0$, i.e. $x = \frac{1-y}{b_i}$.
The second order condition is $-b_i < 0$, so an optimum is obtained.
Substitution yields
$\tilde{U}_i(y) = [\frac{(1-y)}{b_i} + \frac{1}{2}\frac{(1-y)^2}{b_i} - \frac{(1-y)}{b_i}y] = \frac{1}{2}\frac{(1-y)^2}{b_i}$. This completes the proof.

17.4 Portfolio Optimization

This section solves the investor's optimization problem. The analysis applies to either a complete or incomplete market. To obtain a solution, we need to assume that the market is arbitrage free.

Assumption (NA) $\mathfrak{M} \neq \emptyset$.

The investor's optimization problem is

Problem 17.1 (Utility Optimization (Choose the Optimal Trading Strategy))

$$\sup_{(\alpha_0,\alpha)\in\mathbb{R}^{n+1}} E\left[X - \frac{b_i}{2}X^2\right], \qquad \text{where}$$

$$X \in \left(0, \frac{1}{b_i}\right), \quad X = \alpha_0 + \sum_{j=1}^{n} \alpha_j S_j, \quad \text{and} \quad x_i = \alpha_0 + \sum_{j=1}^{n} \alpha_j s_j.$$

To solve this problem, we first simplify it by using the second constraint to remove α_0 as a decision variable.

Problem 17.2 (Utility Optimization (Choose the Optimal Trading Strategy))

$$\sup_{\alpha\in\mathbb{R}^n} E\left[X - \frac{b_i}{2}X^2\right],$$

where

$$X \in \left(0, \frac{1}{b_i}\right) \quad \text{and} \quad X = x_i + \sum_{j=1}^{n} \alpha_j \left[S_j - s_j\right].$$

As in the dynamic problem, we divide the solution into two steps. Step 1 solves for the optimal wealth X. Step 2 solves for the optimal trading strategy (α_0, α) that achieves this wealth. Hence, Step 1 is to solve

Problem 17.3 (Utility Optimization (Choose the Optimal Derivative X))

$$\sup_{X \in \tilde{\mathscr{C}}(x_i)} E\left[X - \frac{b_i}{2} X^2\right],$$

where

$$\tilde{\mathscr{C}}(x_i) = \{X \in L_S : \exists \alpha \in \mathbb{R}^n, X = x_i + \sum_{j=1}^{n} \alpha_j [S_j - s_j]\} \quad \text{and}$$

$$L_S \equiv \{X \in \text{span}(1, S_1, \ldots, S_n), \frac{1}{b_i} > X > 0\}.$$

Without loss of generality, we consider the above optimization problem with free disposal.

Problem 17.4 (Utility Optimization (Choose the Optimal Derivative X))

$$\sup_{X \in \mathscr{C}(x_i)} E\left[X - \frac{b_i}{2} X^2\right],$$

where

$$\mathscr{C}(x_i) = \{X \in L_S : \exists \alpha \in \mathbb{R}^n, \ x_i + \sum_{j=1}^{n} \alpha_j [S_j - s_j] \geq X\}.$$

As the next lemma proves, the solutions to the two problems are equivalent.

Lemma 17.4 (Free Disposal) *X_T is an optimal solution to Problem 17.3 if and only if X_T is an optimal solution to Problem 17.4.*

Proof Given $X < \frac{1}{b_i}$, $U'(X) > 0$ for all $X \in \mathscr{C}(x_i)$. Let X^* be the solution to Problem 17.4 and X^{**} the solution to Problem 17.3. Since $\tilde{\mathscr{C}}(x_i) \subset \mathscr{C}(x_i)$, $E\left[U_i(X^*)\right] \geq E\left[U_i(X^{**})\right]$. Suppose $E\left[U_i(X^*)\right] > E\left[U_i(X^{**})\right]$. This implies that $\exists \alpha \in \mathbb{R}^n$, $Z \equiv x_i + \sum_{j=1}^{n} \alpha_j [S_j - s_j] \geq X^*$. Then $Z \in \mathscr{C}(x_i)$ and $E\left[U_i(Z)\right] > E\left[U_i(X^*)\right]$, contradicting the optimality of X^*. Thus, $E\left[U_i(X^*)\right] = E\left[U_i(X^{**})\right]$. This completes the proof.

To solve this problem, we rewrite the budget constraint in a more convenient form. Let $Y = \frac{d\mathbb{Q}}{d\mathbb{P}} \in M$ denote the Radon–Nikodym derivative of a martingale measure $\mathbb{Q} \in \mathfrak{M}$. Y is a martingale deflator and $E[Y] = 1$. Theorem 17.1 implies that

$$s_j = E\left[S_j Y\right] = E^{\mathbb{Q}}\left[S_j\right] \quad \text{for } j = 1, \ldots, n.$$

Since we are dealing with finite asset portfolios, any $X \in \text{span}(1, S_1, \ldots, S_n)$ with $E[XY] < \infty$ satisfies risk-neutral valuation (see Sect. 2.6 in the fundamental theorems Chap. 2), i.e.

$$x = E[XY] = E^{\mathbb{Q}}[X],$$

where $x = \alpha_0 + \sum_{j=1}^n \alpha_j s_j$ is the cost of constructing the portfolio X.

Proof Given an $X \in \text{span}(1, S_1, \ldots, S_n))$, there exists a unique $(\alpha_0, \alpha) \in \mathbb{R}^{n+1}$ such that $X = \alpha_0 + \sum_{j=1}^n \alpha_j S_j$. Note that (α_0, α) is a trading strategy (it is trivially self-financing and there is no admissibility condition). We have $E[XY] = E\left[\left(\alpha_0 + \sum_{j=1}^n \alpha_j S_j\right) Y\right] = \alpha_0 + \sum_{j=1}^n \alpha_j E\left[S_j Y\right] = \alpha_0 + \sum_{j=1}^n \alpha_j s_j = x$. This completes the proof.

Remark 17.6 (Choosing $\mathbb{Q} \in \mathfrak{M}$) Since we only consider the attainable securities (portfolios of the traded assets, i.e. $X \in \text{span}(1, S_1, \ldots, S_n)$), all martingale measures give the same expectation for any $X \in \text{span}(1, S_1, \ldots, S_n)$. Hence, we can arbitrarily choose a $\mathbb{Q} \in \mathfrak{M}$ for the subsequent analysis. Indeed, suppose that the market is incomplete, so that $|\mathfrak{M}| = \infty$. Consider $Y_1, Y_2 \in \mathscr{M}$, both martingale probability densities with respect to $\mathbb{P}$. Then as in the proof immediately above, $x = E[XY_1] = E[XY_2]$. This implies that for any $Y^* \in M$, $\sup_{Y \in M} E[XY] = E\left[XY^*\right]$. This differs from the continuous-time model. In the continuous-time setting the supremum is taken over the set of supermartingale deflators $\sup_{Y \in D_s} E[XY]$, and it is not true that $E[XY] = x$ for all $Y \in D_s$ because some of the supermartingale deflators may not be martingale deflators. This completes the remark.

Using these insights, we can now obtain

Lemma 17.5 (Budget Constraint Equivalence) *Fix a $\mathbb{Q} \in \mathfrak{M}$ and let $Y = \frac{d\mathbb{Q}}{d\mathbb{P}} \in M$. Then,*

$$\mathscr{C}(x_i) = \{X \in L_S : E[XY] \leq x_i\}.$$

Proof Define $\mathscr{C}_1(x_i) = \{X \in L_S : E[XY] \leq x_i\}$.

(Step 1) Show $\mathscr{C}_1(x_i) \subset \mathscr{C}(x_i)$. Take an $X \in \mathscr{C}_1(x_i)$. Then, there exists an $X \in \text{span}(1, S_1, \ldots, S_n)$ such that $E[XY] \leq x_i$. Thus, $\exists\, (\alpha_0, \alpha) \in \mathbb{R}^{n+1}$, $X = \alpha_0 + \sum_{j=1}^n \alpha_j S_j$. Multiplying by Y and taking expectations yields

$$E[XY] = \alpha_0 + \sum_{j=1}^{n} \alpha_j E[S_j Y].$$

Subtracting gives $X - E[XY] = \sum_{j=1}^{n} \alpha_j [S_j - s_j]$. Since $E[XY] \le x_i$, we get $X = E[XY] + \sum_{j=1}^{n} \alpha_j [S_j - s_j] \le x_i + \sum_{j=1}^{n} \alpha_j [S_j - s_j]$. Hence, $X \in \mathscr{C}(x_i)$. This completes the proof of Step 1.

(Step 2) Show $\mathscr{C}(x_i) \subset \mathscr{C}_1(x_i)$. Take an $X \in \mathscr{C}(x_i)$. Then, $\exists \alpha \in \mathbb{R}^n$, $X \le x_i + \sum_{j=1}^{n} \alpha_j [S_j - s_j]$. Multiplying by Y and taking expectations yields $E[XY] \le x_i$ since $E[S_j] = s_j$. Hence, $X \in \mathscr{C}_1(x_i)$. This completes the proof of Step 2. This completes the proof.

Remark 17.7 (Correspondence to the portfolio optimization Chap. 11) In the portfolio optimization Chap. 11, which studies an incomplete market, the corresponding budget constraint is

$$\mathscr{C}(x) = \{X_T \in L_+^0 : \sup_{Y_T \in D_s} E[X_T Y_T] \le x\}.$$

Making the simplifications, the equivalent expression is

$$\mathscr{C}(x_i) = \{X \in L_S : \sup_{Y \in M} E[XY] \le x_i\}.$$

But here, as noted in Remark 17.6 above, $\sup_{Y \in M} E[XY] = E[XY^*]$ for any $Y^* \in M$. Hence, the formula as given in the previous Lemma is an exact correspondence for an incomplete market (as well as for a complete market). This completes the remark.

Using this lemma, we rewrite the investor's optimization problem in its final form.

Problem 17.5 (Utility Optimization (Choose the Optimal Derivative X))

$$\sup_{X \in \mathscr{C}(x_i)} E\left[X - \frac{b_i}{2} X^2\right],$$

where

$$\mathscr{C}(x_i) = \{X \in L_S : E[XY] \le x_i\}.$$

To solve this problem, we add the following assumptions.

Assumption (The Martingale Deflator Trades)

$$\textit{There exists a } Y \in M \textit{ such that } Y \in \mathrm{span}(1, S_1, \ldots, S_n).$$

This assumption states that there exists a martingale deflator $Y \in M$ that trades.

Assumption ("Small" Initial Wealth) $0 < x_i < \frac{1}{b_i}$.

This assumption states that the investor's initial wealth is strictly positive and small enough so that the investor's marginal utility is strictly positive if evaluated at x_i.

To solve the investor's optimization problem, we define the *Lagrangian*

$$\mathscr{L}(X, y) = E\left[U_i(X)\right] + y\left(x_i - E\left[XY\right]\right), \tag{17.13}$$

where $U_i(X) = X - \frac{b_i}{2}X^2$.

Using the Lagrangian, the *primal problem* is

$$\text{primal}: \quad \sup_{X \in L_S}\left(\inf_{y>0} \mathscr{L}(X, y)\right) = \sup_{X \in \mathscr{C}(x_i)} E\left[U_i(X)\right] = v_i(x_i),$$

where $L_S \equiv \{X \in \text{span}(1, S_1, \ldots, S_n),\ \frac{1}{b_i} > X > 0\}$.

Proof Note that

$$\inf_{y>0} \{y\left(x_i - E\left[XY\right]\right)\} = \begin{cases} 0 & \text{if } \ x_i - E\left[YX\right] \geq 0, \\ -\infty & \text{otherwise.} \end{cases}$$

Hence,

$$\inf_{y>0} \{E\left[U_i(X)\right] + y\left(x_i - E\left[XY\right]\right)\} = \begin{cases} E\left[U_i(X)\right] & \text{if } \ E\left[YX\right] \leq x_i, \\ -\infty & \text{otherwise.} \end{cases}$$

$$\inf_{y>0} \mathscr{L}(X, y) = \begin{cases} E\left[U_i(X)\right] & \text{if } \ E\left[YX\right] \leq x_i, \\ -\infty & \text{otherwise.} \end{cases}$$

So,

$$\sup_{X \in L_S}\left(\inf_{y>0} \mathscr{L}(X, y)\right) = \sup_{X \in L_S} \begin{cases} E\left[U_i(X)\right] & \text{if } \ E\left[YX\right] \leq x_i, \\ -\infty & \text{otherwise.} \end{cases}$$

$$= \sup_{X \in C(x_i)} E\left[U_i\,(X)\right] = v_i(x_i).$$

This completes the proof.

The *dual problem* is

$$\text{dual}: \quad \inf_{y>0}\left(\sup_{X \in L_S} \mathscr{L}(X, y)\right).$$

For optimization problems we know that

$$\inf_{y>0}\left(\sup_{X \in L_S} \mathscr{L}(X, y)\right) \geq \sup_{X \in L_S}\left(\inf_{y>0} \mathscr{L}(X, y)\right). \tag{17.14}$$

Proof We have $\mathcal{L}(X, y) = E\left[U_i(X)\right] + y\left(x_i - E\left[XY\right]\right)$.

Note that $\sup_{X\in L_S} \mathcal{L}(X, y) \geq \mathcal{L}(X, y) \geq \inf_{y>0} \mathcal{L}(X, y)$ for all (X, y).

Taking the infimum over y on the left side and the supremum over X on the right side gives

$\inf_{y>0}\left(\sup_{X\in L_S} \mathcal{L}(X, y)\right) \geq \sup_{X\in L_S}\left(\inf_{y>0} \mathcal{L}(X, y)\right)$. This completes the proof.

To prove a solution exists to the primal problem, we show that a solution exists to the dual problem, and then that the dual solution is a feasible solution to the primal problem. Hence, by expression (17.14), the dual solution solves the primal problem and there is no duality gap.

17.4.1 The Dual Problem

This section provides the solution to the dual problem.

$$\inf_{y>0}\left(\sup_{X\in L_S} \mathcal{L}(X, y)\right)$$

$$= \inf_{y>0}\left(\sup_{X\in L_S} E\left[X - \frac{b_i}{2}X^2 - yXY\right] + x_i y\right).$$

To solve, we first exchange the sup and expectation operator

$$\inf_{y>0}\left(E\left[\sup_{X\in L_S}\left(X - \frac{b_i}{2}X^2 - yXY\right)\right] + x_i y\right).$$

The justification for this step is proven in the appendix to this chapter.

We solve this problem working from the inside out. Fixing y, we solve for the optimal X

$$\sup_{X\in L_S}\left(X - \frac{b_i}{2}X^2 - yXY\right). \tag{17.15}$$

The first-order condition for an optimum gives

$$1 - b_i X_i - yY = 0 \qquad \text{or}$$

$$X_i = \frac{1 - yY}{b_i} = I_i(yY).$$

Note that since $y > 0$ and $Y > 0$, $X_i < \frac{1}{b_i}$. In addition, a feasible wealth is x_i (corresponding to holding only the mma). We have, by the definition of a maximum, that $U_i(X_i) \geq U_i(x_i)$. Since $U_i'(\cdot)$ is increasing, this implies $X_i \geq x_i > 0$ a.s. $\mathbb{P}$. Finally, since $Y \in \text{span}(1, S_1, \ldots, S_n)$, $X \in \text{span}(1, S_1, \ldots, S_n)$. Thus, $X_i \in L_S$ and X_i is the optimum solution to expression (17.15). Note that this implies $yY < 1$ a.s. $\mathbb{P}$.

Given this optimal $X_i \in L_S$, we next solve for y. The conjugate function is

$$\tilde{U}_i(yY) = \sup_{X \in L_S} \left[X - \frac{b_i}{2} X^2 - yXY \right] = \left[X_i - \frac{b_i}{2} X_i^2 - yX_iY \right],$$

where $0 < yY < 1$. This transforms the problem to

$$\inf_{y>0} \left(E\left[\sup_{X \in L_S} \left(X - \frac{b_i}{2} X^2 - yXY \right) \right] + x_i y \right) = \inf_{y>0} \left(E[\tilde{U}_i(yY)] + x_i y \right)$$

for $0 < yY < 1$. Taking the derivative of the right side with respect to y yields

$$E[\tilde{U}_i'(yY)Y] + x_i = 0.$$

This step requires taking the derivative underneath the expectation operator. The justification for this step is proven in the appendix to this chapter.

Noting that $\tilde{U}_i'(y) = -\frac{(1-y)}{b_i}$, we get that the optimal y_i satisfies

$$E\left[\frac{(1 - y_iY)}{b_i} Y \right] = x_i. \tag{17.16}$$

Since $Y \in \text{span}(1, S_1, \ldots, S_n)$, by Lemma 17.3 we have that $E\left[Y^2\right] < \infty$. Algebra and the fact that $E[Y] = 1$ gives

$$y_i = \frac{1 - b_i x_i}{E[Y^2]} > 0.$$

y_i is strictly positive because by assumption $x_i < \frac{1}{b_i}$. We note that expression (17.16) shows that X_i satisfies the budget constraint with an equality, i.e.

$$E\left[X_i Y\right] = x_i.$$

Finally, substitution yields the optimal wealth

$$0 < X_i = \frac{1}{b_i} - \frac{y_i Y}{b_i} = \frac{1}{b_i} - \frac{(1 - b_i x_i)}{b_i E[Y^2]} Y. \tag{17.17}$$

This completes the solution to the dual problem.

17.4.2 The Primal Problem

This section provides the solution to the primal problem. To complete the proof, we need to show that the solution X_i obtained above is a feasible solution to the primal problem, i.e. $X_i \in \mathscr{C}(x_i)$. Then by expression (17.14), since $\inf_{y>0}\left(\sup_{X\in L_S}\mathscr{L}(X,y)\right) \geq v_i(x_i)$, X_i is the optimal solution to the primal problem as well. In this regard, we note that: (i) since $Y \in \text{span}(1, S_1, \ldots, S_n)$, $X \in \text{span}(1, S_1, \ldots, S_n)$, (ii) by expression (17.17), $\frac{1}{b_i} > X_i > 0$, and (iii) $E[XY] = x_i$. This completes the proof that $X_i \in \mathscr{C}(x_i)$.

17.4.3 The Optimal Trading Strategy

This section determines the optimal trading strategy, which is Step 2 in the solution procedure. This is the trading strategy (α_0, α) such that

$$\alpha_0 + \sum_{j=1}^{n} \alpha_j S_j = X_i = \frac{1}{b_i} - \frac{(1-b_i x_i)}{b_i E[Y^2]} Y.$$

Such a solution exists and is unique since

(i) $X_i \in \text{span}(1, S_1, \ldots, S_n)$ and
(ii) $\dim\{\text{span}(1, S_1, \ldots, S_n)\} = n+1$.

To obtain this trading strategy, by Lemma 17.3 expression (17.7), we have

$$\alpha_0 + \sum_{j=1}^{n} \alpha_j S_j = \frac{1}{b_i} - \frac{(1-b_i x_i)}{b_i E[Y^2]}\left(\gamma_0 + \sum_{j=1}^{n} \gamma_j S_j\right).$$

Algebra gives

$$\alpha_0 + \sum_{j=1}^{n} \alpha_j S_j = \left(\frac{1}{b_i} - \frac{(1-b_i x_i)}{b_i E[Y^2]}\right)\gamma_0 + \sum_{j=1}^{n}\left(\frac{1}{b_i} - \frac{(1-b_i x_i)}{b_i E[Y^2]}\right)\gamma_j S_j. \tag{17.18}$$

Uniqueness of the coordinates to the basis $(1, S_1, \ldots, S_n)$ gives that the optimal trading strategy is

$$\begin{aligned} \alpha_0 &= \left(\tfrac{1}{b_i} - \tfrac{(1-b_i x_i)}{b_i E[Y^2]}\right)\gamma_0, \\ \alpha_j &= \left(\tfrac{1}{b_i} - \tfrac{(1-b_i x_i)}{b_i E[Y^2]}\right)\gamma_j \quad \text{for} \quad j = 1, \ldots, n \qquad \text{where} \end{aligned} \tag{17.19}$$

$$\begin{pmatrix} \gamma_1 \\ \gamma_2 \\ \vdots \\ \gamma_n \end{pmatrix} = -\begin{pmatrix} \text{cov}[S_1, S_1] & \text{cov}[S_1, S_2] & \cdots & \text{cov}[S_1, S_n] \\ \text{cov}[S_2, S_1] & \text{cov}[S_2, S_2] & \cdots & \text{cov}[S_2, S_n] \\ \vdots & \vdots & \ddots & \vdots \\ \text{cov}[S_n, S_1] & \text{cov}[S_n, S_2] & \cdots & \text{cov}[S_n, S_n] \end{pmatrix}^{-1} \begin{pmatrix} E[S_1 - s_1] \\ E[S_2 - s_2] \\ \vdots \\ E[S_n - s_n] \end{pmatrix},$$

$$\gamma_0 = 1 - \sum_{j=1}^{n} \gamma_j E[S_j].$$

Remark 17.8 (Partial Derivatives with respect to S_j) From expression (17.18) we have

$$\begin{aligned} X_i &= \alpha_0 + \textstyle\sum_{j=1}^{n} \alpha_j S_j \\ &= \left(\tfrac{1}{b_i} - \tfrac{(1-b_i x_i)}{b_i E[Y^2]}\right) \gamma_0 + \textstyle\sum_{j=1}^{n} \left(\tfrac{1}{b_i} - \tfrac{(1-b_i x_i)}{b_i E[Y^2]}\right) \gamma_j S_j. \end{aligned} \tag{17.20}$$

We can alternatively obtain the trading strategy with respect to the risky assets by taking partial derivatives. Indeed

$$\frac{\partial X_i}{\partial S_j} = \alpha_j = \left(\frac{1}{b_i} - \frac{(1 - b_i x_i)}{b_i E[Y^2]}\right) \gamma_j.$$

The position in the money market account follows by substituting this last equality into expression (17.18) and solving for α_0.

We note the correspondence between this alternative derivation of the trading strategy and that given in the portfolio optimization Chap. 10, Sect. 10.7. This completes the remark.

To obtain the next theorem, we first decompose the initial wealth into the position in the mma α_0 and the wealth in the risky assets,

$$z_i \equiv x_i - \alpha_0 = \sum_{j=1}^{n} \alpha_j s_j. \tag{17.21}$$

Then,

$$Z_i \equiv X_i - \alpha_0 = \sum_{j=1}^{n} \alpha_j S_j \tag{17.22}$$

is the wealth in the risky asset portfolio at time 1. Assume that $z_i > 0$ and define $R_{Z_i} \equiv \frac{Z_i}{z_i} - 1$ to be the return on the risky asset portfolio of investor i. The risky asset portfolio's return $R_{Z_i} = \frac{Z_i}{z_i} - 1$ can be written in an alternative form.

Lemma 17.6 (Risky Asset Portfolio Representation) *Let* $X_i = \alpha_0 + \sum_{j=1}^{N} \alpha_j S_j$ *with* $x_i = \alpha_0 + \sum_{j=1}^{N} \alpha_j s_j$. *Then,*

$$X_i = x_i + z_i R_{Z_i}, \tag{17.23}$$

where $\theta_j \equiv \frac{\alpha_j s_j}{z_i}$ *for* $j = 1, \ldots, n$, $R_{Z_i} = \sum_{j=1}^{n} \theta_j R_j$, *and* $z_i = x_i - \alpha_0$. *Note that*

$$\sum_{j=1}^{n} \theta_j = \sum_{j=1}^{n} \frac{\alpha_j s_j}{z_i} = 1.$$

Proof

$$X_i = x_i + \sum_{j=1}^{n} \alpha_j [S_j - s_j]$$

$$= x_i + z_i \sum_{j=1}^{n} \frac{\alpha_j s_j}{z_i} \frac{[S_j - s_j]}{s_j}.$$

Finally, $R_{Z_i} \equiv \frac{Z_i}{z_i} - 1 = \frac{\sum_{j=1}^{n} \alpha_j S_j - z_i}{z_i} = \sum_{j=1}^{n} \frac{\alpha_j s_j}{z_i} \frac{S_j}{s_j} - 1.$
Given $\sum_{j=1}^{n} \frac{\alpha_j s_j}{z_i} = 1$, we get $R_{Z_i} = \sum_{j=1}^{n} \frac{\alpha_j s_j}{z_i} \frac{S_j}{s_j} - \sum_{j=1}^{n} \frac{\alpha_j s_j}{z_i}$
$= \sum_{j=1}^{n} \frac{\alpha_j s_j}{z_i} \frac{[S_j - s_j]}{s_j}$. Substitution yields $X_i = x_i + z_i R_{Z_i}$.
Finally, using the definition of θ_j completes the proof.

Theorem 17.6 (All Traders Hold the Same Optimal Risky Asset Portfolio)

$$R_{Z_i} = \frac{\sum_{j=1}^{n} \gamma_j S_j}{\sum_{j=1}^{n} \gamma_j s_j} - 1 = \eta R_Y \qquad \text{for all } i = 1, \ldots, \mathscr{I}, \tag{17.24}$$

where $\eta = \frac{\gamma_0 + \sum_{j=1}^{n} \gamma_j s_j}{\sum_{j=1}^{n} \gamma_j s_j}$. *Alternatively stated, the risky asset portfolio return*

$$R_{Z_i} = \sum_{j=1}^{n} \theta_j R_j,$$

where $\theta_j = \frac{\gamma_j s_j}{\sum_{j=1}^{n} \gamma_j s_j}$ *for all* $j = 1, \ldots, n$ *and* $\sum_{j=1}^{n} \theta_j = 1$.

Proof (Step 1) By expression (17.20), using the fact that

$$\alpha_0 = \left(\frac{1}{b_i} - \frac{(1 - b_i x_i)}{b_i E[Y^2]} \right) \gamma_0$$

gives

$$Z_i = \sum_{j=1}^{n} \left(\frac{1}{b_i} - \frac{(1 - b_i x_i)}{b_i E[Y^2]} \right) \gamma_j S_j = \left(\frac{1}{b_i} - \frac{(1 - b_i x_i)}{b_i E[Y^2]} \right) \sum_{j=1}^{n} \gamma_j S_j$$

and

$$z_i = \left(\frac{1}{b_i} - \frac{(1 - b_i x_i)}{b_i E[Y^2]} \right) \sum_{j=1}^{n} \gamma_j s_j.$$

Hence,

$$\begin{aligned} R_{Z_i} &= \frac{Z_i}{z_i} - 1 \\ &= \frac{\left(\frac{1}{b_i} - \frac{(1-b_i x_i)}{b_i E[Y^2]} \right) \sum_{j=1}^{n} \gamma_j S_j}{\left(\frac{1}{b_i} - \frac{(1-b_i x_i)}{b_i E[Y^2]} \right) \sum_{j=1}^{n} \gamma_j s_j} - 1 \\ &= \frac{\sum_{j=1}^{n} \gamma_j S_j}{\sum_{j=1}^{n} \gamma_j s_j} - 1. \end{aligned}$$

But,

$$Y = \gamma_0 + \sum_{j=1}^{n} \gamma_j S_j$$

and

$$Y_0 = \gamma_0 + \sum_{j=1}^{n} \gamma_j s_j.$$

Hence,

$$R_Y = \frac{Y - Y_0}{Y_0} = \frac{\sum_{j=1}^{n} \gamma_j S_j - \sum_{j=1}^{n} \gamma_j s_j}{\sum_{j=1}^{n} \gamma_j s_j} \cdot \frac{\sum_{j=1}^{n} \gamma_j s_j}{\gamma_0 + \sum_{j=1}^{n} \gamma_j s_j}.$$

Or,

$$R_Y = R_{Z_i} \cdot \frac{\sum_{j=1}^{n} \gamma_j s_j}{\gamma_0 + \sum_{j=1}^{n} \gamma_j s_j}.$$

Algebra completes the proof of (Step 1).

(Step 2) By Lemma 17.6, $R_{Z_i} = \sum_{j=1}^{n} \theta_j R_j$ where $\theta_j \equiv \frac{\alpha_j s_j}{z_i}$. From Step 1 we have $z_i = \sum_{j=1}^{n} \left(\frac{1}{b_i} - \frac{(1-b_i x_i)}{b_i E[Y^2]} \right) \gamma_j s_j$. Using this and expression (17.19) gives

$$\theta_j = \frac{\left(\frac{1}{b_i} - \frac{(1-b_i x_i)}{b_i E[Y^2]} \right) \gamma_j s_j}{\left(\frac{1}{b_i} - \frac{(1-b_i x_i)}{b_i E[Y^2]} \right) \sum_{j=1}^{n} \gamma_j s_j} = \frac{\gamma_j s_j}{\sum_{j=1}^{n} \gamma_j s_j}.$$

This completes the proof.

This is an important theorem. It shows that all traders hold the same optimal risky asset portfolio $(\theta_1, \ldots, \theta_n)$. Their optimal portfolio differs only due to amount of wealth held in the risky asset portfolio z_i. The remainder $x_i - z_i$, which also differs across traders, is held in the mma.

17.5 Beta Model (Revisited)

This section revisits the beta model using the characterization of the martingale deflator obtained from solving the investor's optimization problem to obtain a modified version of the beta model. Since the martingale deflator $Y \in M$ trades (by assumption), we get

Theorem 17.7 (Beta Model Revisited)

$$E[R_j] = \frac{\text{cov}[R_j, R_Z]}{\text{var}[R_Z]} E[R_Z] \qquad \textit{for } j = 1, \ldots, n, \tag{17.25}$$

where $R_Z = \sum_{j=1}^{n} \theta_j R_j$ *with* $\theta_j = \frac{\gamma_j s_j}{\sum_{j=1}^{n} \gamma_j s_j}$ *for all* j *and* $\sum_{j=1}^{n} \theta_j = 1$.

Proof Using Theorem 17.5, we get $E[R_j] = \frac{\text{cov}[R_j, R_Y]}{\text{var}[R_Y]} E[R_Y]$.

Expression (17.24) gives $R_Y = \frac{1}{\eta} R_Z$. Substitution yields

$E[R_j] = \frac{\text{cov}[R_j, \frac{1}{\eta} R_Z]}{\text{var}[\frac{1}{\eta} R_Z]} E\left[\frac{1}{\eta} R_Z\right] = \frac{\frac{1}{\eta}\text{cov}[R_j, R_Z]}{\left(\frac{1}{\eta}\right)^2 \text{var}[R_Z]} \frac{1}{\eta} E[R_Z]$, algebra completes the proof.

Remark 17.9 (Non-normalized Economy) Using expression (17.1) in expression (17.25) gives

$$E\left[\frac{\tilde{R}_j - r}{1+r}\right] = \frac{\text{cov}\left[\frac{\tilde{R}_j - r}{1+r}, \frac{\tilde{R}_Z - r}{1+r}\right]}{\text{var}\left[\frac{\tilde{R}_Z - r}{1+r}\right]} E\left[\frac{\tilde{R}_Z - r}{1+r}\right].$$

Algebra gives

$$\frac{1}{1+r}E\left[\frac{\tilde{R}_j - r}{1+r}\right] = \frac{\left(\frac{1}{1+r}\right)^2 \operatorname{cov}\left[\tilde{R}_Z - r, \tilde{R}_Z - r\right]}{\left(\frac{1}{1+r}\right)^2 \operatorname{var}\left[\tilde{R}_Z - r\right]}\left(\frac{1}{1+r}\right)E\left[\tilde{R}_Z - r\right].$$

More algebra and the properties of expectations, variances, and covariances yields the final result

$$E[\tilde{R}_j] = r + \frac{\operatorname{cov}[\tilde{R}_j, \tilde{R}_Z]}{\operatorname{var}[\tilde{R}_Z]}\left(E[\tilde{R}_Z] - r\right). \tag{17.26}$$

This completes the remark.

17.6 The Efficient Frontier

This section introduces the notion of an efficient frontier. To do this, we rewrite the optimization Problem 17.2 in an alternative form and solve it again.

17.6.1 The Solution (Revisited)

Using Lemma 17.6, we rewrite the optimization problem as follows.

Problem 17.6 (Utility Optimization)

$$\sup_{\theta}\left\{\sup_{z} E\left[x_i + zR_\theta - \frac{b_i}{2}\left[x_i + zR_\theta\right]^2\right]\right\}, \qquad \text{where } \sum_{j=1}^{n}\theta_j = 1.$$

We solve the inner problem first, yielding the solution

$$z = \frac{E[R_\theta](1 - b_i x_i)}{E[R_\theta^2] b_i}.$$

Proof

$$\begin{aligned} E\left[x_i + zR_\theta - \frac{b_i}{2}\left[x_i + zR_\theta\right]^2\right] &= E\left[x_i + zR_\theta - \frac{b_i}{2}\left[x_i^2 + 2x_i z R_\theta + z^2 R_\theta^2\right]\right] \\ &= x_i - \frac{b_i}{2}x_i^2 + zE[R_\theta](1 - b_i x_i) - \frac{b_i}{2}z^2 E[R_\theta^2]. \end{aligned}$$

Taking the first derivative and setting it equal to zero yields $E[R_\theta](1 - b_i x_i) - z b_i E[R_\theta^2] = 0$. Algebra completes the proof.

Substituting the optimal z into the original problem, the remaining optimization problem is equivalent to the following optimization problem

$$\sup_\theta \left\{ \frac{1}{2} \frac{(1 - b_i x_i)^2}{b_i} \frac{E[R_\theta]^2}{E[R_\theta^2]} \right\}, \qquad \text{where} \quad \sum_{j=1}^{n} \theta_j = 1.$$

Proof The objective function is

$$x_i - \frac{b_i}{2} x_i^2 + z E[R_\theta](1 - b_i x_i) - \frac{b_i}{2} z^2 E[R_\theta^2],$$

or

$$x_i - \frac{b_i}{2} x_i^2 + \frac{E[R_\theta](1 - b_i x_i)}{E[R_\theta^2] b_i} E[R_\theta](1 - b_i x_i) - \frac{b_i}{2} \frac{E[R_\theta]^2 (1 - b_i x_i)^2}{E[R_\theta^2]^2 b_i^2} E[R_\theta^2].$$

Simplification yields

$$x_i - \frac{b_i}{2} x_i^2 + \frac{E[R_\theta]^2 (1 - b_i x_i)^2}{E[R_\theta^2] b_i} - \frac{1}{2} \frac{E[R_\theta]^2 (1 - b_i x_i)^2}{E[R_\theta^2] b_i}.$$

Algebra completes the proof.

Finally, this is equivalent to solving

$$\sup_\theta \left\{ \frac{E[R_\theta]}{\sqrt{E[R_\theta^2]}} \right\}, \qquad \text{where} \quad \sum_{j=1}^{n} \theta_j = 1. \tag{17.27}$$

The solution to this optimization problem does not depend on the ith trader. This yields another proof of Theorem 17.6 that *all traders hold the same optimal risky asset portfolio*. In symbols, $R_{Z_i} = R_\theta \equiv \sum_{j=1}^{n} \theta_j R_j$ is independent of investor i. The solution $\theta = (\theta_1, \ldots, \theta_n)$ to expression (17.27) is the optimal risky asset portfolio.

17.6.2 Summary

The investor's optimization problem is

(i) to choose z to maximize $E[U_i(x_i + z R_\theta)]$, where $U_i(x) = x - \frac{b_i}{2} x^2$, $R_\theta = \sum_{j=1}^{n} \theta_j R_j$, and $\sum_{j=1}^{n} \theta_j = 1$. Then,

(ii) to choose θ to maximize

$$\sup_{\theta}\left\{\frac{E[R_\theta]}{\sqrt{E[R_\theta^2]}}\right\}, \qquad \text{where } \sum_{j=1}^{n}\theta_j = 1.$$

Both (i) and (ii) determine $\alpha_j = \frac{z\theta_j}{s_j}$ for $j = 1, \ldots, n$.

(iii) Finally, α_0 is determined such that

$$x_i = \alpha_0 + \sum_{j=1}^{n}\alpha_j s_j.$$

The solution θ is independent of the trader's initial wealth x_i.

17.6.3 The Risky Asset Frontier and Efficient Frontier

This section defines and characterizes both the risky asset frontier and the efficient frontier. Consider the following optimization problem

$$\inf_{\theta}\sqrt{E[R_\theta^2]} \quad \text{where } E[R_\theta] = \mu.$$

Denote the solution by θ_μ with the minimum being $\sqrt{E[R_{\theta_\mu}^2]}$.

Solving this problem for different μ maps out what is defined to be the *risky asset frontier* in ("x" $-$ $axis$, "y" $-$ $axis$) $= (\sqrt{E[R^2]}, E[R])$ space. The tangent line from $(\sqrt{E[r^2]} = 0, r = E[r] = 0)$, the money market account's coordinates, to the risky asset frontier is determined by that μ^* such that the slope of this tangent line is maximized, i.e.

$$\frac{E[R_{\theta_{\mu^*}}]}{\sqrt{E[R_{\theta_{\mu^*}}^2]}} = \frac{\mu^*}{\sqrt{E[R_{\theta_{\mu^*}}^2]}} = \sup_{\mu}\left\{\frac{\mu}{\sqrt{E[R_{\theta_\mu}^2]}}\right\}.$$

This tangent line is defined to be the *efficient frontier* (including both the money market account and the risky assets). To relate the optimal portfolio to the efficient frontier, let θ^* denote the optimal risky asset portfolio, i.e. the solution to expression (17.27). Then, θ^* yields a (square root of the expected return squared, mean return) that is on the efficient frontier.

Proof $\sup_{\theta}\left\{\frac{E[R_\theta]}{\sqrt{E[R_\theta^2]}}\right\} \geq \sup_{\mu}\left\{\frac{E[R_{\theta_\mu}]}{\sqrt{E[R_{\theta_\mu}^2]}}\right\}$ since R_{θ_μ}, for different μ, is a subset of R_θ.

But, letting $\tilde{\mu} = E[R_{\theta^*}]$, we have that

$$\frac{\tilde{\mu}}{\sqrt{E[R_{\theta^*}^2]}} \leq \frac{\tilde{\mu}}{\inf_\theta \sqrt{E[R_\theta^2]}} = \frac{\tilde{\mu}}{\sqrt{E[R_{\theta_{\tilde{\mu}}}^2]}} \leq \sup_\mu \left\{ \frac{E[R_{\theta_\mu}]}{\sqrt{E[R_{\theta_\mu}^2]}} \right\}.$$

Thus, $\left\{ \frac{E[R_{\theta^*}]}{\sqrt{E[R_{\theta^*}^2]}} \right\} = \sup_\mu \left\{ \frac{E[R_{\theta_\mu}]}{\sqrt{E[R_{\theta_\mu}^2]}} \right\}$. This completes the proof.

17.7 Equilibrium

This section studies economic equilibrium for the static CAPM economy

$$\left\{ (\mathscr{F}, \mathbb{P}), (N_0 = 0, N), \left(\mathbb{P}, U_i, \left(e_0^i, e^i\right)\right)_{i=1}^{\mathscr{I}} \right\},$$

where the ith trader's endowment is

$$x_i = e_0^i + e^i \cdot s = e_0^i + \sum_{j=1}^n e_j^i s_j, \tag{17.28}$$

and traders' utility functions $U_i(\cdot)$ satisfy expression (17.10). We add the following assumption.

Assumption (Liquidating Cash Flows)
There exists an exogenous random payout vector

$$\xi = (\xi_1, \ldots, \xi_n) : \Omega \to \mathbb{R}^n \text{ at time 1 such that } (S_1, \ldots, S_n) = \xi > 0.$$

Definition 17.3 (Equilibrium) The economy is in *equilibrium* if there exists time 0 prices $(s_1, \ldots, s_n)$ such that

(i) $\left(\alpha_0^i, \alpha^i\right)$ are optimal for $\left(\mathbb{P}, U_i, \left(e_0^i, e^i\right)\right)$ for $i = 1, \ldots, \mathscr{I}$,
(ii) $N_0 = 0 = \sum_{i=1}^{\mathscr{I}} \alpha_0^i$, and
(iii) $N_j = \sum_{i=1}^{\mathscr{I}} \alpha_j^i$ for $j = 1, \ldots, n$.

In an equilibrium, the time 0 asset prices are such that all traders' portfolios are optimal and supply equals demand. The next theorem proves an equilibrium exists for this economy. The proof uses the existence of a representative trader as in Theorem 14.7 in the representative trader Chap. 14. For alternative CAPM equilibrium existence proofs, see Hart [68] and Nielsen [146].

Theorem 17.8 (Existence of a CAPM Equilibrium) *Given is an economy* $\left\{(\mathscr{F}, \mathbb{P}), (N_0 = 0, N), \left(\mathbb{P}, U_i, (e_0^i, e^i)\right)_{i=1}^{\mathscr{I}}\right\}$.

Assume $0 < x_i < \frac{1}{b_i}$ *for all* i.

Define the aggregate utility function on $x \in (0, \sum_{i=1}^{\mathscr{I}} \frac{1}{b_i})$ *by*

$$U(x, \lambda) = \sup_{x = x_1 + \cdots + x_{\mathscr{I}}} \left\{ \sum_{i=1}^{\mathscr{I}} \lambda_i \left[x_i - \frac{b_i}{2} x_i^2 \right] \right\}.$$

Assume that $E\left[U'(N \cdot \xi, \tilde{\lambda})\right] < \infty$, $E\left[U(N \cdot \xi, \tilde{\lambda})\right] < \infty$, *and*

$$0 < \frac{d\mathbb{Q}_{\tilde{\lambda}}}{d\mathbb{P}} \equiv \frac{U'(N \cdot \xi, \tilde{\lambda})}{E\left[U'(N \cdot \xi, \tilde{\lambda})\right]} \in \text{span}(1, S_1, \ldots, S_n)$$

with

$$s^{\tilde{\lambda}} = E_{\tilde{\lambda}}[\xi], \tag{17.29}$$

where $E_{\tilde{\lambda}}[\cdot]$ *is expectation with respect to* $\mathbb{Q}_{\tilde{\lambda}}$ *and*

$$\tilde{\lambda}_i = \frac{1 - b_1 \left(e_0^1 + \left(\frac{E\left[\xi \left(\sum_{j=1}^{\mathscr{I}} \frac{1}{b_j} - N \cdot \xi \right)\right]}{E\left[\sum_{j=1}^{\mathscr{I}} \frac{1}{b_j} - N \cdot \xi\right]} \right) \cdot e^1 \right)}{1 - b_i \left(e_0^i + \left(\frac{E\left[\xi \left(\sum_{j=1}^{\mathscr{I}} \frac{1}{b_j} - N \cdot \xi \right)\right]}{E\left[\sum_{j=1}^{\mathscr{I}} \frac{1}{b_j} - N \cdot \xi\right]} \right) \cdot e^i \right)} \quad \text{for all } i.$$

Then,

$s^{\tilde{\lambda}}$ *is an equilibrium price for the economy.*

Proof The proof involves showing that the sufficient conditions of Lemma 17.10 in the appendix to this chapter are satisfied.

Note that hypotheses (i) is satisfied by assumption. Condition (ii) need only be satisfied by $\tilde{\lambda}$ since we now show that $\tilde{\lambda}$ is the solution to hypothesis (iii).

By definition using Lemma 17.7,

$$s^{\tilde{\lambda}} = \frac{E\left[\xi U'(N \cdot \xi)\right]}{E\left[U'(N \cdot \xi)\right]} = \frac{E\left[\left(\frac{\xi \left(\sum_{j=1}^{\mathscr{I}} \frac{1}{b_j} - N \cdot \xi\right)}{\sum_{i=1}^{\mathscr{I}} \frac{1}{\lambda_i b_i}}\right)\right]}{E\left[\frac{\sum_{j=1}^{\mathscr{I}} \frac{1}{b_j} - N \cdot \xi}{\sum_{i=1}^{\mathscr{I}} \frac{1}{\lambda_i b_i}}\right]}.$$

The shadow price to the ith trader's optimization problem is

$$y_i\left(s^{\tilde{\lambda}}\right) = \frac{1 - b_i x_i}{E[Y^2]} = \frac{1 - b_i(e_0^i + s^{\tilde{\lambda}} \cdot e^i)}{E[Y^2]}$$

$$= \frac{1 - b_i \left(e_0^i + \left(\frac{E\left[\left(\frac{\xi\left(\sum_{j=1}^{\mathscr{I}} \frac{1}{b_j} - N\cdot\xi\right)}{\sum_{i=1}^{\mathscr{I}} \frac{1}{\tilde{\lambda}_i b_i}}\right)\right]}{E\left[\frac{\sum_{j=1}^{\mathscr{I}} \frac{1}{b_j} - N\cdot\xi}{\sum_{i=1}^{\mathscr{I}} \frac{1}{\tilde{\lambda}_i b_i}}\right]} \right) \cdot e^i \right)}{E[Y^2]}.$$

The set of equations in hypothesis (iii) to be solved are

$$\tilde{\lambda}_i = \frac{y_1(s^{\tilde{\lambda}})}{y_i(s^{\tilde{\lambda}})} = \frac{1 - b_1 \left(e_0^1 + \left(\frac{E\left[\left(\frac{\xi\left(\sum_{j=1}^{\mathscr{I}} \frac{1}{b_j} - N\cdot\xi\right)}{\sum_{i=1}^{\mathscr{I}} \frac{1}{\tilde{\lambda}_i b_i}}\right)\right]}{E\left[\frac{\sum_{j=1}^{\mathscr{I}} \frac{1}{b_j} - N\cdot\xi}{\sum_{i=1}^{\mathscr{I}} \frac{1}{\tilde{\lambda}_i b_i}}\right]} \right) \cdot e^1 \right)}{1 - b_i \left(e_0^i + \left(\frac{E\left[\left(\frac{\xi\left(\sum_{j=1}^{\mathscr{I}} \frac{1}{b_j} - N\cdot\xi\right)}{\sum_{i=1}^{\mathscr{I}} \frac{1}{\tilde{\lambda}_i b_i}}\right)\right]}{E\left[\frac{\sum_{j=1}^{\mathscr{I}} \frac{1}{b_j} - N\cdot\xi}{\sum_{i=1}^{\mathscr{I}} \frac{1}{\tilde{\lambda}_i b_i}}\right]} \right) \cdot e^i \right)}.$$

Note that $\tilde{\lambda}_i$ is a constant because $\frac{y_1}{y_i}$ is non-random, hence the terms involving $\tilde{\lambda}_i$ come outside the expectation operators and the denominator's in $s^{\tilde{\lambda}}$ cancel. This yields

$$\tilde{\lambda}_i = \frac{1 - b_1 \left(e_0^1 + \left(\frac{E\left[\xi\left(\sum_{j=1}^{\mathscr{I}} \frac{1}{b_j} - N\cdot\xi\right)\right]}{E\left[\sum_{j=1}^{\mathscr{I}} \frac{1}{b_j} - N\cdot\xi\right]} \right) \cdot e^1 \right)}{1 - b_i \left(e_0^i + \left(\frac{E\left[\xi\left(\sum_{j=1}^{\mathscr{I}} \frac{1}{b_j} - N\cdot\xi\right)\right]}{E\left[\sum_{j=1}^{\mathscr{I}} \frac{1}{b_j} - N\cdot\xi\right]} \right) \cdot e^i \right)}.$$

Since the right side is independent of $\tilde{\lambda}$, this is the solution to this system of equations. This completes the proof.

To characterize the equilibrium risk return relation, we first show that in equilibrium each investor's optimal risky asset portfolio is the market portfolio.

Theorem 17.9 (Equilibrium Portfolio) *In equilibrium, each investor's optimal risky asset portfolio is the market portfolio, i.e.*

$$\theta_j = \frac{N_j s_j}{\sum_{j=1}^{n} N_j s_j} \qquad \text{for } j = 1, \ldots, n. \tag{17.30}$$

Proof In equilibrium, we have $N_j = \sum_{i=1}^{\mathcal{I}} \alpha_j^i$ for all j. Hence, using Lemma 17.6,

$$N_j s_j = \sum_{i=1}^{\mathcal{I}} \alpha_j^i s_j = \sum_{i=1}^{\mathcal{I}} \theta_j^i z_i = \theta_j \sum_{i=1}^{\mathcal{I}} z_i,$$

where θ_j are the holdings in the optimal risky asset portfolio, which are independent of investor i by Theorem 17.6. Algebra gives $\theta_j = \frac{N_j s_j}{\sum_{i=1}^{\mathcal{I}} z_i}$. Next, $\sum_{i=1}^{\mathcal{I}} z_i = \sum_{i=1}^{\mathcal{I}} \left(\sum_{j=1}^{n} \alpha_j^i s_j \right) = \sum_{j=1}^{n} \left(\sum_{i=1}^{\mathcal{I}} \alpha_j^i \right) s_j = \sum_{j=1}^{n} N_j s_j$. Substitution completes the proof.

Define $r_m \equiv \sum_{j=1}^{n} \theta_j R_j$ where θ satisfies expression (17.30). This is the return on the market portfolio. Since in equilibrium $R_{Z_i} = r_m$ for all i, using Theorem 17.7 expression (17.25), we have just proven the key step in obtaining the static CAPM risk return relation.

Theorem 17.10 (Static CAPM Model) *Suppose the economy is in equilibrium. Then,*

$$E[R_j] = \frac{\operatorname{cov}\left[R_j, r_m\right]}{\operatorname{var}\left[r_m\right]} E\left[r_m\right], \tag{17.31}$$

where $r_m = \sum_{j=1}^{n} \theta_j R_j$ with θ satisfying expression (17.30) is the return on the market portfolio.

Remark 17.10 (Non-normalized Economy) In the non-normalized economy this is

$$E[\tilde{R}_j] = r + \frac{\operatorname{cov}[\tilde{R}_j, \tilde{r}_m]}{\operatorname{var}[\tilde{r}_m]} \left(E[\tilde{r}_m] - r \right). \tag{17.32}$$

This completes the remark.

Corollary 17.1 (Mutual Fund Theorem) *The optimal portfolio is a linear combination of the money market account and the market portfolio.*

Remark 17.11 (Alternate Derivation of CAPM) An alternate derivation of Theorem 17.10 can be obtained using the existence of a representative trader economy that reflects the original economy's equilibrium, analogous to the characterization of the systematic risk return relation in Chap. 15 on the characterization of equilibrium, expression (15.2). Indeed, assume that s is an equilibrium price for the economy. Then, by Lemma 17.9 Part 2 in the appendix, the representative trader economy

$$\left\{(\mathscr{F}, \mathbb{P}), (N_0 = 0, N), \left(\mathbb{P}, U(\lambda^*), (N_0 = 0, N)\right)\right\}$$

reflects the economy's equilibrium. Here, the representative trader's martingale deflator

$$Y = U'(N \cdot \xi, \lambda^*) = \frac{\sum_{j=1}^{\mathscr{I}} \frac{1}{b_j} - N \cdot \xi}{\sum_{i=1}^{\mathscr{I}} \frac{1}{\lambda_i^* b_i}}$$

trades because $N \cdot \xi$ equals the value of the market portfolio at time 1.

It can be shown that

$$R_Y = \eta r_m$$

for some constant $\eta \neq 0$.

Proof Define $X_m = N \cdot \xi$ and $x_m = N \cdot s$. These represent the value of the market portfolio at times 1 and 0, respectively. Define $A = \frac{\sum_{j=1}^{\mathscr{I}} \frac{1}{b_j}}{\sum_{i=1}^{\mathscr{I}} \frac{1}{\lambda_i^* b_i}}$ and $B = \frac{-1}{\sum_{i=1}^{\mathscr{I}} \frac{1}{\lambda_i^* b_i}}$.

Then, $Y = A + BX_m$ and

$$R_Y = \frac{A + BX_m - (A + Bx_m)}{A + Bx_m} = \frac{B(X_m - x_m)}{A + Bx_m} = \frac{Bx_m r_m}{A + Bx_m}.$$

Defining $\eta = \frac{Bx_m}{A+Bx_m}$ completes the proof.

Then, substitution into the beta model Theorem 17.5, and algebra gives the final result

$$E[R_j] = \frac{\operatorname{cov}[R_j, \eta r_m]}{\operatorname{var}[\eta r_m]} E\left[\eta r_m\right] = \frac{\operatorname{cov}[R_j, r_m]}{\operatorname{var}[r_m]} E\left[r_m\right].$$

This completes the remark.

17.8 Notes

There has been much written on the static CAPM and the mean-variance efficient frontier. Excellent presentations of the traditional approach to this model include Back [5], Bjork [14], Dana and Jeanblanc [42], Duffie [52], Huang and Litzenberger [72], Ingersoll [74], Merton [140], and Skiadas [171].

Appendix

Proof (Exchange of sup and $E[\cdot]$ operator)
It is trivial that

$$\sup_{X \in L_S} E\left[U_i(X) - yXY\right] \leq E[\sup_{X \in L_S}(U_i(X) - yXY)].$$

We want to prove the opposite inequality.
Since U_i is strictly concave, there exists a unique solution X^* to

$$\sup_{X_T \in L_S}\left[U_i(X) - yXY\right] = \left[U_i(X^*) - yX^*Y\right].$$

But,

$$\sup_{X_T \in L_S} E\left[U_i(X) - yXY\right] \geq E\left[U_i(X^*) - yX^*Y\right]$$

$$= E[\sup_{X \in L_S}(U_i(X) - yXY)],$$

which completes the proof.

Proof (Exchange of $E[\cdot]$ and Derivative)

$$\lim_{\Delta \to 0} \frac{\tilde{U}_i((y + \Delta)Y) - \tilde{U}_i(yY)}{\Delta} = \lim_{\Delta Y \to 0} \frac{\tilde{U}_i((y + \Delta)Y) - \tilde{U}_i(yY)}{(y + \Delta)Y - yY} \cdot \lim_{\Delta \to 0} \frac{(y + \Delta)Y - yY}{\Delta}$$

$$= \lim_{\Delta Y \to 0} \frac{\tilde{U}_i((y + \Delta)Y) - \tilde{U}_i(yY)}{(y + \Delta)Y - yY} \cdot Y = \lim_{\Delta \to 0} \tilde{U}_i'(\kappa Y)Y$$

a.s. $\mathbb{P}$.
The last equality follows from the mean value theorem (Guler [66, p. 3]), i.e. there exists a $\kappa \in [y, y + \Delta]$ such that

$$\tilde{U}_i((y + \Delta)Y) - \tilde{U}_i(yY) = \tilde{U}_i'(\kappa Y)\left[(y + \Delta)Y - yY\right].$$

Thus,

$$\frac{\partial E[\tilde{U}_i(yY)]}{\partial y} = \lim_{\Delta \to 0} \frac{E\left[\tilde{U}_i((y + \Delta)Y) - \tilde{U}_i(yY)\right]}{\Delta} = \lim_{\Delta \to 0} E[\tilde{U}_i'(\kappa Y)Y].$$

But, $0 \leq -Y\tilde{U}'(\kappa Y) = Y\left[\frac{1}{b_i}(1 - \kappa Y)\right] \leq \frac{Y}{b_i} + \frac{\kappa}{b_i}Y^2$ is dominated by this $\mathbb{P}$-integrable random variable.

Using the dominated convergence theorem,

$$\lim_{\Delta \to 0} E[\tilde{U}_i'(\kappa Y)Y] = E[\lim_{\Delta \to 0} \tilde{U}_i'(\kappa Y)Y] = E[\tilde{U}_i'(yY)Y].$$

The last equality follows from the continuity of $\tilde{U}_i'(\cdot)$. The continuity of $\tilde{U}_i'(\cdot)$ follows because $\tilde{U}_i'(\cdot)$ is a strictly increasing function, which is therefore differentiable a.s. $\mathbb{P}$ (see Royden [160, Theorem 2, p. 96]), and hence continuous. This completes the proof.

Lemma 17.7 (Existence and Characterization of the Aggregate Utility Function) *Assume* $0 < x_i < \frac{1}{b_i}$ *for all* i*. Then, for* $x \in (0, \sum_{i=1}^{\mathcal{I}} \frac{1}{b_i})$*,*

$$U(x, \lambda) = \sup_{x = x_1 + \cdots + x_{\mathcal{I}}} \left\{ \sum_{i=1}^{\mathcal{I}} \lambda_i \left[x_i - \frac{b_i}{2} x_i^2 \right] \right\} \tag{17.33}$$

exists with

$$U'(x, \lambda) = \frac{\sum_{j=1}^{\mathcal{I}} \frac{1}{b_j} - x}{\sum_{i=1}^{\mathcal{I}} \frac{1}{\lambda_i b_i}} > 0.$$

Proof Note that $0 < x_i < \frac{1}{b_i}$ for all i implies that $0 < \sum_{i=1}^{\mathcal{I}} x_i < \sum_{i=1}^{\mathcal{I}} \frac{1}{b_i}$.

Also note that $\sum_{i=1}^{\mathcal{I}} \lambda_i \left[x_i - \frac{b_i}{2} x_i^2 \right]$ is strictly increasing and strictly concave on $0 < x < \sum_{i=1}^{\mathcal{I}} \frac{1}{b_i}$.

Define the Lagrangian

$$\mathcal{L} \equiv \sum_{i=1}^{\mathcal{I}} \lambda_i \left[x_i - \frac{b_i}{2} x_i^2 \right] + \mu \left(x - x_1 - \cdots - x_{\mathcal{I}} \right).$$

The necessary and sufficient first-order conditions are
$\lambda_i [1 - b_i x_i] = \mu$ for all i, i.e. $\lambda_1 [1 - b_1 x_1] = \lambda_i [1 - b_i x_i]$ for all i.
Algebra generates $\frac{\lambda_1 - \lambda_i}{\lambda_i b_i} - \frac{\lambda_1 b_1}{\lambda_i b_i} x_1 = -x_i$.
Using the constraint $x - x_1 - \cdots - x_{\mathcal{I}} = 0$, substitution and algebra yields

$$x + \sum_{i=2}^{\mathcal{I}} \frac{\lambda_1 - \lambda_i}{\lambda_i b_i} - x_1 \sum_{i=1}^{\mathcal{I}} \frac{\lambda_1 b_1}{\lambda_i b_i} = 0.$$

Note that

$$\sum_{i=2}^{\mathcal{I}} \frac{\lambda_1 - \lambda_i}{\lambda_i b_i} = \lambda_1 \sum_{i=1}^{\mathcal{I}} \frac{1}{\lambda_i b_i} - \sum_{i=1}^{\mathcal{I}} \frac{1}{b_i}.$$

Substitution and algebra implies that

$$\frac{x + \lambda_1 \sum_{i=1}^{\mathcal{I}} \frac{1}{\lambda_i b_i} - \sum_{i=1}^{\mathcal{I}} \frac{1}{b_i}}{\lambda_1 b_1 \sum_{i=1}^{\mathcal{I}} \frac{1}{\lambda_i b_i}} = x_1.$$

This is true for any j, hence algebra yields

$$\left(\frac{1}{\sum_{i=1}^{\mathcal{I}} \frac{1}{\lambda_i b_i}}\right) \frac{1}{\lambda_j b_j} x + \frac{1}{b_j} - \left(\frac{\sum_{i=1}^{\mathcal{I}} \frac{1}{b_i}}{\sum_{i=1}^{\mathcal{I}} \frac{1}{\lambda_i b_i}}\right) \frac{1}{\lambda_j b_j} = x_j$$

for all j.

Substitution of the optimal solution into the objective function yields the value function

$$\begin{aligned} U(x,\lambda) &= \sum_{j=1}^{\mathcal{I}} \lambda_j x_j - \sum_{j=1}^{\mathcal{I}} \frac{\lambda_j b_j}{2} x_j^2 \\ &= \sum_{j=1}^{\mathcal{I}} \lambda_j \left[\left(\frac{1}{\sum_{i=1}^{\mathcal{I}} \frac{1}{\lambda_i b_i}}\right) \frac{1}{\lambda_j b_j} x + \frac{1}{b_j} - \frac{\sum_{i=1}^{\mathcal{I}} \frac{1}{b_i}}{\sum_{i=1}^{\mathcal{I}} \frac{1}{\lambda_i b_i}} \frac{1}{\lambda_j b_j} \right] \\ &\quad - \sum_{j=1}^{\mathcal{I}} \frac{\lambda_j b_j}{2} \left[\left(\frac{1}{\sum_{i=1}^{\mathcal{I}} \frac{1}{\lambda_i b_i}}\right) \frac{1}{\lambda_j b_j} x + \frac{1}{b_j} - \frac{\sum_{i=1}^{\mathcal{I}} \frac{1}{b_i}}{\sum_{i=1}^{\mathcal{I}} \frac{1}{\lambda_i b_i}} \frac{1}{\lambda_j b_j} \right]^2. \end{aligned}$$

Algebra gives

$$\begin{aligned} U(x,\lambda) = &\left[\frac{\sum_{j=1}^{\mathcal{I}} \frac{1}{b_j}}{\sum_{i=1}^{\mathcal{I}} \frac{1}{\lambda_i b_i}} x + \sum_{j=1}^{\mathcal{I}} \frac{\lambda_j}{b_j} - \frac{\left[\sum_{i=1}^{\mathcal{I}} \frac{1}{b_i}\right]^2}{\sum_{i=1}^{\mathcal{I}} \frac{1}{\lambda_i b_i}} \right] \\ &- \frac{1}{\left[\sum_{i=1}^{\mathcal{I}} \frac{1}{\lambda_i b_i}\right]^2} \sum_{j=1}^{\mathcal{I}} \frac{1}{2} \frac{1}{\lambda_j b_j} \left[x - \sum_{i=1}^{\mathcal{I}} \frac{1}{b_i} + \left(\sum_{i=1}^{\mathcal{I}} \frac{1}{\lambda_i b_i}\right) \lambda_j \right]^2. \end{aligned}$$

Taking the first and second derivative shows that the value function is strictly increasing and strictly concave on its domain.

Indeed,

$$U'(x,\lambda) = \frac{\sum_{j=1}^{\mathcal{I}} \frac{1}{b_j}}{\sum_{i=1}^{\mathcal{I}} \frac{1}{\lambda_i b_i}} - \frac{1}{\left[\sum_{i=1}^{\mathcal{I}} \frac{1}{\lambda_i b_i}\right]^2} \sum_{j=1}^{\mathcal{I}} \frac{1}{\lambda_j b_j} \left[x - \sum_{i=1}^{\mathcal{I}} \frac{1}{b_i} + \left(\sum_{i=1}^{\mathcal{I}} \frac{1}{\lambda_i b_i}\right) \lambda_j \right].$$

Algebra gives

$$U'(x, \lambda) = \frac{\sum_{j=1}^{\mathscr{I}} \frac{1}{b_j} - x}{\sum_{i=1}^{\mathscr{I}} \frac{1}{\lambda_i b_i}}.$$

This completes the proof.

Lemma 17.8 (Representative Trader Optimization Problem) *Assume*

(i) $\mathfrak{M} \neq \emptyset$,
(ii) *there exists a* $Y \in M$ *such that* $Y \in \text{span}(1, S_1, \ldots, S_n)$, and
(iii) $0 < x < \sum_{i=1}^{\mathscr{I}} \frac{1}{b_i}$.

Then, for any $\lambda = (\lambda_1, \ldots, \lambda_{\mathscr{I}}) \in \mathbb{R}_{++}^{\mathscr{I}}$*, the representative trader's optimization problem*

$$\sup_{\alpha \in \mathbb{R}^n} E\left[U(X, \lambda)\right],$$

where $U(X, \lambda)$ *is the aggregate utility function in expression (17.33),*

$$X \in \left(0, \sum_{i=1}^{\mathscr{I}} \frac{1}{b_i}\right), \quad \textit{and} \quad X = x_i + \sum_{j=1}^{n} \alpha_j \left[S_j - s_j\right]$$

has a solution.

Proof By Lemma 17.7, given hypotheses (i)–(iii), $U(x, \lambda)$ satisfies the identical hypotheses of the individual trader's optimization problem in Sect. 17.4 above. Hence, the arguments provided therein prove that the representative trader's optimization problem has a solution. This completes the proof.

Lemma 17.9 (Reflects Economy's Optimal Wealths and Equilibrium) *Let s be a price for the economy* $\left\{(\mathscr{F}, \mathbb{P}), (N_0 = 0, N), \left(\mathbb{P}, U_i, \left(e_0^i, e^i\right)\right)_{i=1}^{\mathscr{I}}\right\}$.

Assume that

(i) $\mathfrak{M} \neq \emptyset$,
(ii) *there exists a* $Y \in M$ *such that* $Y \in \text{span}(1, S_1, \ldots, S_n)$, and
(iii) $0 < x_i < \frac{1}{b_i}$ *for all i.*

Consider the aggregate utility function $U(x, \lambda^*)$*, where*

$$\lambda_i^* = \frac{y_1(s)}{y_i(s)}$$

with $y_i(s)$ *trader i's shadow price of the budget constraint given the price s for* $i = 1, \ldots, \mathscr{I}$.

Then,

(Part 1) the representative trader economy

$$\left\{(\mathscr{F}, \mathbb{P}), (N_0 = 0, N), \left(\mathbb{P}, U(\lambda^*), (N_0 = 0, N)\right)\right\}$$

reflects the economy's optimal wealths, and
(Part 2) if s is an equilibrium price, then the representative trader economy

$$\left\{(\mathscr{F}, \mathbb{P}), (N_0 = 0, N), \left(\mathbb{P}, U(\lambda^*), (N_0 = 0, N)\right)\right\}$$

reflects the economy's equilibrium.

Proof (Part 1: Reflecting the economy's optimal wealth)

This is the proof of Theorem 14.4 in the representative trader Chap. 14, modified for the current setting. For the given price s, from Sect. 17.4, for each trader i the optimal wealth X^i is characterized by the expression

$$U_i'\left(X^i\right) = y_i(s)Y(s), \tag{17.34}$$

where $Y(s)$ depends on the price s.

For a given λ, optimizing the representative trader's wealth, we have using Lemma 17.8 that

$$U'(X^\lambda, \lambda) = y^\lambda(s)Y(s),$$

where from Lemma 17.7, the individual components $\tilde{X}^i$ satisfy

$$\lambda_i U_i'(\tilde{X}^i) = \mu \quad \text{for all } i, \qquad \text{and} \tag{17.35}$$

$$X^\lambda = \sum_{i=1}^{\mathscr{I}} \tilde{X}^i.$$

We want to show $X^i = \tilde{X}^i$ for all i. Using expression (17.34), this will be satisfied if and only if

$$U_i'(\tilde{X}^i) = y_i(s)Y(s) \quad \text{for all } i.$$

Using expression (17.35), this will be satisfied if and only if

$$\lambda_1 y_1(s)Y(s) = \lambda_i y_i(s)Y(s) \quad \text{for } i = 2, \ldots, \mathscr{I}.$$

The solution $\lambda^* = (\lambda_1^*, \ldots, \lambda_{\mathscr{I}}^*)$ to this system of $(\mathscr{I} - 1)$ equations in $\mathscr{I}$ unknowns is

$$\lambda_i^* = \frac{y_1(s)}{y_i(s)} \quad \text{for } i = 1, 2, \ldots, \mathscr{I}.$$

Note that the solution $\lambda^* = (\lambda_1^*, \ldots, \lambda_{\mathscr{I}}^*)$ is a constant. This completes the proof of (Part 1).

(Part 2: Reflecting the economy's equilibrium)

This is the proof of Theorem 14.5 in the representative trader Chap. 14, modified for the current setting. By Part 1, given the *equilibrium* price process s for the original economy, the representative trader economy

$$\left\{(\mathscr{F}, \mathbb{P}), (N_0 = 0, N), \left(\mathbb{P}, U(\lambda^*), (N_0 = 0, N)\right)\right\}$$

reflects the original economy's optimal wealths, i.e. $X^{\lambda^*} = \sum_{i=1}^{\mathscr{I}} X^i$. A trading strategy that achieves this wealth is

$$\alpha_0^{\lambda^*} = \sum_{i=1}^{\mathscr{I}} \alpha_0^i \quad \text{and} \quad \alpha^{\lambda^*} = \sum_{i=1}^{\mathscr{I}} \alpha^i.$$

In the original economy's equilibrium

$$0 = \sum_{i=1}^{\mathscr{I}} \alpha_0^i \quad \text{and} \quad N = \sum_{i=1}^{\mathscr{I}} \alpha^i.$$

This implies that the trading strategy $\alpha_0^{\lambda^*} = 0$ and $\alpha^{\lambda^*} = N$ generates the representative trader's optimal wealth. By uniqueness of the trading strategy generating any terminal wealth, this is the representative trader's optimal trading strategy. Hence, the representative trader economy that reflects the original economy's optimal demands is in equilibrium with price process S. This completes the proof of (Part 2), which completes the proof of the lemma.

Lemma 17.10 (Sufficient Conditions for an Equilibrium) *Given is an economy*

$$\left\{(\mathscr{F}, \mathbb{P}), (N_0 = 0, N), \left(\mathbb{P}, U_i, \left(e_0^i, e^i\right)\right)_{i=1}^{\mathscr{I}}\right\}.$$

Assume that

(i) $0 < x_i < \frac{1}{b_i}$ *for all* i, *and*

(ii) *for all* λ, *the aggregate utility function* $E\left[U'(N \cdot \xi, \lambda)\right] < \infty$, $E\left[U(N \cdot \xi, \lambda)\right] < \infty$, *and* $0 < \frac{d\mathbb{Q}_\lambda}{d\mathbb{P}} \equiv \frac{U'(N \cdot \xi, \lambda)}{E[U'(N \cdot \xi, \lambda)]} \in \text{span}(1, S_1, \ldots, S_n)$.

Define

$$s^\lambda = E_\lambda[\xi], \tag{17.36}$$

where $E_\lambda[\cdot]$ *is expectation with respect to* $\mathbb{Q}_\lambda$, *(iii)* *there exists a $\tilde{\lambda}$ such that*

$$\tilde{\lambda}_i = \frac{y_1(s^{\tilde{\lambda}})}{y_i(s^{\tilde{\lambda}})} \tag{17.37}$$

with $y_i(s^{\tilde{\lambda}})$ trader i's shadow price of the budget constraint given the price $s^{\tilde{\lambda}}$ for $i = 1, \ldots, \mathscr{I}$.

Then, $s^{\tilde{\lambda}}$ is an equilibrium price for the economy.

Proof This is the proof of Theorem 14.7 in the representative trader Chap. 14, modified for the current setting.

(Step 1) Consider the collection of representative trader economies $\{(\mathscr{F}, \mathbb{P}), (N_0 = 0, N), (\mathbb{P}, U(\lambda), (N_0 = 0, N))\}$ indexed by λ.

Note that hypothesis (i) implies $0 < \sum_{i=1}^{\mathscr{I}} x_i = x < \sum_{i=1}^{\mathscr{I}} \frac{1}{b_i}$. By the definition of the price s^{λ}, $\mathfrak{M} \neq \emptyset$. Hence, $U(x, \lambda)$ has an optimal portfolio with initial endowment $(0, N)$, see Lemma 17.8.

We first prove using the price s^{λ} that the representative trader's optimal trading strategy has zero units in the mma and N shares in the risky assets, i.e. $\alpha_0^{\lambda}(s^{\lambda}) = 0$, and $\alpha^{\lambda}(s^{\lambda}) = N$, where $X^{\lambda}(s^{\lambda})$ denotes the representative trader's optimal wealth. Note the explicit dependence of all the previous quantities on λ and s^{λ}. This implies that the representative trader does not trade.

We note that any nonnegative wealth trading strategy for the representative trader $X \in L_{s^{\lambda}}$ is a martingale under $\mathbb{Q}_{\lambda}$.

Consider the trading strategy $(0, N)$. Note that $N \cdot s^{\lambda}$ is a $\mathbb{Q}_{\lambda}$-martingale. This implies that

$$x = E\left[\frac{U'(N \cdot \xi, \lambda)}{E\left[U'(N \cdot \xi, \lambda)\right]} N \cdot \xi\right] = E\left[\frac{U'(N \cdot s, \lambda)}{E\left[U'(N \cdot s, \lambda)\right]} N \cdot S\right].$$

We now show that this trading strategy is optimal. Consider an arbitrary $X \in L_{s^{\lambda}}$. Then,

$$E\left[U(X, \lambda) - U(N \cdot S, \lambda)\right] \leq E\left[U'(N \cdot S, \lambda)(X_T - N \cdot S)\right]$$

by the concavity of U. But,

$$E\left[U'(N \cdot S, \lambda)(X - N \cdot S)\right] = E\left[U'(N \cdot S, \lambda)\right] E\left[\frac{d\mathbb{Q}_{\lambda}}{d\mathbb{P}}(X - N \cdot S)\right] = 0$$

since

$$E\left[\frac{d\mathbb{Q}_{\lambda}}{d\mathbb{P}}(X - N \cdot S)\right] = x - x = 0.$$

This completes the proof of Step 1.

(Step 2) Fix a λ. Consider an economy

$$\left\{(\mathscr{F}, \mathbb{P}), (N_0 = 0, N), \left(\mathbb{P}, U_i, \left(e_0^i, e^i\right)\right)_{i=1}^{\mathscr{I}}\right\}$$

with the price s^λ.

Hypothesis (i) in conjunction with $\mathfrak{M} \neq \emptyset$ for the price s^λ implies that the individual traders have an optimal wealth. For each λ, let $X^i(s^\lambda)$ denote the optimal demands for trader i in the economy with the price s^λ. Note the explicit dependence of the optimal demands on the price s^λ.

(Step 3) Fix the $\tilde{\lambda}$ that satisfies hypothesis (iii).

Jointly consider the economy

$$\left\{(\mathscr{F}, \mathbb{P}), (N_0 = 0, N), \left(\mathbb{P}, U_i, \left(e_0^i, e^i\right)\right)_{i=1}^{\mathscr{I}}\right\}$$

and the collection of representative trader economies

$$\{(\mathscr{F}, \mathbb{P}), (N_0 = 0, N), (\mathbb{P}, U(\lambda), (N_0 = 0, N))\}$$

indexed by λ using the *same* price $\tilde{s} \equiv s^{\tilde{\lambda}}$. Note that this is the identical price process for all the representative trader economies indexed by λ.

Since $\tilde{s} = s^{\tilde{\lambda}}$, $\mathfrak{M} \neq \emptyset$ for the price $\tilde{s}$.

By Lemma 17.9, $\tilde{\lambda}$ is such that the representative trader economy

$$\left\{(\mathscr{F}, \mathbb{P}), (N_0 = 0, N), (\mathbb{P}, U(\tilde{\lambda}), (N_0 = 0, N))\right\}$$

reflects the aggregate optimal wealths in the original economy given the price $\tilde{s}$, i.e.

$$X^{\tilde{\lambda}}(\tilde{s}) = \sum_{i=1}^{\mathscr{I}} X^i(\tilde{s}). \tag{17.38}$$

We note that the trading strategy $\left(\sum_{i=1}^{\mathscr{I}} \alpha_0{}^i(\tilde{s}), \sum_{i=1}^{\mathscr{I}} \alpha^i(\tilde{s})\right)$ generates $\sum_{i=1}^{\mathscr{I}} X^i(\tilde{s})$ and, therefore, $X^{\tilde{\lambda}}(\tilde{s})$.

(Step 4) We claim that $\tilde{s} = s^{\tilde{\lambda}}$ is an equilibrium price for the economy and the representative trader economy

$$\left\{(\mathscr{F}, \mathbb{P}), (N_0 = 0, N), \left(\mathbb{P}, U(\tilde{\lambda}), (N_0 = 0, N)\right)\right\}.$$

(Part a) First, by (Step 1) for the price $s^{\tilde{\lambda}}$ the representative trader with aggregate utility $U(\tilde{\lambda})$'s trading strategy has zero units in the mma and N shares in the risky assets, i.e. $\alpha_0^{\tilde{\lambda}}(s^{\tilde{\lambda}}) = 0$, and $\alpha^{\tilde{\lambda}}(s^{\tilde{\lambda}}) = N$. This implies that $s^{\tilde{\lambda}}$ is an equilibrium price for the representative trader economy

$$\left\{(\mathscr{F}, \mathbb{P}), (N_0 = 0, N), \left(\mathbb{P}, U(\tilde{\lambda}), (N_0 = 0, N)\right)\right\}.$$

(Part b) By expression (17.38), we get

$$X^{\tilde{\lambda}}(s^{\tilde{\lambda}}) = \sum_{i=1}^{\mathscr{I}} X^i(s^{\tilde{\lambda}}).$$

And, by uniqueness of the representative trader's optimal trading strategy generating $X^{\tilde{\lambda}}(s^{\tilde{\lambda}})$,

$$0 = \sum_{i=1}^{\mathscr{I}} \alpha_0{}^i(s^{\tilde{\lambda}}) \quad \text{and} \quad N = \sum_{i=1}^{\mathscr{I}} \alpha^i(s^{\tilde{\lambda}}),$$

which implies that aggregate demand equals aggregate supply in the original economy. This proves that $s^{\tilde{\lambda}}$ is an equilibrium price process for the economy

$$\left\{(\mathscr{F}, \mathbb{P}), (N_0 = 0, N), \left(\mathbb{P}, U_i, \left(e_0^i, e^i\right)\right)_{i=1}^{\mathscr{I}}\right\}.$$

This completes the proof.

Part IV
Trading Constraints

Overview

This Part of the book adds trading constraints to the previous markets, examples of which include short sale restrictions and margin requirements. Trading constraints are a reality in financial markets and it is instructive to understand how the previous results are modified by their inclusion. The key insight of this analysis is that

a market with trading constraints is isomorphic to an incomplete market with no trading constraints.

Once this isomorphism is identified, a market with trading constraints can be analyzed using the theorems from Parts II and III for an incomplete market without trading constraints. In addition, this isomorphism implies that any result generated in an incomplete market without trading constraints will have an analogue in a market with trading constraints and vice-versa.

This Part extends the trading constrained hedging and portfolio optimization analysis in Karatzas and Shreve [118], Chapter 5 to discontinuous sample path processes. We follow the approach of He and Pearson [69] and Karatzas and Zitkovic [119] in modeling trading constraints as a restriction on the number of shares traded as opposed to restrictions on portfolio weights [40, 118] or dollar holdings [37, 38]. For simplicity of presentation, this Part assumes that there is no intermediate consumption in the trader's portfolio optimization problem. This restriction can be easily removed along the lines discussed in Parts II and III.

There are two important implications of trading constraints, in contrast to an unconstrained economy.

1. Under NFLVR in the constrained economy, more types of price bubbles exist. Price bubbles exist not only due to the fact that the retrade value of an asset exceeds its fundamental value (the present value of holding the asset until liquidation), but also because trading constraints inhibit trading strategies that would otherwise eliminate these bubbles.

2. In equilibrium, asset expected returns reflect additional risk premiums due to trading constraints. This is because trading constraints reduce a trader's ability to maximize their preferences. Constraints make trading assets more risky than they would otherwise be. These additional risk premiums imply that more risk factors have nonzero risk premium in a trading constrained economy. Hence, when fitting multiple-factor models to risky asset returns, there will be more priced risk factors in a trading constrained economy than in an unconstrained economy.

Since this part of the book is an extension of Parts I–III, no end of chapter notes with additional textbooks on the topic are provided. The end of chapter notes in Parts I–III corresponding to the chapters extended in this part provide the appropriate references. This Part is based on Jarrow [93].

Chapter 18
The Trading Constrained Market

This chapter introduces trading constraints to the markets studied in Parts I–III of this book. Most of the structure extends in a straightforward fashion.

18.1 The Set-Up

Given is a normalized market $(S, (\mathscr{F}_t), \mathbb{P})$ where the money market account $B_t \equiv 1$. Trading strategies are represented by the number of shares held for each time and state $(t, \omega) \in [0, T] \times \Omega$, i.e. $\alpha_0(t, \omega)$, $\alpha_t = (\alpha_1(t, \omega), \ldots, \alpha_n(t, \omega))' \in \mathbb{R}^n$. For the purposes of this chapter, we need to add more arguments to some of the sets used previously in Parts I–III. Recall from the fundamental theorems Chap. 2 the following definitions.

The set of admissible s.f.t.s.'s

$$\mathscr{A}(x, S) = \{(\alpha_0, \alpha) \in (\mathscr{O}, \mathscr{L}(S)) : X_t = \alpha_0(t) + \alpha_t \cdot S_t, \ \exists c \leq 0, \\ X_t = x + \textstyle\int_0^t \alpha_u \cdot dS_u \geq c, \text{ for all } t \in [0, T]\Big\}.$$

The set of wealth processes generated by the admissible s.f.t.s.'s

$$\mathscr{X}^e(x, S) = \left\{X \in \mathscr{L}^0_+ : \exists (\alpha_0, \alpha) \in \mathscr{A}(x, S), \ X_t = x + \int_0^t \alpha_u \cdot dS_u, \ \forall t \in [0, T]\right\}.$$

The terminal wealths generated by the admissible s.f.t.s.'s

$$\mathscr{C}^e(x, S) = \left\{X_T \in L^0_+ : \exists Z \in \mathscr{X}^e(x, S), \ Z_T = X_T\right\}.$$

R. A. Jarrow, *Continuous-Time Asset Pricing Theory*, Springer Finance,
https://doi.org/10.1007/978-3-319-77821-1_18

The sets of terminal wealths generated by the admissible s.f.t.s.'s with free disposal

$$\mathscr{X}(x,S) = \left\{X \in \mathscr{L}^0_+ : \exists(\alpha_0,\alpha) \in \mathscr{A}(x,S), x + \int_0^t \alpha_u \cdot dS_u \geq X_t, \ \forall t \in [0,T]\right\}$$

$$\mathscr{C}(x,S) = \left\{X_T \in L^0_+ : \exists Z \in \mathscr{X}(x,S), \ Z_T = X_T\right\}.$$

The sets of equivalent probability measures

$$\mathfrak{M}(S) = \{\mathbb{Q} \sim \mathbb{P} : \ S \text{ is a } \mathbb{Q}\text{-martingale}\},$$

$$\mathfrak{M}_l(S) = \{\mathbb{Q} \sim \mathbb{P} : \ S \text{ is a } \mathbb{Q}\text{-local martingale}\}$$
$$= \left\{\mathbb{Q} \sim \mathbb{P} : \ X \text{ is a } \mathbb{Q}\text{-local martingale}, \ X = 1 + \textstyle\int \alpha \cdot dS, \ (\alpha_0,\alpha) \in \mathscr{A}(1,S)\right\},$$

$$\mathfrak{M}_s(S) = \{\mathbb{Q} \sim \mathbb{P} : \ S \text{ is a } \mathbb{Q}\text{-supermartingale}\}.$$

The last set of equivalent probability measures that makes S a supermartingale was not used in Parts I–III, but it is relevant below. Note that these definitions make explicit the dependence of these sets on the asset price process in the market.

18.2 Trading Constraints

This section introduces the trading constraints imposed in the market. The trading constraints restrict the admissible s.f.t.s.'s that an investor can employ to lie in a nonempty, closed, and convex cone.

Assumption (Trading Constraints) *The admissible s.f.t.s.* $(\alpha_0,\alpha) \in \mathscr{A}(x,S)$ *must satisfy*

$$(\alpha_0(t), \alpha_t) \in K \subset \mathbb{R}^{n+1} \textit{for all } t,$$

where $K \subset \mathbb{R}^{n+1}$ *is a nonempty, closed, convex cone with* $(0,0) \in K$.

As shown, it is also assumed that the constraint set K is such that $(0,0) \in K$, which implies that zero trades in the mma and the risky assets satisfy the trading constraint.

Remark 18.1 (Stochastic Trading Constraint Set) The above assumption can be generalized to allow the trading constraint set $K : [0,T] \times \Omega \to 2^{\mathbb{R}^{n+1}}$ to be a set function mapping $(t,\omega) \in [0,T] \times \Omega$ into the (space of) subsets of $\mathbb{R}^{n+1}$ denoted $2^{\mathbb{R}^{n+1}}$. In this case, we need to assume that this set function is $\mathscr{F}_t$-measurable in the following sense, $\{\omega \in \Omega : K_t(\omega) \cap A \neq \emptyset\} \in \mathscr{F}_t$ for every open set $A \subset \mathbb{R}^{n+1}$. All of the subsequent theorems generalize to this stochastic extension of the constraint set. This completes the remark.

We define the trading constrained set of admissible s.f.t.s.'s as

$$\mathscr{A}(x, S, K) = \{(\alpha_{0,}\alpha) \in \mathscr{A}(x, S) : (\alpha_0(t), \alpha_t) \in K \text{ for all } t\}.$$

Given the constrained set of admissible s.f.t.s.'s $\mathscr{A}(x, S, K)$, these are the corresponding constrained sets of wealth processes

$$\mathscr{X}^e_{\mathfrak{C}}(x, S) = \Big\{X \in \mathscr{L}^0_+ : \exists(\alpha_0, \alpha) \in \mathscr{A}(x, S, K),\ X_t = x + \int_0^t \alpha_u \cdot dS_u,$$
$$\forall t \in [0, T]\}$$

and terminal wealths

$$\mathscr{C}^e_{\mathfrak{C}}(x, S) = \Big\{X_T \in L^0_+ : \exists Z \in \mathscr{X}^e_{\mathfrak{C}}(x, S),\ Z_T = X_T\Big\}.$$

The set of terminal wealths generated by the admissible s.f.t.s.'s with free disposal are denoted as follows

$$\mathscr{X}_{\mathfrak{C}}(x, S) = \Big\{X \in \mathscr{L}^0_+ : \exists(\alpha_0, \alpha) \in \mathscr{A}(x, S, K),\ x + \int_0^t \alpha_u \cdot dS_u \geq X_t,$$
$$\forall t \in [0, T]\}$$

$$\mathscr{C}_{\mathfrak{C}}(x, S) = \Big\{X_T \in L^0_+ : \exists Z \in \mathscr{X}_{\mathfrak{C}}(x, S,),\ Z_T = X_T\Big\}.$$

Using these constrained sets of wealth processes the definitions of NA, NUPBR, NFLVR, and ND in the fundamental theorems Chap. 2 extend in a straightforward fashion. We now state these extensions.

Definition 18.1 (No Arbitrage Constrained $(NA_{\mathfrak{C}})$**)** An admissible s.f.t.s. $(\alpha_0, \alpha) \in \mathscr{A}(x, S, K)$ with wealth process $X \in \mathscr{X}^e_{\mathfrak{C}}(x, S)$ is a (simple) *arbitrage opportunity* if

(i) $X_0 = x = 0$ (zero investment),
(ii) $X_T \geq 0$ with $\mathbb{P}$ probability one, and
(iii) $\mathbb{P}(X_T > 0) > 0$.

A market satisfies $NA_{\mathfrak{C}}$ if there are no trading strategies that are arbitrage opportunities.

Definition 18.2 (No Unbounded Profit with Bounded Risk Constrained $(NUPBR_{\mathfrak{C}})$**)** A sequence of admissible s.f.t.s.'s $(\alpha_0, \alpha)_n \in \mathscr{A}(x, S, K)$ with wealth processes $X^n \in \mathscr{X}^e_{\mathfrak{C}}(x, S)$, where $X^n \geq 0$, generates *unbounded profits* *with bounded risk constrained* ($UPBR_{\mathfrak{C}}$) if

(i) $X_0^n = x > 0$,
(ii) $X_T^n \geq 0$ a.s. $\mathbb{P}$, and
(iii) $\lim_{m\to\infty}\left(\sup_n \mathbb{P}\left(X_T^n > m\right)\right) > 0$.

If no such sequence of trading strategies exists, then the market satisfies NUPBR$_{\mathfrak{C}}$.

Definition 18.3 (No Free Lunch with Vanishing Risk Constrained ($NFLVR_{\mathfrak{C}}$)**)** A *free lunch with vanishing risk constrained* ($FLVR_{\mathfrak{C}}$) is a sequence of zero initial investment admissible s.f.t.s.'s $(\alpha_0, \alpha)_n \in \mathscr{A}(0, S, K)$ with initial value $x > 0$, wealth processes $X_t^n \in \mathscr{X}_{\mathfrak{C}}^e(x, S)$, $X^n \geq 0$, and an $\mathscr{F}_T$-measurable random variable $f \geq x$ with $\mathbb{P}(f > x) > 0$ such that $X_T^n \to f$ in probability.

If no such sequence of trading strategies exists, then the market satisfies NFLVR$_{\mathfrak{C}}$.

Definition 18.4 (No Dominance Constrained ($ND_{\mathfrak{C}}$)**)** The ith security $S_i(t)$ is *constrained undominated* if there exists no admissible s.f.t.s. $(\alpha_0, \alpha) \in \mathscr{A}(x, S, K)$ with wealth process $X \in \mathscr{X}_{\mathfrak{C}}^e(x, S)$ and initial wealth $x = S_i(0)$ such that

$$\mathbb{P}\{X_T \geq S_i(T)\} = 1 \quad \text{and} \quad \mathbb{P}\{X_T > S_i(T)\} > 0.$$

A market $(S, (\mathscr{F}_t), \mathbb{P})$ satisfies no dominance constrained (ND$_{\mathfrak{C}}$) if each S_i, $i = 0, \ldots, n$, are constrained undominated.

18.3 Support Functions

This section characterizes the trading constraint set K using support functions. Define the indicator function of the set K, $\delta : \mathbb{R}^{n+1} \to \mathbb{R} \cup \{\infty\}$, by

$$\delta(a) = \begin{cases} 0 & \text{if } a \in K \\ \infty & \text{otherwise.} \end{cases} \tag{18.1}$$

This indicator function is proper, convex, and lower semicontinuous. It is proper because K is nonempty. It is convex because its epigraph is a convex set, and it is lower semicontinuous because the epigraph is closed (see Ruszczynski [162, p. 79]).

Define the *conjugate function* $\delta^* : \mathbb{R}^{n+1} \to \mathbb{R} \cup \{\infty\}$ by

$$\begin{aligned} \delta^*(v) &= \sup_{a\in\mathbb{R}^{n+1}} \left(\langle v, a\rangle - \delta(a)\right) \\ &= \sup_{a\in K} \langle v, a\rangle, \end{aligned} \tag{18.2}$$

where $\langle v, a\rangle = \sum_{i=1}^{n+1} v_i a_i$ is the inner product on $\mathbb{R}^{n+1}$. This is also called the *support function* of the set K. Define the *polar cone*

$$K^o = \{v \in \mathbb{R}^{n+1} : \delta^*(v) < \infty\}. \tag{18.3}$$

This is the set of elements where the support function is finite. We note the following fact. Since K is a cone (see Ruszczynski [162, p. 77]),

$$\delta^*(v) = \begin{cases} 0 & \text{if } v \in K^o \\ \infty & \text{otherwise.} \end{cases} \tag{18.4}$$

Hence, we can show that

$$K^o = \{v \in \mathbb{R}^{n+1} : \langle v, a\rangle \le 0 \text{ for all } a \in K\}.$$

Proof

$$K^o = \left\{v \in \mathbb{R}^{n+1} : \sup_{a \in K} (\langle v, a\rangle) < \infty\right\} = \left\{v \in \mathbb{R}^{n+1} : \langle v, a\rangle \le 0 \text{ for all } a \in K\right\}.$$

The last equality follows because if $\langle v, a\rangle > 0$ for any $a \in K$, then since $\lambda a \in K$ for all $\lambda > 0$ (because K is a cone), we have

$$\sup_{\lambda > 0} (\langle v, \lambda a\rangle) = \sup_{\lambda > 0} (\lambda \langle v, a\rangle) = \infty,$$

contradicting $v \in K^o$. This completes the proof.

This shows that K^o is a cone, i.e. if $v \in K^o$, then $\lambda v \in K^o$ for all $\lambda \ge 0$. It is important to note that $0 \in K^o$. Given these observations, we can prove the following characterization theorem.

Theorem 18.1 (Characterization of the Trading Constraint K)

$$\delta(a) = 0 \quad \Leftrightarrow \quad a \in K \quad \Leftrightarrow \quad \langle v, a\rangle \le 0 \text{ for all } v \in K^o. \tag{18.5}$$

Proof Define the *biconjugate* function $\delta^{**} : \mathbb{R}^{n+1} \to \mathbb{R} \cup \{\infty\}$ by

$$\delta^{**}(a) = \sup_{v \in \mathbb{R}^{n+1}} \big(\langle v, a\rangle - \delta^*(v)\big) = \sup_{v \in K^o} \langle v, a\rangle .$$

It can be shown that for the constraint set K conjugate duality holds (Ruszczynski [162, p. 80]), i.e. $\delta(a) = \delta^{**}(a)$. This gives the following characterization of the constraint set K

$$\big[\delta(a) = 0 \quad \Leftrightarrow \quad a \in K \quad \Leftrightarrow \quad \delta^{**}(a) = 0\big].$$

Define the bipolar set

$$\begin{aligned} K^{oo} &= \{a \in \mathbb{R}^{n+1} : \delta^{**}(a) < \infty\} = \{a \in \mathbb{R}^{n+1} : \sup_{v \in K^o} \langle v, a \rangle < \infty\} \\ &= \{a \in \mathbb{R}^{n+1} : \langle v, a \rangle < \infty \text{ for all } v \in K^o\} \\ &= \{a \in \mathbb{R}^{n+1} : \langle v, a \rangle \leq 0 \text{ for all } v \in K^o\}. \end{aligned}$$

The last inequality follows because if $\langle v, a \rangle > 0$ for any $v \in K^o$, then since $\lambda v \in K^o$ for all $\lambda > 0$ (because K^o is a cone), we have

$$\sup_{\lambda > 0} (\langle \lambda v, a \rangle) = \sup_{\lambda > 0} (\lambda \langle v, a \rangle) = \infty,$$

contradicting $a \in K^{oo}$. Using the bipolar theorem (see Hiriart-Urruty and Lemarechal [71, p. 57]) gives $K = K^{oo}$. Substitution completes the proof.

Lemma 18.1 (Nonnegative Polar Elements) *If* $(0, 1^{(n)}), (1, 0^{(n)}) \in K$, *then all* $(v_0, v^{(n)}) \in K^o$ *satisfy* $v_0 \leq 0$ *and* $v^{(n)} \leq 0$, *where* $x^{(n)} = (x, \ldots, x) \in \mathbb{R}^n$.

Proof By definition, $K^o = \{v \in \mathbb{R}^{n+1} : \langle v, a \rangle \leq 0 \text{ for all } a \in K\}$.

Because K is a cone:

(i) given $(1, 0^{(n)}) \in K$, the only way this can happen for all $(v_0, v^{(n)}) \in K^o$ is if $v_0 \leq 0$;
(ii) given $(0, 1^{(n)}) \in K$, the only way this can happen for all $(v_0, v^{(n)}) \in K^o$ is if $v^{(n)} \leq 0$.

This completes the proof.

Remark 18.2 (No Trading Constraints) When there are no trading constraints, $K = \mathbb{R}^{n+1}$. Here,

$$\delta(a) = \begin{Bmatrix} 0 & \text{if } a \in \mathbb{R}^{n+1} \\ \infty & \text{otherwise} \end{Bmatrix} = 0. \tag{18.6}$$

Simple inspection shows that

$$K^o = \{v \in \mathbb{R}^{n+1} : \langle v, a \rangle \leq 0 \text{ for all } a \in \mathbb{R}^{n+1}\} = \{0\}. \tag{18.7}$$

18.4 Examples (Trading Constraints and Their Support Functions)

This section gives some examples of trading constraint sets K. We study short sale constraints, borrowing constraints, and margin requirements. To do this, fix a pair $(t, \omega) \in [0, T] \times \Omega$. We consider a trading strategy $(\alpha_0(t, \omega), \alpha(t, \omega)) \in K \subset \mathbb{R}^{n+1}$ and an element in its polar cone $(\nu_0(t, \omega), \nu(t, \omega)) \in K^o \subset \mathbb{R}^{n+1}$.

18.4.1 No Trading Constraints

The case of no trading constraints is where

$$K = \left\{(\alpha_0, \alpha) \in \mathbb{R}^{n+1}\right\}.$$

In this case, from Remark 18.2 we have that

$$K^o = \{(\nu_0, \nu) \in \mathbb{R}^{n+1} : (\nu_0, \upsilon) = (0, 0)\}.$$

18.4.2 Prohibited Short Sales

The trading constraint that characterizes a market where there are no short sales is where the risky assets are constrained to be nonnegative, and the mma holdings are unrestricted, i.e.

$$K = \{(\alpha_0, \alpha) \in \mathbb{R}^{n+1} : \alpha_0 \in \mathbb{R}, \alpha \geq 0\}.$$

By inspection, K is a closed and convex cone. The polar cone is

$$K^o = \{(\nu_0, \nu) \in \mathbb{R}^{n+1} : \nu_0 = 0,\ \nu \leq 0\}.$$

We will show later that the elements in the polar cone $(\nu_0, \nu) \in K^o$ correspond to the "shadow prices" of the trading constraints. This implies that there are no costs to holding the mma, but there are costs to holding all of the remaining risky assets.

Proof Note that

$$\delta^*(\upsilon) = \begin{cases} 0 & \text{if } \upsilon \in K^o \\ \infty & \text{otherwise.} \end{cases}$$

Thus, determining when δ^* is finite characterizes K^o. We have

$$\delta^*(\nu_0, \nu) = \sup_{\alpha_0 \in \mathbb{R}, \alpha \geq 0} (\nu_0\alpha_0 + \langle \nu \cdot \alpha \rangle)$$

is finite if and only if $\nu_0 = 0$ and $\nu \leq 0$, in which case

$$\delta^*(\nu_0, \nu) = \sup_{\alpha_0 \in \mathbb{R}, \alpha \geq 0} (\nu_0\alpha_0 + \langle \nu \cdot \alpha \rangle) = 0.$$

Thus, $K^o = \{(\nu_0, \nu) \in \mathbb{R}^{n+1} : \nu_0 = 0,\ \nu \leq 0\}$. This completes the proof.

18.4.3 No Borrowing

The trading constraint that characterizes a market where there is no borrowing is where the mma is constrained to be nonnegative, and the risky asset holdings are unrestricted, i.e.

$$K = \{(\alpha_0, \alpha) \in \mathbb{R}^{n+1} : \alpha_0 \geq 0, \alpha \in \mathbb{R}^n\}.$$

By inspection, K is a closed and convex cone. The polar cone is

$$K^o = \{(\nu_0, \nu) \in \mathbb{R}^{n+1} : \nu_0 \leq 0, \ \nu = 0\}.$$

This implies that there are no costs to holding the risky assets, but there are costs to holding the mma.

Proof Note that

$$\delta^*(v) = \begin{cases} 0 & \text{if } v \in K^o \\ \infty & \text{otherwise.} \end{cases}$$

Thus, determining when δ^* is finite characterizes K^o. We have

$$\delta^*(\nu_0, \nu) = \sup_{\alpha_0 \geq 0, \alpha \in \mathbb{R}} (\nu_0 \alpha_0 + \langle \nu \cdot \alpha \rangle)$$

is finite if and only if $\nu_0 \leq 0$ and $\nu = 0$, in which case

$$\delta^*(\nu_0, \nu) = \sup_{\alpha_0 \geq 0, \alpha \in \mathbb{R}} (\nu_0 \alpha_0 + \langle \nu \cdot \alpha \rangle) = 0.$$

Thus,

$$K^o = \{(\nu_0, \nu) \in \mathbb{R}^{n+1} : \nu_0 \leq 0, \ \nu = 0\}.$$

This completes the proof.

18.4.4 Margin Requirements

The trading constraint that characterizes a market with margin requirements is given by

$$K_t(\omega) = \left\{ (\alpha_0, \alpha) \in \mathbb{R}^{n+1} : \alpha_0 + \sum_{j=1}^{n} \mathfrak{m} \left[(1 + 1_{\alpha_j < 0}) \right] S_j \alpha_j \geq 0 \right\},$$

where the constant $\mathfrak{m} \in [0, 1]$ is the maximum percentage of an asset's wealth that can be borrowed.

As specified, this trading constraint set $K : [0, T] \times \Omega \to 2^{\mathbb{R}^{n+1}}$ is necessarily stochastic, see Remark 18.1.

To understand why this set characterizes margin requirements, we first consider the set

$$\left\{ (\alpha_0, \alpha) \in \mathbb{R}^{n+1} : \alpha_0 \geq (1 + \mathfrak{m}) \left(\sum_{j=1}^{n} \alpha_j^- S_j(t) \right) - \mathfrak{m} \left(\sum_{j=1}^{n} \alpha_j^+ S_j(t) \right) \right\},$$

where $x^+ = \max(x, 0)$ and $x^- = \max(-x, 0)$. Note that $x^+ - x^- = x$. This set is easily seen to reflect the following restrictions on a trading strategy, which characterize margin requirements. Consider a trading strategy involving only the mma and the jth risky asset.

1. If asset j is shorted ($\alpha_j < 0$), then the trading strategy must hold at least $(1 + \mathfrak{m})\alpha_j^-(t)S_j(t)$ units in the mma as margin/collateral. This is the original proceeds from the short plus the margin $\mathfrak{m}$ to protect against losses if the asset price declines.
2. If asset j is purchased ($\alpha_j > 0$), then the trading strategy can borrow no more than $\mathfrak{m}$ percent of its value to finance the purchase. So the units borrowed (shorted) using the mma can be no lower than $-\mathfrak{m}\alpha_j^+(t)S_j(t)$.

Simplifying this set gives the trading constraint $K_t(\omega)$.

Proof

$$\alpha_0 \geq -\mathfrak{m}\,(\alpha \cdot S) + \sum_{j=1}^{n} \max(-\alpha_j, 0) S_j.$$

$$\alpha_0 + \mathfrak{m}\,(\alpha \cdot S) + \sum_{j=1}^{n} min(\alpha_j, 0) S_j \geq 0.$$

$$\alpha_0 + \sum_{j=1}^{n} \mathfrak{m} \left[(1 + 1_{\alpha_j < 0}) \right] S_j \alpha_j \geq 0.$$

This completes the proof.

For fixed $(t, \omega) \in [0, T] \times \Omega$, the constraint set $K_t(\omega)$ is a closed and convex cone.

Proof (Closed Cone)
This follows by inspection.

(Convexity)

To prove that $K_t(\omega)$ is convex, define the function

$$f(\alpha_0, \alpha) = \alpha_0 + \mathfrak{m}(\alpha \cdot S) + \sum_{j=1}^{n} min(\alpha_j, 0)S_j.$$

This is the left side of the inequality in the definition of $K_t(\omega)$. This function is linear (hence concave) in α_0 and concave in α. Thus concave in (α_0, α). For a concave function, we know that $f(\lambda\gamma + (1-\lambda)\beta) \geq \lambda f(\gamma) + (1-\lambda) f(\beta)$ for $\lambda \in [0, 1]$ where $\gamma, \beta \in \mathbb{R}^{n+1}$.

Choose γ and $\beta \in K$. To complete the proof, we need to show $\lambda\gamma + (1-\lambda)\beta \in K_t(\omega)$.

By the definition of $K_t(\omega)$ we have that both $f(\gamma) \geq 0$ and $f(\beta) \geq 0$.

Using the concavity of f gives $f(\lambda\gamma+(1-\lambda)\beta) \geq 0$, i.e. $\lambda\gamma+(1-\lambda)\beta \in K_t(\omega)$. This completes the proof.

The polar cone is

$$K_t(\omega)^o = \{(\nu_0, \nu) \in \mathbb{R}^{n+1} : (\nu_0, \nu) \leq 0\}.$$

This implies that there are costs to holding both the mma and the risky assets.

Proof Note that

$$\delta^*(v) = \begin{cases} 0 & \text{if } v \in K_t(\omega)^o \\ \infty & \text{otherwise.} \end{cases}$$

Determining when δ^* is finite characterizes $K_t(\omega)^o$. We have

$$\delta^*(\nu_0, \nu) = \sup_{\alpha \in K} \sum_{i=0}^{n} \nu_i \alpha_i.$$

If any $\nu_i > 0$, δ^* can be made unboundedly large by choosing $\alpha_i \to \infty$. Note the constraint is always satisfied. Hence, $\nu_i \leq 0$ all i for δ^* to be finite.

Next, given $\nu_i \leq 0$ all i, suppose $\nu_j < 0$ for some j, then δ^* if finite. Indeed, δ^* increases as $\alpha_j \to -\infty$ for α_0, α_j with $j \neq i$ fixed, but the constraint keeps α_j bounded and therefore δ^* is finite. This completes the proof.

18.5 Wealth Processes

This section proves some equivalent characterizations of the wealth processes generated by constrained admissible s.f.t.s.'s. We do this through a few lemmas.

Lemma 18.2 (Equivalence 1)

$$\mathscr{X}_{\mathfrak{C}}(x,S) = \{X \in \mathscr{L}^0_+ : \exists(\alpha_0,\alpha) \in \mathscr{A}(x,S),\ x + \textstyle\int_0^t \alpha_u \cdot dS_u - \int_0^t \delta(\alpha_0(u),\alpha_u)du \geq X_t,\ \forall t \in [0,T]\},$$

$$\mathscr{C}_{\mathfrak{C}}(x,S) = \{X_T \in L^0_+ : \exists Z \in \mathscr{X}_{\mathfrak{C}}(x,S),\ Z_T = X_T\}.$$

Proof Define

$$\mathscr{X}_1(x,S) = \left\{X \in \mathscr{L}^0_+ : \exists(\alpha_0,\alpha) \in \mathscr{A}(x,S), x + \int_0^t \alpha_u \cdot dS_u - \int_0^t \delta(\alpha_0(u),\alpha_u)du \geq X_t,\ \forall t \in [0,T]\right\}$$

and

$$\mathscr{C}_1(x,S) = \left\{X_T \in L^0_+ : \exists Z \in \mathscr{X}_2(x,S),\ Z_T = X_T\right\}.$$

(Step 1) Show $\mathscr{X}_{\mathfrak{C}}(x,S) \subset \mathscr{X}_1(x,S)$.
Take $X \in \mathscr{X}_{\mathfrak{C}}(x,S)$. Then,

$$\exists(\alpha_0,\alpha) \in \mathscr{A}(x,S,K),\ x + \int_0^t \alpha_u \cdot dS_u \geq X_t \text{ for all } t.$$

Hence, $(\alpha_0,\alpha) \in K$, for all t a.s. $\mathbb{P}$, which implies

$$\int_0^t \delta(\alpha_0(u),\alpha_u)du = 0 \text{ for all } t,$$

i.e.

$$x + \int_0^t \alpha_u \cdot dS_u - \int_0^t \delta(\alpha_0(u),\alpha_u)dt \geq X_t \quad \text{for all } t.$$

Thus, $X \in \mathscr{X}_1(x,S)$.

(Step 2) Show $\mathscr{X}_1(x,S) \subset \mathscr{X}_{\mathfrak{C}}(x,S)$.
Take $X \in \mathscr{X}_1(x,S)$. This implies

$$\exists(\alpha_0,\alpha) \in \mathscr{A}(x,S),\ x + \int_0^t \alpha_u \cdot dS_u - \int_0^t \delta(\alpha_0(u),\alpha_t)du \geq X_t \quad \text{for all } t.$$

This implies

$$\int_0^t \delta(\alpha_0(u),\alpha_u)du < \infty \quad \text{for all } t \text{ a.s. } \mathbb{P}.$$

That is,

$$(\alpha_0, \alpha) \in K, \text{ for all } t \text{ a.s. } \mathbb{P}.$$

Thus,

$$X_T \in \mathscr{X}_{\mathfrak{C}}(x, S).$$

(Step 3) $\mathscr{X}_{\mathfrak{C}}(x, S) = \mathscr{X}_1(x, S)$ implies that $\mathscr{C}_{\mathfrak{C}}(x, S) = \mathscr{C}_1(x, S)$.
This completes the proof.

Note that in these sets of wealth processes, an admissible s.f.t.s. is no longer explicitly constrained by the set K. The trading constraint K is implicitly included through the support function δ, which depends on K. Define

$$\mathscr{K}^o = \left\{(\nu_0, \nu) \in \mathscr{L}^0 : (\nu_0(t), \nu(t)) \in K^o \text{ for all } t \in [0, T]\right\}.$$

This is the set of K^o-valued, cadlag (right continuous with left limits existing), $\mathscr{F}_t$-measurable stochastic processes with $(0, 0) \in \mathscr{K}^o$. We note two facts about this set of stochastic processes that will be needed below.

1. If $\{(1, 0), (0, 1)\} \subset K$, then $\mathscr{V}_0(t) = \int_0^t \nu_0(u)du$ and $\mathscr{V}_j(t) = \int_0^t \nu_j(u)du$ for $j = 1, \ldots, n$ are of finite variation since they are nonincreasing processes. This follows directly from Lemma 18.1 since $\{(1, 0), (0, 1)\} \subset K$.
2. $(\nu_0, \nu) \in \mathscr{K}^o$ and $(\alpha_0, \alpha) \in \mathscr{A}(x, S, K)$ implies that $\alpha_0(t)\nu_0(t) + \alpha_t \cdot \nu_t \leq 0$ for all t from expression (18.5). Hence, $\int_0^T (\alpha_0(t)\nu_0(t) + \alpha_t \cdot \nu_t)dt$ is well defined, although it may take the value $-\infty$.

Lemma 18.3 (Equivalence 2)

$$\begin{gathered}\mathscr{X}_{\mathfrak{C}}(x, S) = \{X \in \mathscr{L}_+^0 : \exists(\alpha_0, \alpha) \in \mathscr{A}(x, S), \ \forall t \in [0, T], \\ x + \int_0^t \alpha_u \cdot dS_u - \int_0^t (\alpha_0(u)\nu_0(u) + \alpha_u \cdot \nu_u)du \geq X_t \text{ for all } (\nu_0, \nu) \in \mathscr{K}^o\},\end{gathered}$$

$$\mathscr{C}_{\mathfrak{C}}(x, S) = \left\{X_T \in L_+^0 : \exists Z \in \mathscr{X}_{\mathfrak{C}}(x, S), \ Z_T = X_T\right\}.$$

Proof Recall $\mathscr{X}_1(x, S), \mathscr{C}_1(x, S)$ from the previous lemma's proof.
Define

$$\begin{aligned}\mathscr{X}_2(x, S) = \{X \in \mathscr{L}_+^0 : \exists(\alpha_0, \alpha) \in \mathscr{A}(x, S), \ \forall t \in [0, T], \ x + \int_0^t \alpha_u \cdot dS_u \\ - \int_0^t (\alpha_0(u)\nu_0(u) + \alpha_u \cdot \nu_u)du \geq X_t \text{ for all } (\nu_0, \nu) \in \mathscr{K}^o\}\end{aligned}$$

and

$$\mathscr{C}_2(x,S) = \left\{X_T \in L^0_+ : \exists Z \in \mathscr{X}_2(x,S),\ Z_T = X_T\right\}.$$

Recall that

$$\delta(\alpha_0(t),\alpha_t) = 0, \text{ for all } t \text{ a.s. } \mathbb{P}$$

if and only if

$$\alpha_0(t)\nu_0(t) + \alpha_t \cdot \nu_t \leq 0, \text{ for all } t \text{ a.s. } \mathbb{P}, \text{ for all } (\nu_0,\nu) \in \mathscr{K}^o.$$

(Step 1) Show $\mathscr{X}_1(x,S) \subset \mathscr{X}_2(x,S)$.
Take $X \in \mathscr{X}_1(x,S)$. This implies

$$\exists(\alpha_0,\alpha) \in \mathscr{A}(x,S),\ \forall t \in [0,T],\ x + \int_0^t \alpha_u \cdot dS_u - \int_0^t \delta(\alpha_0(u),\alpha_u)du \geq X_t.$$

This implies

$$\int_0^t \delta(\alpha_0(u),\alpha_u)du < \infty \text{ for all } t \text{ a.s. } \mathbb{P}.$$

That is, $\delta(\alpha_0(t),\alpha_t) < \infty$, for all t a.s. $\mathbb{P}$.
Equivalently, by the definition of δ, $\delta(\alpha_0(t),\alpha_t) = 0$, for all t a.s. $\mathbb{P}$.
Thus,

$$x + \int_0^t \alpha_u \cdot dS_u \geq X_t \text{ for all } t$$

and

$$-(\alpha_0(t)\nu_0(t) + \alpha_t \cdot \nu_t) \geq 0, \quad \text{for all } t \text{ a.s. } \mathbb{P}, \text{ for all } (\nu_0,\nu) \in \mathscr{K}^o.$$

Hence, for all t,

$$x + \int_0^t \alpha_u \cdot dS_u - \int_0^t (\alpha_0(u)\nu_0(u) + \alpha_u \cdot \nu_u)du \geq X_t \text{ for all } (\nu_0,\nu) \in \mathscr{K}^o.$$

Thus, $X \in \mathscr{X}_2(x,S)$.

(Step 2) Show $\mathscr{X}_2(x, S) \subset \mathscr{X}_1(x, S)$.
Take $X \in \mathscr{X}_2(x, S)$. This implies

$$\exists(\alpha_0, \alpha) \in \mathscr{A}(x, S),\ \forall t \in [0, T],$$

$$x + \int_0^t \alpha_u \cdot dS_u - \int_0^t (\alpha_0(u)\nu_0(u) + \alpha_u \cdot \nu_u)du \geq X_t \text{ for all } (\nu_0, \nu) \in \mathscr{K}^o.$$

By definition of $(\nu_0, \nu) \in \mathscr{K}^o$, $(\alpha_0(t)\nu_0(t) + \alpha_t \cdot \nu_t) \leq 0$, for all t a.s. $\mathbb{P}$.
This implies $(\alpha_0(t), \alpha_t) \in K^{oo}$ for all t. But $K = K^{oo}$.
Hence, $(\alpha_0(t), \alpha_t) \in K$, which implies $\delta(\alpha_0(t), \alpha_t) = 0$, for all t a.s. $\mathbb{P}$.
Given $(0, 0) \in \mathscr{K}^o$, we have from the definition of $\mathscr{X}_2(x, S)$ that

$$\forall t \in [0, T],\ x + \int_0^t \alpha_u \cdot dS_u \geq X_t.$$

Thus,

$$x + \int_0^t \alpha_u \cdot dS_u - \int_0^t \delta(\alpha_0(u), \alpha_u)du \geq X_t \text{ for all } t, \text{ and } X \in \mathscr{X}_1(x, S).$$

(Step 3) $\mathscr{X}_1(x, S) = \mathscr{X}_2(x, S)$ implies that $\mathscr{C}_1(x, S) = \mathscr{C}_2(x, S)$.
This completes the proof.

Note that in these sets of wealth processes, the support function δ which depends on K is replaced by a constraint involving the polar set of random processes $\mathscr{K}^o$.

Remark 18.3 (No Trading Constraints) When there are no trading constraints, see Sect. 18.4.1, $K^o = \{(\nu_0, \nu) \in \mathbb{R}^{n+1} : (\nu_0, \upsilon) = 0\}$. In this case the wealth processes collapse to those in the unconstrained market, i.e. $\mathscr{X}_{\mathfrak{C}}(x, S) = \mathscr{X}(x, S)$ and $\mathscr{C}_{\mathfrak{C}}(x, S) = \mathscr{C}(x, S)$. This completes the remark.

Chapter 19
Arbitrage Pricing Theory

This chapter studies the modifications needed due to the introduction of trading constraints in the arbitrage pricing theory of the fundamental theorems Chap. 2. Most, but not all of the three fundamental theorems of asset pricing extend with trading constraints. The set-up begins with a normalized market $(S, (\mathscr{F}_t), \mathbb{P})$ subject to trading constraints where the money market account $B_t \equiv 1$.

19.1 No Unbounded Profits with Bounded Risk (NUPBR$_{\mathfrak{C}}$)

We first study constrained no unbounded profits with bounded risk (NUPBR$_{\mathfrak{C}}$). Recall the definitions of the set of local martingale and supermartingale deflators for an unconstrained market:

$$\mathscr{D}_l(S) = \left\{ Y \in L^0_+ : Y_0 = 1,\ XY \text{ is a } \mathbb{P}\text{-local martingale for } X \in \mathscr{X}^e(x, S) \right\},$$

$$\mathscr{D}_s(S) = \left\{ Y \in L^0_+ : Y_0 = 1,\ XY \text{ is a } \mathbb{P}\text{-supermartingale for } X \in \mathscr{X}^e(x, S) \right\}.$$

Note that these definitions use the sets of wealth processes *without free disposal*, i.e. using $\mathscr{X}^e$ instead of $\mathscr{X}$ (see the trading constraints Chap. 18 for the definitions). Next, we define the set of supermartingale deflators for the constrained markets:

$$\mathscr{D}_l(S, \mathfrak{C}) = \left\{ Y \in L^0_+ : Y_0 = 1,\ XY \text{ is a } \mathbb{P}\text{-local martingale for } X \in \mathscr{X}^e_{\mathfrak{C}}(x, S) \right\},$$

$$\mathscr{D}_s(S, \mathfrak{C}) = \left\{ Y \in L^0_+ : Y_0 = 1,\ XY \text{ is a } \mathbb{P}\text{-supermartingale for } X \in \mathscr{X}^e_{\mathfrak{C}}(x, S) \right\}.$$

We state the next theorem without proof (see Karatzas and Kardaras [116, Theorem 4.12, p. 473]).

R. A. Jarrow, *Continuous-Time Asset Pricing Theory*, Springer Finance,
https://doi.org/10.1007/978-3-319-77821-1_19

Theorem 19.1 (Characterization of $NUPBR_{\mathfrak{C}}$)

$$NUPBR_{\mathfrak{C}} \quad \textit{if and only if} \quad \mathcal{D}_s(S, \mathfrak{C}) \neq \emptyset.$$

This theorem gives the equivalence between $\text{NUPBR}_{\mathfrak{C}}$ and the set of supermartingale deflators being nonempty. It is the analogue of Theorem 2.1 in the fundamental theorems Chap. 2, which is NUPBR if and only if $\mathcal{D}_l(S) \neq \emptyset$. The difference in a constrained market is that in the characterization of NUPBR the set of local martingale deflators is replaced by the set of supermartingale deflators.

19.2 No Free Lunch with Vanishing Risk ($\text{NFLVR}_{\mathfrak{C}}$)

This section studies the First Fundamental Theorem of asset pricing in a trading constrained market. We state the next theorems without proof.

Theorem 19.2 (Characterization of $NFLVR_{\mathfrak{C}}$)

$$NA_{\mathfrak{C}} \text{ and } NUPBR_{\mathfrak{C}} \quad \textit{if and only if} \quad NFLVR_{\mathfrak{C}}.$$

This theorem (see Karatzas and Kardaras [116, Proposition 4.2, p. 466]) is the analogue of Theorem 2.2 in the fundamental theorems Chap. 2, which gives a characterization of no free lunch with vanishing risk, i.e. NA and NUPBR if and only if NFLVR. As with the unconstrained economy, this theorem shows that $\text{NFLVR}_{\mathfrak{C}}$ is a stronger assumption than either $\text{NA}_{\mathfrak{C}}$ or $\text{NUPBR}_{\mathfrak{C}}$. We can now state the First Fundamental Theorem of asset pricing in the constrained market.

Theorem 19.3 (The First Fundamental Theorem)

$$NFLVR_{\mathfrak{C}} \quad \textit{if and only if} \quad \mathfrak{M}_s(S) \neq \emptyset.$$

This theorem (see Karatzas and Kardaras [116, Theorem 4.4, p. 467]) is the generalization of the First Fundamental Theorem 2.3 of asset pricing in Chap. 2 which is NFLVR if and only if $\mathfrak{M}_l(S) \neq \emptyset$. The difference in a trading constrained market in the characterization of NFLVR is that the set of local martingale measures is replaced by the larger set of supermartingale deflators.

Remark 19.1 (The Third Fundamental Theorem of Asset Pricing)
Recall that in an unconstrained market, the Third Fundamental Theorem 2.5 of asset pricing in Chap. 2 states that

$$ND \text{ and } NFLVR \quad \text{if and only if} \quad \mathfrak{M}(S) \neq \emptyset.$$

Unfortunately, due to the existence of asset price bubbles and dominated trading strategies, there appears to be no analogue of this theorem in a market with trading constraints. For some research studying an extension of the Third Fundamental Theorem in constrained markets, see Jarrow and Larsson [97]. This completes the remark.

19.3 Asset Price Bubbles

This section studies asset price bubbles in a trading constrained market. For this section, we impose the simplifying assumption that there is only *one risky asset* trading denoted S_t. The restriction to one risky asset is without loss of generality and it is imposed to simplify the notation (to avoid subscripts for the jth risky asset). The key insight of this section is that

trading constraints may introduce asset price bubbles into the market.

To prove this, we start by assuming that the market with trading constraints is arbitrage free.

Assumption $(NFLVR_{\mathfrak{C}})$ $\mathfrak{M}_s(S) \neq \emptyset$.

Under this assumption the risky asset price process S_t is a $\mathbb{Q}$-supermartingale. Fix a probability measure $\mathbb{Q}$ chosen by the market, and as in the asset price bubbles Chap. 3, we define the asset's *fundamental value as* $E^{\mathbb{Q}}[S_T | \mathscr{F}_t]$. The difference between the asset's market price and fundamental value is the asset's price bubble β_t, i.e.

$$S_t = E^{\mathbb{Q}}[S_T | \mathscr{F}_t] + \beta_t. \tag{19.1}$$

Exactly as in the asset price bubbles Chap. 3, one can prove that this asset price bubble satisfies the following properties.

Theorem 19.4 (Properties of Bubbles)

1. $\beta_t \geq 0$.
2. $\beta_T = 0$.
3. *If* $\beta_s = 0$ *for some* $s \in [0, T]$, *it is zero thereafter.*

Proof

1. This follows since S_t is a supermartingale.
2. Since $S_T = E^{\mathbb{Q}}[S_T | \mathscr{F}_T] + \beta_T$, $\beta_T = 0$.
3. Suppose $\beta_s = 0$. Then $0 = \beta_s \geq E^{\mathbb{Q}}[\beta_t | \mathscr{F}_s] = E^{\mathbb{Q}}[\beta_T | \mathscr{F}_s] \geq 0$. It is a nonnegative martingale after time s, hence it is identically zero after time s.

This completes the proof.

Condition (1) states that asset price bubbles are always nonnegative, $\beta_t \geq 0$. That is, the risky asset is always worth at least its fundamental value. Condition (2) states that the asset price bubble may burst before, but no later than time T. This is because at time T, the risky asset is worth its liquidation value, which also equals its fundamental value at that time by construction. Finally, condition (3) states that if the asset price bubble disappears before time T, it is never reborn.

Remark 19.2 (Bubbles Due To Trading Constraints and Not Retrade Value) The asset price bubbles in Chap. 3 are generated by the asset's retrade value exceeding its value if held until maturity (see Theorem 3.2 in Chap. 3). With trading constraints this need not be the case. With trading constraints, S_t need not be a $\mathbb{Q}$-local martingale. Under NFLVR$_{\mathfrak{C}}$, S_t is only a $\mathbb{Q}$-supermartingale. Note that $\mathfrak{M}_l(S) \subset \mathfrak{M}_s(S)$ and the containment is usually strict. Being a $\mathbb{Q}$-supermartingale is a weaker property than a $\mathbb{Q}$-local martingale. This weaker property is a result of the trading constraints reducing the admissible s.f.t.s.'s in the market. Alternatively stated, NFLVR$_{\mathfrak{C}}$ is a weaker assumption than is NFLVR. With trading constraints, however, S can be a $\mathbb{Q}$-local martingale and even a strict local martingale, in which case the trading constraint induced asset price bubble reduces to those studied in the asset price bubbles Chap. 3. This completes the remark.

19.4 Systematic Risk

This section studies the characterization of systematic risk in a trading constrained market. Given is a normalized market with trading constraints $(S, (\mathscr{F}_t), \mathbb{P})$, where the money market account $B_t \equiv 1$. We start by assuming that the market with trading constraints is arbitrage free.

Assumption $(NFLVR_{\mathfrak{C}})$ $\mathfrak{M}_s(S) \neq \emptyset$.

From the previous section, we know this implies that risky assets may have price bubbles, i.e.

$$S_t = E^{\mathbb{Q}}[S_T | \mathscr{F}_t] + \beta_t, \tag{19.2}$$

where $\beta_t \geq 0$.

As mentioned in the multiple-factor model Chap. 4, using the insights of Jarrow [91], one can now derive: (1) a multiple-factor model, (2) a systematic risk relation, and (3) a beta model for the market with trading constraints. These extensions necessarily include price bubbles, but are otherwise analogous to those contained in the asset price bubbles Chap. 3. We leave the derivation and description of these extensions to Jarrow's [91] paper.

Chapter 20
The Auxiliary Markets

This chapter studies how to transform a trading constrained market into an "equivalent" market which is incomplete, but without trading constraints. The transformed market is called the auxiliary unconstrained market. The goal of this transformation is to use the theorems from an incomplete but unconstrained market to understand trading constrained markets.

Given is a normalized trading constrained market $(S, (\mathscr{F}_t), \mathbb{P})$, where the money market account $B_t \equiv 1$. The motivation for constructing auxiliary unconstrained markets comes from the wealth set equivalence from Lemma 18.3 in the trading constraints Chap. 18, repeated here for convenience, i.e.

$$\mathscr{X}_{\mathfrak{C}}(x, S) = \{X \in \mathscr{L}_+^0 : \exists(\alpha_0, \alpha) \in \mathscr{A}(x, S),\ \forall t \in [0, T],$$
$$x - \int_0^t \alpha_0(u)\nu_0(u)du + \int_0^t \alpha_u \cdot (dS_u - \nu_u du) \geq X_t \text{ for all } (\nu_0, \nu) \in \mathscr{K}^o\}.$$

In this wealth process characterization, it is clear that by adjusting the evolution of the risky asset and the mma, one can view this wealth process evolution as that in an unconstrained market.

To simplify the analysis, we add the following assumption.

Assumption (Buy and Hold Trading Strategies are Unconstrained)

$$\{(1, 0), (0, 1)\} \in K.$$

Assuming that $\{(1, 0), (0, 1)\} \in K$ means that a buy and hold s.f.t.s. involving only the mma, or only the risky assets, or (by convex combinations) all of the assets are contained in K. In conjunction, this assumption implies that the trading constraints only affect portfolios with negative positions in the mma or risky assets. We impose this assumption for all of the remaining chapters in Part IV of this book.

R. A. Jarrow, *Continuous-Time Asset Pricing Theory*, Springer Finance,
https://doi.org/10.1007/978-3-319-77821-1_20

20.1 The Auxiliary Markets

This section introduces the non-normalized hypothetical auxiliary markets

$$\big((B^\nu, S^\nu), (\mathscr{F}_t), \mathbb{P}\big) \qquad \text{for all } (\nu_0, \nu) \in \mathscr{K}^o.$$

Given a $(\nu_0, \nu) \in \mathscr{K}^o$, define a new market with the mma evolving as

$$\begin{aligned} B_t^\nu &= 1 - \textstyle\int_0^t \nu_0(u)du \geq 1 \qquad \text{or} \\ dB_t^\nu &= -\nu_0(t)dt, \end{aligned} \tag{20.1}$$

where $\int_0^t \nu_0(u)du \leq 0$ for all t.

Proof (Negativity of $\nu_0(t)$)

Fix a $(\nu_0, \nu) \in \mathscr{K}^o$.

Since $(1, 0) \in K$ for all t, by Lemma 18.1 in the trading constraints Chap. 18, $\nu_0(t) \leq 0$ for all t.

This completes the proof.

Let the risky assets evolve as

$$\begin{aligned} S_t^\nu &= S_t - \textstyle\int_0^t \nu_u du \geq S_t \qquad \text{or} \\ dS_t^\nu &= dS_t - \nu_t dt, \end{aligned} \tag{20.2}$$

where $\int_0^t \nu_u du \leq 0$ for all t.

Proof (Negativity of $\nu_u(t)$)

Fix a $(\nu_0, \nu) \in \mathscr{K}^o$.

Since $(0, 1) \in K$ for all t, by Lemma 18.1 in the trading constraints Chap. 18, $\nu_u(t) \leq 0$ for all t.

This completes the proof.

The evolution of an unconstrained admissible s.f.t.s. trading strategy $(\alpha_0, \alpha) \in \mathscr{A}(x, (B^\nu, S^\nu))$ in this auxiliary market is

$$X_t = \alpha_0(t)B_t^\nu + \alpha_t \cdot S_t^\nu \qquad \text{or} \tag{20.3}$$

$$dX_t = -\alpha_0(t)\nu_0(t)dt + \alpha_t \cdot (dS_t - \nu_t dt). \tag{20.4}$$

Remark 20.1 (No Trading Constraints) When there are no trading constraints, the polar of the constraint set $K^o = \{(\nu_0, \nu) \in \mathbb{R}^{n+1} : (\nu_0, \upsilon) = (0, 0)\}$. In this case, by substitution into the previous expressions, we see that the auxiliary market collapses to the original market. This completes the remark.

20.2 The Normalized Auxiliary Markets

Consider the normalized auxiliary market using the mma as the numeraire, i.e. $\left(\frac{S^\nu}{B^\nu}, (\mathscr{F}_t), \mathbb{P}\right)$ for all $(\nu_0, \nu) \in \mathscr{K}^o$. Then, we have the following evolutions for the mma and the normalized risky assets.

$$\frac{B_t^\nu}{B_t^\nu} = 1 \tag{20.5}$$

$$\frac{S_t^\nu}{B_t^\nu} = \frac{S_t - \int_0^t \nu_u du}{1 - \int_0^t \nu_0(u) du} \geq 0, \quad \text{or} \tag{20.6}$$

$$d\left(\frac{S_t^\nu}{B_t^\nu}\right) = \frac{1}{B_t^\nu}(dS_t - \nu_t dt) + \frac{\left(S_t - \int_0^t \nu_u du\right)}{B_t^\nu} \frac{\nu_0(t)}{B_t^\nu} dt. \tag{20.7}$$

Proof By the integration by parts formula Theorem 1.1 in Chap. 1 (dropping the t's),

$$d\left(\frac{S - \int \nu_u du}{B^\nu}\right) = \frac{1}{B^\nu}(dS - \nu dt) + \left(S - \int \nu_u du\right) d\left(\frac{1}{B^\nu}\right).$$

The first equality uses $\left[S_t^\nu, \frac{1}{B^\nu}\right] = 0$, since B^ν is continuous and of finite variation (see Lemma 1.6 in Chap. 1).

$$d\left(\frac{1}{B^\nu}\right) = -\frac{1}{B^\nu}\frac{dB^\nu}{B^\nu} = \frac{1}{B^\nu}\frac{\nu_0}{B^\nu} dt$$

Substitution completes the proof.

The evolution of an unconstrained admissible s.f.t.s. $(\alpha_0, \alpha) \in \mathscr{A}\left(x, \frac{S^\nu}{B^\nu}\right)$ in this auxiliary market is

$$\frac{X_t}{B_t^\nu} = \alpha_0(t) + \alpha_t \cdot \frac{S_t^\nu}{B_t^\nu}. \tag{20.8}$$

The wealth process evolution is

$$\left(\frac{X_t}{B_t^\nu}\right) = x + \int_0^t \alpha_u \cdot d\left(\frac{S_u^\nu}{B_u^\nu}\right) \quad \text{or}$$
$$d\left(\frac{X_t}{B_t^\nu}\right) = \alpha_t \cdot d\left(\frac{S_t^\nu}{B_t^\nu}\right).$$

Proof

$$d\left(\frac{X}{B^\nu}\right) = \frac{1}{B^\nu}dX + Xd\left(\frac{1}{B^\nu}\right) = \frac{dX}{B^\nu} - \frac{X}{B^\nu}\frac{dB^\nu}{B^\nu}.$$

The first equality uses the integration by parts formula Theorem 1.1 in Chap. 1 with the fact that $\left[X, \frac{1}{B^\nu}\right] = 0$ since B^ν is continuous and of finite variation (Lemma 1.6 in Chap. 1).

Substitution yields

$$= \frac{-\alpha_0\nu_0 dt + \alpha\cdot(dS - \nu dt)}{B^\nu} + \frac{\alpha_0 B^\nu + \alpha\cdot(S - \int \nu_u du)}{B^\nu}\frac{\nu_0}{B^\nu}dt.$$

Algebra gives

$$= -\alpha_0\frac{\nu_0}{B^\nu}dt + \alpha\cdot(\frac{dS - \nu dt}{B^\nu}) + \alpha_0\frac{\nu_0}{B^\nu}dt - \alpha\cdot\frac{(S - \int \nu_u du)}{B^\nu}\frac{\nu_0}{B^\nu}dt$$

$$= \alpha\cdot(\frac{dS - \nu dt}{B^\nu}) - \alpha\cdot\frac{(S - \int \nu_u du)}{B^\nu}\frac{\nu_0}{B^\nu}dt = \alpha\cdot d\left(\frac{S^\nu}{B^\nu}\right).$$

This completes the proof.

The following change of numeraire lemma will prove useful in subsequent chapters.

Lemma 20.1 (Change of Numeraire)

$$\int_0^t \alpha_u \cdot d\left(\frac{S_u^\nu}{B_u^\nu}\right) = \frac{\int_0^t \alpha_u \cdot dS_u - \int_0^t(\alpha_0(u)\nu_0(u) + \alpha_u\cdot\nu_u)du}{B_u^\nu} \tag{20.9}$$

Proof

$$\left(\frac{X_T}{B_T^\nu}\right) = x + \int_0^T \alpha_t \cdot d\left(\frac{S_t^\nu}{B_t^\nu}\right).$$

Hence,

$$d\left(\frac{X_t}{B_t^\nu}\right) = \alpha_t \cdot d\left(\frac{S_t^\nu}{B_t^\nu}\right),$$

$$d\left(\frac{X_t}{B_t^\nu}\right) = \alpha_t \cdot \frac{1}{B_t^\nu}dS_t^\nu + \alpha_t\cdot\frac{S_t^\nu}{B_t^\nu}\frac{\nu_0(t)}{B_t^\nu}dt,$$

$$d\left(\frac{X_t}{B_t^\nu}\right) = \alpha_t \cdot \frac{1}{B_t^\nu}(dS_t - v_t dt) + \left(\left(\frac{X_t}{B_t^\nu}\right) - \alpha_0\right)\frac{\nu_0(t)}{B_t^\nu}dt,$$

$$d\left(\frac{X_t}{B_t^\nu}\right) = \alpha_t \cdot \frac{1}{B_t^\nu}(dS_t - v_t dt) - \alpha_0\frac{\nu_0(t)}{B_t^\nu}dt + \left(\frac{X_t}{B_t^\nu}\right)\frac{\nu_0(t)}{B_t^\nu}dt,$$

$$d\left(\frac{X_t}{B_t^\nu}\right) - \left(\frac{X_t}{B_t^\nu}\right)\frac{\nu_0(t)}{B_t^\nu}dt = \alpha_t \cdot \frac{1}{B_t^\nu}(dS_t - \upsilon_t dt) - \alpha_0 \frac{\nu_0(t)}{B_t^\nu}dt,$$

$$B_t^\nu d\left(\frac{X_t}{B_t^\nu}\right) - \left(\frac{X_t}{B_t^\nu}\right)\nu_0(t)dt = \alpha_t \cdot (dS_t - \upsilon_t dt) - \alpha_0\nu_0(t)dt,$$

$$B_t^\nu d\left(\frac{X_t}{B_t^\nu}\right) + \left(\frac{X_t}{B_t^\nu}\right)dB_t^\nu = \alpha_t \cdot (dS_t - \upsilon_t dt) - \alpha_0\nu_0(t)dt,$$

$$B_t^\nu \left(d\left(\frac{X_t}{B_t^\nu}\right) + \left(\frac{X_t}{B_t^\nu}\right)\frac{dB_t^\nu}{B_t^\nu}\right) = \alpha_t \cdot (dS_t - \upsilon_t dt) - \alpha_0\nu_0(t)dt.$$

But, $d\left(\frac{X}{B^\nu}\right) = \frac{dX}{B^\nu} - \frac{X}{B^\nu}\frac{dB^\nu}{B^\nu}$. Hence,

$$B_t^\nu \frac{(dX_t)}{B_t^\nu} = \alpha_t \cdot (dS_t - \upsilon_t dt) - \alpha_0\nu_0(t)dt,$$

$$dX_t = \alpha_t \cdot (dS_t - \upsilon_t dt) - \alpha_0\nu_0(t)dt.$$

Finally,

$$X_T = x + \int_0^T \alpha_t \cdot dS_t - \int_0^T (\alpha_0(t)\nu_0(t) + \alpha_t \cdot \upsilon_t)dt.$$

This completes the proof.

This lemma implies the following characterization of the constraint set.

$$\mathscr{X}_{\mathfrak{C}}(x, S) = \{X \in \mathscr{L}_+^0 : \exists(\alpha_0, \alpha) \in \mathscr{A}(x, S),\ \forall t \in [0, T],$$
$$x + \textstyle\int_0^t \alpha_u \cdot d\left(\frac{S_u^\nu}{B_u^\nu}\right) \geq \left(\frac{X_t}{B_t^\nu}\right) \text{ for all } (\nu_0, \nu) \in \mathscr{K}^o\}.$$

Proof The proof of the previous lemma showed that

$$\left(\frac{X_T}{B_T^\nu}\right) = x + \int_0^T \alpha_t \cdot d\left(\frac{S_t^\nu}{B_t^\nu}\right)$$

is equivalent to

$$X_T = x + \int_0^T \alpha_t \cdot dS_t - \int_0^T (\alpha_0(t)\nu_0(t) + \alpha_t \cdot \upsilon_t)dt.$$

Substitution of this fact into the definition of $\mathscr{X}_{\mathfrak{C}}(x, S)$ completes the proof.

Chapter 21
Super- and Sub-replication

This chapter studies super- and sub-replication in a trading constrained market. In a trading constrained market, not all derivatives can be synthetically constructed using constrained admissible s.f.t.s.'s. To price derivatives, we can obtain upper and lower bounds using super- and sub-replication. This is analogous to super- and sub-replication in an incomplete but unconstrained market, see the super- and sub-replication Chap. 8. To obtain the desired results, we need to add more structure to the trading constrained normalized market $(S, (\mathscr{F}_t), \mathbb{P})$, where the money market account $B_t \equiv 1$.

21.1 The Set-Up

First we consider the auxiliary market $(0, 0) \in \mathscr{K}^o$ and then we consider all the remaining auxiliary markets $(\nu_0, \nu) \in \mathscr{K}^o$.

21.1.1 *Auxiliary Market* (0, 0)

We consider the auxiliary market (0, 0) to impose some additional structure on the price process S. This is the original market, but with no trading constraints imposed. In this hypothetical market, we assume that the price process S is such that there are no arbitrage opportunities, i.e.

Assumption (NFLVR) $\quad \mathfrak{M}_l(S) \neq \emptyset.$

Remark 21.1 (NFLVR Stronger than $NFLVR_{\mathfrak{C}}$) Since $\mathfrak{M}_l(S) \subset \mathfrak{M}_s(S)$ we see that $\mathfrak{M}_l(S) \neq \emptyset$ is a stronger assumption than assuming $\mathfrak{M}_s(S) \neq \emptyset$. The assumption $\mathfrak{M}_s(S) \neq \emptyset$ only guarantees NFLVR$_{\mathfrak{C}}$, i.e. there are NFLVR constrained

R. A. Jarrow, *Continuous-Time Asset Pricing Theory*, Springer Finance,
https://doi.org/10.1007/978-3-319-77821-1_21

trading strategies. Assuming $\mathfrak{M}_l(S) \neq \emptyset$ implies the much stronger condition that NFLVR holds, ignoring the trading constraints. The unconstrained set of admissible s.f.t.s.'s is a larger set of trading strategies that can be utilized to construct an FLVR, and even within this larger set, we assume that no FLVRs exist. This completes the remark.

21.1.2 Auxiliary Markets (ν_0, ν)

For the auxiliary markets (ν_0, ν), we consider the normalized version of these markets, i.e. $\left(\frac{S^\nu}{B^\nu}, (\mathscr{F}_t), \mathbb{P}\right)$ for all $(\nu_0, \nu) \in \mathscr{K}^o$. We require that all of these auxiliary markets are arbitrage free.

Assumption (NFLVR for (ν_0, ν)**)**

$$\mathfrak{M}_l\left(\frac{S^\nu}{B^\nu}\right) \neq \emptyset \textit{ for all } (\nu_0, \nu) \in \mathscr{K}^o.$$

This assumption imposes implicit constraints on the stochastic processes $(\nu_0, \nu) \in \mathscr{K}^o$ as they affect the evolution of the price processes $\left(\frac{S^\nu}{B^\nu}\right)$ in expression (20.7). Necessary and sufficient conditions on the price process $\left(\frac{S^\nu}{B^\nu}\right)$ for the NFLVR assumption to be satisfied are contained in Protter and Shimbo [154].

21.2 Local Martingale Deflators

Recall the definitions of the set of local martingale deflators for an unconstrained market

$$\mathscr{D}_l(S) = \left\{Y \in \mathscr{L}^0_+ : Y_0 = 1,\ XY \text{ is a } \mathbb{P}\text{-local martingale for } X \in \mathscr{X}^e(x, S)\right\},$$

$$\mathscr{M}_l(S) = \left\{Y \in \mathscr{D}_l(S) : \exists \mathbb{Q} \sim \mathbb{P},\ Y_T = \frac{d\mathbb{Q}}{d\mathbb{P}}\right\}.$$

Define

$$Y_T^\nu = \frac{d\mathbb{Q}_\nu}{d\mathbb{P}} \text{ with } Y_t^\nu = E\left[Y_T^\nu \,|\mathscr{F}_t\right] \in \mathscr{M}_l\left(\frac{S^\nu}{B^\nu}\right) \text{ for } \mathbb{Q}_\nu \in \mathfrak{M}_l\left(\frac{S^\nu}{B^\nu}\right).$$

We have that

$$\mathfrak{M}_l(S) = \mathfrak{M}_l\left(\frac{S^\nu}{B^\nu}\right) \text{ when } (\nu_0, \nu) = (0, 0) \in \mathscr{K}^o,$$

which implies that

$$Y_T^\nu = Y_T \in M_l(S) \text{ for } (\nu_0, \nu) = (0, 0).$$

We can prove the following equivalences.

Lemma 21.1 (Equivalences)

$$\left(\frac{S^\nu}{B^\nu}\right) \textit{ is a } \mathbb{Q}_\nu\textit{-local martingale if and only if}$$

$$\left(\frac{S^\nu}{B^\nu}\right) Y^\nu \textit{ is a } \mathbb{P}\textit{-local martingale if and only if}$$

$$\left(\frac{X}{B^\nu}\right) Y^\nu \textit{ is a } \mathbb{P}\textit{-local martingale, for all}$$

$$X = 1 + \int_0^t \alpha_u \cdot d\left(\frac{S_u^\nu}{B_u^\nu}\right), \ (\alpha_0, \alpha) \in \mathscr{A}\left(1, \frac{S^\nu}{B^\nu}\right).$$

Proof The first equivalence follows from the definition of Y^ν.

Next we prove the second equivalence.

(Step 1) Suppose Y^ν satisfies the conditions on the right side of this equivalence. Then,

$Y^\nu\left(\frac{X}{B^\nu}\right)$ is a $\mathbb{P}$ local martingale with

$$\left(\frac{X_t}{B_t^\nu}\right) = 1 + \int_0^t \alpha_u \cdot d\left(\frac{S_u^\nu}{B_u^\nu}\right),$$

for all $(\alpha_0, \alpha) \in \mathscr{A}\left(1, \left(\frac{S^\nu}{B^\nu}\right)\right)$.

Choose $\alpha = (1, 0)$, which is a buy and hold s.f.t.s. using only the mma. Since $\left(\frac{B^\nu}{B^\nu}\right) = 1$, this implies Y^ν is a probability density.

Choose $\alpha = (0, 1)$, which is a buy and hold s.f.t.s. using only one of the risky assets. This implies Y^ν is such that $Y\left(\frac{S^\nu}{B^\nu}\right)$ is a $\mathbb{P}$-local martingale.

(Step 2) Suppose Y^ν satisfies the conditions on the left side of this equivalence.

Then,

$$\left(\frac{X_t}{B_t^\nu}\right) = 1 + \int_0^t \alpha_u \cdot d\left(\frac{S_u^\nu}{B_u^\nu}\right)$$

is a $\mathbb{Q}_\nu$-local martingale for all $(\alpha_0, \alpha) \in \mathscr{A}\left(1, \left(\frac{S^\nu}{B^\nu}\right)\right)$ by Lemma 1.5 in Chap. 1 since the admissibility condition implies that $\left(\frac{S^\nu}{B^\nu}\right)$ is bounded below.

Hence $\left(\frac{X}{B^\nu}\right) Y$ is a $\mathbb{P}$ local martingale. This is true for all $(\alpha_0, \alpha) \in \mathscr{A}\left(1, \left(\frac{S^\nu}{B^\nu}\right)\right)$. This completes the proof.

21.3 Wealth Processes Revisited

To obtain the cost of super- and sub-replication, we need to revisit the wealth processes again. We have already proven that

$$\mathscr{C}_{\mathfrak{C}}(x, S) = \{X_T \in L^0_+ : \exists(\alpha_0, \alpha) \in \mathscr{A}(x, S),\\ x + \int_0^T \alpha_t \cdot d\left(\frac{S^\nu_t}{B^\nu_t}\right) \geq \frac{X_T}{B^\nu_T}, \text{ for all } (\nu_0, \nu) \in \mathscr{K}^o\}.$$

We need one last characterization of this set of time T wealths.

Lemma 21.2 (Equivalence 3)

$$\mathscr{C}_{\mathfrak{C}}(x, S) = \left\{ X_T \in L^0_+ : \sup_{\mathbb{Q}_\nu \in \mathfrak{M}_l\left(\frac{S^\nu}{B^\nu}\right)} E_\nu\left[\frac{X_T}{B^\nu_T}\right] \leq x \text{ for all } (\nu_0, \nu) \in \mathscr{K}^o \right\},$$

where $E_\nu[\cdot]$ is expectation with respect to $\mathbb{Q}_\nu$.

Proof Define

$$\mathscr{C}_3(x, S) = \left\{ X_T \in L^0_+ : \sup_{\mathbb{Q}_\nu \in \mathfrak{M}_l\left(\frac{S^\nu}{B^\nu}\right)} E_\nu\left[\frac{X_T}{B^\nu_T}\right] \leq x \text{ for all } (\nu_0, \nu) \in \mathscr{K}^o \right\}.$$

(Step 1) Show $\mathscr{C}_{\mathfrak{C}}(x, S) \subset \mathscr{C}_3(x, S)$.

Take $X_T \in \mathscr{C}_{\mathfrak{C}}(x, S)$. Fix a $\mathbb{Q}_\nu \in \mathfrak{M}_l\left(\frac{S^\nu}{B^\nu}\right)$ and $(\nu_0, \nu) \in \mathscr{K}^o$.

This implies

$$\exists(\alpha_0, \alpha) \in \mathscr{A}(x, S), x + \int_0^T \alpha_t \cdot d\left(\frac{S^\nu_t}{B^\nu_t}\right) \geq \frac{X_T}{B^\nu_T}.$$

We have $\int_0^T \alpha_t \cdot d\left(\frac{S^\nu_t}{B^\nu_t}\right)$ is a $\mathbb{Q}_\nu$-local martingale, hence a $\mathbb{Q}_\nu$-supermartingale by Lemma 1.3 in Chap. 1.

This implies

$$E_\nu\left[\int_0^T \alpha_t \cdot d\left(\frac{S_t^\nu}{B_t^\nu}\right)\right] \leq 0.$$

Or,

$$E_\nu\left[\frac{X_T}{B_T^\nu}\right] \leq x + E_\nu\left[\int_0^T \alpha_t \cdot \left(\frac{S_t^\nu}{B_t^\nu}\right)\right] \leq x.$$

This is true for all $(\nu_0, \nu) \in \mathscr{K}^o$ and $\mathbb{Q}_\nu \in \mathfrak{M}_l\left(\frac{S^\nu}{B^\nu}\right)$.

Thus $X_T \in \mathscr{C}_3(x, S)$.

(Step 2) Show $\mathscr{C}_3(x, S) \subset \mathscr{C}_{\mathfrak{C}}(x, S)$.

Take $X_T \in \mathscr{C}_3(x, S)$. Fix an arbitrary $(\nu_0, \nu) \in \mathscr{K}^o$.

Define

$$W_t \equiv \operatorname*{ess\ sup}_{\mathbb{Q}_\nu \in \mathfrak{M}_l} E_\nu\left[\frac{X_T}{B_T^\nu} | \mathscr{F}_t\right] \geq 0.$$

(Step 2, Part 2a) From Theorem 1.4 in Chap. 1, W_t is a supermartingale for $\mathbb{Q}_\nu \in \mathfrak{M}_l\left(\frac{S^\nu}{B^\nu}\right)$.

(Step 2, Part 2b) The optional decomposition Theorem 1.5 in Chap. 1 implies that

$$\exists \alpha \in \mathscr{L}\left(\frac{S_u^\nu}{B_u^\nu}\right)$$

and an adapted right continuous nondecreasing process A_t with $A_0 = 0$ such that

$$\frac{X_t}{B_t^\nu} \equiv \frac{X_0}{B_0^\nu} + \int_0^t \alpha_u \cdot d\left(\frac{S_u^\nu}{B_u^\nu}\right) - A_t = W_t \text{ for all } t.$$

Let α_t be the position in the risky assets. Define the position in the mma $\alpha_0(t)$ by the expression

$$\frac{X_t}{B_t^\nu} = \alpha_0(t) + \alpha_t \cdot \left(\frac{S_u^\nu}{B_u^\nu}\right) \text{ for all } t.$$

By the definition of $\frac{X_t}{B_t^\nu}$ we get that

$$\frac{X_0}{B_0^\nu} + \int_0^t \alpha_u \cdot d\left(\frac{S_u^\nu}{B_u^\nu}\right) = \alpha_0(t) + \alpha_t \cdot \left(\frac{S_t^\nu}{B_t^\nu}\right) + A_t.$$

Hence, the trading strategy is self-financing with cash flow A_t and initial value $x = \frac{X_0}{B_0^\nu}$.

We have

$$\frac{X_0}{B_0^\nu} + \int_0^t \alpha_u \cdot d\left(\frac{S_u^\nu}{B_u^\nu}\right) = \frac{X_t}{B_t^\nu} + A_t \geq 0$$

since $A_t \geq 0$ and $\frac{X_t}{B_t^\nu} \geq 0$. Hence, the trading strategy's value process plus cash flow $\frac{X_t}{B_t^\nu} + A_t$ is bounded below by zero, and is therefore admissible, i.e. $(\alpha_0, \alpha) \in \mathscr{A}(x, S)$ with $x = \frac{X_0}{B_0^\nu}$. Thus, $X_T \in \mathscr{C}_2(x, S)$.

This completes the proof.

For future reference, we rewrite this using the local martingale deflators.

$$\mathscr{C}_{\mathfrak{C}}(x, S) = \left\{ X_T \in L_+^0 : \sup_{Y_T \in M_l\left(\frac{S^\nu}{B^\nu}\right)} E\left[X_T \frac{Y_T}{B_T^\nu}\right] \leq x \text{ for all } (\nu_0, \nu) \in \mathscr{K}^o \right\}.$$

Remark 21.2 (No Trading Constraints) When there are no trading constraints, the polar set $K^o = \{(\nu_0, \nu) \in \mathbb{R}^{n+1} : (\nu_0, \upsilon) = 0\}$. In this case the set of local martingale deflators collapses to those in the unconstrained market because $Y_T^\nu = Y_T^0 = Y_T$ and $B_T^\nu = B_T^0 = 1$, i.e.

$$\mathscr{C}(x, S) = \left\{ X_T \in L_+^0 : \sup_{Y_T \in M_l(S)} E\left[X_T Y_T\right] \leq x \right\}.$$

This completes the remark.

21.4 Super-Replication

This section derives the super-replication cost of a derivative in a trading constrained market. The super-replication cost $\bar{c}_0$ is the minimum investment required if one sells the derivative and forms an admissible s.f.t.s. to cover the derivative's payoffs for all $\omega \in \Omega$ a.s. $\mathbb{P}$, i.e.

$$\bar{c}_0 = \inf\{x \in \mathbb{R} : \exists(\alpha_0, \alpha) \in \mathscr{A}(x, S, K),\ x + \int_0^T \alpha_t \cdot dS_t \geq X_T\} \tag{21.1}$$

for $X_T \in L_+^0$.

We now characterize the super-replication cost of an arbitrary derivative security using the set of equivalent local martingale measures. To make the notation simpler, define the set

$$M_l(\mathcal{K}^o) = \bigcup_{(\nu_0,\nu)\in\mathcal{K}^o} M_l\left(\frac{S^\nu}{B^\nu}\right).$$

Theorem 21.1 (Super-Replication Cost) *Let* $X_T \frac{Y_T}{B_T^\nu} \in L_+^1(\mathbb{P})$ *for all* $Y_T \in M_l(\mathcal{K}^o)$.

The super-replication cost is

$$\bar{c}_0 = \sup_{Y_T\in M_l(\mathcal{K}^o)} E\left[X_T \frac{Y_T}{B_T^\nu}\right] \geq 0. \tag{21.2}$$

At time t, the super-replication cost is

$$\bar{c}_t = \text{ess} \sup_{Y_T\in M_l(\mathcal{K}^o)} E\left[X_T \frac{Y_T}{B_T^\nu}\middle| \mathcal{F}_t\right] \geq 0. \tag{21.3}$$

Furthermore, there exists an admissible s.f.t.s. $(\alpha_0, \alpha) \in \mathcal{A}(\bar{c}_0, S, K)$ *with cash flows, an adapted right continuous nondecreasing process* A_t *with* $A_0 = 0$*, and a value process* $X_t \equiv \alpha_0(t) + \alpha_t \cdot S_t$ *such that*

$$X_t = \bar{c}_0 + \int_0^t \alpha_s \cdot dS_s - A_t = \bar{c}_t$$

for all $t \in [0, T]$.

The admissible s.f.t.s. $(\alpha_0, \alpha) \in \mathcal{A}(\bar{c}_0, S, K)$ *with cash flows is the super-replication trading strategy.*

The self-financing condition with cash flows implies that $\bar{c}_0 + \int_0^t \alpha_s \cdot dS_s = \alpha_0(t) + \alpha_t \cdot S_t + A_t$.

The adapted process A corresponds to the additional cash one can withdraw from the trading strategy across time and still obtain the derivative's payoff at time T.

Proof

(Step 1) Recall that

$$\mathcal{C}_{\mathfrak{C}}(x, S) = \left\{X_T \in L_+^0 : \sup_{Y_T\in M_l\left(\frac{S^\nu}{B^\nu}\right)} E\left[X_T \frac{Y_T}{B_T^\nu}\right] \leq x \text{ for all } (\nu_0, \nu) \in \mathcal{K}^o\right\}$$

$$= \left\{X_T \in L_+^0 : \exists(\alpha_0, \alpha) \in \mathcal{A}(x, S, K),\ x + \int_0^T \alpha_u \cdot dS_u \geq X_T\right\}.$$

(Step 2)

$$\bar{c}_0 = \inf\{x \in \mathbb{R} : \exists(\alpha_0, \alpha) \in \mathscr{A}(x, S, K),\ x + \int_0^T \alpha_t \cdot dS_t \geq X_T\}$$

$$= \inf\{x \in \mathbb{R} : X_T \in \mathscr{C}_{\mathfrak{C}}(x, S)\}$$

$$= \inf\{x \in \mathbb{R} : \sup_{Y_T \in M_l(\mathscr{K}^o)} E\left[X_T \frac{Y_T}{B_T^\nu}\right] \leq x\} = \sup_{Y_T \in M_l(\mathscr{K}^o)} E\left[X_T \frac{Y_T}{B_T^\nu}\right].$$

(Step 3) The existence of the trading strategy is obtained by using the supermartingale decomposition theorem as in the proof of Lemma 21.2 above, see also the super- and sub-replication Chap. 8. This completes the proof.

21.5 Sub-replication

This section derives the sub-replication cost of a derivative in a trading constrained market. The sub-replication cost $\underline{c}_0$ is the maximum amount one can borrow to buy the derivative and form an admissible s.f.t.s. that covers the borrowing for all $\omega \in \Omega$ a.s. $\mathbb{P}$, i.e.

$$\underline{c}_0 = \sup\{x \in \mathbb{R} : \exists(\alpha_0, \alpha) \in \mathscr{A}(x, S, K),\ x + \int_0^T \alpha_t \cdot dS_t \leq X_T\} \tag{21.4}$$

for $X_T \in L^0_+$.

As before, we can characterize the sub-replication cost using the set of equivalent local martingale measures. We omit the proof because it is identical to that in the super- and sub-replication Chap. 8.

Theorem 21.2 (Sub-replication Cost) *Let $X_T \frac{Y_T^\nu}{B_T^\nu} \in L^1_+(\mathbb{P})$ for all $Y_T^\nu \in M_l(\mathscr{K}^o)$ be bounded from above.*

The sub-replication cost is

$$\underline{c}_0 = \inf_{Y_T \in M_l(\mathscr{K}^o)} E\left[X_T \frac{Y_T}{B_T^\nu}\right] \geq 0. \tag{21.5}$$

At time t, the sub-replication cost is

$$\underline{c}_t = \text{ess} \inf_{Y_T \in M_l(\mathscr{K}^o)} E\left[X_T \frac{Y_T}{B_T^\nu} \middle| \mathscr{F}_t\right] \geq 0. \tag{21.6}$$

Furthermore, there exists an admissible s.f.t.s. $(\alpha_0, \alpha) \in \mathscr{A}(\underline{c}_0, S, K)$ with cash flows, an adapted right continuous nondecreasing process A_t with $A_0 = 0$, and a value process $X_t \equiv \alpha_0(t) + \alpha_t \cdot S_t$ such that

$$X_t = \underline{c}_0 + \int_0^t \alpha_s dS_s + A_t = \underline{c}_t$$

for all $t \in [0, T]$.

This admissible s.f.t.s. $(\alpha_0, \alpha) \in \mathscr{A}(\underline{c}_0, S, K)$ with cash flows is the sub-replication trading strategy.

The self-financing condition with cash flows implies that $\bar{c}_0 + \int_0^t \alpha_s \cdot dS_s = \alpha_0(t) + \alpha_t \cdot S_t - A_t$.

The adapted process A corresponds to the additional cash one must input into the trading strategy across time to obtain the derivative's payoff at time T.

Remark 21.3 (No Trading Constraints) When there are no trading constraints, the polar set $K^o = \{(v_0, v) \in \mathbb{R}^{n+1} : (v_0, v) = (0, 0)\}$. In this case the super-replication and sub-replication prices simplify to the incomplete market in Chap. 8, i.e.

$$\left[\underline{c}_0 = \inf_{\mathbb{Q} \in \mathfrak{M}_l} E^{\mathbb{Q}}[Z_T], \quad \sup_{\mathbb{Q} \in \mathfrak{M}_l} E^{\mathbb{Q}}[Z_T] = \bar{c}_0\right].$$

This completes the remark.

Chapter 22
Portfolio Optimization

This chapter studies a trader's portfolio optimization problem under trading constraints. Given is the normalized trading constrained market $(S, (\mathscr{F}_t), \mathbb{P})$, where the money market account $B_t \equiv 1$.

22.1 The Set-Up

First, we impose the same assumptions used in the previous Chap. 21 on super- and sub-replication. Consider the normalized auxiliary markets, i.e. $\left(\frac{S^\nu}{B^\nu}, (\mathscr{F}_t), \mathbb{P}\right)$ for all $(\nu_0, \nu) \in \mathscr{K}^o$. We want these auxiliary markets to be arbitrage free.

Assumption (NFLVR for (ν_0, ν)**)**

$$\mathfrak{M}_l \left(\frac{S^\nu}{B^\nu}\right) \neq \emptyset \textit{ for all } (\nu_0, \nu) \in \mathscr{K}^o.$$

When $(\nu_0, \nu) = (0, 0)$, this is just $\mathfrak{M}_l (S) \neq \emptyset$, which is equivalent to NFLVR.

As done in Part II, we change the admissibility lower bound from $c < 0$ to zero. That is, the investor can only trade using nonnegative wealth s.f.t.s.'s. This replaces $(\alpha_0, \alpha) \in \mathscr{A}(x, S, K)$ with $(\alpha_0, \alpha) \in \mathscr{N}(x, S, K)$, where

$$\mathscr{N}(x, S, K) = \{(\alpha_0, \alpha) \in (\mathscr{O}, \mathscr{L}(S)) : X_t = \alpha_0(t) + \alpha_t \cdot S_t,$$
$$X_t = x + \int_0^t \alpha_u \cdot dS_u \geq 0,\ (\alpha_0(t), \alpha_t) \in K \text{ for all } t \in [0, T]\Big\}.$$

We note that all of the equivalences between the sets of wealth processes proved using $\mathscr{A}(x, S, K)$ are still valid (with the same proofs) for $\mathscr{N}(x, S, K)$.

R. A. Jarrow, *Continuous-Time Asset Pricing Theory*, Springer Finance,
https://doi.org/10.1007/978-3-319-77821-1_22

We assume that a trader has preferences $\overset{\rho}{\succeq}$ with a state dependent EU representation with a utility function U defined over terminal wealth that satisfies the properties in the utility function Chap. 9, repeated here for convenience.

Assumption (Utility Function) *A utility function is a $U : (0, \infty) \times \Omega \to \mathbb{R}$ such that for all $\omega \in \Omega$ a.s. $\mathbb{P}$,*

(i) $U(x, \omega)$ is $\mathscr{B}(0, \infty) \otimes \mathscr{F}_T$-measurable,
(ii) $U(x, \omega)$ is continuous and differentiable in $x \in (0, \infty)$,
(iii) $U(x, \omega)$ is strictly increasing and strictly concave in $x \in (0, \infty)$, and
(iv) (Inada Conditions) $\lim_{x \downarrow 0} U'(x, \omega) = \infty$ and $\lim_{x \to \infty} U'(x, \omega) = 0$.

The trader's optimization problem is therefore

Problem 22.1 Utility Optimization (Choose the Optimal Trading Strategy (α_0, α))

$$v(x, S) = \sup_{(\alpha_0, \alpha) \in \mathscr{N}(x, S, K)} E\left[U\left(X_T\right)\right], \qquad \text{where}$$

$$\mathscr{N}(x, S, K) = \{(\alpha_0, \alpha) \in (\mathscr{O}, \mathscr{L}(S)) : X_t = \alpha_0(t) + \alpha_t \cdot S_t,$$
$$X_t = x + \textstyle\int_0^t \alpha_u \cdot dS_u \geq 0,\ (\alpha_0(t), \alpha_t) \in K \text{ for all } t \in [0, T]\Big\}.$$

To solve this optimization problem, as in Part II, we decompose the solution procedure into two steps. First, we solve the static problem of choosing the optimal time T wealth (a "derivative"). Then, we solve the dynamic problem by finding the nonnegative wealth s.f.t.s. that generates this optimal time T wealth using the methodology from the fundamental theorems Chap. 2 for constructing a synthetic derivative.

Problem 22.2 Utility Optimization (Choose the Optimal derivative X_T)

$$v(x, S) = \sup_{X_T \in \mathscr{C}^e_{\mathfrak{C}}(x, S)} E\left[U\left(X_T\right)\right], \qquad \text{where}$$

$$\mathscr{C}^e_{\mathfrak{C}}(x, S) = \{X_T \in L^0_+ : \exists(\alpha_0, \alpha) \in \mathscr{N}(x, S, K),\ x + \int_0^T \alpha_t \cdot dS_t = X_T\}.$$

Analogous to the portfolio optimization Chap. 11, the free disposal Lemma 11.1 is valid (the same proof applies). This yields the following equivalent optimization problem.

Problem 22.3 Utility Optimization (Choose the Optimal derivative X_T)

$$v(x, S) = \sup_{X_T \in \mathscr{C}_{\mathfrak{C}}(x, S)} E\left[U\left(X_T\right)\right], \qquad \text{where}$$

$$\mathscr{C}_{\mathfrak{C}}(x, S) = \{X_T \in L^0_+ : \exists(\alpha_0, \alpha) \in \mathscr{N}(x, S, K),\ x + \int_0^T \alpha_t \cdot dS_t \geq X_T\}.$$

To solve this optimization problem, we use an alternative characterization of the budget constraint. This is provided in the next section.

22.2 Wealth Processes (Revisited)

This section provides an equivalent characterization of the investor's budget constraint. We have already proven that

$$\mathscr{C}_{\mathfrak{C}}(x,S)=\left\{X_T\in L^0_+:\sup_{Y_T\in M_l\left(\frac{S^\nu}{B^\nu}\right)}E\left[X_T\frac{Y_T}{B^\nu_T}\right]\le x\ \text{ for all } (\nu_0,\nu)\in\mathscr{K}^o\right\}.$$

We note that $Y_T=\frac{d\mathbb{Q}}{d\mathbb{P}}\in\mathfrak{M}_l\left(\frac{S^\nu}{B^\nu}\right)$ is a probability density with respect to $\mathbb{P}$. Define

$$H^\nu_T=\frac{Y^\nu_T}{B^\nu_T}\in L^0_+,$$

and the local martingale deflator (stochastic) process

$$H^\nu_t=\frac{Y^\nu_t}{B^\nu_t}=E\left[\frac{Y^\nu_T}{B^\nu_T}\middle|\mathscr{F}_t\right]\in\mathscr{L}^0_+.$$

With some abuse of terminology, we also call H^ν_T *a local martingale deflator*. It is sometimes called a *state price deflator*.

We have the following equivalences. The same proof as in the trading constraints Chap. 21 on super- and sub-replication applies.

Lemma 22.1 (Equivalences) $\left(\frac{S^\nu}{B^\nu}\right)Y^\nu$ *is a $\mathbb{P}$-local martingale if and only if $S^\nu H^\nu$ is a $\mathbb{P}$-local martingale if and only if*

XH^ν is a $\mathbb{P}$-local martingale, for all $X=1+\int_0^t\alpha_u\cdot dS^\nu$, $(\alpha_0,\alpha)\in\mathscr{N}(1,S^\nu)$.

Using this lemma, we have that

$$\mathscr{C}_{\mathfrak{C}}(x,S)=\left\{X_T\in L^0_+:\sup_{H_T\in M_l(S^\nu)}E\left[X_TH_T\right]\le x\ \text{ for all } (\nu_0,\nu)\in\mathscr{K}^o\right\}.$$

If H^ν makes S^ν into a $\mathbb{P}$ martingale, then Y^ν_T is a *martingale* deflator and the density of a martingale measure $\mathbb{Q}_\nu\in\mathfrak{M}\left(\frac{S^\nu}{B^\nu}\right)$ with respect to $\mathbb{P}$.

As in the portfolio optimization Chap. 11, to get the existence of an optimal wealth process, we need to expand the set of local martingale deflators to the set of supermartingale deflators. In this regard, define

$$\mathscr{D}_s(S^\nu) = \Big\{H \in \mathscr{L}_+^0 : H_0 = 1,\ HX \text{ is a } \mathbb{P}\text{-supermartingale},\\ \text{for all } X = 1 + \int_0^t \alpha_u \cdot dS^\nu,\ \alpha \in \mathscr{N}(1, S^\nu)\Big\}, \quad \text{and}$$

$$D_s\left(S^\nu\right) = \{H_T \in L_+^0 : \exists Z \in \mathscr{D}_s\left(S^\nu\right),\ Z_T = H_T\}. \tag{22.1}$$

We note that $D_s(S^\nu)$ is the smallest convex, solid, closed subset of L_+^0 (in the $\mathbb{Q}_\nu$-convergence topology) that contains $M_l(S^\nu)$. Solid means that if $\hat{H}_T \le H_T$ and $H_T \in D_s(S^\nu)$, then $\hat{H}_T \in D_s(S^\nu)$. It can be shown that $D_s(S^\nu)$ can be rewritten as (see Pham [149, p. 186])

$$D_s\left(S^\nu\right) = \{H_T \in L_+^0 : H_0 = 1,\ \exists (Z(T)_n)_{n\ge 1} \in M_l\left(S^\nu\right),\ H_T \le \lim_{n\to\infty} Z_n(T)\ a.s.\}. \tag{22.2}$$

Proof (Also See Schachermayer [164]) This follows because by the bipolar theorem in Brannath and Schachermayer [21] the right side of expression (22.1) is the smallest convex, solid, closed subset of L_+^0 (in the $\mathbb{P}$-convergence topology) that contains $M_l(S^\nu) \subset L_+^0$. Since $D_s(S^\nu)$ is also the smallest convex, solid, closed subset of L_+^0 containing $M_l(S^\nu)$ we get the result. This completes the proof.

For simplicity of the notation, define the set of *supermartingale deflators* across all possible $(\nu_0, \nu) \in \mathscr{K}^o$ as

$$D_s(\mathscr{K}^o) = \bigcup_{(\nu_0,\nu)\in\mathscr{K}^o} D_s\left(S^\nu\right).$$

It is important to note that each supermartingale deflator $H_T \in D_s(\mathscr{K}^o)$ is identified with a unique $(\nu_0, \nu) \in \mathscr{K}^o$.

We can now state the final characterization of the budget constraint. The proof is identical to that of Lemma 11.2 in the portfolio optimization Chap. 11.

Lemma 22.2 (Equivalence 4)

$$\mathscr{C}_{\mathfrak{C}}(x, S) = \{X_T \in L_+^0 :\ E\left[X_T H_T\right] \le x \ \text{ for all } H_T \in D_s(\mathscr{K}^o)\}.$$

22.3 The Optimization Problem

Given the characterization of the budget constraint from the previous section, the next theorem holds.

Theorem 22.1 (Equivalence of a Trading Constrained Optimization with an Incomplete Market Optimization)

$$v(x,S) = \sup_{X_T \in \mathscr{C}_{\mathfrak{C}}(x,S)} E\left[U\left(X_T\right)\right], \qquad \textit{where}$$

$$\mathscr{C}_{\mathfrak{C}}(x,S) = \{X_T \in L_+^0 : \exists(\alpha_0,\alpha) \in \mathscr{N}(x,S,K),\ x + \int_0^T \alpha_t \cdot dS_t \geq X_T\}$$

is equivalent to

$$v(x,S) = \sup_{X_T \in \mathscr{C}_{\mathfrak{C}}(x,S)} E\left[U\left(X_T\right)\right], \qquad \textit{where}$$

$$\mathscr{C}_{\mathfrak{C}}(x,S) = \{X_T \in L_+^0 : E\left[X_T H_T\right] \leq x \ \textit{for all}\ H_T \in D_s(\mathscr{K}^o)\}.$$

Note that in the statement of the second optimization problem, the constraint set K enters via the definition of the set of supermartingale deflators

$$D_s(\mathscr{K}^o) = \bigcup_{(\nu_0,\nu)\in\mathscr{K}^o} D_s\left(S^\nu\right).$$

This is an important theorem. It implies that to solve the optimization problem in a market with trading constraints, one can solve an equivalent problem without trading constraints in an incomplete market. Since the incomplete market optimization problem has already been solved in Part II, the portfolio optimization Chap. 11, to solve the trading constrained optimization problem we can use the previous solutions. This is the task to which we now turn.

Define the Lagrangian

$$\mathscr{L}(X_T, y, H_T) = E\left[U(X_T)\right] + y\left[x - E\left[X_T H_T\right]\right]. \tag{22.3}$$

The *primal problem* can be written as

$$\begin{aligned}\text{primal}: \quad & \sup_{X_T \in L_+^0} \left(\inf_{y>0, H_T \in D_s(\mathscr{K}^o)} \mathscr{L}(X_T, y, H_T)\right) \\ & = \sup_{X_T \in \mathscr{C}_{\mathfrak{C}}(x,S)} E\left[U(X_T)\right] = v(x,S).\end{aligned}$$

The *dual problem* is

$$\text{dual}: \quad \inf_{y>0, H_T \in D_s(\mathscr{K}^o)} \left(\sup_{X_T \in L_+^0} \mathscr{L}(X_T, y, H_T)\right).$$

The solutions to the primal and dual problems are equal if there is no duality gap. We show there is no duality gap and then we solve the dual problem to obtain the primal solution.

22.4 Existence of a Solution

We need to show that there is no duality gap and that the optimum is attained, or equivalently, a saddle point exists. To prove this existence, we need three assumptions. The first was already assumed. We repeat it here for emphasis.

Assumption (NFLVR) $\mathfrak{M}_l(S) \neq \emptyset$.

Assumption (Non-trivial Optimization Problem) *There exists an* $x > 0$ *such that* $v(x, S) < \infty$.

Assumption (Asymptotic Elasticity)

$$AE(U, \omega) \equiv \limsup_{x \to \infty} \frac{xU'(x, \omega)}{U(x, \omega)} < 1 \quad a.s.\ \mathbb{P}.$$

Applying the Theorems from the portfolio optimization Chap. 11 directly, we get

Theorem 22.2 (Existence of a Unique Saddle Point and Characterization of $v(x, S)$) *Given the above assumptions, there exists a unique saddle point, i.e. there exists a* $(\hat{X}_T, \hat{H}_T, \hat{y})$ *such that*

$$\inf_{H_T \in D_s(\mathscr{K}^o),\, y>0} \left(\sup_{X_T \in L^0_+} \mathscr{L}(X_T, H_T, y)\right) = \mathscr{L}(\hat{X}_T, \hat{H}_T, \hat{y})$$

$$= \sup_{X_T \in L^0_+} \left(\inf_{H_T \in D_s(\mathscr{K}^o),\, y>0} \mathscr{L}(X_T, H_T, y)\right) = v(x, S). \tag{22.4}$$

Define

$$\tilde{v}(y, S) = \inf_{H_T \in D_s(\mathscr{K}^o)} E\left[\tilde{U}(yH_T)\right] \text{ where } \tilde{U}(y, \omega) = \sup_{x>0}[U(x, \omega) - xy], \quad y > 0.$$

Then, v *and* $\tilde{v}$ *are in conjugate duality, i.e.*

$$v(x, S) = \inf_{y>0} \left(\tilde{v}(y) + xy\right), \ \forall x > 0, \tag{22.5}$$

$$\tilde{v}(y, S) = \sup_{x>0} \left(v(x, S) - xy\right), \ \forall y > 0. \tag{22.6}$$

In addition,

(i) $\tilde{v}$ *is strictly convex, decreasing, and differentiable on* $(0, \infty)$,
(ii) v *is strictly concave, increasing, and differentiable on* $(0, \infty)$,
(iii) *defining* $\hat{y}$ *to be where the infimum is attained in expression (22.5),*

$$v(x, S) = \tilde{v}(\hat{y}, S) + x\hat{y}. \tag{22.7}$$

22.5 Characterization of the Solution

This section characterizes the solution to the investor's optimization problem. It is a direct application of the portfolio optimization Chap. 11's results.

$$v(x, S) = \sup_{X_T \in \mathscr{C}_{\mathfrak{C}}(x,S)} E\left[U(X_T)\right]$$

has the solution for terminal wealth

$$X_T = I(yH_T) \quad \text{with} \quad I = (U')^{-1} = -\tilde{U}',$$

where $H_T \in D_s(\mathscr{K}^o)$ is the solution to

$$\tilde{v}(y, S) = \inf_{H_T \in D_s(\mathscr{K}^o)} E[\tilde{U}(yH_T)],$$

$$\tilde{U}(y, \omega) = \sup_{x>0} \left[U(x, \omega) - xy\right] \text{ for } \quad y > 0,$$

and y is the solution to the budget constraint

$$E\left[H_T I(yH_T)\right] = x.$$

Recall that each $H_T \in D_s(\mathscr{K}^o)$ is identified with a unique $(\nu_0, \nu) \in \mathscr{K}^o$. This implies that the optimal $H_T \in D_s(\mathscr{K}^o)$ identifies a unique (B^ν, S^ν) auxiliary market, i.e. $H_T = D_s(S^\nu)$ for some $(\nu_0, \nu) \in \mathscr{K}^o$. We will use this observation below when characterizing the trader's risk return tradeoff.

The *optimal portfolio wealth* when multiplied by the supermartingale deflator, $X_t H_t$, is a $\mathbb{P}$-martingale. This implies that for an arbitrary time $t \in [0, T]$ we have

$$X_t = E\left[X_T \frac{H_T}{H_t} \middle| \mathscr{F}_t \right].$$

It is important to note, for subsequent use, that the solution $H_T \in D_s(\mathscr{K}^o)$ depends on the utility function U.

22.6 The Shadow Price of the Budget Constraint

This section derives the shadow price of the budget constant. The shadow price is the benefit, in terms of expected utility, of increasing the initial wealth by 1 unit (of the mma). The shadow price equals the Lagrangian multiplier.

Theorem 22.3 (The Shadow Price of the Budget Constraint)

$$y = v'(x, S) \geq E[U'(X_T)] = yE[H_T].$$

These are equalities if and only if $E[H_T] = 1$.

Proof The first equality is obtained by taking the derivative of $v(x, S) = \tilde{v}(y, S) + xy$. The second equality follows from the first-order condition for an optimum $U'(X_T) = yH_T$. Taking expectations gives $E[U'(X_T)] = E[yH_T] \leq y$ because $E[H_T(T)] \leq H_0 = 1$. This completes the proof.

22.7 The Supermartingale Deflator

This section derives the optimal supermartingale deflator. The key insight of the individual optimization problem is the identification of the supermartingale deflator in terms of the individual's utility function and initial wealth. First, we rewrite the first-order condition for the optimal wealth X_T as follows

$$H_T = \frac{U'(X_T)}{y} = \frac{U'(X_T)}{v'(x, S)}.$$

The supermartingale deflator does not need to be a martingale, i.e. $H_t \neq E\left[H_T \,|\mathscr{F}_t\right]$. In fact, it can be proven that H is a $\mathbb{P}$-supermartingale.

Lemma 22.3 $H_t \geq E\left[H_T \,|\mathscr{F}_t\right]$ for all t, i.e. H_t is a $\mathbb{P}$-supermartingale.

Proof H_t is a $\mathbb{P}$-supermartingale because $H_t = H_t X_t$, where X_t represents the wealth process from an admissible s.f.t.s. consisting of a unit holding in the mma. This completes the proof.

Recall that each $H_T \in D_s(\mathscr{K}^o)$ is identified with a unique $(v_0, v) \in \mathscr{K}^o$, i.e. $H_T \in D_s(S^v)$ for some $(v_0, v) \in \mathscr{K}^o$. If $H_T \in D_l(S^v)$ is a local martingale deflator, then $H_t = \frac{Y_t^v}{B_t^v} \in \mathscr{M}_l\left(\frac{S^v}{B^v}\right)$ for that $(v_0, v) \in \mathscr{K}^o$. Assuming that $Y_T^v = \frac{d\mathbb{Q}_v}{d\mathbb{P}}$ is the density of the probability measure $\mathbb{Q}_\nu$ with respect to $\mathbb{P}$, then

$$Y_t^\nu = E\left[Y_T^\nu \,|\mathscr{F}_t\right] = E\left[\frac{U'(X_T)B_T^\nu}{v'(x,S)}\middle|\mathscr{F}_t\right],$$

$$H_t = \frac{Y_t^\nu}{B_t^\nu} = \frac{E\left[Y_T^\nu \,|\mathscr{F}_t\right]}{B_t^\nu} = E\left[\frac{U'(X_T)B_T^\nu}{v'(x,S)B_t^\nu}\middle|\mathscr{F}_t\right], \quad \text{and}$$

$$\frac{H_{t+\Delta}}{H_t} = \frac{E\left[\frac{U'(X_T)B_T^\nu}{v'(x,S)B_{t+\Delta}^\nu}\middle|\mathscr{F}_{t+\Delta}\right]}{E\left[\frac{U'(X_T)B_T^\nu}{v'(x,S)B_t^\nu}\middle|\mathscr{F}_t\right]} = \frac{E\left[U'(X_T)B_T^\nu \,|\mathscr{F}_{t+\Delta}\right]B_t^\nu}{E\left[U'(X_T)B_T^\nu \,|\mathscr{F}_t\right]B_{t+\Delta}^\nu}. \tag{22.8}$$

This last expression will prove useful in a subsequent section.

22.8 The Shadow Prices of the Trading Constraints

This section derives the shadow prices of the trading constraints.

If $(\nu_0, \upsilon) = (0, 0)$, as determined by the trader's optimal supermartingale deflator, then the price processes

$$S^\nu = S \quad \text{and} \quad B^\nu = 1$$

are the same as those in the unconstrained market. In this case, the solution to the trader's optimization problem will be the same as that in an unconstrained market. Hence, when $(\nu_0, \upsilon) = (0, 0)$, trading constraints are non-binding and the shadow prices of the trading constraint are all zero.

If the optimal $(\nu_0, \nu) \neq (0, 0) \in \mathscr{K}^o$, then the trader's optimization problem has a different solution from that in an unconstrained market. This changes the "effective" price processes of the mma and risky assets. The trader behaves as if they trade in the auxiliary market (ν_0, ν) with the price processes

$$B_t^\nu = B_t - \int_0^t \nu_0(u)du \text{ and } S_t^\nu = S_t - \int_0^t \nu_u du,$$

where $\int_0^t \nu_0(u)du \leq 0$ and $\int_0^t \nu_u du \leq 0$ for all t. Hence, $(\nu_0, \nu) \in \mathscr{K}^o$ through these expressions reflect the shadow prices of the trading constraints.

22.9 Asset Price Bubbles

This section studies the existence of asset price bubbles in a trading constrained economy. For this section, we impose the simplifying assumption that there is only *one risky asset* trading denoted S_t. The restriction to one risky asset is without loss of generality and it is imposed to simplify the notation (to avoid subscripts for the jth risky asset). This section relates the shadow prices of the trading constraints to the existence of asset price bubbles in the market. These asset price bubbles are determined by the trader's beliefs and preferences.

Let $(\nu_0, \nu) \in \mathscr{K}^o$ be such that the optimal supermartingale deflator is $H_T^\nu \in D_s(S^\nu)$. We define the risky asset's fundamental value using the trader's supermartingale deflator, i.e.

$$\text{Fundamental Value } = E\left[\frac{H_T^\nu}{H_t^\nu} S_T \middle| \mathscr{F}_t\right].$$

The risky asset's price bubble is the difference between the asset's market price and its fundamental value, i.e.

$$\beta_t = S_t - E\left[\frac{H_T^\nu}{H_t^\nu} S_T \middle| \mathscr{F}_t\right] \geq 0.$$

Note that the asset bubble is nonnegative.

Proof First, $H^\nu S^\nu$ is a $\mathbb{P}$-supermartingale. This follows because S^ν represents the wealth process of a buy and hold trading strategy with a unit holding in the risky asset and $H^\nu X$ is a supermartingale for this trading strategy, see Lemma 21.1 in the trading constraints Chap. 21 on super- and sub-replication. Then, $H_t^\nu S_t = H_t^\nu S_t^\nu + H_t^\nu \int_0^t \nu_u du$ is a supermartingale as well since $\int_0^t \nu_u du \leq 0$ and $H^\nu > 0$. This completes the proof.

The next theorem relates the asset price bubble to the shadow prices of the trading constraints.

Theorem 22.4 (Bubbles and Shadow Prices of the Trading Constraints) *Assume that $d\mathbb{Q}_\nu = Y_T^\nu d\mathbb{P}$ is a probability density for the optimal $(\nu_0, \nu) \in \mathscr{K}^o$. Then,*

$$\beta_t = \left(\int_0^t \nu_u du\right) E_\nu\left[\frac{-\int_t^T \nu_0(u)du}{1-\int_0^T \nu_0(u)du} \middle| \mathscr{F}_t\right]$$
$$-E_\nu\left[\frac{1-\int_0^t \nu_0(u)du}{1-\int_0^T \nu_0(u)du}\left(\int_t^T \nu_u du\right) \middle| \mathscr{F}_t\right] \geq 0.$$

Proof Note that

$$E\left[\frac{H_T^\nu}{H_t^\nu} S_T \middle| \mathscr{F}_t\right] = E_\nu\left[\frac{B_t^\nu}{B_T^\nu} S_T \middle| \mathscr{F}_t\right].$$

To simplify the notation, set $E_\nu[\cdot] = E_\nu[\cdot|\mathscr{F}_t]$.

Also note that

$$B_T^\nu - B_t^\nu = \left(1 - \int_0^T \nu_0(u)du\right) - \left(1 - \int_0^t \nu_0(u)du\right) = -\int_t^T \nu_0(u)du \geq 0$$

and

$$S_t^\nu = E_\nu\left[\frac{B_t^\nu}{B_T^\nu} S_T^\nu\right].$$

Then,

$$\begin{aligned}
\beta_t &= S_t - E_\nu\left[\frac{B_t^\nu}{B_T^\nu} S_T\right] \\
&= S_t - E_\nu\left[\frac{B_t^\nu}{B_T^\nu} S_T^\nu\right] - E_\nu\left[\frac{B_t^\nu}{B_T^\nu}\int_0^T \nu_u du\right] \\
&= S_t - S_t^\nu - E_\nu\left[\frac{B_t^\nu}{B_T^\nu}\int_0^T \nu_u du\right] \\
&= \int_0^t \nu_u du - E_\nu\left[\frac{B_t^\nu}{B_T^\nu}\int_0^T \nu_u du\right] \\
&= \left(\int_0^t \nu_u du\right) E_\nu\left[1 - \frac{B_t^\nu}{B_T^\nu}\right] - E_\nu\left[\frac{B_t^\nu}{B_T^\nu}\int_t^T \nu_u du\right] \\
&= -\left(\int_0^t \nu_u du\right) E_\nu\left[\frac{\int_t^T \nu_0(u)du}{1 - \int_0^T \nu_0(u)du}\right] - E_\nu\left[\frac{1 - \int_0^t \nu_0(u)du}{1 - \int_0^T \nu_0(u)du}\left(\int_t^T \nu_u du\right)\right].
\end{aligned}$$

This completes the proof.

This theorem shows that an asset's price bubble is completely determined by the shadow prices of the trading constraints. If the trading constraints are non-binding so that $(\nu_0(t), \nu_t) = (0, 0)$ for all t, then the asset's price bubble is identically zero.

22.10 Systematic Risk

This section characterizes the trader's risk return tradeoff. For this section, we also impose the simplifying assumption that there is only *one risky asset* trading denoted S_t. Let $(\nu_0, \nu) \in \mathscr{K}^o$ be such that the optimal supermartingale deflator is $H_T^\nu \in D_s(S^\nu)$. Assume that $H_T^\nu = \frac{Y_T^\nu}{B_T^\nu} \in M(S^\nu)$ is a martingale deflator, and not just a supermartingale deflator, where $Y_T^\nu = \frac{d\mathbb{Q}_\nu}{d\mathbb{P}}$ is a probability density. Then,

$Y_T^\nu \in M\left(\frac{S^\nu}{B^\nu}\right)$ and $\frac{S^\nu}{B^\nu}$ is a $\mathbb{Q}_\nu$-martingale. Under this assumption we can apply the theorems from the multiple-factor model Chap. 4 that characterize systematic risk.

Define the returns for the risky asset and the mma in the auxiliary market over $[t, t + \Delta]$ by

$$R_t^\nu = \frac{S_{t+\Delta}^\nu - S_t^\nu}{S_t^\nu} \qquad \text{and}$$

$$r_t^\nu = \frac{B_{t+\Delta}^\nu - B_t^\nu}{B_t^\nu} \geq 0$$

since B_t^ν is increasing. Note that $\frac{B_t^\nu}{B_{t+\Delta}^\nu} = \frac{1}{(1+r_t^\nu)}$.

We assume that traded is a zero-coupon bond that matures at time. Let the time t price of this zero coupon bond in the auxiliary market be

$$p^\nu(t, t+\Delta) = E\left[p^\nu(t+\Delta, t+\Delta) \frac{H_{t+\Delta}^\nu}{H_t^\nu} \middle| \mathscr{F}_t \right] = E\left[1 \cdot \frac{H_{t+\Delta}^\nu}{H_t^\nu} \middle| \mathscr{F}_t \right],$$

where by definition, $p^\nu(t+\Delta, t+\Delta) = 1$. Then, its return over $[t, t+\Delta]$ is

$$r_t^\nu = \frac{1 - p^\nu(t, t+\Delta)}{p^\nu(t, t+\Delta)}.$$

This corresponds to the return on the discrete mma in the auxiliary market $(\nu_0, \nu) \in \mathscr{K}^o$.

From Theorem 4.4 in the multiple-factor model Chap. 4 we have

$$E\left[R_t^\nu | \mathscr{F}_t\right] = r_t^\nu - \operatorname{cov}\left[R_t^\nu, \frac{H_{t+\Delta}^\nu}{H_t^\nu}(1 + r_t^\nu) \middle| \mathscr{F}_t \right]. \tag{22.9}$$

We need to transform this risk return relation from the auxiliary market to the actual market. Define

$$R_t = \frac{S_{t+\Delta} - S_t}{S_t},$$

$$\Upsilon_t = \frac{\int_0^{t+\Delta} \nu_u du - \int_0^t \nu_u du}{\int_0^t \nu_u du} \geq 0$$

since $\nu(u) \leq 0$ for all t, and

$$\Upsilon_0(t) = \frac{\int_0^{t+\Delta} \nu_0(u) du - \int_0^t \nu_0(u) du}{\int_0^t \nu_0(u) du} \geq 0$$

since $\nu_0(u) \leq 0$ for all t. The last two expressions are the returns to the shadow prices of the trading constraints for the risky assets and mma, respectively.

We prove the following lemma.

Lemma 22.4 (Return Relationships)

$$\frac{B^\nu_{t+\Delta} - B^\nu_t}{B^\nu_t} = \eta \Upsilon_0(t) \qquad \textit{and} \tag{22.10}$$

$$R^\nu_t = wR_t + (1-w)\Upsilon_t, \tag{22.11}$$

where $w = \frac{S_t}{S_t - \int_0^t \nu_u du} \geq 0$ *and* $1 \geq \eta = -\frac{\int_0^t \nu_0(u)du}{1-\int_0^t \nu_0(u)du} \geq 0$.

Proof

$$R^\nu_t = \frac{S^\nu_{t+\Delta} - S^\nu_t}{S^\nu_t} = \frac{\left(S_{t+\Delta} - \int_0^{t+\Delta} \nu_u du\right) - \left(S_t - \int_0^t \nu_u du\right)}{S_t - \int_0^t \nu_u du}$$

$$= \frac{S_t}{S_t - \int_0^t \nu_u du} \frac{S_{t+\Delta} - S_t}{S_t} + \frac{\int_0^t \nu_u du}{S_t - \int_0^t \nu_u du} \frac{\int_0^{t+\Delta} \nu_u du - \int_0^t \nu_u du}{\int_0^t \nu_u du}$$

$$= \frac{S_t}{S_t - \int_0^t \nu_u du} R_t + \frac{\int_0^t \nu_u du}{S_t - \int_0^t \nu_u du} \Upsilon_t$$

$$= wR + (1-w)\Upsilon \text{ where } w = \frac{S_t}{S_t - \int_0^t \nu_u du} \geq 0.$$

$$\frac{B^\nu_{t+\Delta} - B^\nu_t}{B^\nu_t} = \frac{\left(1 - \int_0^{t+\Delta} \nu_0(u)du\right) - \left(1 - \int_0^t \nu_0(u)du\right)}{1 - \int_0^t \nu_0(u)du}$$

$$= -\frac{\int_0^t \nu_0(u)du}{1 - \int_0^t \nu_0(u)du} \frac{\int_0^{t+\Delta} \nu_0(u)du - \int_0^t \nu_0(u)du}{\int_0^t \nu_0(u)du}$$

$$= -\frac{\int_0^t \nu_0(u)du}{1 - \int_0^t \nu_0(u)du} \Upsilon_0(t).$$

This completes the proof.

Using this lemma, we can get the risk return relation in the market with trading constraints.

Theorem 22.5 (Risk Return Relation)

$$E\left[R_t \mid \mathscr{F}_t\right] = -\text{cov}\left[R_t, \frac{E\left[U'(X_T)\left(1-\int_0^T \nu_0(u)du\right)|\mathscr{F}_{t+\Delta}\right]}{E\left[U'(X_T)\left(1-\int_0^T \nu_0(u)du\right)|\mathscr{F}_t\right]} \frac{B_t^\nu}{B_{t+\Delta}^\nu}(1+r_t^\nu) \middle| \mathscr{F}_t \right]$$
$$+\frac{1}{w}r_t^\nu$$
$$-\frac{(1-w)}{w}E\left[\Upsilon_t \mid \mathscr{F}_t\right] - \frac{(1-w)}{w}\text{cov}\left[\Upsilon_t, \frac{E\left[U'(X_T)\left(1-\int_0^T \nu_0(u)du\right)|\mathscr{F}_{t+\Delta}\right]}{E\left[U'(X_T)\left(1-\int_0^T \nu_0(u)du\right)|\mathscr{F}_t\right]}\right.$$
$$\left.\frac{B_t^\nu}{B_{t+\Delta}^\nu}(1+r_t^\nu) \middle| \mathscr{F}_t\right], \tag{22.12}$$

where $w = \frac{S_t}{S_t - \int_0^t \nu_u du} \geq 0$ *and* $1 \geq \eta = -\frac{\int_0^t \nu_0(u)du}{1-\int_0^t \nu_0(u)du} \geq 0$.

Proof To simplify the notation, define $E[\cdot] = E[\cdot \mid \mathscr{F}_t]$.

$$E\left[R_t^\nu\right] = r_t^\nu - \text{cov}\left[R_t^\nu, \frac{H_{t+\Delta}^\nu}{H_t^\nu}(1+r_t^\nu)\right],$$

$$E\left[wR_t + (1-w)\Upsilon_t\right] = r_t^\nu - \text{cov}\left[wR_t + (1-w)\Upsilon_t, \frac{H_{t+\Delta}^\nu}{H_t^\nu}(1+r_t^\nu)\right],$$

$$E\left[wR_t\right] = r_t^\nu - E\left[(1-w)\Upsilon_t\right] - wcov\left[R_t, \frac{H_{t+\Delta}^\nu}{H_t^\nu}(1+r_t^\nu)\right]$$
$$-(1-w)\,\text{cov}\left[\Upsilon_t, \frac{H_{t+\Delta}^\nu}{H_t^\nu}(1+r_t^\nu)\right],$$

$$E\left[R_t\right] = \frac{1}{w}r_t^\nu - \frac{(1-w)}{w}E_t\left[\Upsilon_t\right] - \text{cov}\left[R_t, \frac{H_{t+\Delta}^\nu}{H_t^\nu}(1+r_t^\nu)\right]$$
$$-\frac{(1-w)}{w}\text{cov}\left[\Upsilon_t, \frac{H_{t+\Delta^\nu}}{H_t^\nu}(1+r_t^\nu)\right].$$

But,

$$\frac{H_{t+\Delta}^\nu}{H_t^\nu} = \frac{E\left[U'(X_T)B_T^\nu \mid \mathscr{F}_{t+\Delta}\right]B_t^\nu}{E\left[U'(X_T)B_T^\nu \mid \mathscr{F}_t\right]B_{t+\Delta}^\nu}$$

from expression (22.8).

Recall that

$$\frac{B_t^\nu}{B_{t+\Delta}^\nu} = \frac{1}{(1+r_t^\nu)},$$

hence

$$\frac{H_{t+\Delta}^{\nu}}{H_t^{\nu}}(1+r_t^{\nu}) = \frac{E\left[U'(X_T)B_T^{\nu}\,|\mathscr{F}_{t+\Delta}\right]}{E\left[U'(X_T)B_T^{\nu}\,|\mathscr{F}_t\right]}.$$

Algebra and substitution for B_T^{ν} completes the proof.

When the time step is small, i.e. $\Delta \approx dt$, $\frac{\mathbb{B}_t^{\nu}}{\mathbb{B}_{t+\Delta}^{\nu}}(1+r_t^{\nu}) \approx 1$ and $r_t^{\nu} \approx \frac{\mathbb{B}_{t+\Delta}^{\nu}}{\mathbb{B}_t^{\nu}} - 1 = \eta\Upsilon_0(t)$. Using these approximations, this relation simplifies to

$$\begin{aligned} E\left[R_t\,|\mathscr{F}_t\right] \approx &-\mathrm{cov}\left[R_t, \frac{E\left[U'(X_T)\left(1-\int_0^T \nu_0(u)du\right)|\mathscr{F}_{t+\Delta}\right]}{E\left[U'(X_T)\left(1-\int_0^T \nu_0(u)du\right)|\mathscr{F}_t\right]}\middle|\mathscr{F}_t\right] \\ &+\frac{\eta}{w}\Upsilon_0(t) - \frac{(1-w)}{w}E\left[\Upsilon_t\,|\mathscr{F}_t\right] \\ &-\frac{(1-w)}{w}\mathrm{cov}\left[\Upsilon_t, \frac{E\left[U'(X_T)\left(1-\int_0^T \nu_0(u)du\right)|\mathscr{F}_{t+\Delta}\right]}{E\left[U'(X_T)\left(1-\int_0^T \nu_0(u)du\right)|\mathscr{F}_t\right]}\middle|\mathscr{F}_t\right]. \end{aligned} \tag{22.13}$$

The last three terms are the adjustment for the shadow prices of the trading constraints. Note the following changes from the traditional systematic risk return relation.

1. The supermartingale deflator

$$\frac{E\left[U'(X_T)\left(1-\int_0^T \nu_0(u)du\right)|\mathscr{F}_{t+\Delta}\right]}{E\left[U'(X_T)\left(1-\int_0^T \nu_0(u)du\right)|\mathscr{F}_t\right]}$$

is modified due to $\left(1-\int_0^T \nu_0(u)du\right) \geq 1$.

2. The default-free spot rate is modified from zero (in the normalized market) to

$$\frac{\eta}{w}\Upsilon_0(t) - \frac{(1-w)}{w}E\left[\Upsilon_t\,|\mathscr{F}_t\right].$$

3. There is an extra risk premium equal to

$$-\frac{(1-w)}{w}\mathrm{cov}\left[\Upsilon_t, \frac{E\left[U'(X_T)\left(1-\int_0^T \nu_0(u)du\right)|\mathscr{F}_{t+\Delta}\right]}{E\left[U'(X_T)\left(1-\int_0^T \nu_0(u)du\right)|\mathscr{F}_t\right]}\middle|\mathscr{F}_t\right].$$

Note that if the trading constraints are non-binding, then $(\nu_0, \nu) \equiv (0, 0)$ and this Theorem collapses to that in the portfolio optimization Chap. 11.

Chapter 23
Equilibrium

This section studies equilibrium in a trading constrained economy.

23.1 The Set-Up

The set-up is as in Chap. 13 on equilibrium, but modified for trading constraints. Given is a normalized trading constrained market $(S, (\mathscr{F}_t), \mathbb{P})$ with $B \equiv 1$. There are $\mathscr{I}$ traders in the economy with preferences $\overset{\rho}{\succeq}_i$ assumed to have a state dependent EU representation with utility functions defined over terminal wealth that satisfy the properties of the utility function in Chap. 9, repeated here for convenience.

Assumption (Utility Function) *A utility function is a* $U_i : (0, \infty) \times \Omega \to \mathbb{R}$ *such that for all* $\omega \in \Omega$ *a.s.* $\mathbb{P}$,

(i) $U_i(x, \omega)$ *is* $\mathscr{B}(0, \infty) \otimes \mathscr{F}_T$*-measurable,*
(ii) $U_i(x, \omega)$ *is continuous and differentiable in* $x \in (0, \infty)$,
(iii) $U_i(x, \omega)$ *is strictly increasing and strictly concave in* $x \in (0, \infty)$, *and*
(iv) (Inada Conditions) $\lim_{x \downarrow 0} U_i'(x, \omega) = \infty$ *and* $\lim_{x \to \infty} U_i'(x, \omega) = 0$.

In addition, we assume that preferences satisfy the reasonable asymptotic elasticity condition $AE(U_i) < 1$ for all i.

We let the trader's beliefs $\mathbb{P}_i$ be equivalent to $\mathbb{P}$ for all i, i.e. the traders agree on zero probability events and these zero probability events are the same as under the statistical probability measure. Traders have symmetric information, represented by the filtration $(\mathscr{F}_t)$. Traders are also endowed with strictly positive wealth at time 0, denoted $x_i > 0$, and their time T wealth is denoted $X_T^i \geq 0$. Each trader is endowed at time 0 with shares (number of units) of the money market account and the risky assets. These are denoted e_0^i and e_j^i for $j = 1, \ldots, n$, respectively.

R. A. Jarrow, *Continuous-Time Asset Pricing Theory*, Springer Finance,
https://doi.org/10.1007/978-3-319-77821-1_23

We assume that the risky assets pay a liquidating dividend at time T and the risky assets are non-redundant.

Assumption (Liquidating Cash Flows) *There exists an exogenous random payout vector* $\xi = (\xi_1, \ldots, \xi_n) : \Omega \to \mathbb{R}^n$ *at time* T *such that* $S_T = \xi > 0$.

Assumption (Non-redundant Assets) *Given a price process* S*, the trading strategy* $(\alpha_0^i, \alpha^i) \in \mathcal{N}(x, S, K)$ *generating any* $X_T^i \in L_+^0$ *is unique.*

We impose the same assumptions on the prices processes imposed in the trading constraints Chap. 22 on portfolio optimization. Consider the normalized auxiliary markets $\left(\frac{S^\nu}{B^\nu}, (\mathscr{F}_t), \mathbb{P}\right)$ for all $(\nu_0, \nu) \in \mathscr{K}^o$. We assume that the auxiliary markets are arbitrage free.

Assumption (NFLVR for (ν_0, ν)**)**

$$\mathfrak{M}_l\left(\frac{S^\nu}{B^\nu}\right) \neq \emptyset \textit{ for all } (\nu_0, \nu) \in \mathscr{K}^o.$$

The definition of an equilibrium (Chap. 13, Definition 13.1) is the same, with the exception that the trading strategies must lie in the constraint set K. An economy is a collection

$$\left\{((\mathscr{F}_t), \mathbb{P}), (N_0 = 0, N), \left(\mathbb{P}_i, U_i, \left(e_0^i, e^i\right)\right)_{i=1}^{\mathscr{I}}\right\}$$

under the above assumptions. Note that the price process is excluded from the definition of an economy because the price process S will be endogenously determined in equilibrium.

23.2 Representative Trader

The definition of a representative trader is as in the representative trader Chap. 14. The representative trader has the aggregate utility function with weightings λ defined by

$$U(x, \lambda, \omega) \equiv \sup_{\{x_1, \ldots, x_{\mathscr{I}}\} \in \mathbb{R}^{\mathscr{I}}} \left\{\sum_{i=1}^{\mathscr{I}} \lambda_i(\omega) \frac{d\mathbb{P}_i}{d\mathbb{P}} U_i(x_i, \omega) : x = \sum_{i=1}^{\mathscr{I}} x_i\right\},$$

where $\lambda_i : \Omega \to \mathbb{R}$ is $\mathscr{F}_T$-measurable, $\lambda_i > 0$ for all $i = 1, \ldots, \mathscr{I}$, $\lambda \equiv (\lambda_1, \ldots, \lambda_{\mathscr{I}})$, and with an endowment equal to the aggregate market wealth, i.e. zero shares in the mma and all the market's shares in the risky assets, i.e. $(N_0 = 0, N)$.

Given a price process S, the representative trader's portfolio optimization problem is

$$v(x, \lambda_{\mathfrak{C}}, S) = \sup_{(\alpha_0,\alpha)\in\mathcal{N}(x,S,K)} E\left[U\left(X_T, \lambda_{\mathfrak{C}}\right)\right], \qquad \text{where}$$

$$\mathcal{N}(x, S, K) = \Big\{(\alpha_0, \alpha) \in (\mathcal{O}, \mathcal{L}(S)) : X_t = \alpha_0(t) + \alpha_t \cdot S_t,$$
$$X_t = x + \textstyle\int_0^t \alpha_u \cdot dS_u \geq 0,\ (\alpha_0(t), \alpha_t) \in K,\ \forall t \in [0, T]\Big\}.$$

Solving the static problem first, this reduces to

$$v(x, \lambda_{\mathfrak{C}}, S) = \sup_{X_T\in\mathscr{C}_{\mathfrak{C}}(x,S)} E\left[U\left(X_T, \lambda_{\mathfrak{C}}\right)\right], \qquad \text{where}$$

$$\mathscr{C}_{\mathfrak{C}}(x, S) = \{X_T \in L^0_+ : E[X_T H_T] \leq x,\ \forall H_T \in D_s(\mathscr{K}^o)\} \qquad \text{and}$$

$$D_s(\mathscr{K}^o) = \bigcup_{(\nu_0,\nu)\in\mathscr{K}^o} D_s(S^\nu).$$

23.2.1 The Solution

Given a price process S, assuming

(i) *(reasonable asymptotic elasticity)*

$$AE(U, \lambda_{\mathfrak{C}}, \omega) = \limsup_{x\to\infty} \frac{xU'(x, \lambda_{\mathfrak{C}}, \omega)}{U(x, \lambda_{\mathfrak{C}}, \omega)} < 1 \text{ a.s. } \mathbb{P}, \text{ and}$$

(ii) *(non-trivial optimization)*

$$v(x, \lambda_{\mathfrak{C}}, S) < \infty \quad \text{for some } x > 0,$$

the solution to the representative trader's optimization problem exists, and the optimal wealth is characterized by

$$X_T = I(yY_T, \lambda_{\mathfrak{C}}) \quad \text{with} \quad I = (U')^{-1} = -\tilde{U}',$$

where there exists a $(\nu_0, \nu) \in \mathscr{K}^o$ such that $H_T^\nu \in D_s(S^\nu)$ is the solution to

$$\tilde{v}(y, \lambda_{\mathfrak{C}}, S) = \inf_{H_T\in D_s(S^\nu)} E[\tilde{U}(yH_T, \lambda_{\mathfrak{C}})],$$

$$\tilde{U}(y, \omega) = \sup_{x>0} [U(x, \omega) - xy] \text{ for } \ y > 0,$$

and y is the solution to the budget constraint

$$E\left[H_T^\nu I(yH_T^\nu, \lambda_{\mathfrak{C}})\right] = x.$$

The *optimal portfolio wealth* when multiplied by the supermartingale deflator, $X_t H_t^\nu$, is a $\mathbb{P}$-martingale. This implies that for an arbitrary time $t \in [0, T]$,

$$X_t = E\left[X_T \frac{H_T^\nu}{H_t^\nu} \middle| \mathscr{F}_t\right].$$

This characterization is a direct application of the solution to the representative trader's optimization problem in the incomplete market setting given in the representative trader Chap. 14.

23.2.2 Buy and Hold Trading Strategies

If the representative trader's optimal trading strategy is a buy and hold in the mma with positive holdings in every risky asset, then his optimal trading strategy lies within the trading constraints set K, i.e. the trading constraints are nonbinding. In this case the representative trader's optimal solution can also be obtained by solving the unconstrained problem where

$$K^o = \{(\nu_0, \nu) \in \mathbb{R}^{n+1} : (\nu_0, \upsilon) = (0, 0)\}$$

and $\mathscr{C}_{\mathfrak{C}}(x, S) = \mathscr{C}(x, S)$, i.e.

$$\upsilon(x, \lambda_{\mathfrak{C}}, S) = \sup_{X_T \in \mathscr{C}(x,S)} E\left[U\left(X_T, \lambda_{\mathfrak{C}}\right)\right], \qquad \text{where}$$

$$\mathscr{C}(x, S) = \{X_T \in L_+^0 : \sup_{Y_T \in D_s(S)} E\left[X_T Y_T\right] \leq x\}.$$

Here, the supermartingale deflator process is

$$H^0 = \frac{Y^0}{B^0} = Y \in \mathscr{D}_s(S).$$

From the representative trader Chap. 14, Theorem 14.3, if the representative trader's supermartingale deflator is a probability density with respect to $\mathbb{P}$, it is a martingale deflator, i.e.

$$Y_T = \frac{d\mathbb{Q}}{d\mathbb{P}} \in M(S),$$

where

$$S_j(t) = E^{\mathbb{Q}}\left[S_j(T)\,|\mathscr{F}_t\right]$$

for $j = 1, \ldots, n$. We will use this observation below.

Remark 23.1 (Buy and Hold Supermartingale Deflators) When the representative trader's optimal trading strategy is a buy and hold with strictly positive holdings in every risky asset, the representative trader's optimization problem is equivalent to that obtained from an optimization problem in an unconstrained market. There is a key difference, however, between the two optimization problems. Recall that the representative trader's martingale deflator depends on the individual traders' optimization problems and their supermartingale deflators with respect to $\mathbb{P}_i$ through the weightings λ in the aggregate utility function. These weightings will differ across the two economies. Hence, the representative trader's aggregate utility function $U(x, \lambda)$ will differ between the trading constrained and unconstrained economies due the different supermartingale deflators for the individual traders across the constrained and unconstrained economies.

Alternatively stated, in the constrained market, the weights of the representative trader's aggregate utility function $\lambda_{\mathfrak{C}}$ reflect the individual traders' optimal demands from solving the optimization problem with trading constraints. In the unconstrained market, the weights $\lambda_{\mathfrak{U}}$ reflect the individual traders' optimal demands from solving the *unconstrained* optimization problem. These weights will differ, hence, the representative trader's optimal supermartingale deflator process $H_{\mathfrak{C}}$ will differ from the representative trader's optimal supermartingale deflator process $H_{\mathfrak{U}}$ in the unconstrained economy. This implies, of course, that the equilibrium risk premia and prices will differ across the two economies. This completes the remark.

23.3 Existence of Equilibrium

We have the following direct application of the equilibrium existence Theorem 14.7 in the representative trader Chap. 14.

Theorem 23.1 (Sufficient Conditions for an Equilibrium) *Given is an economy*

$$\left\{((\mathscr{F}_t), \mathbb{P}), (N_0 = 0, N), \left(\mathbb{P}_i, U_i, \left(e_0^i, e^i\right)\right)_{i=1}^{\mathscr{I}}\right\}.$$

Assume that

(i) for all λ, $U(x, \lambda, \omega)$ has an asymptotic elasticity strictly less than one a.s. $\mathbb{P}$, $E\left[U'(N \cdot \xi, \lambda)\right] < \infty$, and $E\left[U(N \cdot \xi, \lambda)\right] < \infty$, where ξ is the exogenous liquidating dividends for the risky assets.

Define

$$S_t^{\lambda} = E_{\lambda}\left[\xi \mid \mathscr{F}_t\right] \qquad where \qquad \frac{d\mathbb{Q}_{\lambda}}{d\mathbb{P}} = \frac{U'(N \cdot \xi, \lambda)}{E\left[U'(N \cdot \xi, \lambda)\right]} > 0, \tag{23.1}$$

where $E_{\lambda}[\cdot]$ *is expectation with respect to* $\mathbb{Q}_{\lambda}$,

(ii) for all λ *and i,* $v_i(x_i, S^{\lambda}) < \infty$ *for some* $x_i > 0$*, and*

(iii) there exists a $\tilde{\lambda}$ *such that*

$$\tilde{\lambda}_i = \frac{v_1'(x_1, S^{\tilde{\lambda}})H_T^1(S^{\tilde{\lambda}})}{v_i'(x_i, S^{\tilde{\lambda}})H_T^i(S^{\tilde{\lambda}})}$$

with $H^i(S^{\tilde{\lambda}})$ *the optimal solutions to trader i's dual problem and* $v_i'(x_i, S^{\tilde{\lambda}})$ *trader i's value function, given the initial wealth* x_i *and the price process* $S^{\tilde{\lambda}}$*, for all* $i = 1, \ldots, \mathscr{I}$.

Then,

(1) $S^{\tilde{\lambda}}$ *is an equilibrium price process for the economy,*

(2) $S^{\tilde{\lambda}}$ *is an equilibrium price process for the representative trader economy* $\left\{((\mathscr{F}_t), \mathbb{P}), (N_0 = 0, N), \left(\mathbb{P}, U(\tilde{\lambda}), (N_0, N)\right)\right\}$*, and*

(3) this representative trader economy reflects the economy's equilibrium.

23.4 Characterization of Equilibrium

This section characterizes the trading constrained equilibrium. We assume that an equilibrium exists with the price process S. Hence, by Theorem 14.4 in the representative trader Chap. 14 there exists a representative trader economy indexed by $\lambda_{\mathfrak{C}}$,

$$\left\{((\mathscr{F}_t), \mathbb{P}), (N_0 = 0, N), (\mathbb{P}, U(\lambda_{\mathfrak{C}}), (N_0 = 0, N))\right\},$$

such that the representative trader economy equilibrium is equivalent to the original economy's equilibrium. In particular, this implies that the traders' optimal aggregate demands are reflected in the representative trader economy equilibrium and the equilibrium price process S is the same in the representative trader economy equilibrium as in the original economy's equilibrium.

This is a *no trade* equilibrium where the representative trader's optimal portfolio is a buy and hold trading strategy consisting of zero units in the mma and holdings in each risky asset equal to the aggregate market supply. As discussed in Sect. 23.2 above, this implies that the economy's equilibrium is equivalent to that obtained in an otherwise identical market with no trading constraints.

Therefore, all of the theorems in Chap. 15 on the characterization of equilibrium apply. The difference is that the supermartingale deflator is based on the representative trader's aggregate utility function with weightings $\lambda_{\mathfrak{e}}$, which depend on the individual traders' constrained portfolio optimization problems. This implies, of course, that the supermartingale deflator will differ from that in an otherwise similar economy with no trading constraints. For convenience, we recall the important results.

As in Chap. 15 on the characterization of equilibrium, there are many possible characterizations of the equilibrium supermartingale deflator possible. Assuming that $Y_T^{\lambda_{\mathfrak{e}}}$ is a probability density with respect to $\mathbb{P}$, the supermartingale deflators that can be used to characterize equilibrium prices are

$$\text{Representative Trader} \qquad H_T^{\lambda_{\mathfrak{e}}} = \frac{U'(m_T, \lambda_{\mathfrak{e}}, S)}{v'(m_0, \lambda_{\mathfrak{e}}, S)} \in M(S) \subset D_s(S),$$
$$\text{Individual Trader} \qquad H_T^i = \frac{\frac{d\mathbb{P}_i}{d\mathbb{P}} U_i'(X_T^i)}{v_i'(x_i, S)} \in D_s(\mathscr{K}^o),$$

for $i = 1, \ldots, \mathscr{I}$, where the aggregate market wealth is denoted $m_t = \sum_{j=1}^n N_j S_j(t)$ for $t = 0, T$, $U(\cdot, \lambda_{\mathfrak{e}})$ is the representative trader's utility function with market weights $\lambda_{\mathfrak{e}}$, and $D_s(\mathscr{K}^o)$ is the set of supermartingale deflators with respect to $\mathbb{P}$.

Given that $Y_T^{\lambda_{\mathfrak{e}}}$ is a probability density with respect to $\mathbb{P}$, the representative trader's supermartingale deflator is a martingale deflator because his optimal portfolio is a buy and hold trading strategy with zero units in the mma (see Sect. 23.2 above).

The set of supermartingale deflators depends on the probability measure $\mathbb{P}$ (see the portfolio optimization Chap. 11 after the definition of $\mathscr{D}_s$). This is the reason for the change in probability measures in the numerator of the individual trader's supermartingale deflator (see expression (11.29), Sect. 11.9 in the portfolio optimization Chap. 11).

Remark 23.2 (Equilibrium Asset Price Bubbles) It is important to point out that in equilibrium, the individual traders view risky asset prices as having price bubbles as long as their shadow prices for the trading constraints $(\nu_0^i, \nu^i) \neq (0, 0)$. Indeed, in this case SH^i will be a $\mathbb{P}$ (strict) supermartingale, see the trading constraints Chap. 22, Sect. 22.9. In contrast, assuming that $Y_T^{\lambda_{\mathfrak{e}}}$ is a probability density with respect to $\mathbb{P}$, the representative trader sees no asset price bubble because $SH^{\lambda_{\mathfrak{e}}}$ is a $\mathbb{P}$-martingale, i.e.

$$S_j(t) = E\left[S_j(T) \frac{H_T^{\lambda_{\mathfrak{e}}}}{H_t^{\lambda_{\mathfrak{e}}}} \middle| \mathscr{F}_t \right]$$

for $j = 1, \ldots, n$. This completes the remark.

The risk return relation for individual traders, using the equilibrium prices, is contained in the trading constraints Chap. 22, Sect. 22.10. We repeat this result here for convenience. We note that these risk return relations are for the normalized economy ($B = 1, S$) where the mma return is identically zero.

Theorem 23.2 (Equilibrium Risk Return Relation) *Assume that Y_T^i is a probability density with respect to $\mathbb{P}$. For individual i and risky asset j,*

$$\begin{aligned} E\left[R_j(t)\,|\mathscr{F}_t\right] \approx -\mathrm{cov}&\left[R_j(t), \frac{E\left[\frac{d\mathbb{P}_i}{d\mathbb{P}} U_i'(X_T)\left(1-\int_0^T \nu_0(u)du\right)|\mathscr{F}_{t+\Delta}\right]}{E\left[\frac{d\mathbb{P}_i}{d\mathbb{P}} U_i'(X_T)\left(1-\int_0^T \nu_0(u)du\right)|\mathscr{F}_t\right]} \Bigg| \mathscr{F}_t \right] \\ &+\frac{\eta}{w}\Upsilon_0(t) - \frac{(1-w)}{w} E\left[\Upsilon_t\,|\mathscr{F}_t\right] \\ &-\frac{(1-w)}{w}\mathrm{cov}\left[\Upsilon_t, \frac{E\left[\frac{d\mathbb{P}_i}{d\mathbb{P}} U_i'(X_T)\left(1-\int_0^T \nu_0(u)du\right)|\mathscr{F}_{t+\Delta}\right]}{E\left[\frac{d\mathbb{P}_i}{d\mathbb{P}} U_i'(X_T)\left(1-\int_0^T \nu_0(u)du\right)|\mathscr{F}_t\right]} \Bigg| \mathscr{F}_t \right], \end{aligned} \tag{23.2}$$

where

$$R_j(t) = \frac{S_j(t+\Delta) - S_j(t)}{S_j(t)},$$

$$w = \frac{S_t}{S_t - \int_0^t \nu_u du} \geq 0, \quad 1 \geq \eta = -\frac{\int_0^t \nu_0(u)du}{1 - \int_0^t \nu_0(u)du} \geq 0,$$

$$\Upsilon_t = \frac{\int_0^{t+\Delta} \nu_u du - \int_0^t \nu_u du}{\int_0^t \nu_u du} \leq 0, \text{ and}$$

$$\Upsilon_0(t) = \frac{\int_0^{t+\Delta} \nu_0(u)du - \int_0^t \nu_0(u)du}{\int_0^t \nu_0(u)du} \leq 0.$$

The changes from the equilibrium risk return relation in an unconstrained economy are observations 1–3 below.

1. The supermartingale deflator

$$\frac{E\left[\frac{d\mathbb{P}_i}{d\mathbb{P}} U_i'(X_T)\left(1-\int_0^T \nu_0(u)du\right)|\mathscr{F}_{t+\Delta}\right]}{E\left[\frac{d\mathbb{P}_i}{d\mathbb{P}} U_i'(X_T)\left(1-\int_0^T \nu_0(u)du\right)|\mathscr{F}_t\right]}$$

is modified due to $\left(1-\int_0^T \nu_0(u)du\right) \geq 1$.

2. The default-free spot rate is modified from zero (in the normalized economy) to

$$\frac{\eta}{w}\Upsilon_0(t) - \frac{(1-w)}{w} E\left[\Upsilon_t\,|\mathscr{F}_t\right].$$

3. There is an extra risk premium

$$-\frac{(1-w)}{w}\operatorname{cov}\left[\Upsilon_t, \frac{E\left[\frac{d\mathbb{P}_i}{d\mathbb{P}}U_i'(X_T)\left(1-\int_0^T \nu_0(u)du\right)|\mathscr{F}_{t+\Delta}\right]}{E\left[\frac{d\mathbb{P}_i}{d\mathbb{P}}U_i'(X_T)\left(1-\int_0^T \nu_0(u)du\right)|\mathscr{F}_t\right]}\middle|\mathscr{F}_t\right].$$

We can also characterize the equilibrium risk return relation using the representative trader's martingale deflator. For the rest of this section, assume that the representative trader's supermartingale deflator $Y_T^{\lambda_{\mathfrak{C}}}$ is a probability density with respect to $\mathbb{P}$. Using the previous risk return relation for the representative trader, recalling that the representative trader's optimal trading strategy is a buy and hold in equilibrium implying that $(\nu_0, \nu) = (0, 0)$, yields

Theorem 23.3 (The Equilibrium Risk Return Relation)

$$E\left[R_j(t)\,|\mathscr{F}_t\right] \approx -\operatorname{cov}\left[R_j(t), \frac{E[U'(m_T, \lambda_{\mathfrak{C}})\,|\mathscr{F}_{t+\Delta}]}{E[U'(m_T, \lambda_{\mathfrak{C}})\,|\mathscr{F}_t]}\middle|\mathscr{F}_t\right]. \tag{23.3}$$

Remark 23.3 (Trading Constrained Versus Unconstrained Equilibrium Risk Return Relations) Although the appearance of expression (23.3) is the same as that in an unconstrained economy, there is an important difference. The difference is that the representative trader's supermartingale deflator $H_T^{\lambda_{\mathfrak{C}}} = \frac{U'(m_T,\lambda_{\mathfrak{C}})}{v'(m_0,\lambda_{\mathfrak{C}},S)}$ differs from that in the unconstrained economy due to the different weightings in the aggregate utility function. This difference, in turn, implies that the magnitude of a risky asset's risk premium $E[R_j(t)\,|\mathscr{F}_t]$ will differ across the two economies. This difference will be reflected in the CCAPM and ICAPM risk return relations derived immediately below. This completes the remark.

The same proof as in Chap. 15 on the characterization of equilibrium, Theorem 15.3 generates the consumption capital asset pricing model (CCAPM) generalized to include trading constraints.

Theorem 23.4 (CCAPM) *Define* $\mu_m(t) \equiv E[m_T\,|\mathscr{F}_t]$. *Assume* $U(x, \lambda_{\mathfrak{C}}, \omega)$ *is three times continuously differentiable in x. Then,*

$$E\left[R_i(t)\,|\mathscr{F}_t\right] \approx -\frac{U''(\mu_m(t), \lambda_{\mathfrak{C}})}{U'(\mu_m(t), \lambda_{\mathfrak{C}})}\operatorname{cov}\left[R_i(t), (\mu_m(t+\Delta) - \mu_m(t))\,|\mathscr{F}_t\right]. \tag{23.4}$$

Define $r_m(t) \equiv \frac{m_{t+\Delta}-m_t}{m_t}$ to be return on aggregate market wealth, which is the return on the market portfolio. The characterization of equilibrium in Chap. 15, Theorem 15.4 gives the following multiple-factor beta model. For this theorem, we use the non-normalized economy.

Theorem 23.5 (Multiple-Factor Beta Model)

$$\begin{aligned} R_i(t) - r_0(t) = \beta_{im}(t)\,(r_m(t) - r_0(t)) \\ +\textstyle\sum_{j\in\Phi_i} \beta_{ij}(t)\left(r_j(t) - r_0(t)\right), \end{aligned} \tag{23.5}$$

where $\beta_{ij}(t) \neq 0$ *for all* (i, j), $r_0(t) = \frac{1}{p(t,t+\Delta)} - 1$ *is the return on a default-free zero-coupon bond that matures at time* $t + \Delta$, $R_i(t) = \frac{\mathbb{S}_i(t+\Delta)-\mathbb{S}_i(t)}{\mathbb{S}_i(t)}$ *is the return on the* i*th risky asset, and* $r_j(t)$ *is the return on the* j*th risk factor.*

Of course, the betas in this expression differ from those in an economy without trading constraints, but otherwise the expression is identical in appearance.

Taking time t conditional expectations gives the intertemporal capital asset pricing model (ICAPM) extended to include trading constraints.

Corollary 23.1 (ICAPM)

$$\begin{aligned} E[R_i(t)\,|\mathscr{F}_t] - r_0(t) = \beta_{im}(t)\,(E[r_m(t)\,|\mathscr{F}_t] - r_0(t)) \\ +\textstyle\sum_{j\in\Phi_i} \beta_{ij}(t)\left(E[r_j(t)\,|\mathscr{F}_t] - r_0(t)\right). \end{aligned} \tag{23.6}$$

Remark 23.4 (Empirical Implications) The implications of trading constraints for empirical testing are three-fold.

1. Trading constraints change the equilibrium price process S^{λ}, and hence returns. This implies that the number and composition of the set of risk factors (Φ_i) for any individual stock may differ from an economy without trading constraints and the beta coefficients (β_{ij}) may differ.
2. Asset expected returns reflect additional risk premiums due to the trading constraints. This is because trading constraints reduce a trader's ability to maximize their preferences. Constraints make trading assets more risky than they would otherwise be.
3. These additional risk premiums imply that more risk factors may have nonzero risk premium. Hence, when fitting multiple-factor models to risky asset returns, there will be more priced risk factors in a trading constrained economy than in an otherwise equivalent unconstrained economy.

References

1. K. Amin, R. Jarrow, Pricing options on risky assets in a stochastic interest rate economy. Math. Finance **2**(4), 217–237 (1992)
2. J.P. Ansel, C. Stricker, Couverture des actifs contingents et prix maximum. Ann. Inst. H. Poincare Probab. Statist. **30**(2), 303–331 (1994)
3. R. Ash, *Real Analysis and Probability* (Academic, New York, 1972)
4. K. Back, *A Course in Derivative Securities: Introduction to Theory and Computation* (Springer, Berlin, 2010)
5. K. Back, *Asset Pricing and Portfolio Choice Theory* (Oxford University Press, Oxford, 2010)
6. P. Bank, D. Baum, Hedging and portfolio optimization in financial markets with a large trader. Math. Finance **14**(1), 1–18 (2004)
7. N. Barberis, R. Thaler, A survey of behavioral finance, in *Handbook of Economics and Finance*, ed. by G. Constantinides, M. Harris, R. Stulz, vol. 1, Part B. Financial Markets and Asset Pricing (Elsevier B.V., Amsterdam, 2003)
8. S. Basak, D. Cuoco, An equilibrium model with restricted stock market participation. Rev. Financ. Stud. **11**(2), 309–341 (1998)
9. R. Battig, R. Jarrow, The second fundamental theorem of asset pricing: a new approach. Rev. Financ. Stud. **12**(5), 1219–1235 (1999)
10. M. Baxter, A. Rennie, *Financial Calculus: An Introduction to Derivative Pricing* (Cambridge University Press, Cambridge, 1996)
11. J. Berger, *Statistical Decision Theory: Foundations, Concepts, and Methods* (Springer, Berlin, 1980)
12. T. Bielecki, M. Rutkowski, *Credit Risk: Modeling, Valuation and Hedging* (Springer, Berlin, 2002)
13. P. Billingsley, *Probability and Measure*, 2nd edn. (Wiley, New York, 1986)
14. T. Bjork, *Arbitrage Theory in Continuous Time* (Oxford University Press, Oxford, 1998)
15. T. Bjork, Y. Kabanov, W. Runggaldier, Bond market structure in the presence of marked point processes. Math. Finance **7**(2), 211–239 (1997)
16. T. Bjork, G. Di Masi, Y. Kabanov, W. Runggaldier, Towards a general theory of bond markets. Finance Stoch. **1**(2), 141–174 (1997)
17. C. Bluhm, L. Overbeck, *Structured Credit, Portfolio Analysis, Baskets and CDOs* (Chapman & Hall /CRC, New York, 2007)
18. C. Bluhm, L. Overbeck, C. Wagner, *An Introduction to Credit Risk Modeling* (Chapman & Hall /CRC, New York, 2003)
19. J. Borwein, A. Lewis, *Convex Analysis and Nonlinear Optimization: Theory and Examples*, 2nd edn. (Springer, Berlin, 2006)

R. A. Jarrow, *Continuous-Time Asset Pricing Theory*, Springer Finance,
https://doi.org/10.1007/978-3-319-77821-1

20. A. Brace, D. Gatarek, M. Musiela, The market model of interest rate dynamics. Math. Finance **7**(2), 127–147 (1997)
21. W. Brannath, W. Schachermayer, A Bipolar Theorem for $L^0_+(\Omega, \mathscr{F}, P)$. Seminaire de probabilites (Strasbourg) **33**, 349–354 (1999)
22. D. Breeden, An intertemporal asset pricing model with stochastic consumption and investment opportunities. J. Financ. Econ. **7**, 265–296 (1979)
23. P. Bremaud, *Point Processes and Queues: Martingale Dynamics* (Springer, Berlin, 1980)
24. D. Brigo, F. Mercurio, *Interest Rate Models – Theory and Practice* (Springer, Berlin, 2001)
25. J. Campbell, A. Lo, A.C. MacKinlay, *The Econometrics of Financial Markets* (Princeton University Press, Princeton, NJ, 1997)
26. R. Carmona, HJM: a unified approach to dynamic models for fixed income, credit and equity markets, in *Paris–Princeton Lectures in Mathematical Finance*, 2005, ed. by R. Carmona et al. Lecture Notes in Mathematics, vol. 1919 (Springer, Berlin, 2009), pp. 3–45
27. R. Carmona, S. Nadtochiy, Local volatility dynamic models. Finance Stoch. **13**, 1–48 (2009)
28. R. Carmona, M. Tehranchi, *Interest Rate Models: an Infinite Dimensional Stochastic Analysis Perspective* (Springer, Berlin, 2006)
29. P. Carr, R. Jarrow, A discrete time synthesis of derivative security valuation using a term structure of futures prices. Handb. OR&MS **9**, 225–249 (1995)
30. U. Cetin, R. Jarrow, P. Protter, Liquidity risk and arbitrage pricing theory. Finance Stoch. **8**(3), 311–341 (2004)
31. J. Choi, K. Larsen, Taylor Approximation of incomplete Radner equilibrium models. Finance Stoch. **19**, 653–679 (2015)
32. P. Christensen, K. Larsen, Incomplete continuous time securities markets with stochastic income volatility. Rev. Asset Pric. Stud. **4**(2), 247–285 (2014)
33. J. Cochrane, *Asset Pricing*, revised edn. (Princeton University Press, Princeton, NJ, 2005)
34. R. Cont, P. Tankov, *Financial Modelling with Jump Processes*. Chapman & Hall/CRC Financial Mathematics Series (Chapman & Hall/CRC, New York, 2004)
35. J. Cox, M. Rubinstein, *Options Markets* (Prentice Hall, Englewood Cliffs, NJ, 1985)
36. J. Cox, J. Ingersoll, S. Ross, The relation between forward prices and futures prices. J. Financ. Econ. **9**, 321–346 (1981)
37. D. Cuoco, A Martingale characterization of consumption choices and hedging costs with margin requirements. Math. Finance **10**(3), 355–385 (1997)
38. D. Cuoco, Optimal consumption and equilibrium prices with portfolio constraint and stochastic income. J. Econ. Theory **72**, 33–73 (1997)
39. C. Cuoco, H. He, Dynamic equilibrium in infinite-dimensional economies with incomplete financial markets. Working paper, Wharton (1994)
40. J. Cvitanic, I. Karatzas, Convex duality in constrained portfolio optimization. Ann. Appl. Probab. **2**(4), 767–818 (1992)
41. Q. Dai, K. Singleton, Term structure dynamics in theory and reality. Rev. Financ. Stud. **16**(3), 631–678 (2003)
42. R. Dana, M. Jeanblanc, *Financial Markets in Continuous Time* (Springer, Berlin, 2003)
43. M. DeGroot, *Optimal Statistical Decisions* (McGraw Hill, New York, 1970)
44. F. Delbaen, W. Schachermayer, A general version of the fundamental theorem of asset pricing. Math. Ann. **312**, 215–250 (1994)
45. F. Delbaen, W. Schachermayer, The Banach space of workable contingent claims in arbitrage theory. Ann. l'IHP **33**(1), 114–144 (1997)
46. F. Delbaen, W. Schachermayer, The fundamental theorem of asset pricing for unbounded stochastic processes. Math. Ann. **300**, 463–520 (1998)
47. F. Delbaen, W. Schachermayer, *The Mathematics of Arbitrage* (Springer, Berlin, 2000)
48. H. Dengler, R. Jarrow, Option pricing using a binomial model with random time steps (a formal model of gamma hedging). Rev. Deriv. Res. **1**, 107–138 (1997)
49. J. Detemple, R. Garcia, M. Rindisbacher, Simulation methods for optimal portfolios, in *Handbooks in OR & MS*, ed. by J.R. Birge, V. Linetsky, vol. 15 (Elsevier B. V. Amsterdam, 2008)

50. J. Detemple, S. Murthy, Equilibrium asset prices and no-arbitrage with portfolio constraints. Rev. Financ. Stud. **10**(4), 1133–1174 (1997)
51. G. Di Nunno, B. Oksendal, F. Proske, *Malliavin Calculus for Levy Processes with Applications in Finance* (Springer, Berlin, 2009)
52. D. Duffie, *Dynamic Asset Pricing Theory*, 3rd edn. (Princeton University Press, Princeton, NJ, 2001)
53. D. Duffie, R. Kan, A yield factor model of interest rates. Math. Finance **6**, 379–406 (1996)
54. D. Duffie, K. Singleton, *Credit Risk: Pricing, Measurement, and Management* (Princeton University Press, Princeton, NJ, 2003)
55. E. Eberlein, S. Raible, Term structure models driven by general Levy processes. Math. Finance **9**(1), 31–53 (1999)
56. L. Eisenberg, R. Jarrow, Option pricing with random volatilities in complete markets. Rev. Quant. Finance Acc. **4**, 5–17 (1994)
57. E. Fama, Efficient capital markets: a review of theory and empirical work. J. Finance **25**(2), 383–417 (1970)
58. E. Fama, Efficient capital markets: II. J. Finance **46**(5), 1575–1617 (1991)
59. E. Fama, Market efficiency, long-term returns and behavioral finance. J. Financ. Econ. **49**, 283–306 (1998)
60. D. Filipovic, *Consistency Problems for Heath–Jarrow–Morton Interest Rate Models*. Lecture Notes in Mathematics (Springer, Berlin, 2001)
61. P. Fishburn, *Nonlinear Preference and Utility Theory* (Johns Hopkins University Press, Baltimore, 1988)
62. H. Follmer, Y.M. Kabanov, Optional decomposition and Lagrange multipliers. Finance Stoch. **2**, 69–81 (1998)
63. H. Follmer, A. Schied, *Stochastic Finance: An Introduction in Discrete Time*, 2nd edn. (Walter de Gruyter, Berlin, 2004)
64. A. Friedman, *Foundations of Modern Analysis* (Holt, Rinehart and Winston, Austin, 1970)
65. H. Geman, The importance of the forward neutral probability in a stochastic approach of interest rates. Working paper, ESSEC (1989)
66. O. Guler, *Foundations of Optimization* (Springer, New York, 2010)
67. J.M. Harrison, S. Pliska, A stochastic calculus model of continuous trading: complete markets. Stoch. Process. Appl. **15**, 313–316 (1983)
68. O. Hart, On the existence of equilibrium in a securities model. J. Econ. Theory **9**, 293–311 (1974)
69. H. He, N. Pearson, Consumption and portfolio policies with incomplete markets and short sale constraints: the infinite dimensional case. J. Econ. Theory **54**, 259–304 (1991)
70. D. Heath, R. Jarrow, A. Morton, Bond pricing and the term structure of interest rates: a new methodology for contingent claims valuation. Econometrica **60**(1), 77–105 (1992)
71. J. Hiriart-Urruty, C. Lemarechal, *Fundamentals of Convex Analysis* (Springer, New York, 2004)
72. C. Huang, R. Litzenberger, *Foundations for Financial Economics*. Classic Textbook (North-Holland, New York, 1988)
73. J. Hugonnier, Rational asset pricing bubbles and portfolio constraints. J. Econ. Theory **147**, 2260–2302 (2012)
74. J. Ingersoll, *Theory of Financial Decision Making* (Rowman & Littlefield, Lanham, MD, 1987)
75. J. Jacod, P. Protter, *Probability Essentials* (Springer, New York, 2000)
76. J. Jacod, P. Protter, Risk neutral compatibility with option prices. Finance Stoch. **14**, 285–315 (2010)
77. R. Jagannathan, E. Schaumburg, G. Zhou, Cross-sectional asset pricing tests. Annu. Rev. Financ. Econ. **2**, 49–74 (2010)
78. R. Jarrow, An integrated axiomatic approach to the existence of ordinal and cardinal utility functions. Theory Decis. **22**, 99–110 (1987)

79. R. Jarrow, The pricing of commodity options with stochastic interest rates. Adv. Future Options Res. **2**, 15–28 (1987)
80. R. Jarrow, *Finance Theory* (Prentice Hall, Englewood Cliffs, NJ, 1988)
81. R. Jarrow, Market manipulation, bubbles, corners, and short squeezes. J. Financ. Quant. Anal. **27**(3), 311–336 (1992)
82. R. Jarrow, Derivative security markets, market manipulation, and option pricing theory. J. Financ. Quant. Anal. **29**(2), 241–261 (1994)
83. R. Jarrow, *Modeling Fixed Income Securities and Interest Rate Options*, 2nd edn. (Stanford University Press, Stanford, CA, 2002)
84. R. Jarrow, Risky coupon bonds as a portfolio of zero-coupon bonds. Finance Res. Lett. **1**, 100–105 (2004)
85. R. Jarrow, Credit risk models. Annu. Rev. Financ. Econ. **1**, 37–68 (2009)
86. R. Jarrow, The term structure of interest rates. Annu. Rev. Financ. Econ. **1**, 69–96 (2009)
87. R. Jarrow, Active portfolio management and positive alphas: fact or fantasy? J. Portf. Manag. 17–22 (2010)
88. R. Jarrow, Credit market equilibrium theory and evidence: revisiting the structural versus reduced form credit risk model debate. Finance Res. Lett. **8**, 2–7 (2011)
89. R. Jarrow, Asset price bubbles. Annu. Rev. Financ. Econ. **7**, 201–218 (2015)
90. R. Jarrow, Asset market equilibrium with liquidity risk. Ann. Finance **14**, 253–288 (2018)
91. R. Jarrow, Bubbles and multiple-factor asset pricing models. Int. J. Theor. Appl. Finance **19**(1), 19 pp. (2016)
92. R. Jarrow, An equilibrium capital asset pricing model in markets with price jumps and price bubbles. Q. J. Finance (2017, forthcoming)
93. R. Jarrow, A CAPM with trading constraints and price bubbles. Int. J. Theor. Appl. Finance (2017, forthcoming)
94. R. Jarrow, On the existence of competitive equilibrium in frictionless and incomplete stochastic asset markets. Math. Financ. Econ. **11**(4), 455–477 (2017)
95. R. Jarrow, A. Chatterjea, *An Introduction to Derivative Securities, Financial Markets, and Risk Management* (W. W. Norton and Co., New York, 2013)
96. R. Jarrow, M. Larsson, The meaning of market efficiency. Math. Finance **22**(1), 1–30 (2012)
97. R. Jarrow, M. Larsson, Informational efficiency under short sale constraints. SIAM J. Financ. Math. **6**, 804–824 (2015)
98. R. Jarrow, M. Larsson, On aggregation and representative agent equilibria. J. Math. Econ. (2015, forthcoming)
99. R. Jarrow, D. Madan, Option pricing using the term structure of interest rates to hedge systematic discontinuities in asset returns. Math. Finance **5**(4), 311–336 (1995)
100. R. Jarrow, D. Madan, Hedging contingent claims on semimartingales. Finance Stoch. **3**, 111–134 (1999)
101. R. Jarrow, G. Oldfield, Forward contracts and futures contracts. J. Financ. Econ. **4**, 373–382 (1981)
102. R. Jarrow, P. Protter, The martingale theory of bubbles: implications for the valuation of derivatives and detecting bubbles, in *The Financial Crisis: Debating the Origins, Outcomes, and Lessons of the Greatest Economic Event of Our Lifetime*, ed. by A. Berd (Risk Publications, London, 2010)
103. R. Jarrow, P. Protter, Discrete versus continuous time models: local Martingales and singular processes in asset pricing theory. Finance Res. Lett. **9**, 58–62 (2012)
104. R. Jarrow, P. Protter, Positive alphas and a generalized multiple-factor asset pricing model. Math. Financ. Econ. **10**(1), 29–48 (2016)
105. R. Jarrow, A. Rudd, *Option Pricing* (Richard D. Irwin, Inc., Homewood, IL, 1983)
106. R. Jarrow, S. Turnbull, Credit risk: drawing the analogy. Risk Mag. **5**(9), 63–70 (1992)
107. R. Jarrow, S. Turnbull, Pricing derivatives on financial securities subject to credit risk. J. Finance **50**(1), 53–85 (1995)
108. R. Jarrow, S. Turnbull, *Derivative Securities*, 2nd edn. (South-Western College Publishing, Cincinnati, OH, 2000)

109. R. Jarrow, P. Protter, K. Shimbo, Asset price bubbles in a complete market. Adv. Math. Finance 105–130 (2007). In Honor of Dilip B. Madan
110. R. Jarrow, H. Li, F. Zhao, Interest rate caps smile too! But can the LIBOR market models capture the smile? J. Finance **57** (February), 345–382 (2007)
111. R. Jarrow, P. Protter, K. Shimbo, Asset price bubbles in incomplete markets. Math. Finance **20**(2), 145–185 (2010)
112. R. Jarrow, Y. Kchia, P. Protter, How to detect an asset bubble. SIAM J. Financ. Math. **2**, 839–865 (2011)
113. M. Jeanblanc, M. Yor, M. Chesney, *Mathematical Methods for Financial Markets* (Springer, Berlin, 2009)
114. M. Jensen, Some anomalous evidence regarding market efficiency. J. Financ. Econ. **6**, 95–101 (1978)
115. J. Kallsen, P. Kruhner, On a Heath–Jarrow–Morton approach for stock options. Finance Stoch. **19**, 583–615 (2015)
116. I. Karatzas, C. Kardaras, The Numeraire portfolio in semimartingale financial models. Finance Stoch. **11**, 447–493 (2007)
117. I. Karatzas, S. Shreve, *Brownian Motion and Stochastic Calculus* (Springer, Berlin, 1988)
118. I. Karatzas and S. Shreve, *Methods of Mathematical Finance* (Springer, Berlin, 1999)
119. I. Karatzas, G. Zitkovic, Optimal consumption from investment and random endowment in incomplete semimartingale markets. Ann. Probab. **31**(4), 1821–1858 (2003)
120. I. Karatzas, P. Lehoczky, S. Shreve, G. Xu, Martingale and duality methods for utility maximization in an incomplete market. J. Control Optim. **29**(3), 702–730 (1991)
121. C. Kardaras, Market viability via absence of arbitrage of the first kind. Finance Stoch. **16**, 651–667 (2012)
122. C. Kardaras, H. Xing, G. Zitkovic, Incomplete stochastic equilibria with exponential utilities close to Pareto optimality. SSRN (2015)
123. A. Klenke, *Probability Theory: A Comprehensive Course* (Springer, Berlin, 2008)
124. R. Korn, E. Korn, *Option Pricing and Portfolio Optimization: Modern Methods of Financial Mathematics*. Graduate Studies in Mathematics, vol. 31 (American Mathematical Society, Providence, RI, 2001)
125. D. Kramkov, Existence of an endogenously complete equilibrium driven by a diffusion. Finance Stoch. **19**, 1–22 (2015)
126. D. Kramkov, W. Schachermayer, The asymptotic elasticity of utility functions and optimal investment in incomplete markets. Ann. Appl. Probab. **9**(3), 904–950 (1999)
127. D. Kramkov, K. Weston, Muckenhoupt's Ap condition and the existence of the optimal martingale measure. Working paper, Carnegie Mellon University (2015)
128. D. Kreps, *A Course in Microeconomic Theory* (Princeton University Press, Princeton, NJ, 1990)
129. D. Lamberton, B. Lapeyre, *An Introduction to Stochastic Calculus Applied to Finance* (Chapman & Hall, New York, 1996)
130. D. Lando, On Cox Processes and credit risky securities. Rev. Deriv. Res. **2**, 99–120 (1998)
131. D. Lando, *Credit Risk Modeling: Theory and Applications* (Princeton University Press, Princeton, NJ, 2004)
132. K. Larsen, Continuous equilibria with heterogeneous preferences and unspanned endowments. Working paper, Carnegie Mellon University (2009)
133. K. Larsen, T. Sae-Sue, Radner equilibrium in incomplete Levy models. Working paper, Carnegie Mellon University (2016)
134. D. Luenberger, *Optimization by Vector Space Methods* (Wiley, New York, 1969)
135. A. Mas-Colell, M. Whinston, J. Green, *Microeconomic Theory* (Oxford University Press, Oxford, 1995)
136. P. Medvegyev, *Stochastic Integration Theory* (Oxford University Press, New York, 2009)
137. R.C. Merton, An intertemporal capital asset pricing model. Econometrica **41**, 867–888 (1973)
138. R.C. Merton, Theory of rational option pricing. Bell J. Econ. **4**(1), 141–183 (1973)

139. R.C. Merton, On the pricing of corporate debt: the risk structure of interest rates. J. Finance **29**, 449–470 (1974)
140. R.C. Merton, *Continuous Time Finance* (Basil Blackwell, Cambridge, MA, 1990)
141. T. Mikosch, *Elementary Stochastic Calculus with a Finance in View* (World Scientific, Singapore, 1998)
142. K. Miltersen, K. Sandmann, D. Sondermann, Closed form solutions for term structure derivatives with lognormal interest rates. J. Finance **52**, 409–430 (1997)
143. A. Mood, F. Graybill, D. Boes, *Introduction to the Theory of Statistics*, 3rd edn. (McGraw-Hill, New York, 1974)
144. O. Mostovyi, Necessary and sufficient conditions in the problem of optimal investment with intermediate consumption. Finance Stoch. **19**, 135–159 (2015)
145. M. Musiela, M. Rutkowski, *Martingale Methods in Financial Modelling*, 2nd edn. (Springer, Berlin, 2005)
146. L. Nielsen, Existence of equilibrium in CAPM. J. Econ. Theory **52**, 223–231 (1990)
147. M. O'Hara, *Market Microstructure Theory* (Blackwell, Oxford, 1995)
148. S. Perlis, *Theory of Matrices*, 3rd edn (Addison-Wesley, Reading, MA, 1958)
149. H. Pham, *Continuous time Stochastic Control and Optimization with Financial Applications* (Springer, Berlin, 2009)
150. S. Pliska, *Introduction to Mathematical Finance: Discrete Time Models* (Blackwell, Oxford, 1997)
151. P. Protter, *Stochastic Integration and Differential Equations*, 2nd edn., version 2.1 (Springer, Berlin, 2005)
152. P. Protter, A mathematical theory of financial bubbles. Working paper, Columbia University (2012)
153. P. Protter, Strict local martingales with jumps. Stoch. Process. Appl. **125**, 1352–1367 (2015)
154. P. Protter, K. Shimbo, No arbitrage and general semimartingales, in *Markov Processes and Related Topics: A Festschrift for Thomas G. Kurtz*. IMS Collections, vol. 4 (Institute of Mathematical Statistics, Beachwood, OH, 2008)
155. R. Radner, Equilibrium under uncertainty. Econometrica **36**(1), 31–58 (1982)
156. R. Rebanato, *Modern Pricing of Interest Rate Derivatives: the LIBOR Market Model Land Beyond* (Princeton University Press, Princeton, NJ, 2002)
157. L. Rogers, D. Williams, *Diffusions, Markov Processes, and Martingales: Volume 2 Ito Calculus* (Wiley, New York, 1987)
158. S. Ross, The arbitrage theory of capital asset pricing. J. Econ. Theory **13**, 341–360 (1976)
159. S. Ross, *Stochastic Processes*, 2nd edn. (Wiley, New York, 1996)
160. H. Royden, *Real Analysis*, 2nd edn. (MacMillan, New York, 1968)
161. D. Ruppert, *Statistics and Data Analysis for Financial Engineering* (Springer, Berlin, 2011)
162. A. Ruszczynski, *Nonlinear Optimization* (Princeton University Press, Princeton, NJ, 2006)
163. K. Sandmann, D. Sondermann, K. Miltersen, Closed form term structure derivatives in a Heath–Jarrow–Morton model with lognormal annually compounded interest rates, in *Proceedings of the Seventh Annual European Research Symposium*, Bonn, September 1994. Chicago Board of Trade (1995), pp. 145–164
164. W. Schachermayer, Portfolio optimization in incomplete financial markets, in *Mathematical Finance: Bachelier Congress 2000*, ed. by H. Geman, D. Madan, S.R. Pliska, T. Vorst (Springer, Berlin, 2001), pp. 427–462
165. J. Schoenmakers, *Robust Libor Modelling and Pricing of Derivative Products* (Chapman & Hall, New York, 2005)
166. P. Schonbucher, *Credit Derivatives Pricing Models: Models, Pricing and Implementation* (Wiley, Chichester, NJ, 2003)
167. M. Schweizer, J. Wissel, Term structures of implied volatilities: absence of arbitrage and existence results. Math. Finance **18**, 77–114 (2008)
168. S. Shreve, *Stochastic Calculus for Finance I: The Binomial Asset Pricing Model* (Springer, Berlin, 2004)

169. S. Shreve, *Stochastic Calculus for Finance II: Continuous-Time Models* (Springer, Berlin, 2004)
170. G. Simmons, *Topology and Modern Analysis* (McGraw-Hill, New York, 1963)
171. C. Skiadas, *Asset Pricing Theory* (Princeton University Press, Princeton, NJ, 2009)
172. D. Sondermann, *Introduction to Stochastic Calculus for Finance* (Springer, Berlin, 2006)
173. K. Takaoka, M. Schweizer, A note on the condition of no unbounded profit with bounded risk. Finance Stoch. **18**, 393–405 (2014)
174. A. Taylor, D. Lay, *Introduction to Functional Analysis*, 2nd edn. (Wiley, New York, 1980)
175. H. Theil, *Principles of Econometrics* (Wiley, New York, 1971)
176. P. Wakker, H. Zank, State dependent expected utility for savage's state space. Math. Oper. Res. **24**(1), 8–34 (1999)
177. R. Zagst, *Interest-Rate Management* (Springer, Berlin, 2002)
178. G. Zitkovic, Utility maximization with a stochastic clock and an unbounded random endowment. Ann. Appl. Probab. **15**(1B), 748–777 (2005)
179. G. Zitkovic, Financial equilibria in the semimartingale setting: complete markets and markets with withdrawal constraints. Finance Stoch. **10**, 99–119 (2006)
180. G. Zitkovic, An example of a stochastic equilibrium with incomplete markets. Finance Stoch. **16**, 177–206 (2012)

Index

R. A. Jarrow, *Continuous-Time Asset Pricing Theory*, Springer Finance,
https://doi.org/10.1007/978-3-319-77821-1

Printed in the United States
By Bookmasters